Tradeoff Decisions in System

A. Terry Bahill • Azad M. Madni

Tradeoff Decisions in System Design

Springer

A. Terry Bahill
Systems and Industrial Engineering
University of Arizona
Tucson, AZ, USA

Azad M. Madni
Astronautical Engineering Department
University of Southern California
Los Angeles, CA, USA

Springer Geography
ISBN 978-3-319-82892-3 ISBN 978-3-319-43712-5 (eBook)
DOI 10.1007/978-3-319-43712-5

Printed on acid-free paper

This Springer imprint is published by Springer Nature
The registered company is Springer International Publishing AG
The registered company address is: Gewerbestrasse 11, 6330 Cham, Switzerland

Foreword

Tradeoff Decisions in System Design is an easy-to-read, user-oriented book that offers readers a comprehensive treatment of the fundamental processes of system design and development. The volume covers in a thorough but readily understood manner the principles of requirements and specifications discovery, tradeoff studies and the analysis of system risks. Terry A. Bahill and my colleague at the Viterbi School of Engineering of the University of Southern California, Azad Madni, have done an excellent job of capturing the essential elements of system design and presenting them using illustrative examples that aid the reader in understanding the evolution of the conception and creation of complex systems. The authors are especially qualified to write about this subject because of their collective experiences in military systems, space systems, commercial sector systems, sociotechnical systems and fields as varied as education and baseball.

Although the book will be of value to those readers involved in the design of systems that fall outside the realm of what is normally considered engineering, such as business processes, health care, education and other organizational structures, it will be of particular interest to those engaged in systems engineering, the interdisciplinary field of engineering that embraces the design, development and management of complex systems. Originating in the days before WWII, the discipline of systems engineering offers a holistic view of engineering principles and practices, transcending and, yet, incorporating the traditional fields of civil, chemical, electrical, industrial and mechanical engineering as well as the more recent entries of software engineering, information and communication technologies, biotechnology and robotics. Importantly, Bahill and Madni point out to the reader the importance of the user community in the generation of the system requirements, performance measures, the user interfaces to the system operation, the possible human biases and errors in the decision making leading to the planning, design, development and deployment of the system, among other social influences.

The authors describe SIMILAR, an approach to the design and development of system products and processes. SIMILAR is composed of a number of steps that are essential to achieving a higher probability of success in the system development and implementation. These steps are: *S*tate the Problem, *I*nvestigate Alternatives, *M*odel the System, *I*ntegrate the System, *L*aunch the System, *A*ssess Performance, and *R*e-evaluate. Their successful completion ensures the achievement of the system's requirements throughout its useful life.

As the nations of our global society become more interdependent and more beset with complex and threatening environmental, security and social challenges, among others, the systems needed to confront those challenges will become more difficult to define, create, operate and maintain. Their design, development and deployment will require cooperation and commitment from a broad and diverse set of participants who are characterized by a myriad of cultures, nationalities, languages, religions and backgrounds. System designers will have to know more than science and engineering to develop the organizational structures, processes and arrangements that will be needed to serve the needs of this disparate world population and meet the difficult and demanding requirements of the future. It is to these women and men to whom I commend this book for it offers more than a set of rules and equations to solve a range of system problems. It provides an insight into decision making, tradeoff and risk analysis strategies for those system designers who recognize that we are all part of a pluralistic but fractious society—a multi-faceted, global system of people and nations—that is facing daunting issues whose solutions will require talented individuals with the advanced systems thinking abilities and tools put forth in this important book.

University of Southern California
Los Angeles, CA
USA

John Brooks Slaughter

Preface

This book is primarily for seniors and graduate students in engineering and business and professionals interested in acquiring knowledge of three fundamental processes associated with system development—requirements discovery, trade-off studies, and risk analysis. These three important technical areas are covered in a unified manner in this single, easy-to-read volume. This book is organized into seven parts, with each part comprising multiple independent, self-contained chapters. This structure allows both instructors and students to read the chapters in the order of their need, preference, or interest. It can be used as a textbook for a single course covering the three main topic areas or as a reference text for three separate courses, each covering a topic area. This flexibility should appeal to faculty members, students, professional development instructors, and consultants.

This is a learner-centric book. It allows students to access a relevant body of knowledge from a single source rather than scouring multiple volumes to find the needed information. The examples employed to illustrate the use of different methods also discuss potential pitfalls to avoid when using them. The examples are small enough to be cognitively manageable and rich enough to convey key concepts and highlight key aspects of the methods. A strategy that we employed in scoping the illustrative problems was to ensure that each one spanned the process being discussed, rather than resort to multiple, disconnected examples. The continuity provided by having a few illustrative examples that serve as a common thread across the process being discussed is a distinguishing feature of this book that should make it an easy read.

We cover requirements discovery, trade-off studies, and risk analysis—three core processes in system development, product reengineering, and service design and implementation. We specifically address human cognitive biases and heuristics and their implications while executing these processes. We draw attention to the fact that while on the surface these processes appear to be distinctly different, employ different vocabularies, and require different inputs, they have the same underlying structure and process pattern. We exploited this recognition to identify a general process and thereby reduced the cognitive load on the readers. In other words, learning how to perform one process facilitates the learning of the others.

Requirements discovery, trade-off studies, and risk analysis are critical and recurrent processes that come into play when developing, redesigning, and deploying systems. These processes require human judgment and frequent assessments. And we know that human judgment can be fallible and humans tend to exhibit significant variability arising from cognitive biases and occasional misapplication of heuristics that can throw off carefully planned trade-off studies and risk analyses. To circumvent these outcomes, we offer recommendations on how to exploit heuristics to cope with problem and system complexity and mitigate the effects of cognitive biases when conducting trade-off studies and performing risk analyses. This is another key discriminator of this book.

The requirements for deriving maximum value from this book are quite modest. You need to be conversant in using calculators and Microsoft Excel® spreadsheets. Knowledge of modeling languages such as UML and SysML is desirable, but not necessary. Knowledge of basic mathematics is also desirable but not necessary to appreciate the full impact of sensitivity analyses and scoring functions—two key functions employed in trade-off studies. Learners unfamiliar with the underlying mathematics can skip the mathematical material without experiencing lack of continuity. This book invites the reader to explore the richness of requirements discovery, trade-off studies, and risk analyses. The illustrative examples provide a continuous thread through each core process. On rare occasions, the illustrative examples can appear to be somewhat long. However, in those instances that length is necessary to illuminate the different aspects of the methods. Such is the case when making multiple iterations through a process in light of new evidence.

This book reflects our real-world experiences in academia and industry. We have had the rare opportunity to be involved in the early and late stages of development of systems ranging from smart portable devices to missiles, satellite networks, spacecraft, commercial aircraft, and electrical power grids, to command and control of complex system-of-systems. The opportunity to participate in the development of systems of such vastly different scales has given us valuable perspectives that we have shared with you in this volume.

We enjoyed writing and assembling the various chapters of this book with you, the intended audience, in mind. We hope you will find the illustrative examples and homework problems to be informative, *fun*, and insightful. We expect to refine and add to the content based on the feedback we receive from you.

You are about to embark on the journey of unraveling the mysteries and understanding the nuances of requirements discovery, trade-off studies, and risk analysis. We hope that you will find the journey rewarding in understanding and applying the methods. And we hope that the methods will prove to be of value when you are engineering new systems or redesigning existing systems and services.

Tucson, AZ A. Terry Bahill
Los Angeles, CA Azad M. Madni

Acknowledgments

We thank Edwin Ordoukhanian for writing the scoring function toolbox program and for managing the references for this textbook and providing feedback from a student's perspective. Edwin is pursuing his doctorate in astronautical engineering under Professor Madni's direction at the University of Southern California. We thank Ken Cureton at the University of Southern California for reviewing the book.

Contents

Abbreviations and Acronyms

$S_{\alpha}^{F} = \frac{\partial F}{\partial \alpha}\Big\vert_{\mathrm{NOP}}$	Absolute-sensitivity function
$\overline{S}_{\alpha}^{F} = \frac{\partial F}{\partial \alpha}\Big\vert_{\mathrm{NOP}} \frac{\alpha_0}{F_0}$	Relative-sensitivity function
$\widetilde{S}_{\alpha}^{F} = \frac{\partial F}{\partial \alpha}\Big\vert_{\mathrm{NOP}} \alpha_0$	Semirelative-sensitivity function
1B	Single
2B	Double
3B	Triple
AC	Alternating current
act	Activity diagram
AHP	Analytic hierarchy process
ATM	Automated teller machine
B	Baseline value
BA	Batting average
BB	Base on balls
bdd	Block definition diagram
BICS	Bahill Intelligent Computer Systems
BIMS	BICS Illuminance Management System
BiST	Built-in self-test
BSA	Boy Scouts of America
CCB	Change control board
CDR	Critical design review
CDRL	Contract deliverable requirements list
CF	Certainty factor
CFL	Compact fluorescent light bulb
CMMI	Capability Maturity Model Integration
ConOps	Concept of operations
CoR	Cost requirement
COTS	Commercial off the shelf
CS	Caught stealing
CuR	Customer requirement

DAR	Decision analysis and resolution
dB	Decibel
DC	Direct current
DD	Difficulty of detection
DeriveReqt	Derive requirement
DM	Decision-maker
DoE	Design of experiments
EV	Expected value
F	Frequency of occurrence
FEMA	Failure modes and effects analysis
FoM	Figure of merit
FR	Functional requirement
GDP	Grounded into a double play
GIS	Geographic information system
GNC	Guidance, navigation, and control
GUI	Graphical user interface
HOA	Homeowner association
HR	Homerun
HVAC	Heating, ventilation, and air conditioning
Hz	Frequency in Hertz
ibd	Internal block diagram
IC	Illuminance controller
Id	Identification
IJS	Integers
INCOSE	International Council on Systems Engineering
IR	Infrared
IZ	Inputs to a system
L	Lower threshold
LED	Light-emitting diode
LIMPET	Land Installed Marine Power Energy Transmitter
LP	Launch platform
MAUT	Multi-attribute utility technique
MCAS	Marine Corps Air Station
MCR	Mission concept review
MG	Motor generator
MLB	Major League Baseball
NBA	National Basketball Association
NCAA	National Collegiate Athletic Association
NFL	National Football League
NFPR	Nonfunctional performance requirement
NOAO	National Optical Astronomy Observatory
NOP	Nominal operating point
NZ	Next state function for a system
OCD	Operational concept description

OMG	Object Modeling Group
OZ	Outputs from a system
PAL	Process assets library
par	Parametric diagram
PCB	Printed circuit board
PDR	Preliminary design review
PI	Performance index
PIN	Personal identification number
PRR	Production readiness review
PV	Photovoltaic
QFD	Quality function deployment
R	Risk
R	Runs scored
RA	Risk analyst
RAM	Random access memory
RAM	Rolling airframe missile
RBI	Run batted in
req	Requirements diagram
ReqF	Functional requirement
RF	Radio frequency
RLS	Real numbers
RMS	Royal mail ship
ROE	Reached base on error
RR	Risk requirement
RZ	Readout function for a system
S	Severity of consequences
S	Slope of tangent to the curve at B
SB	Stolen base
sc	Score
sd	Sequence diagram
SEMP	Systems engineering management plan
SEU	Subjective expected utility
SH	Successful bunt
SI	Systeme International d'Unites
SO	Strikeout
SR	Schedule requirement
SRR	System requirements review
SSF	Standard scoring function
stm	State machine diagram
SysML.	Systems Modeling Language
SZ	States of a system
TBD	To be determined
TEP	Tucson Electric Power
TPM	Technical performance measure

TRR	Test readiness review
TST	Total system test
uc	Use case diagram
UML	Unified Modeling Language
UofA	University of Arizona
v	Value of the input
V&V	Verification and validation
VLS	Vertical launch system
w	Weight
Z	System

About the Authors

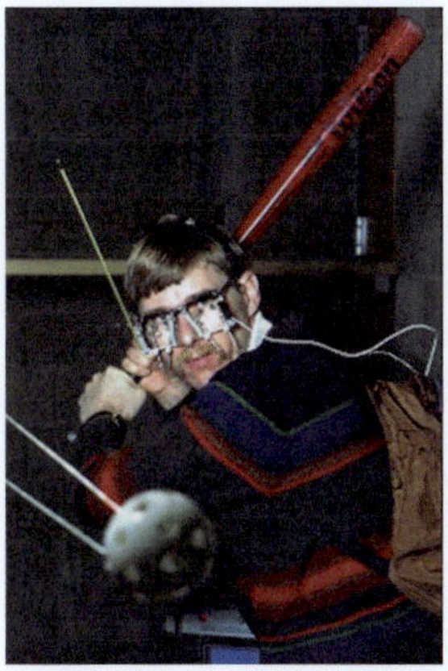

A. Terry Bahill is professor emeritus of systems and industrial engineering at the University of Arizona in Tucson. He received his Ph.D. in electrical engineering and computer science from the University of California, Berkeley, in 1975. He is the author of six engineering books and 250 papers; over 100 of these are in peer-reviewed scientific journals. Bahill has worked with dozens of technical companies presenting seminars on systems engineering, working on system development teams, and helping them to describe their systems engineering processes. He holds a US patent for the Bat Chooser™, a system that computes the Ideal Bat Weight™ for individual baseball and softball batters. He was elected to the Omega Alpha Association, the systems engineering honor society. He received the Sandia National Laboratories Gold President's Quality Award. He is an elected fellow of the Institute of Electrical and Electronics Engineers (IEEE), of Raytheon Missile Systems, of the International Council on Systems Engineering (INCOSE), and of the American Association for the Advancement of Science (AAAS). He is the founding chair emeritus of the INCOSE Fellows Committee. His picture is in the Baseball Hall of Fame's exhibition "Baseball as America." You can view this picture at http://www.sie.arizona.edu/sysengr/. His research interests are in the fields of system design, modeling physiological systems, eye-hand-head coordination, human decision-making, and systems engineering application and theory. He has tried to make the public appreciate engineering research by applying his scientific findings to the sport of baseball.

Azad M. Madni is a Professor of Astronautical Engineering and the Technical Director of the Systems Architecting and Engineering Program in the University of Southern California's Viterbi School of Engineering. He is also a Professor (by courtesy) in USC's Schools of Medicine and Education. He is the founder and Chairman of Intelligent Systems Technology, Inc., a high tech R&D company specializing in game-based educational simulations, and methods, processes and tools for complex systems engineering. He received his B.S., M.S. and Ph.D. degrees from the University of California, Los Angeles. His research has been sponsored by both government research organizations such as DARPA, OSD, ARL, RDECOM, ONR, AFOSR, DHS S&T, DTRA, NIST, DOE and NASA, and aerospace and automotive companies such as Boeing, Northrop Grumman, Raytheon and General Motors. He is an elected Fellow of the American Association for Advancement of Science (AAAS), the American Institute for Aeronautics and Astronautics (AIAA), the Institute for Electrical and Electronics Engineers (IEEE), the Institution for Electronics and Telecommunications Engineers (IETE), the International Council on Systems Engineering (INCOSE) and the Society for Design and Process Science (SDPS). His recent awards include *2016 Boeing Lifetime Accomplishment Award* and *2016 Boeing Visionary Systems Engineering Leadership Award* for contributions to industry and academia (awards received in Boeing's 100th anniversary); the 2016 INCOSE RMC Special Award for pioneering, industry-relevant contributions to Transdisciplinary Systems Engineering; the *2016 Distinguished Engineering Educator Award* from the Engineers' Council; the *2016 Outstanding Educator Award* from the Orange County Engineering Council; the 2014 *Lifetime Achievement Award* from the International Council on Systems Engineering; the 2013 *Innovation in Curriculum Award* from the Institute of Industrial Engineers; the 2012 *Exceptional Achievement Award* from INCOSE and the 2011 *Pioneer Award* from the International Council on Systems Engineering. He serves on the USC's Council of the Center of Cyber-Physical Systems and Internet of Things (CCI), and Steering Committee of USC Provost's STEM Consortium. His research interests include formal and probabilistic methods in systems engineering, model-based architecting and engineering, engineered resilient systems, cyber-physical systems, and exploiting disciplinary and technology convergence in systems engineering. He is listed in Who's Who in Science and Engineering, Who's Who in Industry and Finance and Who's Who in America.

Chapter 1
System Design and the Design Process

1.1 Section 1. The SIMILAR Process

Processes are used by individuals, teams, and organizations to increase predictability of outcomes and ultimately, probability of success. Over the years, these processes have been described in both technical and nontechnical terms. Although they use different vocabularies and levels of detail, it turns out that most process descriptions are surprisingly similar. Bahill and Gissing [1] compared the descriptions of a hundred processes and extracted the similarities. They discovered that common to most processes are seven key activities: **S**tate the problem, **I**nvestigate alternatives, **M**odel the system, **I**ntegrate components, **L**aunch the system, **A**ssess performance, and **Re**-evaluate. These seven activities are conveniently summarized using the acronym, SIMILAR: **S**tate, **I**nvestigate, **M**odel, **I**ntegrate, **L**aunch, **A**ssess, and **Re**-evaluate. They are incorporated into the SIMILAR process of Fig. 1.1. We use this process to provide the overall context for problem solving and system design in this book. At the outset, we note that the functions in the SIMILAR process are performed iteratively and in parallel. Each activity in this process is described next.

1.1.1 State the Problem

The problem statement begins either with a description of the purpose of the system, the top-level functions that the system is expected to perform, or a deficiency that needs to be corrected. The system's *mandatory and preference requirements* should originate in and be traceable to this problem statement [2]. Acceptable system designs are those that satisfy all mandatory requirements. Preference requirements are a special type of requirement that are brought into balance as part of trade-off analysis to find the preferred alternative from among acceptable

A.T. Bahill, A.M. Madni, *Tradeoff Decisions in System Design*,
DOI 10.1007/978-3-319-43712-5_1

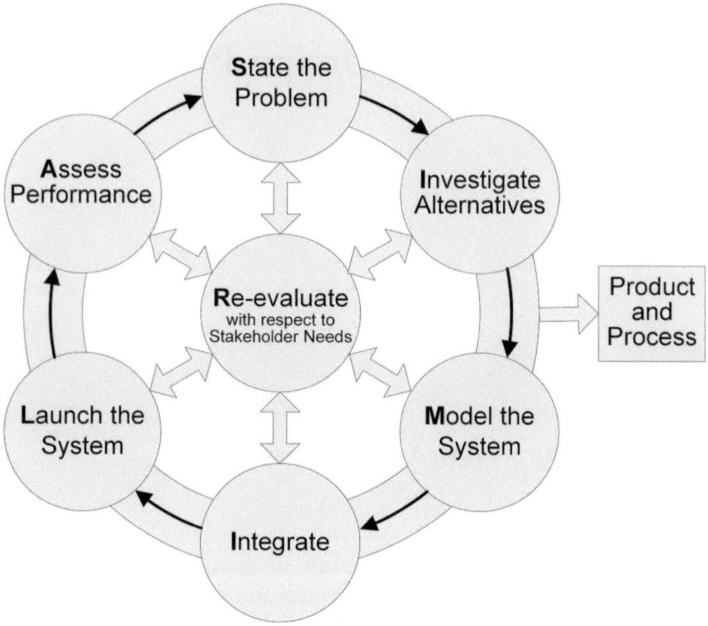

Fig. 1.1 The Similar process

alternatives. The problem should be stated in terms of *what* needs to be done, not *how* it must be done. The problem statement may be in prose form or in the form of a model. Inputs are provided by end users, operators, maintainers, suppliers, acquirers, owners, customers, bill payers, regulatory agencies, affected individuals or organizations, sponsors, manufacturers, and other stakeholders [3]. The Re-evaluate activity will help refine the problem statement.

"The beginning is the most important part of the work," Plato, *The Republic*, Fourth century B.C.

"All the really important mistakes are made the first day," Rechtin, *The Art of System Architecting*, 2000.

1.1.2 Investigate Alternative Designs

Alternative designs are evaluated using criteria such as performance, cost, schedule, and risk. Since no one particular design is likely to be best on all evaluation criteria, multi-criteria decision-making techniques are used to identify the preferred alternatives. This type of analysis is typically repeated as and when additional data become available. For example, evaluation criteria should be computed initially based on estimates made by design engineers. Then, models are constructed and evaluated. Next, simulation data are generated. Thereafter, prototype performance is measured, and ultimately tests are performed on the real system. When designing complex systems, alternative designs are generated to explore design sensitivities to specific parameters of interest and to reduce project risks. Alternative designs range from analysis of alternatives at the system level down to alternative subsystem designs.

As importantly, investigating candidate design alternatives helps to clarify and occasionally reformulate the statement of the problem [4, 5].

1.1.3 Model the System

Abstract models are typically created for most alternative designs. After a design has been selected, the model for the preferred alternative is appropriately elaborated (that is, expanded) and then used to manage system development and use throughout the system's life cycle. A variety of system models can be used to create system designs. These include physical analogs, analytic equations, state machines, block diagrams, functional flow diagrams, object-oriented models, computer simulations, and mental models. The engineering function is typically responsible for creating both the product and the process for producing it. Therefore, models are invariably constructed for both the product and the process, and capture the relationships between them [6]. For example, *process* models can be used to study scheduling changes, create dynamic PERT charts, and perform sensitivity analyses to determine the effects of delaying or accelerating certain subprojects. Executing process models can reveal bottlenecks and fragmented activities, identify opportunities to reduce cost, and expose duplication of effort. *Produc*t models help in understanding and explaining the system, and in conducting design trade-off studies and identifying and managing risks.

1.1.4 Integrate

Integration means bringing elements together so that they work as a whole to accomplish their intended purpose and deliver value. Specifically, systems, enterprises, and people need to be integrated to achieve desired outcomes. To this end, interfaces need to be designed between subsystems. Subsystems are typically defined along natural boundaries in a manner that minimizes the amount of information exchanged between the subsystems. Well-designed subsystems send finished products to other subsystems. Feedback loops around individual subsystems are easier to manage than feedback loops involving densely interconnected subsystems.

1.1.5 Launch the System

System launch means either deploying and running the actual system in the operational environment, or exercising the model (with requisite level of fidelity) in a simulated environment to produce outputs necessary for evaluation. In a manufacturing environment, this might mean modifying and using commercial off-the-shelf hardware and software, writing prototype code, and/or bending metal. In the business environment, launching the systems means the business plan is decomposed into tasks and actions and is deployed throughout the company. The purpose of system launch is to provide an environment that allows the system or its model to do what it is being designed to do.

1.1.6 Assess Performance

Evaluation criteria, technical performance measures and metrics are used to assess performance. Evaluation criteria are also used to quantify requirements in trade-off studies. Technical performance measures are used to mitigate risk during both design and manufacturing. Metrics are used to manage a company's processes. Ultimately, it comes down to measurements. If you cannot measure, you cannot control the system. If you cannot control the system, you cannot improve it [7].

1.1.7 Re-evaluate

Re-evaluation is arguably one of the most important functions. For a century, engineers have used feedback to control systems and improve performance. It is one of the most fundamental engineering concepts. Re-evaluation is a continual process with multiple parallel loops. Re-evaluation means observing outputs and using this information to modify the inputs, the product, and/or the process.

The SIMILAR process (Fig. 1.1) clearly shows the distributed nature of the Re-evaluate function in the feedback loops. However, it is important to realize that not all the loops will always come into play. The loops that are used depend on the problem to be solved, the point in the system life cycle, and the problem context.

1.1.8 Ancillary Materials

We have several web sites that support this textbook. For students we have http://www.springer.com/gp/book/9783319437101#otherversion=9783319437125 and http://sysengr.engr.arizona.edu/forStudents/index.html.

These sites have examples, templates, a scoring function toolbox, and spreadsheets that were too big to put into the textbook that contain the mathematical functions. For instructors we have http://www.springer.com/gp/book/9783319437101#otherversion=9783319437125 and http://sysengr.engr.arizona.edu/forInstructors/index.html. This directory is password protected. Instructors who have adopted this textbook for their class may obtain a solutions manual, addition homework problems and solutions, spreadsheets, hands-on exercises, and PowerPoint slides for lectures by sending a resume and a syllabus for their course that will use this textbook to terry@sie.arizona.edu.

A program written by Edwin Ordoukhanian in Astronautical Engineering at the University of Southern California that creates scoring function graphs is available free at http://sysengr.engr.arizona.edu/forStudents/index.html. It is called the Scoring Function Toolbox.

1.1.9 Homework

1.1.9.1 State the Problem

Why are you taking this course?

1.1.9.2 Investigate Alternatives

List at least half dozen alternatives for you taking this course.

1.1.9.3 Survival

Assume that you have just crash-landed. You were able to salvage the items listed below. You can choose ten items to take with you on your search for survival. Which would you choose? This is a very common team building exercise that we presume many of you have already participated in. If you can remember your previous answers, please use them in this problem.

The book *Edible Animals of the Desert*
Bottle of 1000 salt tablets
Box of matches
Compress kit with gauze
Cosmetic mirror
Fifty feet of nylon rope
First-aid kit with injection needles
Five gallons of water
Flash light (size D batteries)

Food concentrate
Magnetic compass
One case of dehydrated milk
Pair of sunglasses per person
Parachute (red and white)
Plastic raincoat (large size)
Sectional air map of the area
Self-inflating life raft
Signal flares
Solar-powered FM transmitter
Solar-powered portable heating unit
Stellar map (moon's constellation)
Swiss Army knife
Topcoat one per person
Two .45 caliber pistols (loaded)
Two 100-lb tanks of oxygen

Please prioritize your top ten choices.

1.1.9.4 Problem Stating

A politician once asked Wayne Wymore, who founded Systems Engineering at the University of Arizona 1960, if he would help them solve Arizona's water problem. Wymore said, "Sure, what is your water problem?" The politician replied, "We don't have enough water for all of our people." Wymore responded, "A simple solution to your problem is to shoot half your people." The flabbergasted politician exclaimed, "That's ridiculous. That's not a solution to our problem." Wymore replied, "It *is* a solution to the problem that you stated." The key point here is to fully articulate the problem in a manner that captures the intent and all implied constraints when writing a statement for the politician.

At the turn of the century, the US government said that we must reduce our use of fossil fuels and reduce the production of greenhouse gases by running our cars on gasohol. The government spent considerable sum of money on growing and fermenting corn (and sugar cane) into alcohol, which was then blended with gasoline. With the federal subsidies, this process was economically successful. However, it drove the price of corn so high that Mexican peasants could no longer afford to buy tortillas, their staple food. Furthermore, it takes 50 gal of water to grow enough corn to make enough alcohol to drive a car one mile. This only exacerbates the water problem! Thus, it is important to write the problem statement for the US government that takes all key government considerations into account.

In the 1950s, IBM was offered the patents for Xerox's photocopy machine at an attractive price. IBM's market research concluded that no one would pay thousands of dollars for a machine intended to replace carbon paper. Once again, it was important to write a problem statement for IBM that conveyed the value proposition.

1.1.9.5 Lawn Care

A grass lawn needs to be kept neat. You could hire a gardener and tell him to, "cut my grass." Later, you might decide to do it yourself. In that case, you will need a lawn mower. Although inserting an implied point solution into the problem statement is poor practice, you might say, "I need a lawn mower to cut my grass." What is wrong with these statements?

1.1.9.6 Marijuana Surprise

Assume that you live in a country and state in which using marijuana is illegal. You are a father, you have just come home from work and your wife shows you a baggie containing what looks like parsley, sage, rosemary and thyme, and some self-roll cigarettes papers. She says that your 16-year-old son is smoking marijuana. And you must do something about it. Please state the problem. You must do this without any other data.

1.1.9.7 Illegal Drugs

The pattern shown in Fig. 1.2 has been repeated over and over again. The government, spurred by moralists, decides that a certain inexpensive chemical is not good for the people: so they make it illegal. They do not bother to state what the problem actually is or to investigate alternative solutions. Then criminals produce and market this chemical on the black market at high prices. This concentrates a large amount of money into the hands of organized crime. This produces the unintended consequences of murders, burglaries, police and judicial corruption, explosion of prison populations, increased spread of diseases, and increased violence. The government has done this with ethyl alcohol, marijuana, cocaine, methamphetamines, and selling sex. "Those who cannot remember the past are condemned to repeat it [Santayana, 8]." Help the government to state the problem. Use marijuana as your specific example.

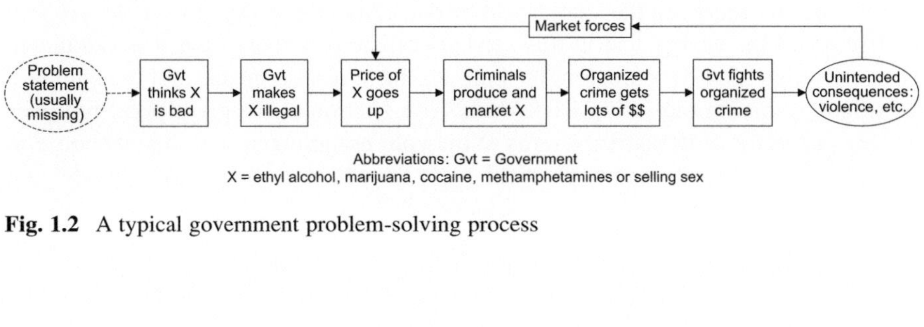

Fig. 1.2 A typical government problem-solving process

1.1.9.8 Carbon Tax Alternatives

To combat global warming, the US government wants to tax carbon emissions. It is also forcing automobile manufacturers to increase their average fleet fuel efficiency. It is estimated that one-third of the fuel used in automobiles is spent to overcome friction. Assume that half of that friction is in the engine and by using superlubricity materials the coefficient of friction could be reduced from 0.4 to 0.004. Presently, pyrolytic graphite surfaces and carbon nanotubes can do this, but only at the nanoscale. If the engine surfaces could be coated with such materials, what would be the fuel efficiency of a car that presently gets 30 miles per gallon of fuel?

1.1.9.9 Predicting the Past

A common way to verify models, particularly those implemented with computer simulations, is to run them backwards in time and see if they can predict the past. For example, can the models that are used to predict global warming be run backwards to predict the earth's temperature for the last millennium? Running backwards in time is easy for the simulation part, but it is harder for the physical world part. Here is an example for a physical event. Let us consider a golfer driving a golf ball off of a tee. The golfer winds up and swings the golf club. This imparts angular momentum and kinetic energy to the club. We determine how much by measuring the angular velocity. When the club head hits the ball, it transfers angular momentum and kinetic energy to the ball. The club-ball collision has a coefficient of restitution (CoR) of 0.8, so 64 % of the kinetic energy will be transferred to the ball. The rest will be lost. From the collision model [9], we determine the velocity, launch angle, and spin of the ball. That is all we need to determine its trajectory and therefore where it lands. We need to make a simplifying assumption here, namely that the ball does not bounce or roll after it hits the ground. Now let us run the model backward. To launch the ball backwards, we use an air cannon, or a similar device, that will give the ball the exact velocity, launch angle, and spin needed to make it go back to the tee. This is a simple calculation [9]. When it gets back to the tee, there will be a club mounted on a pivot point. The ball will strike the club and make it rotate. But in the original collision, energy was lost. So in this backward collision, energy must be gained. Therefore, a small explosive charge will be mounted on the club head. The energy in the charge will be the sum of the energy lost in the original collision and the energy lost in this second collision. If everything was calculated correctly, then after the collision, the ball will be at rest on the tee and the club will be rotating, tracing the swing backwards. The friction at the pivot point removes energy just as the golfer added energy. Now your assignment, should you choose to accept it, is to think of another physical event for which you can make a model and run it backwards in time.

1.2 Section 2. Abstraction

Engineers typically interview customers and subject matter or domain experts to determine the intended use of the system. This information is used to create abstractions of the system architecture and required capabilities. The abstractions are transformed into models which, in turn, are converted into system designs. The system design specifies the components and the interfaces needed to implement the proposed system. Abstraction is difficult to teach, because it requires conceptual thinking that engineers may not be exposed to.

The natural instinct of an inexperienced engineer is to think in terms of a point solution to the specific problem. Let us pause a moment and reflect. Perhaps this specific problem is a specific instance of a class of problems. So, resisting the urge to jump to a point solution and instead thinking in terms of abstractions can potentially produce superior systems. Without conceptualizing the problem as an instance of a general class of problems, important elements of a more general solution to the particular problem can be missed. Furthermore, failure to find more general concepts and solutions can limit the ability to address future similar problems.

Engineers frequently design systems by selecting components, designing interfaces between the components, and then assembling them. Consequently, it is difficult to get engineers to unlearn these deeply ingrained practices and learn to create abstractions of components to create solutions that are more enduring.

It is worth recalling that design is an abstraction. A system model, that employs a state machine diagram representation, is not the system. It is an abstraction of the system. In fact, all such diagrams are abstractions and simplifications that are intentionally created to help with system understanding by suppressing extraneous system details.

Some dictionary definitions of abstract and abstraction are:

"**abstract** *adj.* **1.** Considered apart from concrete existence... **4.** Thought of or stated without reference to a specific instance."

"**abstraction** *n.* The process of extracting the underlying essence, removing dependence on real world objects."

Stated differently, *abstraction* is the process of removing dependence on real-world objects. It allows designers and modelers to describe a problem at a high level by hiding low-level information. In general, the greater the level of abstraction, the more details that are hidden from the user. In light of the foregoing, here is another definition.

"**Abstraction**: A representation in term of presumed essentials, with a corresponding suppression of the non-essential [10]"

The more abstract an image is, the more people it will describe. For example, consider the person depicted in Fig. 1.3. The photo on top is singular: it describes one person. Then progressing downward are drawings with more and more

Fig. 1.3 Six stages of abstraction. Drawings are by Angelo Hammond

abstraction. Finally, the abstract smiley face cartoon on the bottom is universal: it describes nearly all people.

A broader definition of abstraction can go beyond *just* eliminating details to including specialization, i.e., deliberate changes to specific details. Depending on the details addressed, different abstractions can be created of the same object. For example, in software, a sequence of objects {a, b, c, d, e, f} may be of interest. If the intent is to read or replace any of these objects, then this sequence can be abstracted as an array. However, if the intent is to also insert and remove objects, then this sequence should be abstracted as a list. Today abstraction requires human intelligence. Abstraction also includes deliberately leaving out nonessential components.

In 1945, Pablo Picasso made a drawing of a bull. A month later, he finished his 11th stage of abstraction, which comprised six lines. Célestin said, "When you stand before his eleventh bull, it's hard to imagine the work that went into it [11]." Picasso abstracted the essence of the bull (exemplified by the horns and the penis) through progressive analysis and investigation of its form. The development of abstractionism in modern art is attributed to Picasso and Henri Matisse who concurrently developed this art form (Fig. 1.4).

Picasso's 11 stages of abstraction is a metaphor for system design. You go through iteration after iteration until you can present the essence of your design concisely. Then you begin the elaboration part of the design process.

Here is yet another example of abstraction. Start with a pile of data points. Organize them into a scatter graph and then fit the data with a line with variance bars or a statistical distribution (for example, Poison or Gaussian). You have eliminated details, but kept the essence of the data. To abstract further you could include more data sets and generate a set of equations to describe the data. Finally, those equations could be organized into vectors and matrices (for example, $\dot{\mathbf{x}} = \mathbf{A}\mathbf{x} + \mathbf{B}\mathbf{u}$). Each step up removes details and increases the abstraction.

Object-oriented design and the Unified Modeling Language (UML) [12] use the concept of generalization, which is different from abstraction. In generalization, there are several objects and you find what is common to all of them. You create a class that has the important properties of each of the objects. This may require abstraction. Generalization, which is related to inheritance, takes you up the hierarchy and specialization takes you down. Abstraction starts with a concrete detailed object and removes detail. Generalization starts with many objects and finds what is common to all. Consider these examples. "All pretty girls are female" is an abstraction. "All pretty girls put on make-up" is a generalization.

1.2.1 Homework

1.2.1.1 Abstraction

Create a sketch or a drawing showing the essence of a baseball playing field.

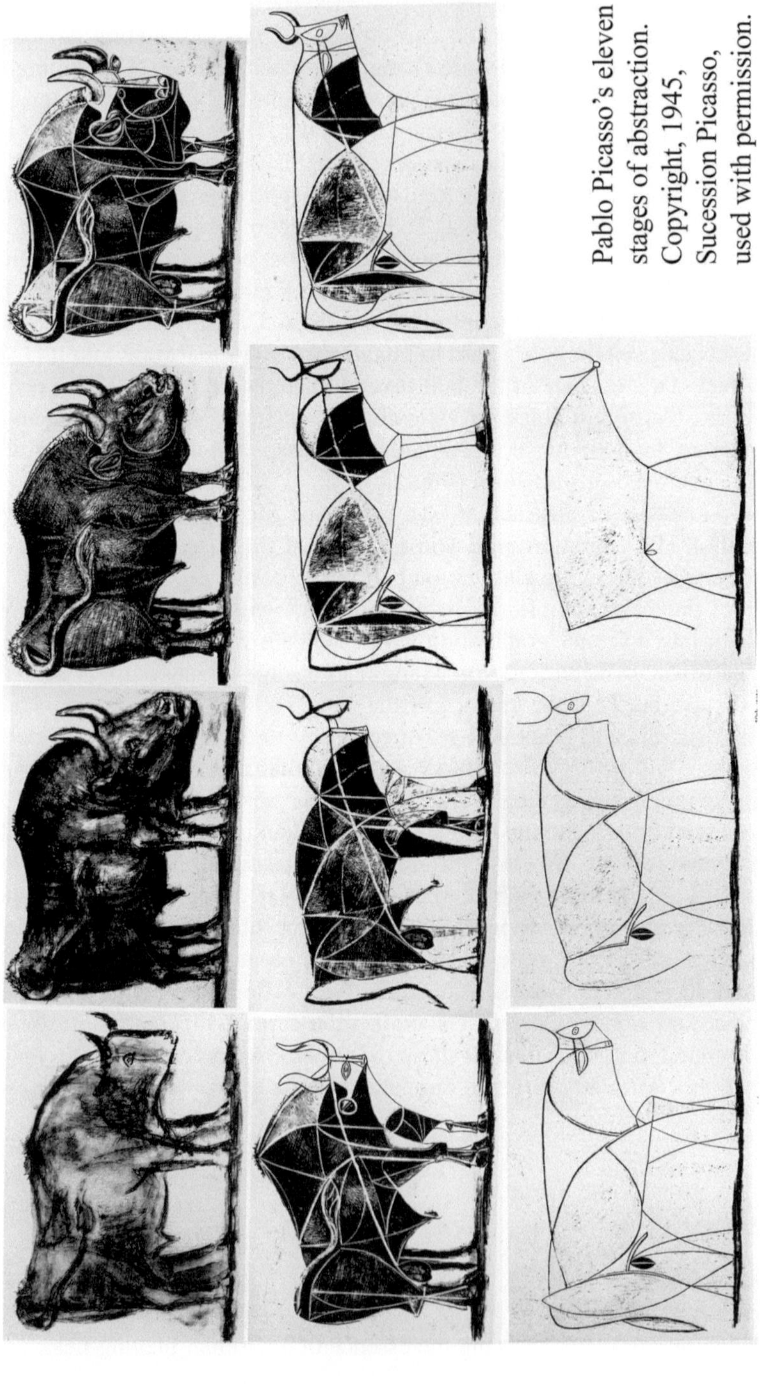

Fig. 1.4 Picasso's 11 stages of abstraction. Copyright ©, 1945, Succession Picasso, used with permission

Fig. 1.5 Gala apple

1.2.1.2 Abstraction and Generalization of Apples

Create an abstraction of the red apple in Fig. 1.5.
Create a generalization of the apples in Fig. 1.6.

1.2.1.3 Baseball in Flight

Create an abstraction for a baseball in flight. Consider spin, drag, direction of travel, gravity, and velocity.

1.2.1.4 Wise Investment

Suppose that you deposit \$1000 into a savings account that pays an annual interest rate of 7 %, compounded monthly. Provide a model (abstraction) for this investment. Use it to calculate the value of your account after 10 years.

$$\text{Value at end of term} = \text{Principal}\left(1 + \frac{\text{Annual interest rate}}{\text{Number of times compounded}}\right)^{\text{Term}\times\text{Number of times compounded}}$$

Fig. 1.6 (**a**) Red apple; (**b**) Golden apple; (**c**) Green apple

1.3 Section 3. The Design Process

Our system design process is a use case-based iterative process [3, 13, 14]. It starts with a simple story that describes what the system is intended to do and how it is expected to operate. This story evolves into a statement of the problem to be solved. Proper understanding of the problem requires exploration of possible solutions. A rough schedule of who does what and when is also created. These are brief activities. Next, use cases are written that describe the expected behavior of the system. While writing use cases, functional and nonfunctional requirements are developed. Historically, requirements documents tended to be voluminous. However, today with the advent of model-based systems engineering (MBSE), these are model-enabled activities that spur the creation of smaller documents. Design of tests typically starts once use cases are written. Thereafter, systems engineers derive technical requirements while test engineers create test requirements. Equipped with partial requirements, preliminary evaluation criteria are created for use in trade-off studies. Design engineers create the system model using UML and SysML diagrams that depict *system behavior* using state machine diagrams, sequence diagrams, and activity diagrams, and *system structure* using block diagrams. As noted earlier, this system design process is iterative and hierarchical. Therefore, creating documents is also necessarily iterative. With multiple iterations, there are multiple opportunities for parallel processing. This process involves identifying various risk aspects (for example, cost, schedule, performance, project, business, safety, environmental). At the beginning of a risk analysis, defining the system and explaining its behavior are the two most important activities. Whereas, near the conclusion of a risk analysis, communicating with decision makers and explaining opportunities and risks are the most important activities. The example in Section 3 is used to explain this design process. If some of the words, phrases, and concepts in this example are unfamiliar, there is no cause for concern. You, the learner, will become totally familiar as you continue your reading (Fig. 1.7).

1.3.1 Design of a Chocolate Chip Cookie Acquisition System

Imagine that while reading this book, you experience an irresistible urge for chocolate chip cookies and a glass of milk. You can simply state the problem as, "I want chocolate chip cookies." You begin to explore how to get hold of chocolate chip cookies to satisfy this urge. You quickly discover that there are no chocolate chip cookies in your house, but you do have yogurt. Unfortunately, yogurt won't do today. You begin to explore your options. You could head over to a bakery and buy chocolate chip cookies. But, wait! That would cut into valuable study time. You simply cannot afford to do that! How about settling for having a pizza delivered? No! You don't want pizza. You want *chocolate chip cookies*. Frantically, you start rummaging the kitchen. Lo and behold, you find a tube of chocolate chip cookie

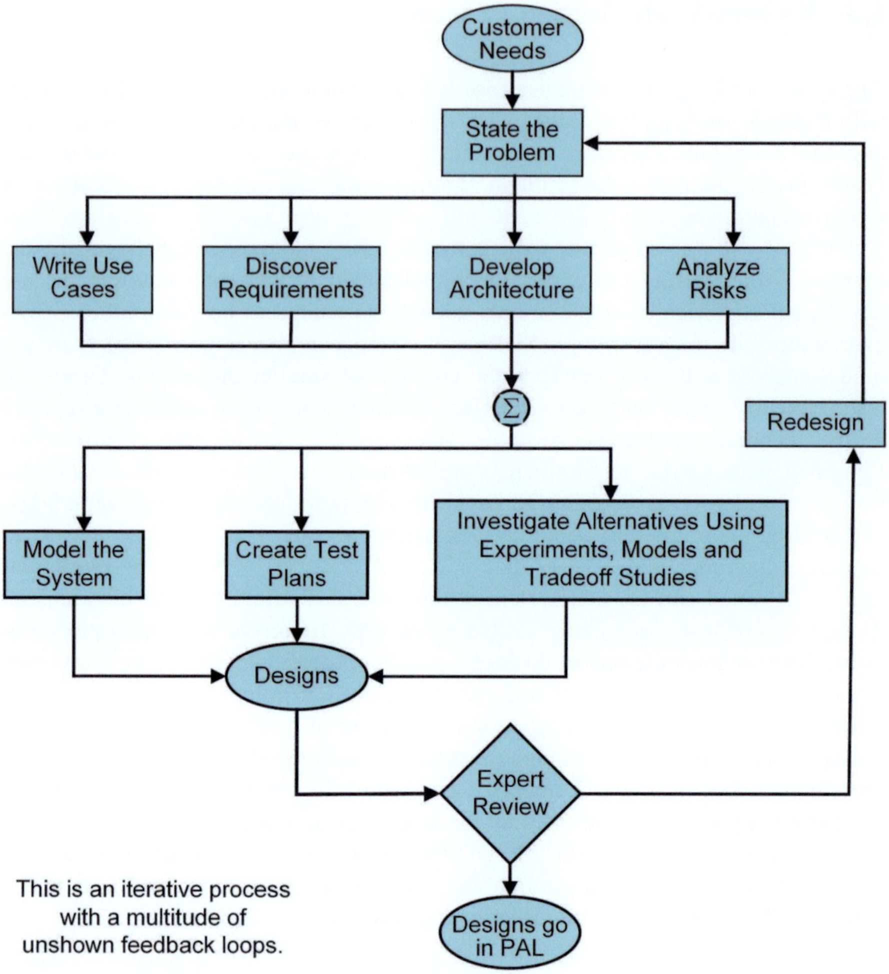

Fig. 1.7 The initial phase of the system design process

dough in your refrigerator! You are going to make chocolate chip cookies! Ah, it will take 20 minutes! Not a problem. That will work fine.

Now assume that you have a typical home kitchen—a stove, an oven, a timer, pots and pans, utensils and, of course, cookie sheets. You have the all-important tube of Pillsbury's Chocolate Chip Cookie dough, with instructions clearly printed on the label:

- Preheat oven to 350 °F.
- Spoon heaping teaspoons of well-chilled dough about 2 in. apart onto a cool ungreased cookie sheet.
- Bake at 350 °F for 10 min.

You are in business!

1.3.1.1 Concept Exploration

Now let us elaborate on the summary description from that last section. To begin with, you need to explore your available options (alternative concepts). You need to describe a few options and specify a few evaluation criteria. So, let's begin with the "buy-reuse-make" decision. You could go to the neighborhood bakery and simply buy the chocolate chip cookies. Or, you could try to reuse partial ingredients already in your house. Perhaps you may find a half-bag of potato chips in the pantry, or some cream puffs in the freezer. Then again, you could collect the ingredients (butter, flour, salt, baking soda, sugar, eggs, milk, vanilla extract, and chocolate chips) and make the cookies from scratch. Of course, there are other options. You could pretend to be on a diet and forgo the cookies! You could bake a cake, or make no-bake oatmeal cookies. Out of desperation, you could eat the cookie dough raw! Or, you could fall back to a trusted standby—ask your mother to bake cookies for you. These alternatives are conceived by the system architect.

How about the evaluation criteria? Potential evaluation criteria could include: Cookie freshness (time in hours), time spent with mother (hours), time to create (hours), cost (US dollars), flavor (survey of a dozen students), aroma (survey of a dozen students), chewiness (survey of a dozen students), and, of course, calories (kCal/cookie).

1.3.1.2 Write a Use Case

Write a use case that will describe how your system should work. Here is a simple use case template. You should use the relevant slots and ignore the rest.

1.3.1.3 A Use Case Template

A *use case* is an abstraction of required functions of a system. A use case produces an observable result of value to the user. Each use case describes a sequence of interactions between one or more actors and the system. Our design process is use case based.

Name: A use case should be named with a verb phrase in active present tense form. It should not relate to any particular solution.
Iteration: This is configuration management. Sometimes we just number them.
Derived from: Explain the source for the use case, for example, the mission statement, concept of operations, business use case or customer requirement.
Brief description: Describe the general sequence that produces an observable result of value to the user.
Level: The amount of detail required in the use case. Do not mix classes of different levels in the same use case.

Priority: The importance of this use case relative to other use cases.
Scope: This defines the boundary of what the use case applies to.
Added value: Describe the benefit (usually) for the primary actor. This is an important slot.
Goal: The goal is the behavior that the primary actor expects to get from the use case. You should have a goal or an added value, but probably not both.
Primary actor: Actors are named with nouns or noun phrases. Actors reflect roles of things outside the system that interact with the system. Primary actors initiate the functions described by use cases.
Supporting actor: Supporting (or secondary) actors are used by the system. They are not a part of the system, and thus cannot be designed or altered. They often represent external systems or commercial off-the-shelf (COTS) components. If your system affects them, then those effects are unintended consequences.
Frequency: How often is the use case likely to be used? When this slot is helpful, it is *very* helpful. When it is not, do not use it.
Precondition: The precondition (and the postcondition) should contain, among other things, the state of the system and values for pertinent attributes before the main success scenario starts.
Trigger: The trigger should contain the event that causes a transition from the precondition state to the first step in the Main Success Scenario.

Main Success Scenario

1. The numbered set of steps that illustrates the usual, successful interactions of actors with the system. Usually the first step states the action of the primary actor that starts the use case.

   2a. The last step tells you where to go next, for example [exit use case].

   Alternate Flows

   2b. Alternate flows describe failure conditions and unsuccessful interactions [exit use case].

Postcondition: Describes the state of the system after exit of use case no matter which flows were executed. This is hard to write.
Specific Requirements
The steps in the main success scenario should suggest the functions that the system is supposed to perform. From these we should be able to write system requirements.
Functional Requirements: Describe the functional requirements with shall statements.
Nonfunctional Requirements: Describe the nonfunctional (often performance) requirements with shall statements.
Author/owner: This is an important field.

Last changed: Use some form of configuration management such as the date of the last change or the revision history.

No standard specifies which slots should be in a use case description. Your minimal set should be based on your company requirements' template. The number of slots and the detail in each slot increase as the design progresses from the requirements model, to the analysis model, to the design model, to the implementation model. Other useful slots contain priority, rules, assumptions, and extension points. Do not use slots that do not help you. If you find that the trigger, precondition, and postcondition do not help you to create state machine diagrams, then do not use them. A use case description is also called a use case report and a use case narrative. A use case package contains a use case description, sequence diagrams, supplementary requirements, and other UML stuff.

1.3.1.4 Use Case for the Design of a Chocolate Chip Cookie Acquisition System

Name: Bake my Cookies (Fig. 1.8)
Iteration: 2.4
Derived from: Problem statement
Brief description: The Cookie acquisition system bakes cookies to perfection. It is named Cookie.
Level: High
Priority: This use case is of the highest priority.
Scope: A typical home kitchen with pots, pans, utensils, etc.
Added value: Students' brains always work better with a tummy full of cookies.
Goal: To produce stupendous freshly baked cookies
Primary actor: Student
Supporting actors: A tube of Pillsbury's Chocolate Chip Cookie dough
Frequency: Once a week
Precondition: All ingredients and cookware are available
Trigger: Student gets "hungry" for cookies.

Fig. 1.8 Use case diagram for the Cookie acquisition system

Main Success Scenario

1. Student decides to bake cookies.
2. Student turns on the oven and sets the desired oven temperature to 350 °F.
3. Cookie increases the temperature in the oven.
4. Student gets a tube of Pillsbury's Chocolate Chip Cookie dough out of the refrigerator and spoons heaping teaspoons of well-chilled dough about 2 in. apart onto an ungreased cookie sheet. Your Mom would probably worry about the expiration date of the product, which is a possible risk.
5. Cookie signals that the oven has preheated to 350 °F.
6. Student puts the cookie sheet full of cookies into the oven and sets the timer for 10 min.
7. Cookie signals that the baking time is over. If this signal is erroneously too late, the cookies could burn.
8. Student takes the cookies out of the oven. The cookie sheet will be hot. Student must wear an oven mitt. Student puts the cookies on a cooling rack and turns the oven off. Failing to turn the oven off creates a risk.
9. Student eats the cookies and notes their quality [exit use case].

Unanchored alternate flow: At any time, student can abort the process and turn off the oven [exit use case].

Postcondition: The kitchen is a mess but the oven is off.

The steps in the main success scenario suggest the functions that the system is supposed to perform. From these we can write the following system requirements [15].

Specific Requirements

Functional Requirements

ReqF1: Cookie shall provide a mechanism for student to enter a desired baking time. The abbreviation ReqF* means a functional requirement.

ReqF2: Cookie shall display the desired baking time entered by student.

ReqF3: Cookie shall heat up the oven from room temperature to 350 °F in less than 5 min.

ReqF4: Cookie shall calculate and display the remaining baking time.

ReqF5: Cookie shall emit an audible signal when the oven is preheated and when the baking time is over.

ReqF6: Cookie shall visually indicate when the oven is preheated and when the baking time is over.

ReqF7: Cookie shall execute Built-in Self-Tests (BiST). (Derived from company policy).

ReqF8: Cookie shall have a hard upper limit on oven temperature at 550°, even during self-cleaning. Rationale: this will help prevent fires.

ReqF9: Cookie shall turn off the oven when it is no longer being used. Implementation could be (1) turning the oven off 20 min after the end of the timer interval, (2) turning on an alarm 20 min after the end of the timer interval, (3) turning the oven off 20 min after the end of the timer interval if there is no food inside of the oven.

Nonfunctional Requirements

ReqNF1: The remaining baking time displayed by Cookie shall be visible to a student with 20/20 vision standing 5 ft from the oven in a room with an illuminance level between 0 and 1000 lx. The abbreviation ReqNF* means a nonfunctional (usually performance) requirement.

ReqNF2: Cookie shall raise the temperature of food in the oven so that temperatures at two distinct locations in the food differ by less than 10 %.

ReqNF3: Cookie shall update the remaining baking time display every minute.

ReqNF4: The audible signal emitted by Cookie shall have an intensity level of 80 ± 2 decibels (dB) at a distance of 30 cm and a frequency of 440 Hz.

ReqNF5: Cookie shall comply with section 1030 of Title 21—Food and Drugs, Chapter I—Food and Drug Administration, Department of Health and Human Services, Subchapter J: Radiological Health.

ReqNF6: The desired baking time shall be adjustable between 1 min and 10 h.

Author/owner: Hungry student
Last changed: November 24, 2016

1.3.1.5 Write a Test Plan for This System

An important part of the incipient system design is describing how the system will be tested. Fortunately, with uses cases this is a simple task. To test means to apply inputs, measure and record outputs, compare outputs to requirements, and finally indicate passing status.

This test plan is based on the main success scenario of the use case.

1. Tester turns on the oven and sets the desired temperature to 350 °F.
2. Tester waits until Cookie signals that the oven has preheated to 350 °F.
3. Tester stands 5 ft from the oven and observes the visual display. He measures the sound intensity and the frequency of the auditory signal from a distance of 30 cm. He measures the actual temperature inside the oven. He records the results.
4. Tester sets the timer for 10 min.
5. Tester waits until Cookie signals that 10 min is over.
6. Tester stands 5 ft from the oven and observes the visual display. He measures the sound intensity and the frequency of the auditory signal from a distance of 30 cm. He measures the actual temperature inside the oven, he notes the desired and actual elapsed time (10 min) and records all of the results.
7. Tester turns the oven off
8. Tester notes that the oven temperature is decreasing [end of test].

1.3.1.6 Risk, Unintended Consequences, and BiST

Risk management starts with the incipient system design. In this section, we (1) construct a risk register identifying and evaluating the major risks, (2) identify early tests or other measures that could be used to mitigate at least some of the

Table 1.1 A risk register

Risk no.	Potential failure event	Consequences	Likelihood per system usage	Severity	Risk
1	Incorrect ingredient quantities when baking from scratch	Off-taste; bad consistency	0.09	0.5	**0.045**
2	Product is out of date	Possible illness	0.002	0.9	0.018
3	Oven temperature too high	Burned cookies	0.009	0.09	0.0008
4	Baking time too long	Burned cookies	0.009	0.09	0.0008
5	Student could gain weight	Student would be unhappy	0.1	0.1	0.01
6	Failure to turn oven off	The oven could start a fire	0.01	0.95	0.0095
7	Student does not wear an oven mitt and burns his or her hand	Student is unhappy	0.04	0.8	0.032

Risk is defined as Likelihood times Severity

major risks, (3) describe a few potential unintended consequences of this system, and (4) describe a few BiST for this system.

Table 1.1 shows some of the risks to the system and its primary actor. Likelihood expresses our feelings about how likely this failure event is, on a scale of 0–1. Severity expresses our feelings about how severe this failure event is, on a scale of 0–1. Risk is the product of likelihood and severity.

The biggest risk is incorrect ingredient quantities, which is why people prefer to use a tube of Pillsbury's Chocolate Chip Cookie dough instead of starting from scratch. Next, we identify the most likely outcome and the most severe outcome. Gaining weight is the most likely event that needs to be mitigated, if convenient. A fire in the oven is the most severe event; therefore, we need to keep an eye on the oven.

Now we should identify early tests or other measures that could be used to mitigate the major risks.

Possible early mitigation tests include

1. Check "use by" dates on ingredients.
2. Use an independent (not a part of the oven) thermometer.
3. Use an independent (not a part of the oven) timer.
4. Cutting back on the quantity or quality of the butter or sugar is not a reasonable mitigation strategy.
5. Build-in a governor that will restrict the maximum temperature to 550 °F.

It is important to identify potential unintended consequences of the system. Two possible negative unintended consequences are (1) the aroma could attract undesirable neighbors to crash into your home and (2) heat generated by the oven could heat up the house (particularly undesirable in the summer). This could ultimately contribute to global warming!

All systems should have Built-in Self-Test (BiST). It is amazing that a simple oven system has at least five BiSTs: (1) the clock indicates that the oven is connected to electricity, (2) the oven-on light indicates that the function select knob is operable, (3) the preheated indicator shows that the oven heats up, (4) an internal temperature probe indicates that the oven is functioning, and (5) the student being vigilant by looking for smoke and smelling for cookies burning protects against catastrophic failures.

1.3.1.7 Model the System

Ports

It is important to recognize that most systems not only have multiple inputs, but they also have many *kinds* of inputs. For example, in the Cookie system the oven door represents one kind of input while the electrical power wires and the temperature knob represent two more. This possibility of different kinds of inputs leads to the idea of *input ports*. Sometimes there is a question as to whether a model should have one input port with multiple values, or multiple input ports with fewer values. To distinguish between these options, we must determine if the inputs could occur simultaneously. If they can, then they must be different input ports. If not, then they can be different values in the allowed set for one port. An input port has a set of allowed values each of which is possible individually, but no two can occur simultaneously. In our Cookie example, the allowed values for the oven door input port might be a cookie sheet or a turkey. Different input ports could have the same set of allowed values. For example, a bank with four tellers on duty would have four input ports. The allowed input values for each port would be the same: {money, checks, withdrawal slips, etc.}.

The allowed values for a port might be called messages. In diagrams, a connector joins two ports and serves as a conceptual communications path between the systems. A connector may be physical or conceptual.

Input ports are often called interfaces. As such, their description should include the originating system, the receiving system, the owner, and the set of allowed values (or messages).

Finally, there are two types of input ports: control and data. Control inputs are the events that can trigger transitions between states in state machine diagrams. Data (or material) inputs are the things that get transformed into outputs by functions. Do not mix control inputs and data (or material) inputs in the same diagram. Differentiating between control and data input ports was probably the most important system design decision of the last 40 years. It was established in 1981 by the U.S. Air Force with IDEF0. However, these valuable distinctions may disappear in future versions of the SysML standard [16].

Similarly, each system can have one or many output ports. The outputs are the actions or activities that are described on transitions between states or in the state boxes of state machine diagrams. Output ports are interfaces. As such, their descriptions should include the originating system, the receiving system, the owner, and the set of allowed values.

The analogous entities for states are called state variables.

Input and Output Ports for Cookie

The Cookie model will include a description of the inputs, outputs, and states of the system.

These are the (data or material) input flow ports and their legal values for Cookie:

Input flow port-1: electric wires or a gas pipe {natural gas or electricity}
Input flow port-2: the oven door {a sheet of Pillsbury's Chocolate Chip Cookie dough}

These are the input control ports and their legal values for the Cookie controller:

Input control port-1: tempKnob {200–550 °F}, desired oven temperature is set using the oven temperature knob
Input control port-2: ovenSetting {Bake, Broil, Clean, Off}, selects the function to be performed
Input control port-3: timerSwitch {Start, Stop, Reset}, controls the timer
Input control port-4: ovenTemp {150–550 °F}, the actual internal oven temperature as measured by a thermostat
Input control port-5: remainingBakeTime {0–600 min}, as indicated by the timer
Input control port-6: desiredBakeTime {1–600 min}

These are the output flow ports and their legal values for Cookie:

Output flow port-1: oven door {baked chocolate chip cookies}
Output flow port-2: top, bottom, and sides of the oven

These are the output control ports and their legal values for Cookie:

Output control port-1: preheatedIndicator (Light or Beeper) {On, Off}, indicates that the oven is heated up to the selected temperature
Output control port-2: Beeper {On, Off}, the beeper can have two functions. It can signal that the baking time is over and it can also signal that the oven is preheated. This is allowable according to Section 6
Output control port-3: ovenTemp {150–550 °F}, actual internal oven temperature as measured by the thermostat
Output control port-4: remainingBakeTime {0–600 min}, indicated by the timer

A block diagram showing interconnections of the control inputs and outputs of the Cookie acquisition system is shown in Fig. 1.9.

These are the states for Cookie:

{Oven is Off,
Oven is Heating Up,
Oven is Preheated,
Baking Cookies,
Baking Time is Over}.

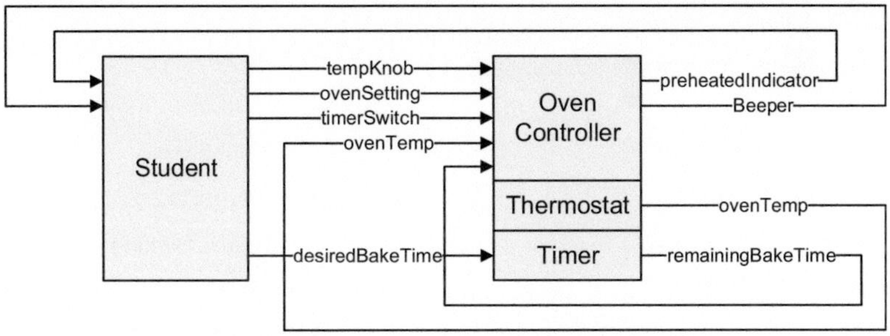

Fig. 1.9 Block diagram showing the interconnections of input and output ports

This is the next state function for Cookie. Each line shows
((Present State, the input condition that causes a transition to), the Next State)

{((Oven is Off, ovenSetting = Bake), Oven is Heating Up)
((Oven is Heating Up, ovenTemp = tempKnob), Oven is Preheated)
((Oven is Preheated, timerSwitch = Start), Baking Cookies)
((Baking Cookies, remainingBakeTime = 0), Baking Time is Over)
((Baking Time is Over, ovenSetting = Off), Oven is Off)}

Finally, this is the readout function for Cookie. Each line shows

(Present State, and the Output for that state)
{(Oven is Off, Nothing)
(Oven is Heating Up, Nothing)
(Oven is Preheated, turnPreheatedIndicatorOn)
(Baking Cookies, Nothing)
(Baking Time is Over, turnOnBeeper)}

A state machine diagram that shows the transitions between states is shown in Fig. 1.10.

In a state machine diagram, each state has the state name in the upper part of the state box and the outputs in the bottom part of the state box. Three types of outputs are supported: entry, exit, and do. The entry outputs are executed upon entry to the state. The exit outputs are executed upon exiting the state. And the do outputs are executed the whole time that the system is in that state. The inputs are labeled on the transitions between states.

When designing a system or a process, it is desirable to have as many elements in parallel as possible. The state machine shown in Fig. 1.11 has parallel concurrent states.

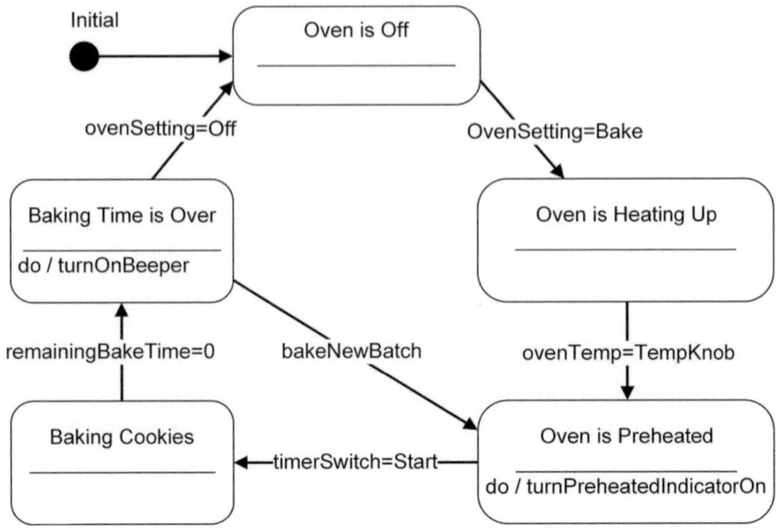

Fig. 1.10 A state machine diagram for the Cookie baking system from the *perspective of the oven*

1.3.1.8 Sensitivity Analysis

Students unfamiliar with the mathematics in this section may skip this section without experiencing loss of continuity. Sensitivity analyses are presented in detail in Section 26.

The instructions tell us to bake the cookies at 350 °F for 10 min. However, some might prefer to bake at 400 °F. It seems intuitive that with a higher temperature the baking time should be less. However, this is not a linear relationship, because baking soda only starts to decompose producing the gas that causes the cookies to rise when the cookies' internal temperature reaches 270 °F. A reasonable equation for the process is

$$\text{Doneness} = \frac{\text{Temperature}}{350} \times \frac{\text{BakingTime}^2}{100}$$

where Doneness is a number between 0 and 1, with 0.5 being just perfect. We will use the following abbreviations.

$$D = \frac{T}{350} \times \frac{BT^2}{100}$$

Alternative models that could be considered are

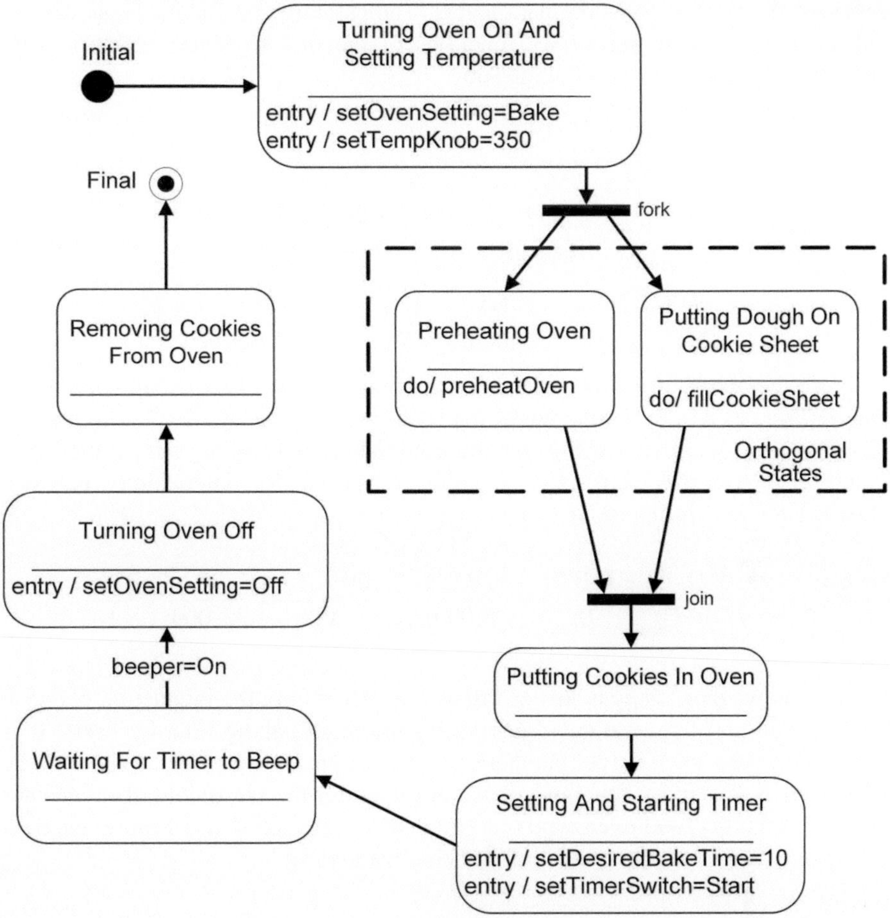

Fig. 1.11 A state machine diagram for the Cookie acquisition system from the *perspective of the student*

$$\text{Doneness} = \frac{\text{Temperature}}{350} \times \frac{\text{BakingTime}}{10} - 0.5,$$

which has a negative y-intercept and

$$\text{Doneness} = \frac{\text{Temperature}}{350} \times \frac{t-5}{10},$$

which has a time delay, because we must wait for the ingredients to heat up.

Oven temperature varies with position inside the oven and with time. Thermostats and timers are not especially accurate. Therefore, we would like to know how sensitive the process is to the parameters, Temperature and BakingTime. So, we calculate the relative sensitivity functions of the first Doneness equation with respect to Temperature and BakingTime.

$$\overline{S}_{\alpha}^{F} = \left.\frac{\partial F}{\partial \alpha}\right|_{\text{NOP}} \frac{\alpha_0}{F_0}$$

$$\overline{S}_{T}^{D} = \left.\frac{\partial D}{\partial T}\right|_{\text{NOP}} \frac{T_0}{D_0} = \left.\frac{BT^2}{35{,}000}\right|_{\text{NOP}} \frac{T_0}{D_0} = \frac{100 \times 350}{35{,}000 \times 1} = 1$$

$$\overline{S}_{\text{BT}}^{D} = \left.\frac{\partial D}{\partial \text{BT}}\right|_{\text{NOP}} \frac{\text{BT}_0}{D_0} = \left.\frac{2T\text{BT}}{35{,}000}\right|_{\text{NOP}} \frac{\text{BT}_0}{D_0} = \frac{2 \times 350 \times 10 \times 10}{35{,}000 \times 1} = 2$$

From the above, it can be seen that, the baking process is twice as sensitive to BakingTime as it is to Temperature. Does this match your intuition? Well maybe. The process is more sensitive to baking time than it is to baking temperature. We could also use this sensitivity function to check for interactions between BakingTime and Temperature.

$$\overline{S}_{T-\text{BT}}^{D} = \left.\frac{\partial^2 D}{\partial T \partial \text{BT}}\right|_{\text{NOP}} \frac{T_0 \text{BT}_0}{D_0^2} = \left.\frac{2\text{BT}}{35{,}000}\right|_{\text{NOP}} \frac{350 \times 10}{1} = \frac{20}{35{,}000} \times \frac{3500}{1} = 2$$

This shows that $\overline{S}_{T-\text{BT}}^{D}$ is just as big as $\overline{S}_{\text{BT}}^{D}$. Therefore, the interaction between baking time and temperature is just as important as baking time by itself. It is unusual to find such large interactions. An alternate sensitivity function, the semirelative sensitivity function, yields the same result. We should also consider altitude, humidity, and oven type (e.g., forced-air convection and heating element type/placement) as alternative environmental variables.

1.3.2 Section Summary

The advantage of use case-based design is that requirements, test plans, risks, unintended consequences, and BiST come right out of the use case, almost for free. When writing and rewriting the steps of the use case, the engineer should be alert for these opportunities and record them appropriately. Here is an example excerpted from the Cookie acquisition system.

Use Case for Cookie
Name: Bake my Cookies

Main Success Scenario

1. Student decides to bake cookies.
2. Student turns on the oven and sets the desired oven temperature to 350 °F.

3. Cookie increases the temperature in the oven.
4. Student takes a tube of Pillsbury's Chocolate Chip Cookie dough out of the refrigerator and spoons heaping teaspoons of well-chilled dough about 2 in. apart onto an ungreased cookie sheet.
5. Cookie signals that the oven has preheated to 350 °F.
6. Student puts the cookie sheet full of cookies into the oven and sets the desired baking time for 10 min.
7. Cookie signals that the desired baking time is over.
8. Student takes the cookies out of the oven, puts them on a cooling rack and turns the oven off.
9. Student eats the cookies [exit use case].

Functional Requirements

ReqF1: Cookie shall provide a means for student to enter a desired baking time.
ReqF2: Cookie shall display the desired baking time entered by student.
ReqF3: Cookie shall heat up the oven and cook the food.
Note: Requirements should contain only one concept. So, later, this requirement will be decomposed into two requirements.
ReqF4: Cookie shall calculate and display the remaining baking time.
ReqF5: Cookie shall emit an audible signal when the baking time is over.
ReqF6: Cookie shall indicate visually when the baking time is over.
ReqF7: Cookie shall execute Built-in Self-Tests (BiST). (Derived from company policy.)

Nonfunctional Performance Requirements

ReqNF1: The remaining baking time displayed by Cookie shall be visible to student with 20/20 vision standing 5 ft from the oven in a room with an illuminance level between 0 and 1000 lx.
ReqNF2: Cookie shall raise the temperature of food in the oven so that temperatures at two distinct locations in the food differ by less than 10 %.
ReqNF3: Cookie shall update the remaining baking time display every minute.
ReqNF4: The audible signal emitted by Cookie shall have an intensity level of 80 ± 2 dB at a distance of 30 cm.
ReqNF6: The desired baking time shall be adjustable between 1 min and 10 h.

Test Plan for Cookie
This test plan is based on the main success scenario of the use case.

1. Tester turns on the oven and sets the desired oven temperature to 350 °F.
2. Tester waits until Cookie signals that the oven has preheated to 350 °F.
3. Tester stands 5 ft from the oven and observes the visual display. He measures the volume and frequency of the auditory signal from a distance of 30 cm. He measures the actual temperature inside the oven. He records the results.
4. Tester sets the timer for 10 min.
5. Tester waits until Cookie signals that 10 min is over.

6. Tester stands 5 ft from the oven and observes the visual display. He measures the volume and frequency of the auditory signal from a distance of 30 cm. He measures the actual temperature inside the oven, He notes the desired and actual elapsed time (10 min) and records all of the results.
7. Tester turns the oven off
8. Tester notes that the oven temperature is decreasing [end of test].

Risks for Cookie

Potential failure event	Consequences	Likelihood	Severity	Risk
Incorrect ingredient quantities when baking from scratch	Cookies taste bad	0.09	0.5	**0.045**
Product is out of date	Possible illness	0.002	0.9	0.018
Oven temperature is too high	Burned cookies	0.009	0.09	0.0008
Baking time is too long	Burned cookies	0.009	0.09	0.0008
Failure to turn off the oven	The oven could start a fire	0.01	0.95	0.0095

Risk is defined as likelihood times severity

Unintended Consequences

To the extent possible, it is important to identify potential unintended consequences of the system. Two possible negative unintended consequences are (1) the aroma could attract intrusive neighbors to knock and seek entry and (2) heat generated by the oven could over-heat up the kitchen (particularly undesirable in the summer). Could this ultimately contribute to global warming?

BiST

All systems should have Built-in Self-Test (BiST). It is rather impressive that a simple oven system already has four BiSTs: (1) the clock indicates that the oven is connected to electricity, (2) the "oven on" light indicates that the function select knob is operable, (3) the preheated indicator shows that the oven does heat up, and (4) an internal temperature probe indicates that the oven is functioning normally.

1.3.3 Homework

1.3.3.1 SIMILAR and Cookie

Map the SIMILAR process to the Cookie acquisition system of Section 3.1.

1.3.3.2 SIMILAR and System Design Process

Map the SIMILAR process to the system design process of Fig. 1.7.

1.3.3.3 Cookie Figures

Figures 1.9, 1.10, and 1.11 are examples of what?

1.3.3.4 Cookie System

These are the (data or material) flow ports for the Cookie acquisition system.

- Input flow port-1: electric wires or a gas pipe
- Input flow port-2: the oven door
- Output flow port-1: the oven door
- Output flow port-2: top, bottom, and sides of the oven

These ports do not seem to appear in any of the figures. How would they be used? Draw a diagram showing an oven with these input and output flow ports.

Similarly, these two output control ports do not seem to be used.

- Output control port-3: ovenTemp {150–550 °F}
- Output control port-4: remainingBakeTime {0–600 min}

Should we eliminate them?

1.3.3.5 Input Ports

How many input ports are needed to model the following? Describe your input ports.

A pushbutton telephone in normal operation?
A human being?
A laboratory door combination lock? Typically, these locks have five buttons that can be pressed alone or in combination, and a doorknob, as in the combination lock homework problem SZ12.
A combination lock for a school locker? It is the type where you must first turn two complete turns clockwise and stop on the correct number, then turn one complete turn counterclockwise and stop on the second number, then turn clockwise and stop on the final number (Fig. 1.12).

1.3.3.6 System Synopsis for the Cookie Acquisition System

A system synopsis is a quick way to explain a system. There are examples in http://sysengr.engr.arizona.edu/forStudents/index.html. Write a system synopsis for the Cookie acquisition system. Use the following template based on [17].

Fig. 1.12 School locker lock

System Synopsis Template

For	(target customer)
who	(statement of need or opportunity)
the (system name)	is a (system category)
that	(provides key benefits or problem solving capability)
unlike	(primary competitive alternative)
(our system)	(statement of primary differentiation)

Put your system-specific information inside the parentheses ().

A System Synopsis is called an *elevator speech*, because of the following scenario. You step into the elevator and see your Vice President. You realize that you will have about one minute alone with her. You want to spark her interest, so that she will read your concept of operations when it shows up on her desk. So, you blurt out your system synopsis.

1.4 Section 4. Systems Engineering Design Documents

In general there are about eight different documents that the system designer must write while designing a system. They are often collectively called a Systems Engineering Management Plan (SEMP). They are living documents: they grow and they are continually updated throughout the entire system life cycle. They are not written sequentially, but they are structured so that they can be read sequentially. Just as Introduction, Methods, Results, Discussion, and Conclusion is not the order in which a scientific paper is written, but it *is* the order in which a scientific paper is usually read. Creating these eight documents is not a serial process. There must be many iterations and there are many opportunities for parallel processing.

> "Begin at the beginning," the King said gravely, "and go on 'til you come to the end; then stop." (From Lewis Carroll, *Alice's Adventures in Wonderland*.)

Document 1: The Problem Situation Document is the executive summary. It specifies the system's mission, explains the customers' needs and expectations, states the goals of the project, defines the business needs, prescribes the system's capabilities, delineates the scope of the system, expresses the concept of operations, describes the stakeholders, presents the key decisions that must be made, shows the suggested architecture, and (in the final version) highlights the preferred alternatives. This document summarizes the four most important program metrics: performance, cost, schedule, and risk. It lists the deliverables: what will be delivered to whom and when. It contains the ubiquitous language glossary. It is written in plain language and is intended for management and the public.

Document 2: The Customer Requirements Document is a detailed description of the problem in plain language. This document must not describe a preconceived solution. It is based on the Specific Requirements sections of the use cases and the Supplementary Requirements Specification. It is intended for management, the customer, and Systems Engineering.

Document 3: The Derived Requirements Document is a technical description (or model) of the problem statement and the requirements of Document 2. Each requirement in this document must trace to Documents 1 or 2 and identify rationale for why the requirement was derived as stated. Alternative names for this document are technical requirements and design requirements. The audience for Document 3 is engineering.

Document 4: The Verification and Validation Document has four parts. **Validating the system**: Building the *right system*: making sure that the system does what it is supposed to do. Validation determines the correctness and completeness of the product, and ensures that the system will satisfy the actual needs of the customer. **Verifying the system**: Building the *system right*: ensuring that the system complies with its requirements and conforms to its design. **Validating requirements**: Ensuring that the *set* of requirements is correct, complete, and consistent, that a model can be created that satisfies the requirements, that a real-world solution can be built that satisfies the requirements, and that this real-world solution can be tested to prove that it satisfies the requirements. If the client has requested a perpetual motion machine, this is the place to stop the project and save the money. **Verifying requirements**: Proving that each requirement has been satisfied. Depending on the system, verification can be done by logical argument, inspection, modeling, simulation, analysis, test, or demonstration**.** The verification matrix must show a one-to-one mapping between the requirements and the test plan. Its audience is Systems Engineering and the customer.

Document 5: The Concept Exploration Document is used to study alternative system designs through brainstorming, modeling, simulation, prototypes, and finally the complete system. In early iterations of document 5, this exploration helps produce the problem statement and the incipient system architecture. This document is rewritten continually as more information becomes available. Evaluation criteria are derived from a subset of the trade-off requirements. They are used in the heart of later iterations, which is a trade-off study where the preferred alternatives are suggested by the data. If your system is not unprecedented, then your alternatives should include the competition. You should explain the significance of the number of digits being printed after the decimal place. If the "do nothing" alternative scores high, you should suggest a cause. Often this is caused by having too many dependent cost figures of merit. A sensitivity analysis identifies the most important parameters in a trade-off study, often these are the cost drivers that are worthy of further investment. A sensitivity analysis completes a trade-off study by gauging the robustness of the preferred alternatives and the rationality of the overall study. This is one of the most popular documents. It should be readable by the public.

Document 6: The Use Case Model contains many use case reports that taken together describe the required behavior of the proposed system. The system design should be based on these use cases. The early use cases should be independent of potential solutions. The intended audience for Document 6 is management, engineering, the customer, the public, and Systems Engineering.

Document 7: The Design Model shows the objects that implement the required functionality of the system. System interfaces and their rationale are characterized in this document. This document should contain the system boundary diagram, a list of assumptions, class diagrams, sequence diagrams, state machine diagrams, block definition diagrams, internal block diagrams, and parametric diagrams. As time progresses the design model will be expanded into the implementation model and then into the operational model. Limited resources will probably constrain development of design models for only the preferred alternatives.

Document 8: The Mappings and Management Document contains a potpourri of vital information, such as a table that shows the mappings between the requirements of Documents 2 and 3, the verification plan of Document 4, the evaluation criteria of Document 5, the functions of Document 6, and the objects of Document 7. Document 8 should also include activity diagrams, the user's manual, the risk analysis, schedule information (Gantt Charts, Pert Charts, etc.), the project work breakdown structure, the business plan, and a log of meetings with the customer.

1.4.1 Example: Disposing of Financial Records

Here is an extremely abbreviated example of the eight system design documents. Your customer has said that she needs a system to get rid of a box of old financial records.

Document 1: Problem Situation
Top-level system function—dispose of a box of paper financial records.
Owner—person who owns the financial records.

Recommended solution—burn the records in the fireplace and put the ashes into a Container.

Performance—no residue should be left in the house.

Cost—our customer is willing to spend about 75 dollars US on this project.

Schedule—the financial records must be gone one week after this design is complete.

Risk—if the system does not destroy the financial records completely, then the customer will be worried about identity theft.

Document 2: Customer Requirements

Deficiency—our customer's home has run out of storage space and it is cluttered with obsolete financial records

Input/output—The input is a box of financial records, the output is an empty box. Yes, that is right: the customer's statement of need is usually incorrect. In this case, the customer does not want to get rid of the box; she probably wants to minimize the contents.

Functional requirement—The system shall destroy financial records.

Operational life—the system shall be reusable and shall have a life of 5 years.

Technology—the customer's house is in a residential section of Pima County and it has all of the technology that is typical in this area.

Cost requirement—customer has about 75 dollars for this project.

Trade-off—the Performance Requirement will be given ¾ of the points and the Cost ¼ of the points.

System Test—the volume of residue will be measured.

Glossary—financial records include things like bank statements, stockbroker reports, and credit card statements.

Document 3: Derived Requirements

Problem statement—our customer needs to free up space in her home and to get rid of obsolete financial records.

Inputs, functions and outputs—

Input—a box full of paper financial records.

High-level function—get the customer's financial records out of her house and on the way to the landfill.

External output—residue taken to landfill.

Input—a box full of paper financial records.

Low-level function—destroy records.

Internal outputs—empty box and residue.

Operational life—the system shall function without replacement of major components for 5 years.

Technology—only commercial off-the-shelf equipment can be used.

Performance-amount—The system shall destroy one cubic foot of paper and leaving no more than one cubic inch of residue in the house. [This sounds like two requirements rolled into one.]

The total performance score will be computed using

2 points for no residue left in house
1 point for one cubic inch of residue left in house
0 points for more residue left in house

Performance-rate—The system shall destroy at least one cubic foot of financial records per hour.
Cost—Total life cycle cost shall be about 75 dollars.
The total cost score will be computed using

2 points for total life cycle cost less than 50 dollars
1 point for total life cycle cost less than 75 dollars
0 points for total life cycle cost greater than 75 dollars

Trade-off—The system score is calculated by

$0.75 \times$ performance-amount score $+ 0.25 \times$ cost score.

The highest score wins.
If the schedule can be traded off, it would go in here, otherwise it is a mandatory requirement and it goes below in Acceptability.

System Test—The customer will observe the disposal of the records and measure the volume of the residue left in the house. The customer will evaluate system cost after reading the catalogs and advertisements that you show her.
Acceptability—(mandatory requirement) the system will be acceptable if it protects our customer's confidentiality in that the financial records are disposed of in a manner so that strangers cannot take and read the financial records.

Document 4: Verification and Validation
These requirements are complete, correct, and consistent. These requirements could be satisfied by taking the box of financial records to the campus police station and shredding them. Therefore, the system requirements are valid. We must still discover if our customer's real need is to get rid of the documents or to prevent identity theft.

Test Procedure

1. Tester puts one cubic foot of financial records into system.
2. System destroys financial records. Tester records the amount and the time.
3. Tester measures the amount of residue and records the result.
4. Tester disposes of residue.

Document 5: Concept Exploration
The following data came from experiments with prototypes.
Concept-1—throw the financial records into a garbage can.
Using the trade-off requirement, the scores are

Performance score—2 points
Cost score—2 points
Overall system score $= 0.75 \times 2 + 0.25 \times 2 = 2.00$

However, this system is *not acceptable*, because people could raid her garbage can and discover her financial secrets.
Concept-2—shred the records and then put them into a garbage can
Using the trade-off requirement, the scores are

Performance score—2 points
Cost score—1 points
Overall system score $= 0.75 \times 2 + 0.25 \times 1 = 1.75$

Concept-3—burn the financial records in the fireplace, and then put the ashes into a garbage can.
Using the trade-off requirement, the scores are

Performance score—2 points
Cost score—2 points
Overall system score $= 0.75 \times 2 + 0.25 \times 2 = 2.0$

Concept-4—do nothing
Using the trade-off requirement, the scores are

Performance score—0 points
Cost score—2 point
Overall system score $= 0.75 \times 0 + 0.25 \times 2 = 0.5$

Recommended alternative—The recommended alternative is Concept-3 (burn the records and put the ashes into a Container) with a score of 2.0.
Sensitivity analysis—Which requirement is the real cost driver? It is the need for confidentiality. Therefore, we should go back to the customer and ask if this is a real requirement. This is important, because it was not listed.
Rationale—Other concepts that were considered: use a trash compactor (paper does not compact well), put the records in a recycle bin (not confidential), shred the records, soak them in water and make new paper (labor intensive), store them in a Tuff Shed (too costly and a fire hazard).
Other considerations—Should we add an evaluation criterion for environmental impact *in* particular for air pollution? Is it hazardous to burn paper in a fireplace? Should the empty box be considered residue left in the house? We assumed the customer already owns a garbage can, thus trash pickup produced no additional cost. We observed that our customer does have a fireplace.

Document 6: Use Case Model
Name: Destroy Financial Records
Iteration: 2.6
Derived from: Problem statement
Brief description: Owner takes a box of financial records, put them into our system to destroy them and then disposes of the residue.
Added value: Owner gains valuable space in the house and feels better about the prospect of identity theft.
Level: User goal

Scope: This use case describes one usage of the system by owner in her house in Tucson Arizona.
Primary actor: Owner of documents and system
Supporting actor: Waste Management Truck and Shredder or Fireplace
Frequency: Once a week
Precondition: Owner has a box of financial records that she wants to dispose of.

Main Success Scenario

1. Owner puts appropriate amount of financial records into system.
2. System destroys financial records [repeat at step 1 until all papers are destroyed].
3. Owner disposes of residue [exit use case].

Alternate Flow
1b-1. System fails to destroy papers.
1b-2. Owner's husband fixes system [repeat at step 1].

Postcondition: The amount of financial records has been reduced.
Specific requirements
Functional requirement: The system shall destroy paper financial records.
Nonfunctional performance requirement 1:
The system shall be capable of destroying at least one cubic foot of financial records per hour.
Nonfunctional performance requirement 2:
The system shall be capable of destroying one cubic foot of paper and leaving no more than one cubic inch of residue in the house.
Author/owner: Paranoid Pack Rat
Last Changed: August 27, 2016

Document 7: Design Model
The system can be decomposed into two classes

1. Machine for destroying the records and
2. Container to hold the residue.

The function Destroy Records is assigned to the machine and the function Hold Residue for Trash Pickup is assigned to the container. The residue will be transferred from the machine to the container in plastic garbage bags.

Document 8: Mappings and Management

Doc 2	Doc 3	Doc 4	Doc 5	Doc 6	Doc 7
The system shall destroy financial records	The system shall be capable of destroying at least one cubic foot of	Owner will time the operation		The system shall destroy financial records	Machine for destroying records

(continued)

Doc 2	Doc 3	Doc 4	Doc 5	Doc 6	Doc 7
	financial records per hour				
	The system shall be capable of destroying one cubic foot of paper and leaving no more than one cubic inch of residue in the house	Owner will measure the volume of residue	Performance amount is an evaluation criterion in the trade-off study	Owner disposes of residue	Container
The system shall cost less than 75 dollars	Total life cycle cost shall be less than 75 dollars	Owner will evaluate system cost	Cost is an evaluation criterion in the trade-off study	Owner must buy machine for destroying records	Machine for destroying records
[Check with customer]	Acceptable if confidentiality is preserved	Owner will observe disposal of records		System destroys records	System for destroying records
Trade-off requirement	Trade-off requirement		Trade-off equation		
Input—box of records *Function*—get rid of records *Output*—empty box	*Input*—box of paper financial records *High-level function*—get customer's financial records on the way to the landfill. *External output*—residue taken to landfill *Input*—box of paper financial records *Low-level function*—destroy records *Internal output*—empty box and residue				

Check—Are there any requirements without functions or functions without requirements? No. That is good. Are any mappings missing? Yes. Confidentiality of Document 3 does not map to a customer requirement in Document 2. Therefore, ask the customer if she requires it.

1.4.2 Homework

1.4.2.1 Cooking Food

Your boss just moved into a new house with an unfinished kitchen but with a stocked refrigerator and pantry. The food needs to be cooked in a sanitary fashion, while satisfying cultural norms. You could order the right system to cook food for your boss over the next year. You want to make sure that your boss is pleased with your choice. Document 2, Customer Requirements, is a good description of customer's needs. At the very least, you should specify the Deficiency and the following requirements: Input/Output, Technology, Performance, Cost, and System Test. List at least two dozen alternative candidate systems that satisfy your requirements at least partially. Your answer should not take more than two pages.

1.4.2.2 Bridge Across the Santa Cruz River

The following is an edited transcript of an apocryphal meeting with a government administrator (G) and an engineer (E).

G: The bridge, across the Santa Cruz River, near Mission San Xavier del Bac has been washed out. We want to replace it. As you know, Father Kino, a Jesuit priest whom the Tohono O'odhams called "the padre on horseback," came to the Tohono O'odham village named Bac in 1692 and started the first mission. The present church was built in the 1770s. The name of this mission is unusual: it is a combination of Spanish (Saint Xavier of the) and Tohono O'odham (place where the water appears) languages. The Santa Cruz River is also very unusual, because it flows south to north. Get me some bids for this bridge.

E: Do you want a wood, steel, or concrete bridge?

G: I don't know. What's the difference?

E: Cost, maximum weight, maximum speed, amount of traffic, system life, and deterioration due to rain, salt, and termites.

E: How wide is the river?

G: About 30 meters.

E: How much water flows in the river?

G: Well, none from October to June. But in half the days of July and August it has one meter: sometimes we see two meters. In September, the average is about 15 cm.

E: So when it rains south of Tucson there will be water in the river. Otherwise, it is dry.

G: Yes

E: Who uses this road?

G: December to March hundreds of tourists use it daily. And throughout the year dozens of Tohonos use it daily to get from their homes near the mission to their Bureau of Indian Affairs Medical Facility.

E: If there were no bridge, what would be their alternative?
G: A six-kilometer detour.
E: Is it needed by emergency vehicles such as fire trucks or ambulances?
G: No.

Assume you are going to do the systems engineering for this project. Write the requirements for this system. Document 2, Customer Requirements, would be a good description of your customer's needs. At the least, you should specify the Deficiency and the following requirements: Input/Output, Technology, Performance, Cost, and System Test. List at least ten alternative candidate systems that you think should be considered. Do not restrict yourself to bridges, because often the customers do not know what they want. As with all real-world problems, if you do not have data, guess. Your answer must not take more than two pages

1.4.2.3 Extinguishing a Candle

Write the first five systems engineering documents for a system to extinguish a candle. You are **strictly** limited to three pages: so be concise.

1.4.2.4 Moving a Sheet of Paper

Write the first five systems engineering documents for a system to move a sheet of paper (8½ by 11 in. or 21 by 29.7 cm) across the room (10 ft or 3 m). You are **strictly** limited to three pages.

1.5 Section 5. Fundamental Principles of Good Design

Design is a creative activity. Consequently, there is no process that *guarantees* a good design. However, there are principles that can increase the likelihood of realizing a good design [18]. This section, which is based on Botta and Bahill [19], presents a set of principles for good system design. Using these principles will make a product better, more reusable for future systems, and cost-effective (i.e., lower redesign costs) when requirements change. Of course, the customer may choose to exclude some or all of these principles.

Principles of good design

Design for the total life cycle of the system
Use models to design systems
Use hierarchical, top-down design
Work on high-risk items first
Prioritize where possible

Control the level of interacting entities
Design interfaces
Produce satisficing designs
Resist premature optimization
Maintain an up-to-date model of the system
Develop stable intermediate configurations
Design resilient systems and processes
Use evolutionary development
Understand the enterprise in which the system will operate
State requirements in terms of *what* not *how*
Enumerate functional requirements in use cases
Allocate each function to just one component
Document all functions
Provide observable states
Conduct rapid prototyping to reduce implementation risks
Develop iteratively and test immediately
Employ a modular architecture
Create libraries of reusable metadata-tagged objects
Use open standards
Identify elements that are likely to change
Write extension points in use cases
Group data and behavior to simplify design
Use data hiding
Write a glossary of relevant terms
Circumscribe requirements with flexibility for revision
Incorporate design margins
Design for testability
Design for producibility
Design for evolvability
Design with human factors in mind
Build in preparation for buying
Create COTS technology-enabled new design process
Create incentives to achieve behavioral change in people
Write a lessons learned document

These design principles come from the experience of hundreds of engineers and managers. The particular references cited in the following paragraphs are not meant to be **the** authority; they are merely examples with entries to the literature.

Design for the total life cycle of the system. Stating the problem correctly is perhaps the most important phase of the system design process shown in Fig. 1.13. After that, we try to discover the system requirements. Now that we know what the system is supposed to do, we produce many alternative concepts that will satisfy the problem and the requirements. One or several (for example, in WWII, two completely different designs for the atomic bomb were pursued all the way through final delivery) of these alternatives will be selected for full-scale engineering

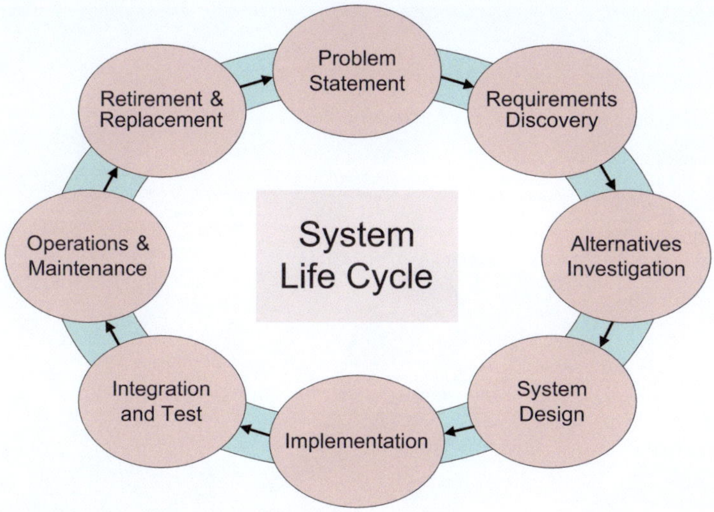

Fig. 1.13 The total life cycle of a system

design. Implementation means build the system: write the code and bend metal. Most systems consist of smaller subsystem. These subsystems must be interfaced and tested to ensure that they do what the system is supposed to do. The operations and maintenance phase of the system life cycle is short for a nuclear decay or a flashbulb and it is long for a building or the afterlife of a supernova. A very important phase of the system design is retirement and replacement. This should be addressed at the very beginning of the design.

There are dozens of life cycle models. Figure 1.14 shows one that emphasizes the iterations that occur. It starts in the center with the customer's need. Then we design requirements, design a model, build the model, simulate the model, and validate the model. Finally, we review all of that work. Then we start a new iteration. In this new iteration, the spiral is wider. Everything is bigger. We refine the old requirements and add new ones. We design a prototype, build the prototype, and test the prototype. Then we review all of that work. Now, we are ready to start another bigger iteration. With every action, the spiral gets wider.

Use models to design systems: System design can be requirements based, function based, or model based. Model-based system engineering and design has an advantage of executable models that improve efficiency and rigor. The earliest development of this technique was in Wymore's [20] book entitled *Model-based System Engineering*. Model-based systems engineering depends on having and using well-structured models that are appropriate for the given problem domain [21, 22]. An ancient Chinese proverb, which was invented by a New York City journalist a century ago, says "A picture is worth a thousand words." In engineering design, this phrase is elaborated to "A model is worth a thousand pictures." This means that models greatly reduce the complexity of a system description.

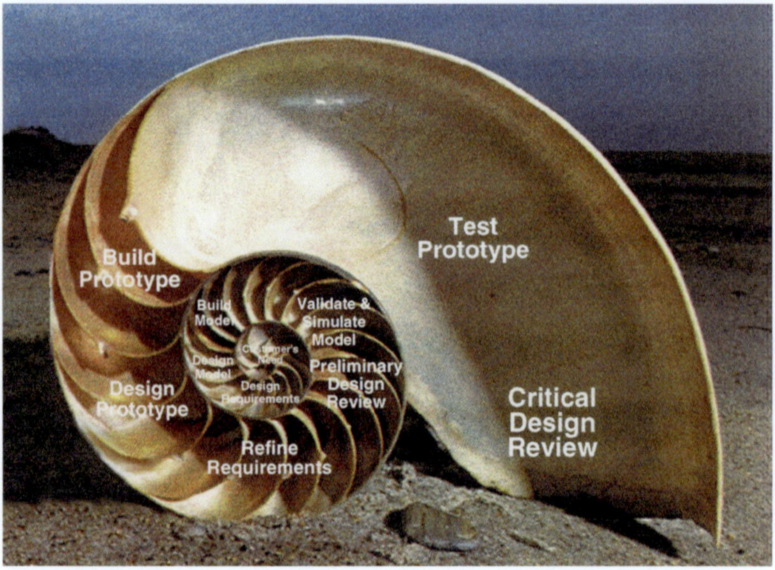

Fig. 1.14 The spiral life cycle model of a system

Use hierarchical, top-down design: Early on, translate the customer's needs into goals, capabilities, and functions: these provide guidance for all future development. Work on high-level functions first, because although high-level functions are less likely to change, when they do change, they cause changes in other functions. Decompose systems into subsystems, subsystems into subsubsystems, and so on [23]. In software, this decomposition is called layered architecture [24]. Implementation is invariably simpler if the dependencies and action initiations between these layers are unidirectional.

Work on high-risk items first: Work on high-risk items first in order to reduce risk; in addition, high-risk items are more likely to change thereby producing changes in other items. So working the high-risk items first will reduce rework due to changing requirements. Furthermore, if it were impossible to satisfy the high-risk capabilities and the project got canceled, you will have saved the money that otherwise would have been squandered in satisfying low-risk requirements [25]. The original spiral model of Boehm [26] advocated risk-driven development.

Prioritize: Requirements, goals, customer needs, capabilities, risks, directives, initiatives, issues, activities, features, functions, use cases, technical performance measures, weights of importance, and evaluation criteria should all be prioritized. Prioritization helps with budget, schedule, system architecture, customer satisfaction, and risk reduction [19].

Control the level of interacting entities: It is desirable that objects exchange inputs and outputs with other objects at the same level, or perhaps at one level above or below [21 2008]. It is a mistake to have objects interact with objects two or more levels above or below [27]. Some inter-level interactions may be specified as unidirectional [24], in which case the implementation is simplified.

Design the interfaces: Be sure to design interfaces [10, 28]. Interfaces between subsystems and between the main system and the external world must be designed. Subsystems should be defined along natural boundaries. When the same information travels back and forth among different subsystems, a natural activity may have been fragmented. In the automobile industry, they do not build a car frame, send it across town to install the motor and then send it back to put on the axels and wheels. Subsystems should be defined to minimize the amount of information to be exchanged between the subsystems [10].

Minimize subsystem interactions. Well-designed subsystems send finished products to other subsystems. Another way of stating this is to minimize the coupling between subsystems. Interfaces should connect things of similar levels [21]. Feedback loops around individual subsystems are easier to manage than feedback loops around interconnected subsystems. When possible, different entities should use the same interface, rather than having a specialized interface for each entity [29]. Browning [30] uses a Design Structure Matrix to analyze and modify the interfaces in a process. Special care should be given to interface design so that the interface does not have to change when its associated systems change. The name and the brief description of the interface should proclaim its intention.

Produce satisficing designs: Engineers should strive for designs that satisfice. They should not pursue the elusive goal of optimal designs, because for complex systems it is impossible to do so. Simon [31] wrote that the key to successful design is "the replacement of the goal of maximization with the goal of satisficing, of finding a course of action that is 'good enough.' ... Since the [designer] ... has neither the senses nor the wits to discover an 'optimal' path—even assuming the concept of optimal to be clearly defined—we are concerned only with finding a choice mechanism that will lead it to pursue a 'satisficing' path, a path that will permit satisfaction at some specified level of all its needs."

Do not optimize early: If optimization is absolutely required, do it late in the design process. If you optimize early in the design process, you will have to reoptimize every time the design changes. At the risk of being too specific, optimization should not be done before Critical Design Review (CDR). This does not mean that you should not strive to find a good architecture early in the design process. It means, for example, not using integer programming to minimize part counts or wire lengths before CDR. It means not using optimization methods for routing problems on networks with stochastic failures before CDR.

Maintain an updated model of the system: Engineering must create a model that simulates the intended system at every level from both the business and technical viewpoints [32]. A single model should underlie analysis, design, implementation, and team communication [24]. Update this model throughout the design process [33]. Update this model to "as built" in the implementation phase and "as modified" in the test, operational and retirement and replacement phases. When the requirements inevitably change, engineering should review all decisions, revise the system model, re-run all simulations, and show the effects of requirement changes on cost, schedule, performance, risk, and the designs of other systems [34].

Every operational system should have an accurate up-to-date model. This model should be run to evaluate both performance enhancements and possible disasters. If the system is a regional electric power grid and it has been suggested that its capacity could be increased by adding solar and wind generators, then the system performance should be evaluated on the model before it is installed on the real system. If the system is a large building, and evacuation because of a hurricane must be planned, then run the model to see how hurricane damage might disrupt the plan. Every Interstate highway bridge should have a model that degrades as the bridge degrades. When the model predicts a potentially dangerous situation, the bridge should be shut down. When a nuclear reactor is to be decommissioned, its model should be run to help formulate the decommissioning plan.

Develop stable intermediates: "A large change is best made through a series of smaller, planned, stable intermediate states" [10]. These intermediate states should be stable enough that the progression could be stopped at predetermined points. With stable intermediates, if a large complex system is canceled, for example, because of loss of political support, then something useful for mankind will still exist. The Central Arizona Project was designed this way, but the Superconductor Supercollider in Texas was not.

Design resilient systems and processes: Anticipate possible failure events and design recovery states that will allow rapid reoperation of the system. There are many ways to accomplish this. One of them is a layered defense, where if one component fails there are others that can perform its functions [34–37].

Use evolutionary development: The Department of Defense (DoD) created the evolutionary acquisition process. Using it, you create a usable system, and later add requirements and money to get a more complex usable system (DoD 5000.1). The B-52 airplane is a quintessential example. Over its 60-year life, every half-dozen years the Air Force added requirements and money and received a new capability. They hope to be just as successful with the F-35.

Understand the enterprise: Understand how the system being designed fits into the enterprise. Frameworks help people organize and assess completeness of integrated enterprise models. Several popular frameworks have been used to architect enterprises. The Zachman framework [38] like several others, considers multiple perspectives and multiple aspects of an enterprise [32].

State what, not how: State what function needs to be performed, not how to implement the solution. For example, do say, "Play music." Do not say, "Play a CD" or "Hire a pianist" or "Turn on the radio and tune into a music station." In the object-oriented software world, this is called polymorphism [39]. However, one man's floor is another man's ceiling. First, state *what* the customer needs. Then figure out *how* to satisfy this need. This *how* then becomes a system feature. State *what* this feature must do. Then figure out *how* to implement it. This implementation then becomes a requirement that states *what* the system is expected to do.

List functional requirements in the use cases: Traditional requirements are typically written as textual statements of imperative. They are the primary means by which systems engineers bound and communicate system functions, capabilities, and constraints. The rapid adoption of use case modeling for capturing functional requirements in the software community has caused systems engineers to adopt use case models for capturing system-level functional requirements. We advocate a hybrid requirements process (to be presented in Section 29) in which use case modeling and traditional shall-statement requirements are applied together to effectively express both functional and nonfunctional requirements [15]. The use cases should contain the main and alternate scenarios. These describe the system functions. The next section of the use case should formally describe the functional requirements.

Allocate each function to only one component: Each function should be allocated to only one physical component, and, therefore, each function would have just one owner. If there were two owners for a function, then one might decide to change his or her requirements, and this would change the system for the other. In the object-oriented world, this would be phrased as, "Do not allow multiple actors to have the same role [40]." If two actors are trying to assume the same role, generalize them into one abstract actor. There are exceptions to this principle, but they are unusual. Violation of this principle is captured in the American proverb, "Too many cooks spoil the soup." In the twentieth century, the functions of gathering intelligence and tracking terrorists were allocated to the FBI, the CIA, the NSA, the Pentagon, etc. Until 9/11/01, the results were mixed. This principle is discussed in more detail in Section 6.

Do not allow undocumented functions: Do not hide the existence of system functions. Make sure that all system functions are identified and described appropriately. Undocumented functions can exist due to developmental testing activities, accidents, playfulness, or malicious intent. The PDP 11-45 computer had secret op codes. Microsoft Excel 97 had a built-in flight simulator. Thousands of such Easter eggs are documented at http://www.eeggs.com/tree/1-1-111.html. Don't try this at home, but starting a text line with "=rand (200,99) (Enter)" (excluding the quotes) in a Word document or a PowerPoint presentation, will give you 235 pages of "The quick brown fox jumps over the lazy dog," and perhaps a computer crash. Microsoft fixed this feature/bug in Word 2013. Now instead 235 pages of the "The quick brown fox jumps over the lazy dog" you get 805 pages of how "To make your document look professionally produced."

Provide observable states: System equivalence cannot be proven using input/output behavior. State behavior must be used to prove equivalence of dynamic systems [41]. It would be best to provide the total system state, but it would still be useful to provide only a significant number of reset states or restore points

[42]. Providing states will allow future designers to reuse existing systems, upgrade systems, use commercial off-the-shelf (COTS) products, replicate field failures, verify that a physical system conforms to its design, and verify evolving systems [41, 43]. It will also allow them to find a system mode behavior function that can be placed in front of the input to the system being reused to produce the desired system behavior [20].

Accelerometers from automobile air bag systems can be reused as g-switches in missile safing systems if the states are observable. If the only indication the accelerometer gives is *Fired* or *NotFired*, then they are not reusable. But the state of the accelerometer can be defined as the g-force being measured. If this g-force can be observed, then they can be reused in missiles.

Rapid prototyping: Develop a prototype quickly. Prototypes can be "throwaway" or reusable. In the interest of time, prototyping is often done to assure early involvement of stakeholders [3]. This strategy helps identify the true needed capabilities of the system. This is particularly useful when delving into unfamiliar design areas. Therefore, the first prototypes are often discarded, particularly in software systems [44].

Develop iteratively and test immediately: Build a large-scale system by constructing it as a series of smaller products of increasing completeness. Because the early entities are small, it is easier to get them right and testing can be completed immediately [33].

Create modules: Design the system with many small independent but interacting modules [45]. The principle of independence means that changing one module should not force changes in other modules [46]. Good examples are the modules of the Unix (or Linux) computer operating system and the libraries of Java. Modules should have low external complexity (loose coupling) and high internal complexity (strong cohesion) [29]. Loose coupling should reduce or eliminate propagation of failures to other modules. When technology changes, you should be able to replace a module, leaving its interfaces intact.

Create libraries of reusable objects: Patterns are exemplar solutions for doing common things. There are many books with software patterns, see, for example, [47] and some with use case patterns [40]. However, this library principle is not restricted to software and use cases. For example, a company could create 100 generalized evaluation criteria (measures of effectiveness) that could be tailored and used in company trade-off studies. Each criterion would contain the name of the criterion, description, weight of importance (priority), basic measure, units, measurement method, input (with expected values or the domain), output, and scoring function (type and parameters) [48]. This criterion library would form the nucleus of a trade-off study, but many customized criteria would be added. This particular library should be a part of the company CMMI DAR process [49]. Libraries could also be created for generic requirements.

Use open standards: Open standards are publicly available specifications for accomplishing a particular task. Open standards are available for all to read and implement. Examples of open standards include ANSI/EIA-632, ANSI/EIA 731, IEEE 1220, ISO 15288, UML, SysML, OMG IDL, CORBA, and USB. In contrast, proprietary standards are controlled by a single entity. Java Server Faces can be the interface between HTML code and logic code. The Java Server Faces is a mass-market product that is highly scalable, but it is still controlled by a single commercial entity, likewise with Microsoft's Source Code Control Interface (SCCI) Specification.

Identify elements that are likely to change: Differentiate between aspects of a system that are likely to change and those that are likely to remain relatively constant. Identify the aspects that are likely to change and put extra effort into designing their interfaces. For example, an air transportation system will always have functions such as load people, load luggage, take-off, land, and comply with FAA regulations. It also has entities that are sure to change such as the type of aircraft, speed, size, airports, routes, and cargo capacity. The system should be designed to accommodate changes in the things that are likely to change [24]. For example, the processing speed of computers is expected to increase. Therefore, hardware and software should be designed so that changes are not necessary when processor speeds do increase.

Write extension points: In use cases, use extension points where change is expected [39, 49–51]. That way the base use case does not need to be changed when the extending use case is changed.

Group data and behavior: Group together data and the behavior (logic) that operates on it, because data and the behavior that accesses that data often change together [47]. Software accomplishes this with header files and with class diagrams that have attributes (data) and operations (behavior). Hardware accomplishes this with distributed control where each board has a processor and memory.

Use data hiding: Hide data from objects that do not have "a need to know" [45]. That way if the data structure is changed, the other objects do not have to be notified about the change. This principle is also called information hiding and function hiding [52]. In the UML, model elements, such as attributes and operations, can be marked with one of these visibility indicators: public, protected, private, or package. In public documents, describe the interfaces and only the public functions and data. This principle has nothing to do with security: its purpose is simply to improve the efficiency of making changes. Architecture hierarchies and levels also influence data hiding.

Write a glossary of relevant terms: Collect and provide definitions for terms used in the domain and design models. Capture the jargon of the domain experts. Include names of functions, classes, methods, modules, and high-level organizing principles. This collection of terms is called a ubiquitous language [24].

Circumscribe requirements: Estimate requirements top-down and fill in details later, for example, we need to test (measure) a DC voltage that surely will be less than 40 V. Later we find that it will probably be between 5 and 20 V. Finally, we get a requirement of 12 ± 2 V. This principle ameliorates changing requirements.

Incorporate design margins: If your estimated need is 2 GHz of bandwidth, specify four. But this is expensive: if everyone did this, the system would be gold-plated. It is better to give all of the flexibility in cost, schedule, and performance to the program manager and allow him or her to distribute it across the project.

Design with human factors in mind: That is, design products, systems, and processes to take account of the interactions among humans and other elements of the system. Design systems that fit the human body and its cognitive abilities.

Design for testability: Early in the design process, it should be determined how the system will be tested. The architecture should be selected to enable built-in self-test (BiST). BiST can be triggered externally or it can be performed whenever the system is not busy performing its normal functions [53].

Design for producibility: Early in the design process, the system designers should talk to manufacturing to determine that would simplify or complicate manufacturing.

Design for evolvability: The capacity of an existing system to successfully adapt to changing requirements throughout its life cycle is called evolvability [54]. A system can adapt to change by reconfiguring existing system entities, by increasing the size of existing system entities, or by adding new entities. Often design, development, and testing take multiple years. Technologies can change during that period. Designs should have adequate flexibility and adaptability to take advantage of this fact rather than fighting it. The classic example is Motorola's satellite phone system that was made obsolete by cell phone technology before it was completed.

In designing the Boeing 747 airliner, Boeing engineers designed weight growth capability into the airplane by giving it bigger wings and tail surfaces than it initially needed. The landing gear was more robust than needed and it was designed so that landing gear that is more capable could be substituted without the expense of a major redesign. This allowed a dozen different versions of the 747 to be built over 30 years [55].

Build in preparation for buying: If a commercial off-the-shelf (COTS) product that suits your needs is available, then of course you should prefer to use it rather than develop a new product in-house. But if a COTS product is not available, then build a custom product and monitor the market place: when a COTS product becomes available, switch to that COTS product. In hardware, sometimes the contractor will use commercial grade products while waiting for military standard parts to come along.

Create a new design process: When using a significant amount of commercial off-the-shelf (COTS) products you must change your design process. These are some of the new activities that must be performed. Search the marketplace for candidate COTS products and technologies. Work with the customer in early project phases to refine the customer needs statement and get high-level, flexible objectives, instead of extensive requirements lists. Iteratively negotiate the requirements and the architecture with the customer. Change contracting so that funding for upgrades and marketplace research does not end with the project's period of performance. Create a satisficing design that may not meet all of the requirements, but one that provides the high priority capabilities and satisfies the customer's operational needs. Create test procedures that will be reused in the Operation and Maintenance phase of the system life cycle whenever upgrades or new COTS products are brought into the system. Plan for insertion of upgrades and technologies throughout the system life cycle. Continually search the marketplace for upgrades and new products that can replace existing COTS products, particularly in the Operation and Maintenance phase. Choosing to design with a significant amount of COTS products is a decision that should not be made lightly. Indeed the preliminary COTS-Based Engineering package of one technology company had a dozen documents containing 3 MB of text.

Write lessons learned: At the end of a project, the design engineers should describe the lessons that were learned on the project. The audience for this document is future design engineers. This document must be stored where it is easily accessible by future engineers, but it must be safeguarded because it may contain embarrassing mistakes.

Get people to change their behavior: This is a last resort, and requires changing incentives and inhibitors that cause people to behave in a certain way. It often means trying to change the bureaucracy and exploiting human nature to arrive at a stable set of requirements up-front.

1.5.1 Design for Reuse

New systems can be designed better, faster, and cheaper if reuse of previously developed artifacts is done correctly. For example, when it comes to software, reuse should pertain to software models, not computer code. The artifacts/products might have been developed in-house, or purchased as commercial off-the-shelf (COTS) components. There are several facets to design for reuse: (1) refining customer needs and negotiating system requirements with the customer, (2) selecting and evaluating potential reuse products, (3) technology planning and insertion, (4) product life cycle planning, and (5) designing products so that they are more reusable. This section concerned the fifth topic, design for reuse. Designing systems with increased likelihood of reuse is not free: it requires extra effort and cost. This produces another facet of the design for reuse process: convincing your boss or customer that it is worth the extra cost to design a system for increased likelihood of future reuse.

1.5.2 Section Summary

This section has presented essential principles for good design. Not all apply to all programs. Importantly, none of these principles contradicts each other. If care is not taken, the principle of "Develop iteratively and test immediately" could confound the principles of "Use hierarchal, top-down design" and "Work on high-risk items first." But in a careful design, the high-level or high-risk items would be decomposed into smaller modules that could be tested immediately. Similarly, "Design for testability" and "Design for evolvability" are orthogonal, not contradictory. There are no other obvious interactions between these principles.

Some of these principles are just good design principles: they help with reuse, changing requirements, testability, and evolvability. Whereas others primarily ameliorate changing requirements, increase reuse potential, or enhance evolvability. Table 1.2 presents these principles and indicates the specific facets that they apply to most. However, the Xs are fuzzy, not crisp. The right column indicates principles that are likely to impact the short-term cost of the system. But, of course, the trade-off between short-term cost and customer value is a complex issue [56].

Table 1.2 Design principles and pertinent facets

Design principle	Reuse potential	Changing requirements	Evolvability	Good design	Impact cost?
Design for total system life cycle	X	x	X	x	
Use models to design systems	x	x		X	
Use hierarchical, top-down design				X	
Work on high-risk items first				X	
Prioritize where possible		x		X	
Control the level of interacting entities				X	
Design the interfaces				X	
Produce satisficing designs				X	
Resist premature optimization				X	
Maintain an up-to-date model of system				X	$
Develop stable intermediate configurations		x	X	x	
Use evolutionary development		x	X		$
Understand larger enterprise context			X	x	$
State requirements as "what," not "how"				X	
Enumerate functional requirements associated with use cases		x		X	$
Allocate each function to just one component		x		X	

(continued)

Table 1.2 (continued)

Design principle	Reuse potential	Changing requirements	Evolvability	Good design	Impact cost?
Ensure all functions are documented	x			X	
Provide observable states	x		x	X	$
Conduct rapid prototyping (implementation risks, user acceptance)		x		X	$
Develop iteratively and test immediately		x		X	
Employ modular architecture, where feasible	X			X	
Create libraries of reusable, metadata-tagged objects	X				$
Use open standards whenever possible	X		x		
Identify elements that are likely to change and make provisions for such changes		x	X		
Write extension points where needed		x		X	
Group data and behavior to simplify design				X	
Use data hiding to reduce cognitive complexity				X	
Write a glossary of relevant terms	x			X	$
Circumscribe requirements with flexibility for revision		X			
Incorporate design margins		X			$
Design for testability				X	$
Design for evolvability			X		$
Build in preparation for buying	X				$
Create COTS-technology-enabled new design process	X				$
Change right incentives to achieve behavior change in people		X			$

Upper case X means the principle has a great affect. Lowercase x means the principle has substantial affect. The dollar sign indicates that use of the principle comes with a cost

The aforementioned principles of good design are concrete, and low-level. They do not cover basic systems engineering principles such as stating the problem, discovering requirements, and conducting trade-off studies [5, 47, 56–58], performing configuration management, and exploring "hotspots" through sensitivity analyses [59–61]. Of course, these other activities must be done, but they are covered adequately elsewhere. Examples of failures that result from violating these system design principles are also given elsewhere [60–61].

While most engineers and managers are probably familiar with the principles of good design, there is value in presenting them all in one place as a checklist, to make sure that they are all considered when designing systems.

1.5.3 Homework

1.5.3.1 Principles of Good Design

For the Cookie acquisition system, did we *Work on high-risk items first*?

Prioritize the functional requirements in the Bake My Cookies use case by listing them in order of importance, the most important first.

What interface of Cookie is hiding complexity from the user?

What optimization technique could have (but has not yet) been applied to the Cookie acquisition system?

How could we have designed the system to satisfy the principle of *Develop stable intermediate configurations*?

Suggest ways to make the Cookie acquisition system resilient.

What is the enterprise context in which the Cookie acquisition system will be deployed?

Modify the Cookie acquisition system so that it violates the principle of *State requirements in terms of what not how*.

Point out the sections of Cookie that *Enumerate functional requirements in use cases*.

Does Cookie provide observable states?

Point out the undocumented functions that might be in the Cookie acquisition system.

Modify the Cookie acquisition system so that it violates the principle of *Use open standards*.

In the Cookie acquisition system, *Identify elements that are likely to change*.

For the Cookie acquisition system, *Write a glossary of relevant terms*.

What incentives does Cookie provide that could change the behavior of people?

1.5.3.2 Defining the Customer

Assume that you have just been hired as a teaching assistant for your college's course in Numerical Methods. Before you start lecturing, you decide to design the system that will teach and administer this course. The set of *all* people who have the right or responsibility to impose requirements on a system is a called the Customer. Describe your customer by filling out the following portion of Systems Engineering Document 1. You are limited to one page.

Owners:
Bill payers:
The Client:
Users:
Operators:
Beneficiaries:
Victims:
Technical representatives:
Social impact:
Economic impact:
Environmental impact:

1.5.3.3 Requirements Specification

Suppose your boss wants you to order a digital clock from a company in Japan. You do not have their catalog, but you are supposed to describe what you want, and obtain their most acceptable item. Write the requirements that you will send to them. Make sure you get it right, because it takes a long time to ship things from Japan, it is too expensive to clarify things on the telephone and your boss will be mad if you waste money. You should at least specify the following requirements: Input-Output, Technology, Performance, Cost, and System Test. Your answer should not take more than two pages.

1.5.3.4 System Life Cycle-1

Assume you were hired on June 1 to teach a two-semester course on Systems Engineering that would run from September to May. Assign the seven phases of the system life cycle to the 12 months.

1.5.3.5 System Life Cycle-2

Like everything in this book, there is no unique definition of the System Life Cycle. If you ask a dozen companies, you will get a dozen definitions. The definition will even differ depending on whether the thing being designed is a product, like a bus, or a project, like a municipal bus system.

Life Cycle 1

Research and development
Market introduction

Growth
Maturity
Deterioration
Death

Life Cycle 2

Conceptual Phase
Definition Phase
Production Phase
Operational Phase Divestment Phase

Which of these is for a product and which is for a project?

1.5.3.6 System Life Cycle-3

The definition of System Life Cycle even varies from industry to industry.

Life Cycle 1

Startup
Definition
Main analysis
Termination

Life Cycle 2

Formation
Buildup
Production
Phase-out
Final audit

Life Cycle 3

Conceptual
Planning
Definition and design
Implementation
or alternatively
Analysis
Design
Programming
Testing
Implementation

Life Cycle 4

Planning and data gathering

Studies and basic engineering
Major review
Detailed engineering
Construction
Testing and commissioning

Which of these is from Construction, Manufacturing, Computer Programming, and Engineering?

1.5.3.7 System Life Cycle-4

The following describes the system life cycle for a type of systems and includes the percentage of the total life cycle cost associated with each phase. What type of a system is being described?

1.5.3.8 Concept Exploration

Assume you were hired on June 1 to teach a two-semester course in Systems Engineering that would run from September to May. What concepts would you explore in the Concept Exploration Document that you write in the Investigate Alternatives Phase of the system life cycle?

1.5.3.9 System Test

At the behest of the federal government many states, counties, and municipalities have mandated that gasoline be laced with ether or alcohol to reduce carbon monoxide emissions. If you are in one of these localities, find out what the performance requirements were (i.e., what was supposed to improve and by how much?) and find out what system tests were designed to see whether or not these additives improved the air quality. Be aware that these gasohols also increase formaldehyde emissions, and aldehydes are injurious to the human respiratory system.

1.5.3.10 Retirement and Replacement of the Hubble Telescope

What should we do with the Hubble Space Telescope when it reaches the end of its life cycle?

1.5.3.11 Retirement and Replacement of a Baseball Stadium

Describe the retirement and replacement of a baseball or a football stadium. For an example, you might use the 1.3 billion dollar New York Yankee's stadium, which opened in 2009 http://newyork.yankees.mlb.com/nyy/ballpark/stadium_history.jsp or any other new stadium

https://ondeckcircle.wordpress.com/2013/01/20/major-league-ballparks-oldest-to-newest/

1.6 Section 6. Principles of Function Design

Functional decomposition is a core system design technique. It begins by identifying the top-level function that the system must perform, then decomposing that function into subfunctions, and then decomposing the subfunctions into subsubfunctions, and so on. This decomposition continues until you get to functions that can be implemented with commercial off-the-shelf products. When using this design technique, creating the functions and allocating them to components should not be arbitrary. The functions should be designed using principles that help with good design.

1.6.1 Allocate Each Function to Only One Component

Each function should be allocated to one and only one physical component and, therefore, each function will have only one owner. In the object-oriented world, this concept is stated as, "Do not allow multiple actors to have the same role [40]." If two actors are trying to assume the same role, generalize them into one abstract actor. In a theater play, it would also be mistake to allow two actors to assume the same role. This principle is easy to implement as long as you pay attention to the principle. Let us now look at a few examples of following and violating this principle.

One function allocated to two components, bad design: Violation of this principle is captured in the proverb, "Too many cooks spoil the broth." In the twentieth century, the function of tracking terrorists was allocated to the FBI, the CIA, the NSA, the Pentagon, etc. Until 9/11/01, the results were mixed. In a baseball game, if the runner on first base is likely to steal second base, then the function of covering second is allocated by agreement to the second baseman or the short stop, but never to both. This agreement prevents a bad design. This principle may be the reason why back-seat drivers are so distained.

One function allocated to two components, exceptions that proof the rule: One valid reason for assigning a function to more than one component is that the function is performed by one component in a certain mode, and by another component in another mode. For example, to decelerate an airplane, on the landing approach the pilot employs air brakes (the big flaps on the top and bottom of the trailing edges of the wings); when the wheels first hit the runway the pilot reverses the thrusters on the engines; later, when preparing to move from the runway to the taxiway he applies the brakes on the wheels. Another reason for assigning one function to two components would be to incorporate deliberate redundancy to enhance reliability. For example, allowing one portion of the system to take on a function if another portion fails to do so, as was the case on the Space Shuttle, which had five flight control computers. Another reason is to truly distribute functionality. For example, you can decelerate a car with either front wheels or back wheels. For decelerating an airplane, the air brakes on the wings, the reverse thrusters on the engines, and the brakes on the wheels might be for redundancy, or distributed functionality. But if this is the case, then everyone should know about it.

One system with two subsystems and two functions, OK design: A claw hammer can be used to pound nails or pulls nails. A pencil is used to make or erases lines. A teapot can heat water and also whistle. A clock radio wakes up people or plays music. A heating, ventilation and air conditioning system both heats and cools a house. Computers perform multiple tasks, often at the same time.

One object with two functions, OK design: A drawbridge allows cars to move over the river or boats to move under the bridge. Mammal sex organs are used for reproduction and waste elimination. An air conditioner cools and dehumidifies the air. A dog can guard the house and be a pet.

One object with two functions, marginal design: Sports stadiums have been built for both football and baseball games. However, these stadiums are not popular now, and this design has fallen out of favor. A combination washer and dryer, washes clothes and dries clothes: these are not currently popular. An item with functions of create and operate in the same object is viewed as a bad design [24]. When you buy a new car and it needs service, you do not take it to the factory that manufactured it. You take it to a separate service center.

One function with two purposes, OK design: An oven cooks food and kills bacteria. It has one function (heat food) but two purposes. A dishwasher cleans dishes and kills bacteria.

1.6.2 Assure Functional Independence

Functions should not depend on each other. This principle assures that when one function is changed, it does not affect others. For example, one may routinely use clock radio to wake up with music. But when in a strange town, one may want to use an alarm buzzer, in case the radio station is off the air, or the radio is mistuned when one wants to wake up. Allowing the wake-up function to depend on the play music function is risky and one needs to avoid it when the wake-up function is critical. There are techniques to help maintain functional independence [62].

A bad design: On most company telephones, dialing 9 gets an outside line, and then dialing 1 starts a long distance dial. At this point, if the 1 button is inadvertently hit again, then the caller is connected to 911, the emergency line in some countries, such as the United States. If you hang up real quick, they send a policeman to save you.

Limit side effects: Functions should do what their names imply and nothing else. For example, in an automobile, the Ignition Switch turns the ignition system on and off, but it also has a side effect of disconnecting electric power to the radio. Renaming this switch the Electric Power Switch would improve this side effects problem.

In the summer of 2015, Fiat Chrysler was fined $105 million and forced to recall a million cars because the ignition switch could wiggle out of the run position and disable the air bag system.

These principles apply to **object-oriented design** as well. In use cases, the functions are described in the Functional Requirements part of the Specific Requirements Section (to be discussed in Section 29) [15]. In activity diagrams, functions are the activities. In state machine diagrams, functions are the actions and activities. In class diagrams, functions are the operations listed in the third part of the class box. In sequence diagrams, functions are the commands. So functions appear in many places.

In an object-oriented design, it would be a mistake to give two functions the same name and it would be a mistake to assign one function to two classes. However, it is very natural to have one function appear in a use case, a class diagram, a state machine diagram, and a sequence diagram. It would be specified in one place and used in multiple diagrams.

It is interesting to note that these principles of good design do not apply to biological systems because biological systems evolve, they are not human-engineered.

1.6.3 Homework

1.6.3.1 Functional Design

Section 1.6.1 tells us to *Allocate Each Function to Only One Component*. Has the Cookie acquisition system violated this principle?

1.6.3.2 Disposing of Financial Records

What were the functions discovered for the Disposing of Financial Records system of Section 4?

1.6.3.3 Functional Decomposition

Assume that you will be teaching a course in systems engineering this coming semester. Do a functional decomposition of your job. Consider administrative as well as academic matters.

1.7 Section 7. Generic Tactical Missile

System design is an iterative process. It uses sequential incremental iterations (model a little, build a little, test a little, repeat). There are multiple iterations and many opportunities for parallel processing. Throughout the design of a system, many use case models are written. Usually, each iteration produces use case models that are larger and more complex. Mappings can be created that show this progression between models.

This section shows the progression of use cases through three iterations for a generic tactical missile. They apply to a missile launched from an airplane, a ship, or a ground station. In these models, Fire Control is an electronic system that communicates with the missile and the operator. It is located on the airplane, ship, or ground platform.

These use cases were written by three different people at three different times. For simplicity they do not show the specific requirements section.

Name: Launch and Fly to Target (Fig. 1.15)
`Iteration: 1`
Brief description: Missile is launched from the platform, flies to the target, and detonates near the target.
Level: `high`
Added value: Depending on the type of missile, it may be to save a pilot's life, save a ship, or destroy a target.
Primary actor: Operator (pilot, crew member, or ship's captain)
Supporting actor: Fire Control and Target
Precondition: Missile is in Track Mode

Main Success Scenario

1. The Target emits energy continuously. This energy is received by the missile and by Fire Control.
2. Fire Control sends information to the missile continuously.

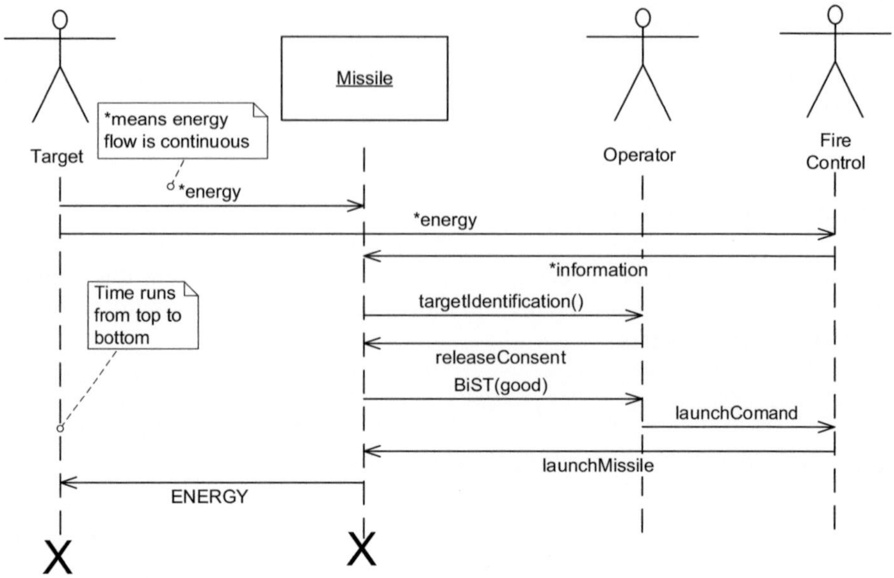

Fig. 1.15 Sequence diagram for the main success scenario of the first iteration of the Launch and Fly to Target use case

3. Missile sends target identification to Operator.
4. Operator sends *Release Consent* to the missile.
5. Missile sends BiST(good) to the Operator.
6. Operator tells Fire Control to launch the missile.
7. Fire Control launches the missile.

8a. Missile delivers ENERGY to the target [exit use case].

Alternate Flow

8b. Missile misses the target and self-destructs [exit use case].

Postcondition: Missile has detonated.
Author/owner: Terry Bahill
Last changed: Summer 2016

Name: Launch and Fly to Target
Iteration: 2
Brief description: Missile is launched from the platform, flies to target, and detonates near the target.
Level: *medium*
Primary actor: Operator
Secondary actor: Target and Fire Control
Precondition: Missile has acquired the Target.

Main Success Scenario

1. Operator sends *Release Consent* to missile.
2a. Missile sends BiST(good) to operator.
3. Operator sends launch signal to Fire Control.
4. Fire Control sends launch signal to missile.
5. Missile is launched.
6. Sensor acquires energy from the Target.
7. Guidance and Control guides missile toward the Target.
8. The *away-switch* tells the Safety and Arming (sometimes called Safing and Arming) device that the missile is away.
9. The *g-switch* sends an acceleration signal to the Safety and Arming device.
10. Motor sensor tells the Safety and Arming device that the boost motor has burned out.
11. Safety and Arming device arms the missile.
12a. Fuze tells the processor that it is time to detonate.
13. Processor detonates the warhead.
14. Fire Control confirms target destruction to the Operator [exit use case].

Postcondition: Missile has detonated or it was not launched because it failed BiST.

Alternate Flows

2b. Missile sends BiST(fail) to operator.
2b1. Missile is not launched [exit use case].
12b. Fuze does not tell processor that it is time to detonate.
12b1. Time out device tells processor to self-destruct the missile.
12b2. Processor self-destructs the missile [exit use case].

Author/owner: Jeff Gurvine
Last changed: Fall 2016

Name: Identify and Track Target
Iteration: 3
Brief description: Operator identifies desired target based on sensed energy (e.g., electro-optical or IR signature) and commands missile to track it.
Level: *low*
Primary actor: Operator
Secondary actor: Target
Precondition: Missile is in Ready Mode (missile has been "boresighted" or aligned to platform gunsights and/or wing pylons; power has been applied; environmental conditioning of sensors is complete).

Main Success Scenario

1a. Operator identifies potential targets, using either the missile seeker or an external targeting system.

2. Operator enables the acquisition sequence by "uncaging" (physically unlocking) the missile seeker and slewing it to establish line of sight (LOS) with the desired Target.
3. Missile seeker detects target energy and provides indication to Operator.
4. Operator commands the missile to track the desired Target.

5a. Missile tracker responds to input from the seeker and begins closed-loop tracking; missile provides a "good lock" indication to Operator [**include** Fly to Target use case].

Postcondition: Missile is in Track mode (i.e., ready to launch) or sortie has been aborted.

Alternate Flows

1b. Operator is unable to identify valid targets. Sortie is aborted [exit use case].

5b. Target signature is too weak to establish lock-on (insufficient to guarantee that tracking would be maintained throughout missile flight). Operator responds by maneuvering the aircraft to achieve adequate signature (e.g., closing in) [repeat at 1 or exit use case].

Author/owner: Kathleen Finkelstein, (c), 2016, used with permission
Last changed: Winter 2016

Name: Launch and Fly to Target (Fig. 1.16)
Iteration: 3
Brief description: Operator commands launch of the missile. Missile is propelled from the launch platform and guides to the target. Missile detonates in proximity to target or upon impact.
Level: *low*
Primary actor: Operator
Secondary actor: Target
Precondition: Missile is in Track Mode (i.e., missile has acquired and "locked on" to a target) from the end of the Identify and Track Target use case.
Trigger: Operator issues a Launch command.
Main Success Scenario

1a. Missile activates its secondary (on board) power source (e.g., thermal battery).

2a. Secondary voltages reach minimum threshold in a specified time.

3. Missile ignites the propulsion subsystem and begins to move forward.

4a. At missile separation from the launch platform (signaled by the *away-switch*) and detection of sufficient voltage from the secondary power source, the arming sequence of the Safe-Arm Fuzing device (SAF) begins.

5. After a specified time, control surfaces are unlocked and the control surface actuation system begins responding to steering commands from the autopilot.

6a. The missile seeker maintains closed-loop tracking of the Target.

7. The autopilot generates steering commands based on inputs from the missile seeker indicating LOS angle, and sensed missile acceleration and angular rate from an on-board sensor package, in accordance with a predetermined guidance law that optimizes flight trajectory.

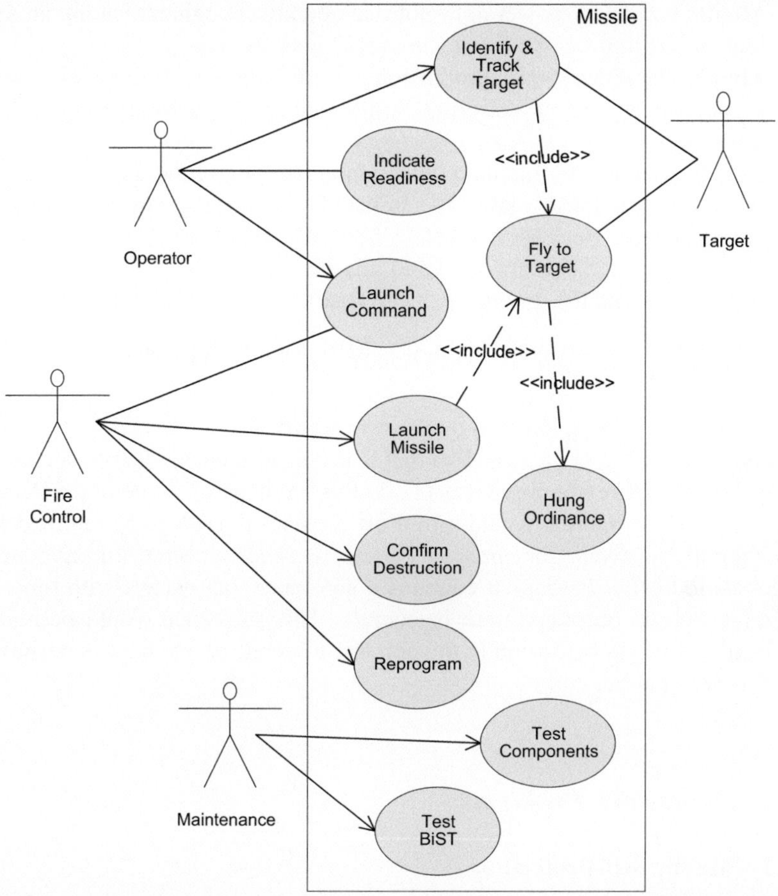

Fig. 1.16 Use case diagram for the third iteration of the Identify and Track Target and Launch and Fly to Target use cases

8. After a specified time duration and a sensed acceleration profile (by the *g-switch*), the SAF arms.
9. The missile propulsion system concludes operation and flight continues based on missile potential and kinetic energy.
10. When the missile reaches the Target, the SAF initiates the warhead detonation sequence in response to a signal from the proximity or contact trigger mechanism.
11. Warhead detonates [exit use case].

Postcondition: System has one less missile available.

Alternate Flows

1b. Missile does not respond to Launch command. Operator issues a second Launch command. If missile still does not respond, [exit use case].

2b. Missile secondary power does not rise quickly enough, indicating an unsafe launch condition, and missile "misfires" [exit use case].
4b. Missile "hangfires" (i.e., propulsion system is ignited, but missile does not separate from launch platform). Sortie is aborted. [**include** Hung Ordnance use case].
6b. Missile is unable to maintain closed-loop tracking of the Target.
6b1. Alternate guidance law is initiated, modifying missile trajectory appropriately.
6b2. Fuzing sequence is terminated and missile "duds."
6b3. Missile impacts the ground [exit use case].

Author/owner: Kathleen Finkelstein, (c), 2016, used with permission
Last changed: Winter 2016

Protection from being hoist on one's own petard

The missile should be designed so that if it falls into enemy hands, it cannot be used against us. There are many ways to do this. Perhaps someone in the Pentagon could send out a signal that would turn it off. Perhaps it needs to be returned to the manufacturer every year for *updates*. Perhaps it needs a secret arming code that changes periodically. Perhaps it contains a substance that decays with time, as it decays the missile becomes more inaccurate. This substance would be replaced periodically. It might be desirable to keep this a secret, or maybe it is desirable to make this *very* public.

1.7.1 Homework Problems

1.7.1.1 Missile Self-Protection

We suggested several techniques to ensure that if a missile fell into enemy hands, it could not be used against us. Suggest another technique. That paragraph ended with the sentence "It might be desirable to keep this a secret, or maybe it is desirable to make this *very* public." Which do you think would be best?

1.7.1.2 Release Consent

What is the release consent message shown in the sequence diagram of the generic missile?

1.7.1.3 Identify Functions

Identify the missile system functions that are implied by iteration 1 of the use case Launch and Fly to Target and the Identify and Track Target use case. For example, step 1 of the main success scenario of iteration 1 of the use case Launch and Fly to Target implies that

1. The missile receives energy from the Target and
2. Fire Control receives energy from the Target.

Do any of these functions violate the principles of function design given in Section 6?

1.7.1.4 Safety and Arming

Draw an electromechanical Safety and Arming (S&A) device that would be appropriate for the Launch and Fly to Target, *Iteration: 2* use case.

1.7.1.5 Little Boy

Figure 1.17 shows the arming, fuzing, and firing (AF&F) system for Little Boy, the first atomic bomb. When the B-29 was near its target, the Electronics Test Officer (ETO) removed the three green safing plugs on the bomb and replaced them with red safing plugs. This armed the bomb. When the bomb cleared the bomb bay doors, six arming wires attached to the bomb bay were pulled out of a bank of six clock-operated switches in the bomb. These clocks then began counting down from 15 s. The end of this interval turned on (powered up) the four radars. Six barometric pressure sensors were already active. When they sensed an altitude of 6600 ft, they activated the radars. When at least two of the four radars detected an altitude of 2000 ft, a signal was sent through the three firing switches that detonated the primers. Those three primers exploded the four silk bags of cordite explosives. This shot a subcritical mass of U-235 shaped as a hollow cylinder (the bullet) down the gun barrel. When the cylinder surrounded the subcritical U-235 solid cylindrical target, a super critical mass was created. Neutrons from four beryllium/polonium initiators, started the nuclear explosion. Discuss the redundancy in this system. Draw a sequence diagram, as in Fig. 1.15, for this system. To help you to identify classes, we have underlined some nouns that are good candidates [63].

1.8 Section 8. Model-Based System Engineering

Model-based systems engineering is transforming systems engineering from a document-driven process to a model-based process capable of generating documents on demand [22]. The models in MBSE are represented using languages such as the Unified Modeling Language (UML) [64] and the Systems Modeling Language (SysML) [16, 65]. It is important to recognize that MBSE does not equate to UML or SysML. However, these languages provide a convenient notation for representing a system from different perspectives. In this section, we use these

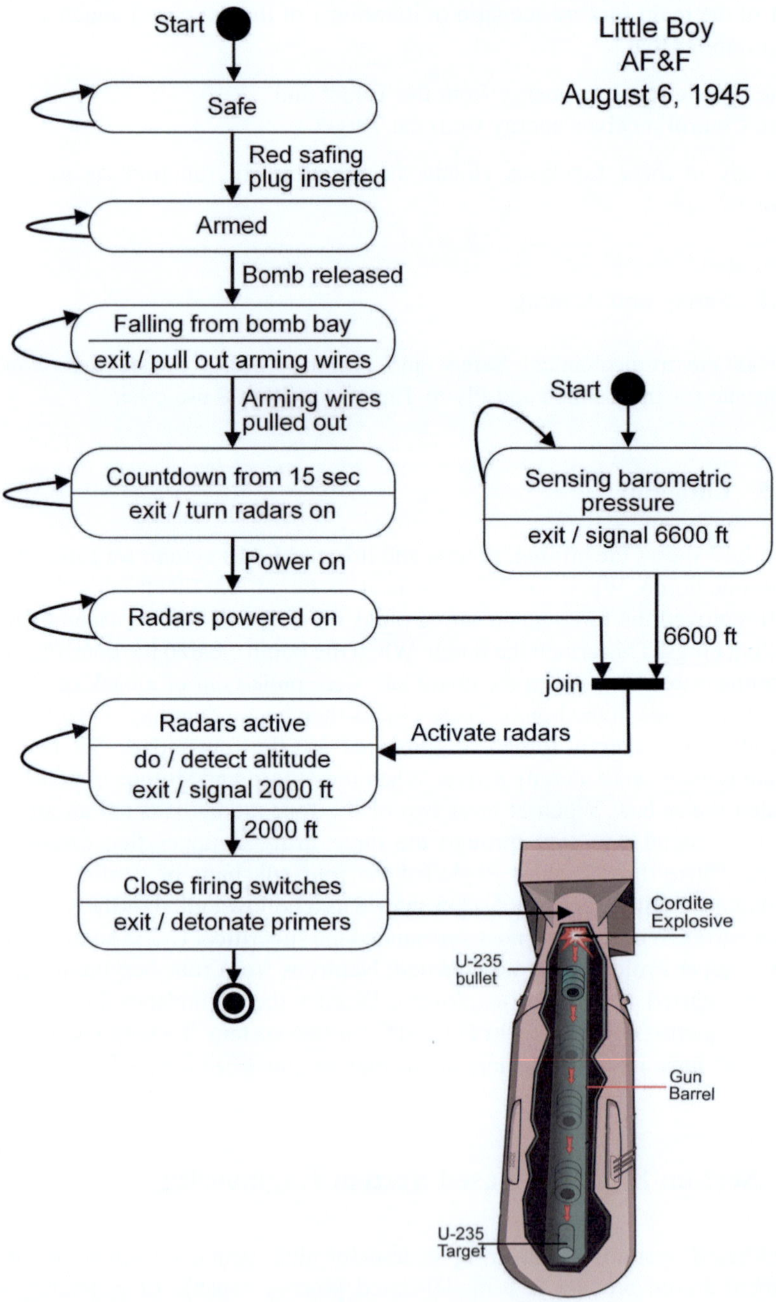

Fig. 1.17 State machine diagram for the fuzing system of Little Boy

languages to depict different models. At the heart of MBSE is state-based modeling. State-based modeling is discussed next.

1.8.1 State-Based Modeling

We begin with a few key definitions. A *system* is a combination of interacting elements that performs a function that is not possible with any of the individual elements. The elements can include hardware, software, humans, facilities, policies, and processes. A system accepts inputs, over which it has no direct control, and transforms them into outputs. A system should have a well-defined boundary.

Systems can be categorized as memoryless, or dynamic. In a *memoryless system*, the outputs depend only on the present values of its inputs. In a *dynamic system*, the outputs depend on the present and past values of its inputs.

The concept of a *state* is an essential aspect of dynamic systems. The *state of a system* encapsulates the system's history. In other words, the state of a system contains all the information needed to calculate responses to present and future inputs without reference to the past inputs and outputs (that is, history). The state of the system, current inputs and the sequence of future inputs allow computation of future states (and outputs). For example, the present balance in your checking account can represent the state of that system. This state could have reached its present value in a number of ways. However, when you write a check, that historical path becomes irrelevant to you. On the other hand, your model depends on how you define the system. Credit agencies would want the state definition to include the number of times you had a negative balance. The bank would want the state definition to include your average daily balance to calculate your interest.

Some dynamic systems are best modeled with *state equations* while others are best modeled with *state machines*. State equation systems are modeled with equations. For example, a baseball's movement can be modeled with state equations that capture position, linear velocity, and angular velocity as functions of time. In contrast, state machine systems focus less on physical variables, and more on logical attributes. Therefore, these systems are said to have memory, and are modeled using the notation of finite state machines. Most computer systems, for example, are modeled with finite state machines using high-level software packages.

Most real-world systems are modeled as either state equation systems or state machine systems. However, there are rare physical systems where both types of models are useful together: for example, systems using water, because the heat flow dynamics are different depending on whether H_2O is in its liquid, solid, or gas state, Friedenthal, Moore, and Steiner [16], as shown in Fig 1.18.

The state variable in this model is temperature T and Q is the heat input and output. The state equation is $Q = Cm\Delta T$. Where Q is heat (J), m is mass (g), T is temperature (°C). C is the specific heat capacity, which is 4.186 J/(g °C) for water, about 2.01 J/(g °C) for steam, and about 2.09 J/(g °C) for ice. For the state

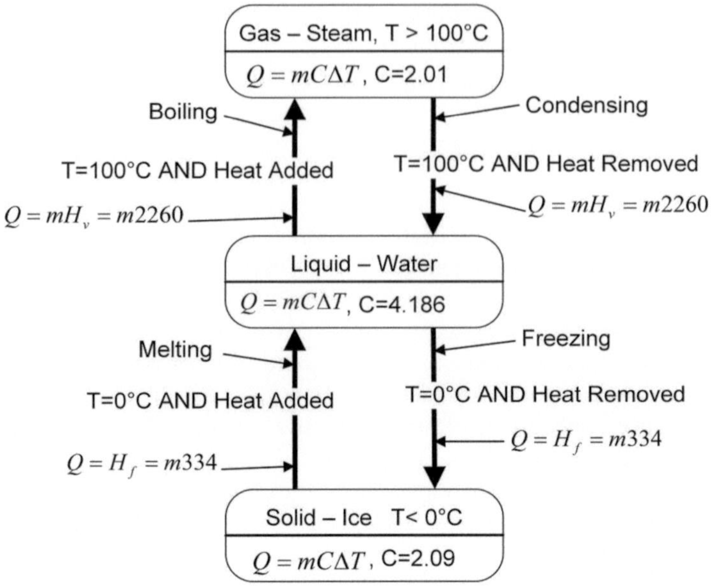

Fig. 1.18 An unusual system that uses state equations in a state machine diagram

transitions H_v is the latent heat of vaporization = 2260 J/g and H_f is the latent heat of fusion = 334 J/g. Besides using state equations in a state machine diagram, another reason that this system model is unusual is that the transitions are not instantaneous. Durations of the transitions depend on the speed of heat application. This could be fixed by adding four more states named Melting, Boiling, Condensing, and Freezing.

At each instant of time, a dynamic system is said to be in a specific state. State equation systems can have one or many state variables: at any time, the system's state is defined as the unique values for each of the state variables. State machine systems can be modeled with one or many concurrent state machines: at any time, each of the concurrent state machines must be in one and only one state. A *state* is a unique snapshot that (1) is specified by values for a set of state variables, (2) characterizes the system for a period of time, and (3) is different from other states. Each state differs from other states in either the inputs it responds to, or the outputs it produces, or the transitions along the way. A *transition* is a response to an input that may cause a change of state. In state diagrams, the inputs are labeled on the transitions that they produce.

1.8.2 Finesse Point

One of the first architectural decisions that the systems designer must make is whether to design the system as memoryless or dynamic. It is imperative that this decision be made and not left to default.

Memoryless systems are also called combinatorial systems because the outputs depend only on *combinations* of present inputs. As an example of a memoryless system, toss two coins at the same time, if one is heads and the other tails, then you win. The output depends only on the present inputs.

Dynamic systems are also called sequential systems because their outputs depend on the *sequence* of previous inputs. For example, toss one coin twice. If the first toss is heads and second toss is tails, then you win. The output depends on the sequence of inputs. This is modeled nicely with a state machine diagram. However, not all dynamic systems are modeled best with state machine diagrams. For example, toss one coin twice. If one toss is heads and the other tails in either order, then you win. The output depends on the present input with memory of the previous input. This merely needs a simple system with a memory element. As another example, it would be cumbersome to model a computer program with a state machine diagram. But computer programs are dynamic systems, where the state is specified by the stack pointer and the status register. Counters can also be used to implement dynamic systems. So, dynamic systems can be implemented with state machines, memory units, registers, and counters. For continuous systems modeled with state equations the presence of an integrator or an energy storage element indicates a dynamic system. Their outputs are functions of time.

1.8.3 Activity Diagram Modeling

As we have just explained, state machine diagrams are used to model the behavior of dynamic systems. However, activity diagrams are also used to model the behavior of systems.

The essence of a state machine diagram is its states. The state of a system encapsulates the system's history. Whereas the activity diagram is characterized by the activities (or functions or operations) that are being performed. These actions are usually named with verbs or verb phrases with nouns separating them.

1.8.3.1 The Activity Diagram

Now we want to discuss a model for a fragment of the brain of a batter in a game of baseball. For the batter, the pitch can be divided into thirds. During the first third of the pitch, the batter gathers sensory information. During the second third, he or she must compute *where* and *when* the ball will make contact with the bat and then decide whether to swing, duck, or take the pitch. During the final third of the pitch, the batter swings the bat and can do almost nothing to alter the bat's trajectory. The activity diagram of Fig. 1.19 is useful for describing the batter's actions during the pitch. Computing where and when has been modeled with equations [66] and the swing of the bat has also been modeled with equations [67]. These would be the only actions in this diagram that are described this way. We do not have equations that describe wait for the pitch or gather sensory information.

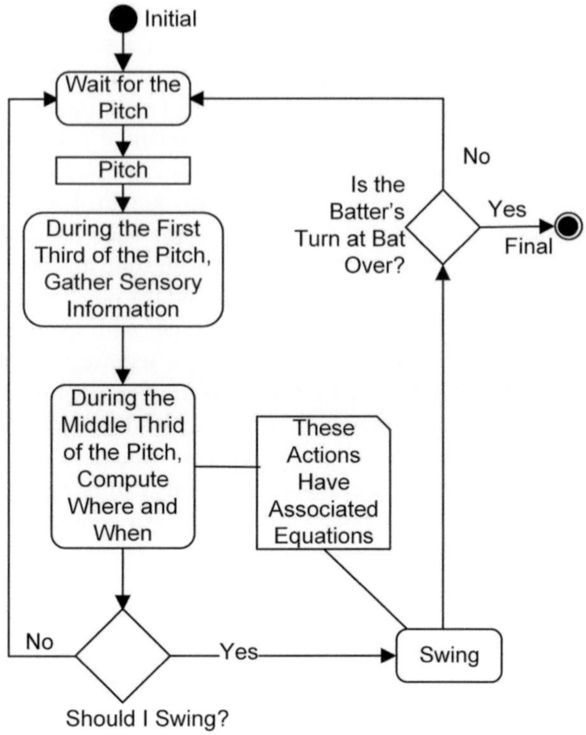

Fig. 1.19 Activity diagram (**act**) for a fragment of the brain of a batter in a baseball game

The activity diagram has evolved. In 1967, Jim Long developed a technique for representing system logic that incorporated functions, inputs, outputs, and control in a single hierarchical set of diagrams [68]. This developed into the extended functional flow block diagram, which, with the addition of the colored Petri-net token concept, was the basis for the SysML activity diagram [69]. Activity diagrams are functional flow block diagrams with verbs in the rounded boxes and nouns in the rectangles in between them.

A good model is a simplified representation of one aspect of a real system: models are successful because they do not consider all the complexity of the real system. As with all successful models, the example of Fig. 1.19 only models one aspect of the batter's behavior. The actual batter is doing many other tasks in parallel [31, 65, 69, 77–75]. However, including many tasks in the same model is detrimental to understanding [78].

1.8.3.2 State Machine Diagram

Now we want to discuss another model for the batter in a game of baseball. For the batter, the pitch can be divided into thirds. During the first third of the pitch, the batter is gathering sensory information. In this state, the batter must block out all

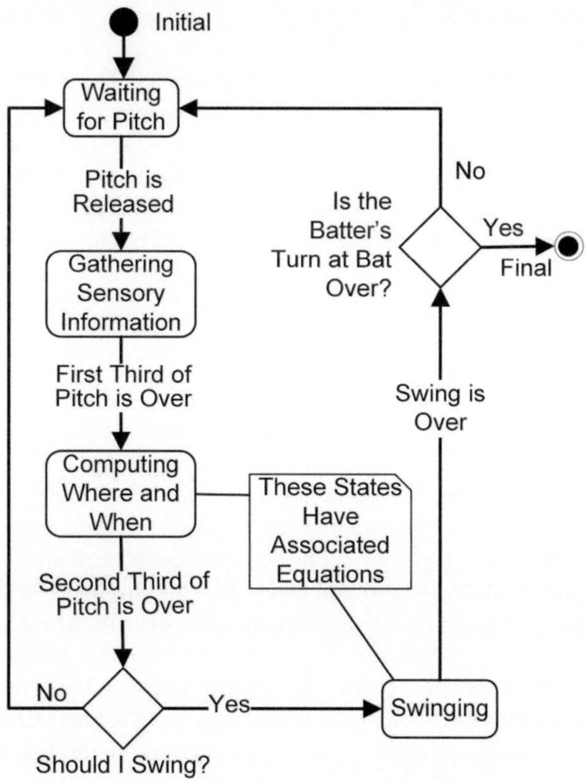

Fig. 1.20 State machine diagram for a fragment of the batter's brain in a baseball game

distractions and concentrate on the ball. During the second third, he or she is computing *where* and *when* the ball will make contact with the bat and then deciding whether to swing, duck, or take the pitch. During the final third of the pitch, the batter is swinging the bat and can do almost nothing to alter its trajectory. The state machine diagram of Fig. 1.20 is useful for describing the batter's behavior during the pitch. Computing where and when the ball will cross the plate has been modeled with equations [66] and the swing of the bat has also been modeled with equations [67]. These would be the only states in this diagram that are described this way. We do not have equations that describe waiting for the pitch or gathering sensory information. The transitions between states are labeled with the events that cause the transitions.

1.8.3.3 Which Is the Best Model, the Activity Diagram of Fig. 1.19 or the State Machine Diagram of Fig. 1.20?

If we were to expand the model upward to include learning how to swing the bat and how to compute where and when the ball will cross the plate, then we might want to use the state machine diagram. However, if we were to expand the model downward to include how the batter predicts time to contact and how he uses the spin on the ball to predict the spin-induced deflection of the ball, then we might want to use the activity diagram. What about the transitions? Transitions between states are triggered by events. Pitch is Released, First Third of Pitch is Over and Second Third of Pitch is Over certainly sounds like an events. The final thing that might influence our choice is that the batter is not likely to make decisions in parallel; therefore, the activity diagram loses its forte. In conclusion, in many cases it will not be clear whether an activity diagram or a state machine diagram would be best. Creating rules for making such a decision is clearly outside the scope of this section.

1.8.4 Homework

1.8.4.1 Wolf, Goat, Cabbage Riddle

In the forest on the East bank of a river, a traveler is making his way with his wolf, goat, and two dozen heads of cabbage. His goal is to reach the town on the West bank of the river with all his charges and possessions. He has available a small boat that at most can hold him and one of his charges. His task is further complicated by the fact that if left to themselves, the wolf will eat the goat, and the goat will eat the cabbage. Furthermore, he can't leave the cabbage alone on the bank by the forest, because the mice in the immediate vicinity in the forest might eat it. (This problem is attributed to Alcuin, a friend of Charlemagne.) Create a diagram that describes the sequence of states that the system must go through to successfully cross the river with all charges and possessions intact (Fig. 1.21).

1.8.4.2 State Variables

With inputs and outputs, we started with single-input, single-output systems and listed their legal values. Then we introduced the concept of input and output ports, where the applied input or the resulting output was the Cartesian product of the input or output legal values. Similarly, for the system states, we can have multiple state variables. Here the states are defined by the Cartesian product of the legal values of the state variables. A system with, for example, three state variables could be modeled as a state machine with three state variables or as three separate state

Fig. 1.21 The Wolf, Goat, Cabbage riddle

Fig. 1.22 SpringMass Dashpot system

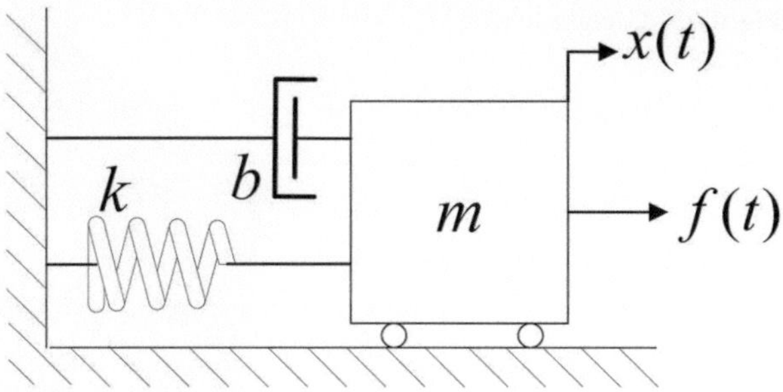

machines. How many state variables should there be in a model for the following everyday systems? Describe the state variables you have envisioned for

The watch on your wrist
A simple calendar
The Odd Parity Detector homework problem of Section 9
The human body, as assessed by a nurse in a simple physical exam
Your TV set
A baseball game, so that you could resume a rain-delayed game

1.8.4.3 Spring Mass Dashpot System

Consider the mechanical system illustrated in Fig. 1.22. Assume an ideal spring, mass and dashpot, and frictionless wheels. Assume initial conditions are zero and do a small signal analysis. Create a state space model for this system.

In general, the systems engineer gets models from the domain experts. For this system, assume we are given these models. $f_k = kx$, $f_b = b\frac{dx}{dt} = b\dot{x}$ and $f_m = ma = m\ddot{x}$. We will use SI units: the spring constant (k) has units of N/m, the damping coefficient (b) has units of N s/m, and the mass (m) has units of kg.

1.8.4.4 Useless Box

Captain Jules' extraordinary useless box is a wooden box with a switch on top. Inside are batteries, wires, gears, and a lever.

http://longhoang.ca/blog/most-pointless-item-ever/

Create a state machine diagram and an activity diagram for this box. Here is how it behaves.

Name: Turn Off (Fig. 1.23)
Iteration: 1.2
Derived from: The physical object
Brief description: The owner turns it on. The Box turns itself off.
Level: Low level

Fig. 1.23 Captain Jules' Useless Box

Priority: Low priority
Scope: The wooden box
Added value: The owner enjoys seeing the look of amazement on the faces of Users of the Box.
Primary actor: The Owner, the User
Supporting actor: None
Frequency: A dozen times on days when there is a party
Precondition: The Box is off and is being held stationary

Main Success Scenario

1. The User pushes the switch to the right. That turns the Box on.
2a. The Box turns its gears, opens its lid, and moves its lever in a manner that turns the switch off.
3. The Box withdraws its lever. [Exit use case]

Alternate Flows

2b. The gears do not turn.
2b1. The owner must troubleshoot the Box. [Exit use case]

Postcondition: The Box is off.
Specific Requirements
Functional Requirements
ReqF1. The system shall open its lid.
ReqF2. The system shall move the switch to its off position.
ReqF3. The system shall return gears and the lever to their original positions.
ReqF4. The system shall close its lid.
Nonfunctional Requirements
ReqNF. The system shall accomplish ReqF1 to ReqF4 within 2 s.
Author/owner: Zach Bahill
Last changed: April 10, 2016

1.8.4.5 Sleep

Most mammals and birds sleep. Their existence can be divided into two states: sleeping and awake. The sleeping state can be divided into two substates: rapid eye movement (REM) sleep and slow wave sleep (SWS). What is the difference between sleep and hibernation?

1.9 Section 9. Automobile Seatbelt System

This section presents a simple example of state-based modeling: an automobile seatbelt system. It is described using the notation of state machines with statecharts of Harel [79] and model-based system engineering of Wymore [20].

Several decades ago, the United States government issued a regulation that required drivers and passengers to wear seatbelts. This regulation required automobile manufactures to mandate that people in an automobile wear seatbelts. In the following paragraphs, we derive a state machine for a system that enables the ignition circuit of an automobile *only* if the driver, after being seated, fastens the seatbelt. We assume that the act of the driver being seated and the seatbelt being fastened signals are simple switch closures. We will also specify a testing program to prove that our system works.

A general technique for creating state machine diagrams is to (1) define input and output ports, (2) start a list of assumptions, (3) define a start state, (4) draw states along the main success scenario of the use case, (5) fill in states along other flows, (6) go back and make sure that there are $m \times n$ arrows (transitions) leaving each state, where n is the number of input ports and m is the number of values each port can have (assuming each port can accept the same number of values), (7) add states as needed so that all arrows terminate on a state, (8) repeat steps 5–7 until all possible transitions have been accounted for, and (9) simplify, which may include removing the start state, minimizing the model, creating hierarchical and/or concurrent state machines.

We will call our Automobile Seatbelt System, Z1. These are the two input ports for Z1 along with their legal values:

I1Z1 = {0, 1}, /* A 1 at this input port indicates that the seat is occupied */ and
I2Z1 = {0, 1}, /* A 1 at this input port means that the seatbelt is fastened */
Z1 has a single output.
OZ1 = {On, Off}, /* An output of On signifies the ignition is enabled */

Our start state is that the driver is outside of the car, the ignition is off, and there is nothing on the drivers' seat. The system is awaiting driver action. Our main success scenario is that from this start state, the driver opens the car door, gets in the car, and settles in the driver's seat. Then the driver fastens the seatbelt, and turns the ignition key on. The car starts up, and the driver drives off. This behavior is captured in the statechart of Fig. 1.24.

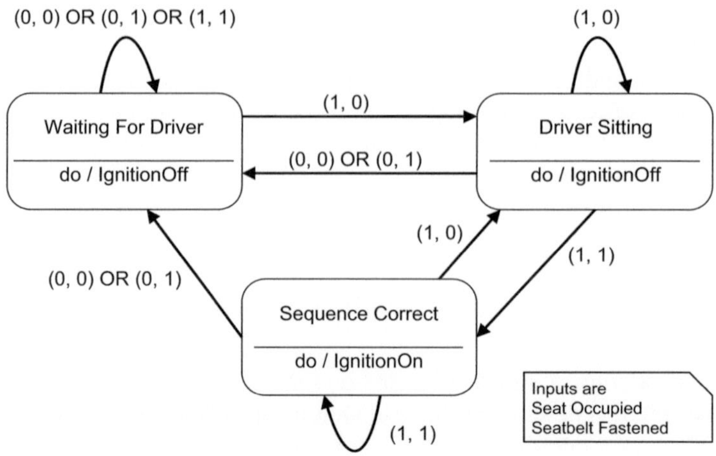

Fig. 1.24 A state machine diagram (or statechart) for the automobile seatbelt system

In a state machine diagram, each state has the state name in the upper part of the state box and the outputs in the bottom part of the state box. Three types of outputs are supported: entry, exit, and do. The entry outputs are executed upon entry to the state. The exit outputs are executed upon exiting the state. And the do outputs are executed the whole time that the system is in that state. The inputs are labeled on the transitions between states.

The main success scenario is as follows. The system is in the Waiting For Driver state. The driver sits on the seat producing an input pair (1, 0), which produces a transition to the Driver Sitting state. Next, the driver fastens the seatbelt producing an input pair (1, 1), which produces a transition to the Sequence Correct state, which has an output IgnitionOn for the whole time while in this state. At the conclusion of the trip, the driver unfastens the seatbelt, and gets out of the car producing an input pair (0, 0), which produces a transition to the Waiting For Driver state, which has an output of IgnitionOff.

In developing state machines, it is best to name the states with historical statements or temporal phrases describing past or present activities, e.g., Driver Seated, Waiting for Seatbelt. Care should be exercised that the names of the states not sound like mere combinations of input conditions. For example, the Sequence Correct state could be renamed Seated *then* Belted, but it should not be renamed Seated *and* Belted.

In this book, whenever possible, we show alternative notations, solutions, and representations to the reader. Here we present the alternative Wymorian set theory notation for a state machine [20]. While this notation is compact and complete, it may appear intimidating to customers. In that case, present a friendly pictorial and/or textual description of inputs, outputs, and states.

The Wymorian state machine description for the Automobile Seatbelt System is as follows. The first line states that the system named Z1 is composed of five artifacts: SZ1 is the set of states in the system, IZ1 is the set of inputs to the system,

OZ1 is the set of outputs of the system, NZ1 is the next state function (for every present state and possible input it gives the next state), and finally the RZ1 is the readout function that gives the output for every state.

Z1 = (SZ1, IZ1, OZ1, NZ1, RZ1), where
the following are the three states of the system
SZ1 = {Waiting For Driver, Driver Sitting, Sequence Correct},
the following are the two inputs to the system and their legal values, which are zero and one
I1Z1 = {0, 1}, /* A 1 at this input port indicates seat occupied */
I2Z1 = {0, 1}, /* A 1 at this input port means seatbelt is fastened */
the following is the output of the system and its legal values
OZ1 = {On, Off}, /* An output of On signifies the ignition is enabled */
the next state function comes next. Each element of the next state function shows ((the present state, the present inputs,) and the next state). If this notation confuses you, then ignore it, because this information is also in the statechart
NZ1 = {((Waiting For Driver, (0, 0)), Waiting For Driver),
((Waiting For Driver, (0, 1)), Waiting For Driver),
((Waiting For Driver, (1, 0)), Driver Sitting),
((Waiting For Driver, (1, 1)), Waiting For Driver),
((Driver Sitting, (0, 0)), Waiting For Driver),
((Driver Sitting, (0, 1)), Waiting For Driver),
((Driver Sitting, (1, 0)), Driver Sitting),
((Driver Sitting, (1, 1)), Sequence Correct),
((Sequence Correct, (0, 0)), Waiting For Driver),
((Sequence Correct, (0, 1)), Waiting For Driver),
((Sequence Correct, (1, 0)), Driver Sitting),
((Sequence Correct, (1, 1)), Sequence Correct)},
each element of the readout function shows (the present state, and its outputs)
RZ1 = {(Waiting For Driver, Off), (Driver Sitting, Off), (Sequence Correct, On)}.

There are two input ports and each can accept two values. Therefore, there are $2^2 = 4$ possible transitions from each state. There are three states, so there are 12 total transitions. In the description above, we explicitly list all 12 transitions. In Fig. 1.25, we show how to combine some transitions to simplify the statechart.

Using the abbreviation DC for "don't care" we can draw the state diagram with fewer transitions, as shown in Fig. 1.25. Three typical reasons why we might not care are (1) it makes no difference, for example, in the state Waiting For Driver, the seatbelt input cannot cause a transition, (2) in a particular state perhaps a particular input is impossible and (3) sometimes one input port has priority over a second; therefore, if the first input port has a certain value, then we don't care about the value of the second input port. For example, if we are in the Driver Sitting state and the driver gets out of the car (maybe he was sitting on his seatbelt) we want to go to the Waiting For Driver state no matter what the seatbelt input indicates.

As with most real systems, there is no unique solution [5]. For example, if we are in the Sequence Correct state when the driver unfastens the seatbelt, then this

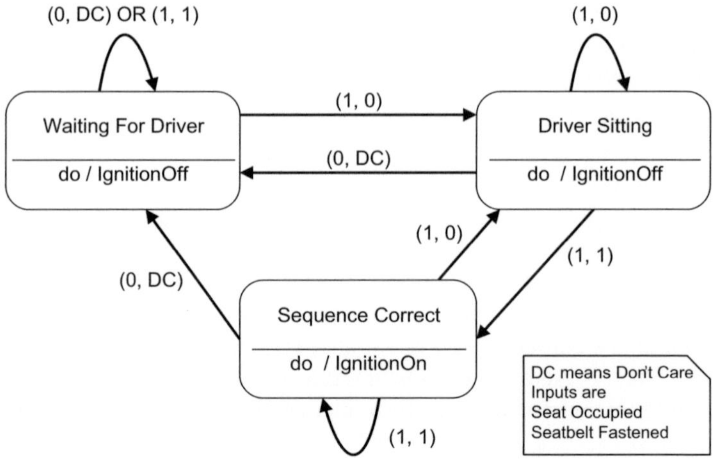

Fig. 1.25 An alternative statechart for the automobile seatbelt system using Don't Cares (DC)

system goes back to the Driver Sitting state. It would be just as reasonable to go back to the Waiting for Driver state, where the system would immediately go to the Driver Sitting state. As another alternative, we note that in the Corvette, when the seat weight sensor determines that the passenger weights less than 80 lb, it turns off the passenger air bags.

This statechart is simple, but incomplete and it could get confused. For we have assumed that (1) If the system is in the Sequence Correct state, and the seatNotOccupied event occurs, then the system will go to the Waiting for Driver state and the driver will have to intervene, for example, by unfastening the seatbelt in order to rectify the system. (2) If the system is in the Waiting for Driver state and the driver fastens the seatbelt, the system will stay in that state and the driver will have to intervene, for example, by unfastening the seatbelt in order to rectify the system. If these assumptions are not acceptable to the customer, then extra mistake states or guard conditions must be added. For example, using a guard condition, the transition from the Waiting for Driver state to the Driver Sitting state could be labeled seatOccupied[seatBeltNotEngaged]. The condition inside the square brackets [] is a guard condition. If the guard condition is false, then a transition cannot occur.

To test this system, first put it in the Waiting for Driver state (using special purpose inputs if necessary). Then, apply all possible inputs and make sure the outputs are acceptable. If it is too much work to do **all** possible inputs, then do only those that are necessary to convince the customer that the system works. This strategy is shown below.

For this testing phase, we introduce an alternative notation.

Inputs

$s_1 = 1$ means seat is occupied
$s_1 = 0$ means seat is not occupied

$s_2 = 1$ means seatbelt is fastened
$s_2 = 0$ means seatbelt is not fastened

Output

$z = 1$ means ignition is enabled = on
$z = 0$ means ignition is not enabled = off

Note the timing relationship between inputs and states (outputs). In System Experiment 1 at time = 3 the system is in the state Driver Sitting, therefore the output is Off. At this time, apply the inputs 1 and 1, the next state will become Sequence Correct and therefore the output at time = 4 is On (Tables 1.3, 1.4, 1.5, 1.6, 1.7, and 1.8)

Table 1.3 System Experiment 1 (A normal sequence)

Time	Inputs		State	Output
	s_1	s_2		
0	0	0	Waiting For Driver	Off
1	1	0	Waiting For Driver	Off
2	1	0	Driver Sitting	Off
3	1	1	Driver Sitting	Off
4	1	1	Sequence Correct	On
5	0	0	Sequence Correct	On
6			Waiting For Driver	Off

Table 1.4 System Experiment 2 (buckle before sitting)

Time	Inputs		State	Output
	s_1	s_2		
0	0	0	Waiting For Driver	Off
1	0	1	Waiting For Driver	Off
2	1	0	Waiting For Driver	Off
3			Waiting For Driver	Off

Table 1.5 System Experiment 3 (test arrow (1, 0) from Sequence Correct State)

Time	Inputs		State	Output
	s_1	s_2		
0	1	0	Waiting For Driver	Off
1	1	1	Driver Sitting	Off
2	1	0	Sequence Correct	On
3	1	1	Driver Sitting	Off
4			Sequence Correct	Off

Table 1.6 System Experiment 4 (test arrow (0, 1) from Driver Sitting)

Time	Inputs		State	Output
	s_1	s_2		
0	1	0	Waiting For Driver	Off
1	0	1	Driver Sitting	Off
2			Waiting For Driver	Off

Table 1.7 System Experiment 5 (test arrow (0, 0) from Driver Sitting)

Time	Inputs		State	Output
	s_1	s_2		
0	1	0	Waiting For Driver	Off
1	0	0	Driver Sitting	Off
2			Waiting For Driver	Off

Table 1.8 System Experiment 6 (test loop (1, 1) from Waiting For Driver)

Time	Inputs		State	Output
	s_1	s_2		
0	1	1	Waiting For Driver	Off
1			Waiting For Driver	Off

Table 1.9 Because we are starting the implementation phase, we introduce another notation for the next state function

Present state		Next state AB($t+1$)				
Name	AB(t)	$s_1s_2 = 00$	$s_1s_2 = 01$	$s_1s_2 = 11$	$s_1s_2 = 10$	Present output Z
Waiting For Driver	00	00	00	01	00	0
Driver Sitting	01	00	00	01	11	0
Sequence Correct	11	00	00	00	11	1

All state transitions (arrows on the state diagram) have been verified.

In the above tables, both the outputs and the states have been shown in order to prove conformance of the real system with its design. Studying inputs and outputs alone is not sufficient [42] (Table 1.9).

1.9.1 Summary

In general, defining input ports and states is usually the most difficult aspect in system design. However, input ports for state machines are quite simple. Typical inputs are switches that can be open or closed, and inputs are memoryless. For example, when trying to detect the presence of two consecutive 1s or two consecutive 0s in a communications line, you should not describe the input as current bit and previous bit were 1s. Also, inputs should not be described as intelligent machines. For example, when designing a system for a space shuttle launch, you need to determine if all prelaunch operations (for example, load crew, close door, load oxygen, load hydrogen, release clamps) have been done in the correct order. For this problem, the input should **not** be

$x = 1$, means all operations were done in correct order, and
$x = 0$, means operations were not done in correct order.

States should be named with historical statements or temporal phrases describing past or present activities. A state name should reflect an interval of time when something is happening. A state name can be a gerund, a noun phrase, a clause, or even a sentence. However, state names should never sound like combinations of input conditions. For example, for the automobile seat belt problem, "Seat Occupied *and* Seatbelt Fastened" would be an inappropriate name for a state. However, "Seat Occupied *then* Seatbelt Fastened" might be appropriate. Past tense might be appropriate for verbs. Note that in general, if a state has the same name as an input, then you have most likely made a mistake. However, if a state has the same name as an output, it may be fine, because you merely used state readout as the output.

In developing state machines, name the states with temporal phrases, for example, "Driver Seated, Waiting for Seatbelt." However, after minimization, it might be the case that a state has a name comprising two incompatible temporal phrases. Therefore, for the final version of the state machine, historical phrases may have to be removed.

Knowing the present state of a system and the future input trajectory, future states can be calculated. Therefore, it does not matter how the present state was reached. For example, look at the wolf-cabbage-goat homework problem of Section 8. If we know the system is in state $\overline{T}W\overline{G}C(0101)$, then we don't care if it got there from TWGC (1111) by TG crossing or if it got there by starting in TWGC (1111) with TG crossing to state $\overline{T}W\overline{G}C(0101)$, followed by T crossing to state $TW\overline{G}C(1101)$, followed by T crossing to state $\overline{T}W\overline{G}C(0101)$.

For most of our problems, you may, if you wish, ignore startup states and assume that the system is operating in steady state.

1.9.2 Homework

1.9.2.1 Set Theory Notation

These problems are intended to help familiarize you with our set theoretic notation.
Let

$A = \text{IJS}[1\text{–}9]$,

/* The notation IJS[1–9] means the integers 1 to 9.*/

$B = \{x: x \in A; x \text{ is odd}\}$,

/* This means B is the set of all x where x is an element of the set A and x is odd.*/

$C = \{x: x \in A; x \text{ is even}\}$,
$D = \text{IJS}[3\text{–}5]$,
$E = \{x: x \in D; x \text{ is odd}\}$.

These sets can also be defined by enumeration as follows:

$A = \{1, 2, 3, 4, 5, 6, 7, 8, 9\}$,
$B = \{1, 3, 5, 7, 9\}$,
$C = \{2, 4, 6, 8\}$,
$D = \{3, 4, 5\}$,
$E = \{3, 5\}$.

Find:

$B \cup C$. This is called B Union C.
$B \times D$. This is the Cartesian product of sets B and D.
$E \cap C$. This is called the Intersection of E and C.
$A\text{–}B$. This is often expressed as $A \cap B$.
Deduce all possible sets X for which $X \cap B = \{\}$, the empty set. Do not limit your consideration to only the sets A through E given above. Find all possible sets X by considering the universe of all possible elements that those sets may contain.
Define F by enumeration, where $F = \{1, 2\}\hat{}3$, that is $F = \{1, 2\} \times \{1, 2\} \times \{1, 2\}$. Please observe this use of the caret: ˆ3 means that there are three input ports.

1.9.2.2 Odd Parity Detector

Computer words are to be transmitted sequentially down a wire. Draw a state diagram for a system whose output will be 1 if and only if the present and previous bits showed odd parity. That is, the output should become a one if the previous two bits were 01 or 10 and should stay at 1 until the next bit is received. Conversely, the output should become zero if they were 00 or 11. Draw a state machine diagram for this system or specify Z3 = (SZ3, IZ3, OZ3, NZ3, RZ3).

1.9.2.3 Invalid BCD Digit Detector

BCD	Decimal
0000	0
0001	1
0010	2
0011	3
0100	4
0101	5
0110	6
0111	7
1000	8
1001	9

In the Binary Coded Decimal (BCD) representation, each decimal digit is represented by four binary bits. The binary numbers 0–9 represent valid BCD digits, and the binary numbers 10–15 constitute invalid BCD digits. Draw a state diagram for a single input, single output system that determines whether an invalid BCD digit has been received. If at the end of the BCD word (the four bits), the system has determined that the BCD digit was invalid, then its output should become a 1 and stay there until another bit is received. The output should be 0 at all other times. Least significant bits of the words appear first in time. You may ignore startup transients and design the system for steady state operation. You need not minimize the system. It will be easier if you treat the input as a sequence of words not as a string of bits, that is test bits 1, 2, 3, and 4, then bits 5, 6, 7, and 8; do not test bits 1, 2, 3, and 4, then bits 2, 3, 4, and 5. Call this system Z7; specify Z7 = (SZ7, IZ7, OZ7, NZ7, RZ7). If you wanted to test your model by applying all possible input trajectories and all reasonable initial states, how many test trajectories would you need and how long would they have to be?

1.9.2.4 Candy Machine

In our geedunk machine, candy cost 15 cents. (OK, it is a generous machine and the portions are small.) The machine sorts coins into three categories: nickels, dimes, and other. Draw a state machine diagram for a system that will drop candy if the customer deposits: either three nickels, or a nickel and a dime. If the customer deposits foreign coins or illegal combinations (for example, two dimes), all deposited coins should be returned. Describe your inputs, outputs, and states. Label your states with meaningful names. State all assumptions you make. Name this system Z9; specify Z9 = (SZ9, IZ9, OZ9, NZ9, RZ9).

1.9.2.5 Telephone System

In the old days, you did not have to dial a 1 to tell the telephone system that you wanted to place a long distance phone call. You merely dialed the area code and the system figured it out. How did it do this? All area codes had a 1 or a 0 as the second digit, and no phone numbers did (except for 911 emergency). Design a system that will turn on a light after the third number, if it is an area code. The light should stay on until the customer hangs up. Describe your input ports, output ports, and states. Label your states with meaningful names. State all assumptions you make. Name this system Z10; specify Z10 = (SZ10, IZ10, OZ10, NZ10, RZ10).

1.9.2.6 Spelling Checker

Design a system to detect spelling errors. Or more simply, create a state machine to implement the spelling rule "i before e except after c." If a word violates this rule your system should stop processing words and turn on an error light. When the human acknowledges the mistake and turns off the error light, your system should resume processing words. For example, the words piece and receive are correct so your system should continue processing words. However, weird violates this rule so your system should stop and wait for human action. You may assume that bizarre sequences such as "ceie" will go undetected and that Professor Lucien Duckstein will not use it. Describe your inputs, outputs, and states. Label your states with meaningful names. Assume the system starts in a reset state. State all assumptions you make. Name this system Z11; specify Z11 = (SZ11, IZ11, OZ11, NZ11, RZ11). How could you test your system?

Spelling Checker Use Case
Name: Check Spelling
Level: user
Scope: a single MS Word document
Brief description: The Spelling Checker detects possible misspelled words, suggests alternative words, and makes corrections.
Primary actor: Student.
Supporting actor: Dictionary, Document.
Precondition: Student has an MS Word document with possible misspelled words.
Frequency: Student uses the Spelling Checker every 10 min.

Main Success Scenario

1. Student requests Spelling Checker to check the spelling in his or her document.
2a. Spelling Checker searches the document, highlights a potentially misspelled word, and suggests some alternatives.
3a. Student accepts one of the suggested alternatives.
4. Spelling Checker changes the word [repeat at 2].

Anchored Alternate Flows

2b. Spelling Checker reaches the end of the document without finding possible mistakes [exit use case].
3b. Student chooses to add the word to the dictionary.
3b1. Spelling Checker adds the word to the dictionary [repeat at 2]
3c. Student chooses to ignore the potential misspelling [repeat at 2].

Postcondition: Student has an MS Word document with no potentially misspelled words identified.

Rules

Rule1:. "i before e except after c"
Rule2:. "other" means any letter other than c, e or i.
Rule3:. For now, our system ignores spaces, numbers, punctuation and exceptions to Rule1.
Rule 4:. It is assumed that Typist corrects or ignores the mistake before resetting the Spelling Checker.

Frequency: Millions of users might use the system hundreds of times per day.
Author/owner: Ms. Dulzer
Date: May 9, 2016

1.9.2.7 Combination Lock

There is a combination lock on the door of a laboratory. It has five buttons that can be pressed individually or in combination, and a doorknob that can be turned. Presume that the correct combination is to push buttons 4 and 2 at the same time, and then button 3. Turning the doorknob clockwise opens the door and resets the lock. (The doorknob cannot be turned in a counterclockwise direction.) Make a model of this system. Describe your input ports, output ports, and states. Draw a state machine diagram. Label your states with meaningful names. State all assumptions you make. Name this system Z12; specify Z12 = (SZ12, IZ12, OZ12, NZ12, RZ12) (Fig. 1.26).

1.9.2.8 Gasoline Pumps

Gasoline stations today have a computer system in each pump. It first asks if you want to pay with cash or a credit card. Assume that if you say cash, it will charge you $3.00 per gallon, but if you say credit card it will charge $3.09 per gallon. Then it asks if you want to buy $5, $20, $45, or fill up the tank. (To simplify this problem, let us ignore the possibility of filling up the tank.) When the total amount you owe equals the amount you wanted, it turns off the pump. Design such a system. Describe your input ports, output ports, and states. Draw a state machine diagram.

Fig. 1.26 Laboratory Door Lock

Label your states with meaningful names. State all assumptions you make. Name this system Z15; specify Z15 = (SZ15, IZ15, OZ15, NZ15, RZ15).

1.9.2.9 Campus Phone System

We only dial five numbers to get someone on campus; all main campus numbers begin with a 1 and all medical school numbers begin with a 6. To get an outside line we dial 9. To call long distance we dial 82, unless it is inside Arizona in which case we dial 81 instead. Of course, 0 will get the operator. Design a system that will monitor a telephone and turn on an error light if a mistake is made. The light should stay on until the person hangs up. You may ignore the * and # buttons. Describe your inputs, outputs, and states. Draw a state diagram. Label your states with meaningful names. State all assumptions you make Name this system Z17; specify Z17 = (SZ17, IZ17, OZ17, NZ17, RZ17).

1.10 Section 10. Parking Garage, First Iteration

The next five sections present five iterations in the design of a parking garage system. In each section, the problem statement and then the design solution are presented. At first glance, parts of the design may appear repetitive, but the design, in fact, grows with each iteration. In addition to drawings, we also present a formal description for each state machine of the design [23]. Students unfamiliar with the Unified Modeling Language (UML) [64] or the Systems Modeling Language (SysML) [16, 65] may skip Sections 10 to 14 without loss of continuity.

The Concept of Operations (ConOps) is usually given to the designers by the customer.

ConOps for a parking garage. Create a design for a university parking garage. This is how the system should work. A person drives up to the entrance gate, the system reads the car's transponder, if it is valid, the gate opens, when the system senses that the car has left the gate area, the gate closes. When that person leaves for the day, he or she drives up to the exit gate, the system reads the car's transponder, if it is valid, the gate opens. When the system senses that the car has left the gate area, the gate closes. The gate shall not close on top of the car or break the radio antenna. To prevent people from using one transponder for two cars, the entrance gate will not open if the system thinks that the car is already inside. The transponder contains only an identification number. All of the other data are stored in the system. To make the system simpler, design it for only one transponder (that is, consider only one car). For simplicity, assume that the system has only two gates: one for entrance and one for exit. Restrict the inputs and outputs to things that could actually be in the parking garage.

The first step in our design process is to write a use case, based on the ConOps. Then we will create a state machine diagram (statechart) for the system. It is important to vigorously describe the inputs, outputs, and states and to state all assumptions that are made.

A use case is an abstraction of the required functionality of a system. A use case produces an observable result of value to the user. A typical system will have many use cases each of which satisfies a goal of a particular user. Each use case describes a sequence of interactions between one or more actors and the system. This sequence of interactions is described in a use case narrative and the relationships between the system and its actors are portrayed in use case diagrams. State machine diagrams describe event-based behavior of a system. They represent a part of the life cycle of a class or a block (Fig. 1.27).

Fig. 1.27 Entrance gate for a parking garage system

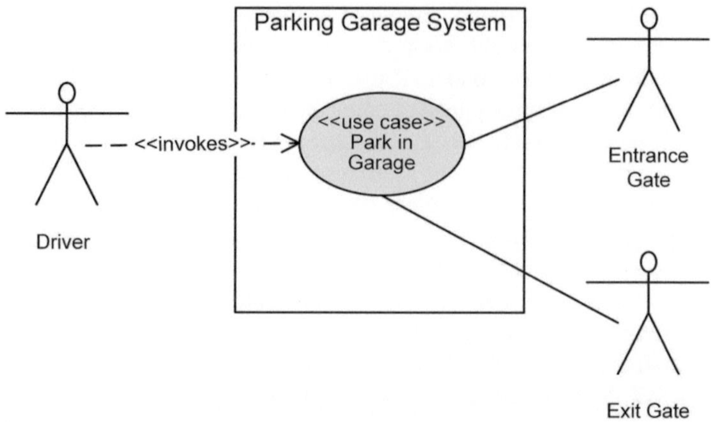

Fig. 1.28 Use case diagram for a high-level model of the parking garage system

1.10.1 First Iteration Design Solution

Use Case (Fig. 1.28)
Name: Park in Garage
Iteration: 3.5
Derived from: Concept of Operations
Brief description: A person drives up to the entrance gate of the parking garage. The system reads the car's transponder: if it is valid, the gate opens. When the car has left the gate area, the gate closes. A similar sequence occurs when the person leaves the garage.
Level: Summary
Scope: Park Avenue and Speedway Boulevard parking garage
Added value: Driver can park his car in a safe, shaded environment within a half mile of his or her classroom or office.
Primary actor: Student, faculty, or staff member who is the Driver of a car. In this use case, we will treat the Driver and the Transponder as one actor.
Supporting actors: Entrance Gate and Exit Gate
Frequency: About 1500 cars use the system each day.
Precondition: Driver's car is outside of the parking garage and the system believes that the car is outside of the parking garage.

Main Success Scenario

1. Driver approaches the Entrance Gate.
2. The system reads the car's transponder.
3a. The system determines that the transponder is valid and that the car is not already in the garage.
4. The system opens the gate.
5. Driver drives through the Entrance Gate.

6. The system detects that the car is clear of the Entrance Gate.
7. The system closes the gate.
8. The system stores the transponder's identification number (ID), the date, and the time.
9. Driver parks the car and goes to class or his/her office.
10. Driver returns to the car and drives up to the Exit Gate.
11. The system reads the car's transponder.

12a. The system determines that the transponder is valid and that the car is actually in the garage.

13. The system opens the gate.
14. Driver drives through the Exit Gate.
15. The system detects that the car is clear of the Exit gate.
16. The system closes the gate.
17. The system stores the transponder's identification number (ID), the date, and the time.
18. Driver drives home [exit use case].

Alternate Flows

3b. The system determines that the transponder is not valid or that the car is already in the garage.

3b1. The system does nothing (This may be stupid, but it is what the system actually does.) [exit use case].

12b. The system determines that the transponder is not valid or that the car is not in the garage.

12b1. The system does nothing (This design flaw must be fixed.) [exit use case].

Postcondition: Driver's car is outside of the parking garage and the system has recorded that the car is outside of the parking garage. (This does not include the present behavior of step 12b.)

Specific Requirements

Functional Requirements

Req1. The system shall determine when a car is waiting at the Entrance or Exit Gate. This requirement (and other similar compound requirements) will be decomposed later.

Req2. The system shall determine the validity of the transponder's ID.

Req3. The system shall store the identity of every car in the garage (from the ConOps).

Req4. The system shall open and close the Entrance and Exit Gates. This requirement will be decomposed later.

Req5. The system shall determine when a car has cleared the Entrance or Exit Gates.

Derived Requirement

Req6. The system shall allow a car to enter the garage only if the transponder's ID is valid and the database shows that the car is outside of the garage.

Nonfunctional Requirements

Req7. When the transponder is valid, the system shall open the gate within 2 s of interrogation.

Author/owner: Parking and Transportation Department.
Last changed: February 10, 2016

For systems that are state dependent, like this garage system, the fundamental diagram that describes its behavior is a state machine diagram a.k.a. a statechart, as shown in Fig. 1.29.

State boxes contain the state name, a dividing line, and then the state outputs. The outputs are differentiated as *entry* meaning the output is executed upon entering the state, *exit* meaning the output is executed upon exiting the state or *do* meaning the output is executed the entire time that the system is in that state. The inputs are labeled on the transition between the states.

It is important to recognize that a statechart is not a stand-alone artifact. It needs to have backup documentation that describes all the input, outputs, and states. These can be simple textual descriptions or precise mathematical descriptions.

The formal state machine description for the parking garage system is presented next. The first line states that the system named Z2 is composed of five artifacts: SZ2 is the set of states in the system, IZ2 is the set of inputs to the system, OZ2 is the set of outputs of the system, NZ2 is the next state function (for every present state and possible input it gives the next state), and finally the RZ2 is the readout function that gives the output for every state.

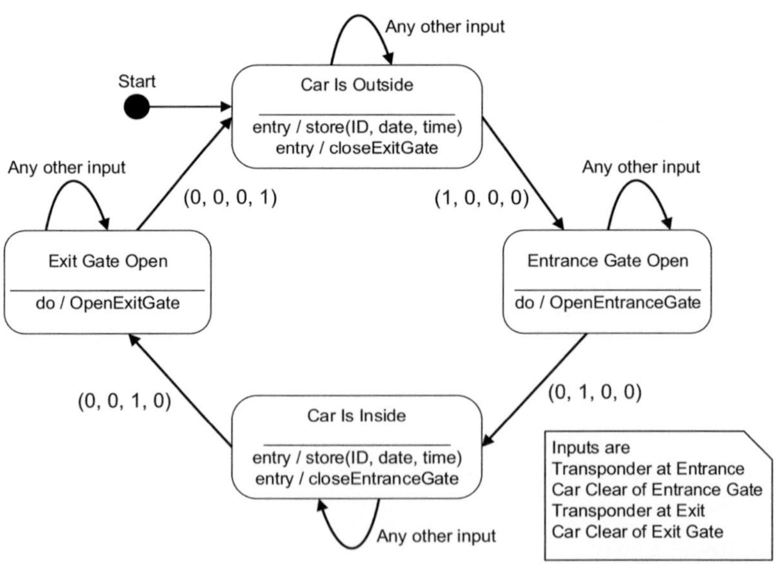

Fig. 1.29 State machine diagram for the high-level behavior of the parking garage system, named Z2

Z2 = (SZ2, IZ2, OZ2, NZ2, RZ2), where
the following are the four states of the system
SZ2 = {Car is Outside, Entrance Gate Open, Car is Inside, Exit Gate Open},
the following are the four inputs to the system and their legal values, which are zero and one
IZ2 = I1Z2 × I2Z2 × I3Z2 × I4Z2,
I1Z2 = {0, 1}, /* 1 means valid transponder at entrance gate */
I2Z2 = {0, 1}, /* 1 means car is clear of entrance gate */
I3Z2 = {0, 1}, /* 1 means valid transponder at exit gate */
I4Z2 = {0, 1}, /* 1 means car is clear of exit gate */
the following are the three outputs of the system and their legal values
OZ2 = O1Z2 × O2Z2 × O3Z2,
O1Z2 = {0, 1}, /* 1 means openEntranceGate*/
O2Z2 = {0, 1}, /* 1 means openExitGate*/
O3Z2 = {0, 1}, /* 1 means store(ID, date, time) */
the next state function comes next. Each element of the next state function shows ((the present state, the inputs,) and the next state). If this notation is too confusing, then ignore it, because this information is in the statechart
NZ2 = {((Car is Outside, I1Z2), Entrance Gate Open),
((Car is Outside, I1Z2 OR I2Z2 OR I3Z2), Car is Outside),
((Entrance Gate Open, I2Z2), Car is Inside),
((Entrance Gate Open, I1Z2 OR I3Z2 OR I4Z2), Entrance Gate Open),
((Car is Inside, I3Z2), Exit Gate Open),
((Car is Inside, I1Z2 OR I2Z2 OR I4Z2), Car is Inside),
((Exit Gate Open, I4Z2), Car is Outside),
((Exit Gate Open, I1Z2 OR I2Z2 OR I3Z2), Exit Gate Open)},
each element of the readout function shows (the present state, and its outputs)
RZ2 = {(Car is Outside, (0, 0, 1)), (Entrance Gate Open, (1, 0, 0)),
(Car is Inside, ((0, 0, 1)), (Exit Gate Open, (0, 1, 0))}.

While this notation is compact and complete, it might not be appropriate for customers, who prefer a friendlier presentation [80].

After a lengthy engineering discussion, it was concluded that a car could only be in one place at one time. Thus, we cannot have simultaneous occurrences of inputs. Therefore, we can simplify the model so that it has just one input port that can take on any of four values:

validCarAtEntrance means a valid transponder is detected at the entrance gate
carClearOfEntrance means the car is clear of the entrance gate
validCarAtExit means a valid transponder is detected at the exit gate
carClearOfExit means the car is clear of the exit gate

Therefore, we define a new model Z2a in Fig. 1.30 such that IZ2a = {validCarAtEntrance, carClearOfEntrance, validCarAtExit, carClearOfExit}

SZ2a, OZ2a, and RZ2a are the same as SZ2, OZ2, and RZ2, respectively. NZ2a will be different from NZ2.

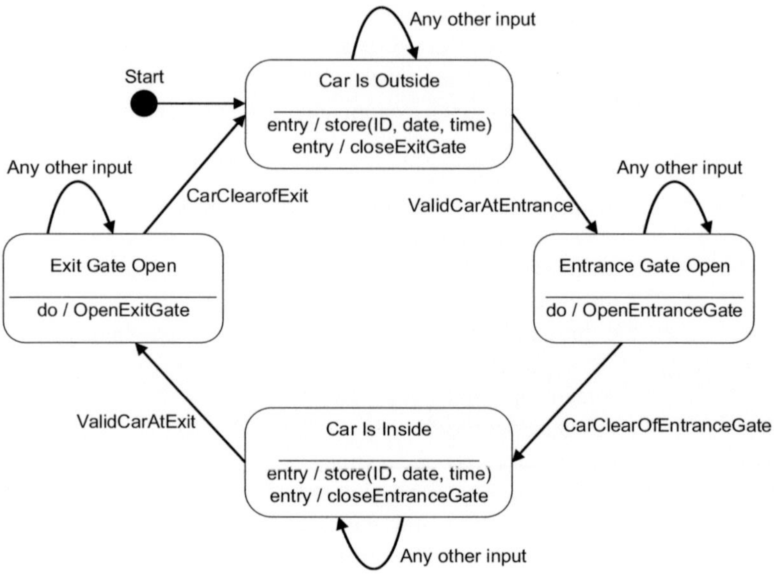

Fig. 1.30 Alternative state machine diagram for the high-level behavior of the parking garage system, Z2a

1.10.2 Homework

1.10.2.1 System Synopsis for the Parking Garage System

A system synopsis is a quick way to explain a system. Write a System Synopsis for the parking garage system. Use the following template.

System Synopsis Template

For	(target customer)
who	(statement of need or opportunity)
the (system name)	is a (system category)
that	(provides key benefits or problem solving capability)
unlike	(primary competitive alternative)
(our system)	(statement of primary differentiation)

Put your system-specific information inside the parentheses ().

A System Synopsis is called an *elevator speech*, because of the following scenario. You step into the elevator and see your Vice President. You realize that you will have about one minute alone with her. You want to spark her interest, so that she will read your concept of operations when it shows up on her desk. So, you blurt out your system synopsis.

1.10.2.2 Fix the Postcondition

For the main success scenario of the Park in Garage use case, the behavior at step 3b1 is stupid, but not disastrous. The driver just gets angry and drives over to the Parking and Transportation building. However, the behavior at step 12b1 is more serious, because the driver is locked in the garage. How could you fix this problem?

1.10.2.3 Difference in State Machines

What is the difference between the state machines stmParkingGarage1 (Fig. 1.29) and stmParkingGarage2 (Fig. 1.30)? Write an expression for the input port of Z2a.

1.10.2.4 The OR Function

The Park in Garage use case has OR functions. OR functions are unusual in use cases. How could you implement an OR function in UML?

1.10.2.5 Make Your Customer Happy

The Director of the Parking and Transportation Department is your customer. As well as designing a wonderful parking garage, you must also make her happy. What does she want? List the things that will make her happy and the things that will make her sad.

1.10.2.6 Principles of Good Design

Apply Section 5 fundamental principles of good system design to the parking garage system. Did we use models to design the system?

Did we work on high-risk items first?

Prioritize the functional requirements of the first iteration of the parking garage (Section 10) use case by listing them in order of importance, the most important first.

What interfaces of parking garage are hiding complexity from the user?

How could we have designed the system to satisfy the principle of develop stable intermediate configurations first?

Suggest ways to make the parking garage system resilient.

Use evolutionary development

What is the enterprise context in which the parking garage system will be deployed?

Modify the requirements of the first iteration of the parking garage (Section 10) use case so that they violate the principle of state requirements in terms of *what* not *how*.
For the parking garage system, did we enumerate functional requirements in use cases?
Does the parking garage provide observable states?
Point out the undocumented functions that might be in the parking garage system.
In the parking garage system, identify elements that are likely to change.
For the parking garage system, write a glossary of relevant terms.

1.10.2.7 Dog Door Problem

This is an extensive homework problem.

Concept of Operation: A family has a Golden Retriever named Sasha. She is always on the wrong side of the door. When she is in, she wants out. When she is out, she wants in. When she wants in, she jumps up and impatiently bangs and scratches on the door. The Dad wants to put in a dog door. But the wife says, "If you put in a door big enough for the dog, then burglars will be able to crawl in." So we would like you to design a *secure* dog-door system. We will have a radio frequency identification (RF ID) tag on her collar. The door lock will only be enabled if the dog is nearby, maybe 2 ft. The door controller will use this ID tag to help unlock the door. But we do not want the door to be unlocked, just because she walks near the door. So we will put switches on the door: one on the inside and one on the outside. We will train her to push a switch if she wants the door to open. The door should lock itself about 5 s after it opens. Model this secure dog-door system at the user level. Please make sure that your names for actions, events, messages, etc. use the exact same words, spelling, punctuation, and grammar throughout this problem. The following functionality is excluded from the system. The system would be more secure if the door could only be unlocked if the inside door switch were pressed when the dog was inside or if the outside door switch was pushed when the dog was outside. But the RF ID reader cannot tell us if the dog is inside or outside. Furthermore, the dog might push the button, let the door open and then back away. So we cannot require alternations of inside and outside accesses.

We desire to use as much commercial off-the-shelf (COTS) equipment as possible. The switches and the door lock might come from automobile remote door locking systems. The door opener might use garage door openers. The RF ID tag and RF ID reader could come from toll bridge equipment or library systems such as in the Seattle Library. The dog door could come from DrsFosterSmith.com item #VD-15182.

Please define the **input and output ports** of your Secure Dog-door System
Write a **use case narrative** for this Secure Dog-door System.
Draw a **use case diagram** for this Secure Dog-door System

Draw a **class diagram** for your Secure Dog-door System. *Candidate* classes include RF ID Tag, RF ID Reader, RF ID Reader Interface, Inside Door Switch, Inside Door Switch Interface, Outside Door Switch, Outside Door Switch Interface, Controller, Controller On/Off Switch, Batteries, COTS Dog Door, COTS Dog Door Lock, a boundary class to interface your controller to the COTS Dog Door Lock, Timer, Built-in Self-Test (BiST), day/night and vacation algorithms, an entity class to store usage data, results of BiST and battery charge.
Draw a **sequence diagram** for your Secure Dog-door System
Draw a **statechart** for the Controller of your Secure Dog-door System
If you have the background, draw the circuit for **Built-in Self-Test.**
Write a **test procedure** for your Secure Dog-door system.
Perform a **risk analysis** for your Secure Dog-door System

What is your biggest risk?

1.10.2.8 Spin Coach

A fragment of the Operational Concept Description (OCD) for the Spin Coach is given below.

The system alerts the player with an auditory or a visual cue that a pitch is about to come. One second later the system displays video images of the selected PitchType for the selected DisplayDuration. The player predicts the direction of spin-induced deflection and indicates this direction using a pointing device (mouse, trackball, or joy stick). The player will have a Window of time in which to indicate his or her predicted direction of spin-induced deflection. This Window of time starts at the beginning of the pitch display. If the player responds within this Window, then the system records the player's response in the player's ResponseDataTable and provides positive or negative feedback to the player. Otherwise, it informs the player that he or she has missed the window of opportunity.

Legal values for PitchType are {four-seam fastball, two-seam fastball, curveball, slider}, legal values for DisplayDuration are {5 s, 1 s, 800 ms, 600 ms, 400 ms, 200 ms}, Legal values for Window are {5 s, 1 s, 800 ms, 600 ms, 400 ms, 200 ms}.

For this OCD, write a **use case description**.

Create **a state machine diagram** for this use case of the Spin Coach.

1.10.2.9 Generic Missile

Create a state machine diagram for the Arming, Fuzing & Firing (AF&F) system of the following use case.

Name: Fly to Target
Iteration: 1
Brief description: Missile is launched from the platform, flies to the target, and detonates near the target.

Level: *high*
Primary actor: Missile AF&F
Supporting actor: Missile Processor
Precondition: Missile is in Ready Mode.

Main Success Scenario

1. Operator sends Release Consent to the missile.
2. AF&F enters the Safe state.
3. The missile is launched.
4. AF&F enters the Launched state.
5. The missile activates its batteries
6. The away-switch detects that the missile has separated from the platform.
7. The AF&F enters the Away state.
8. When the motor fires, the g-switch detects a large acceleration.
9. The missile activates more batteries
10. The AF&F enters its Armed state.

11a. When the Fuze decides that it is time, it sends the Fire command to the explosive [end of use case].

Alternate Flow

11b. The Fire command never comes. What should we do?

Postcondition: Missile has detonated.
Author/owner: Terry Bahill
Last changed: Winter 2016

Can we run the state machine backwards?

1.11 Section 11. Parking Garage, Second Iteration

Each iteration needs to be a complete design. Therefore, things that have not changed from the previous iteration are repeated. And, of course, some items will be changed and new items will be added.

The second design iteration is for the analysis model of the parking garage system. In going from the **high-level model** of the first iteration to the **analysis model** of the second iteration we added a lot of detail. We edited or recreated the use case, the use case diagram and the statechart. *Things that have been added are set in a Italic Sans-Serif font.*

We added a requirements diagram and explained how we verify our requirements. We then identified classes, drew a class diagram, made a sequence diagram, assigned functions to classes, created a new state machine diagram, created SysML block definition diagrams, and created a SysML parametric diagram for the entrance gate controller for this parking garage system. We are continuing to add detail as we work our way to an executable model.

1.11.1 Summary of the Diagrams

The use case diagram is the table of contents for the use case model. It shows all of the use cases that have been written and only those use cases. The individual use cases show the required functionality of the system. The sequence diagram is used to depict the operations or functions of the classes. A sequence diagram captures the behavior of a single scenario. There is a sequence diagram for each flow of each use case. Class diagrams show the components or parts of the system and their relationships. This is the workhorse of the UML. Class diagrams have three compartments: name, attributes, and functions. If more compartments are needed, then block definition diagrams can be used instead. Block definition diagrams show system structure as components along with their properties, constraints, parameters, and relationships. Parametric diagrams show how the parameters from each of the equations can be bound together to create a complex network of equations. Parametric diagrams show the bindings of the constraints: that is, they show where you would go to get a value for a variable in an equation. Requirements diagrams show the requirements and their relationships to each other, to use cases, and to test cases.

1.11.2 Second Iteration Design Solution

1.11.2.1 Analysis Model for the Parking Garage

ConOps for a parking garage. A university parking garage should work like this. A person drives up to the entrance gate, the system interrogates the car's transponder, if it is valid, the wooden arm opens, when the system senses that the car has left the gate area, the wooden arm closes. When that person leaves for the day, he or she drives up to the exit gate, the system interrogates the car's transponder, if it is valid, the wooden arm opens. When the system senses that the car has left the gate area, the wooden arm closes. The wooden arm shall not close on top of the car or break the radio antenna. To prevent people from using one transponder for two cars, the entrance gate will not open if the system thinks that the car is already inside. The transponder contains only an identification number. All of the other data are stored in the system. To make the system simpler, design it for only one transponder (that is, consider only one car). For simplicity, assume that the system has only two gates: one for entrance and one for exit. Restrict the inputs and outputs to things that could actually be in the parking garage.

Glossary

Transponder: a passive electronic device (a radio frequency identification tag, RFID) usually mounted in the lower-left corner of a car 's windshield. It has the user 's identification number stored in it.

Interrogate: the system detects the presence of a transponder (This could be done by continuous scanning or be triggered by the EntranceGateSensor.). Then it sends out electromagnetic signals, in order to read the information stored in the transponder.

Use case

Name: Park in Garage
Iteration: 3.6
Derived from: Concept of Operations
Brief description: A person drives up to the entrance gate of the parking garage. The system interrogates the car's transponder: if it is valid, the system raises the wooden arm. When the system senses that the car has left the gate area, the system lowers the wooden arm. A similar sequence occurs when the person leaves the garage.
Level: User goal
Scope: Parking Garage
Added value: The Driver can park his car in a safe, shaded environment within a half mile of his or her classroom or office.
Primary actor: The student, faculty, or staff member who is the Driver of a car. In this use case, we will treat the Driver and the Transponder as one actor.
Supporting actor: Entrance Gate and Exit Gate
Frequency: About 1500 cars use the system each day.
Precondition: The Driver's car is outside of the parking garage and the system believes that the car is outside of the parking garage.

Main Success Scenario

1. The Driver approaches the Entrance Gate.
2. The system interrogates the car's transponder.
3a. The system determines that the transponder is valid and that the car is not already in the garage.
4. The system raises the wooden arm.
5. The Driver drives through the Entrance Gate.
6. The system detects that the car is clear of the Entrance Gate.
7. The system lowers the wooden arm.
8. The system stores the transponder's identification number (ID), the date, and the time.
9. The Driver parks the car and goes to class.
10. The Driver returns to the car and drives up to the Exit Gate.
11. The system interrogates the car's transponder.
12a. The system determines that the transponder is valid and that the database says that the car is actually in the garage.
13. The system raises the wooden arm.
14. The Driver drives through the Exit Gate.
15. The system detects that the car is clear of the Exit Gate.
16. The system lowers the wooden arm.
17. The system stores the transponder's identification number (ID), the date, and the time.
18. The Driver drives home [exit use case].

Alternate Flows

3b. The system determines that the transponder is not valid or that the car is already in the garage.
3b1. The system does nothing. (This may be stupid, but it is what the real system actually does.) [exit use case].
12b. The system determines that the transponder is not valid or that the car is not in the garage.
12b1. The system does nothing (This flaw must be fixed.) [exit use case].

Postcondition: The Driver's car is outside of the parking garage and the system has recorded that the car is outside of the parking garage. (This does not include the present behavior of step 12b.)

Specific Requirements

Functional Requirements

FR1. The system shall determine that a car is waiting at the Entrance or Exit Gate. This requirement (and other similar compound requirements) will be decomposed later.
FR2. The system shall determine the validity of the transponder's ID.
FR3. The system shall store the identity of every car in the garage (from the ConOps).
FR4. The system shall open and close the Entrance and Exit Gates.
FR5. The system shall determine when a car has cleared the Entrance or Exit Gate areas.

Derived Requirement

FR6. The system shall allow a car to enter the garage only if the transponder's ID is valid and the database shows that the car is outside of the garage. This prevents two cars from using the same transponder.

Nonfunctional Requirements

NFPR1. Electronic response time. When the transponder is valid, the system shall begin raising the wooden arm within 2 s of interrogation.
NFPR2. Mechanical movement time. The system shall raise and lower the wooden arms in 1 to 3 s.

Author/owner: Parking and Transportation.
Last changed: April 24, 2016

The use case diagram for the Park in Garage use case is the same as Fig. 1.28. No change was necessary for this diagram from the first iteration

The system requirements document would go here. Because of its length, we will not include it in this textbook.

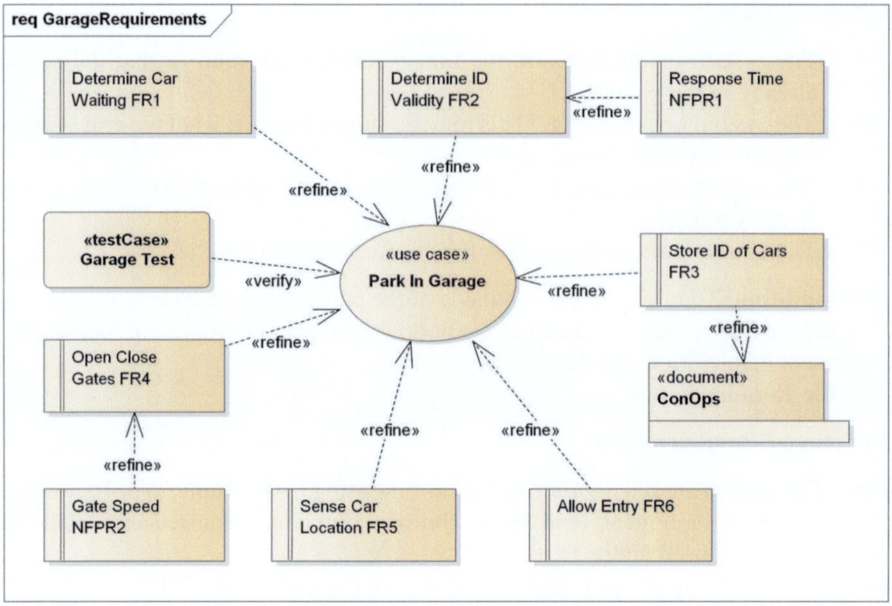

Fig. 1.31 Requirements diagram for parking garage entrance gate

In this second iteration, we are becoming more formal. Requirements 1–6 map to functional requirements FR1–FR6 and requirements 7 and 8 become nonfunctional performance requirements NFPR1 and NFPR2.

Our design process starts with the use cases. The requirements and everything else follow. Therefore, the dependencies in Fig. 1.31 show that, for example, the requirement SenseCarLocation is more detailed and thus refines the use case Park in Garage. If instead, the customer gave us all of the requirements and then we wrote the use cases, then the arrow's direction would show that the use case, being more detailed, refines the requirements.

1.11.2.2 *Parking Garage Test Plan*

This test plan is based on the use case.

Precondition: Tester must ensure that (1) Tester's car is outside of the parking garage, (2) the system believes that Tester's car is outside of the parking garage (He can verify this by calling Parking and Transportation.), and (3) Tester's car transponder is valid.

Main Success Scenario

1. *Tester approaches the Entrance Gate.*
2. *The system interrogates the car's transponder.*

3. *The system determines that the transponder is valid and that the car is not already in the garage.*
4. *The system opens the gate. Tester records that the gate opened. This tests the following requirements:*

   FR1. The system shall determine that a car is waiting at the Entrance Gate.
   FR2. The system shall determine the validity of the transponder 's ID.
   FR4a. The system shall open the Entrance Gate.
   FR6. The system shall allow a car to enter the garage only if the transponder 's ID is valid and the database shows that the car is outside of the garage.

   Tester shall use a stopwatch to time function execution and he shall record the execution times. This tests the following requirements:

   NFPR1. Electronic response time. When the transponder is valid, the system shall begin raising the wooden arm within 2 s of interrogation.
   NFPR2. Mechanical movement time. The system shall raise the gates in 1–3 s.

5. *Tester drives through the Entrance Gate.*
6. *The system detects that the car is clear of the Entrance Gate.*
7. *The system closes the gate. Tester records that the wooden arm went down. This tests the following requirements:*

   FR4b. The system shall close the Entrance Gate.
   FR5. The system shall determine when a car has cleared the Entrance Gate area.

   Tester shall use a stopwatch to time function execution and he shall record the execution times. This tests the following requirements:

   NFPR2. The system shall lower the wooden arms in 1–3 s.

8. *The system stores the transponder 's identification number (ID), the date, and the time. Tester calls Parking and Transportation on 6th Street and Highland and asks if his car is in the garage. If they confirm that his car is in the garage, then Tester records that the following requirement has passed its test.*

   FR3. The system shall store the identity of every car in the garage.

9. *Tester parks the car and gets out.*
10. *Tester returns to the car and drives up to the Exit Gate.*
11. *The system interrogates the car 's transponder.*
12. *The system determines that the transponder is valid and that the database says that the car is actually in the garage.*
13. *The system opens the gate. Tester records that the wooden arm went up. This tests the following requirements:*

    FR1. The system shall determine that a car is waiting at the Exit Gate.
    FR2. The system shall determine the validity of the transponder 's ID.
    FR4c. The system shall open the Exit Gate.

FR6. The system shall allow a car to exit the garage only if the transponder 's ID is valid and the database shows that the car is inside of the garage.

Tester shall use a stopwatch to time function execution and he shall record the execution times. This tests the following requirements:

NFPR1. When the transponder is valid, the system shall begin raising the wooden arm within 2 s of interrogation.
NFPR2. The system shall raise the wooden arm in 1–3 s.

14. *Tester drives through the Exit Gate.*
15. *The system detects that the car is clear of the Exit Gate.*
16. *The system closes the gate. Tester records that the wooden arm went down. This tests the following requirements:*

    FR4d. The system shall close the Exit Gate.
    FR5. The system shall determine when a car has cleared the Exit Gate area.

    Tester shall use a stopwatch to time function execution and he shall record the execution times. This tests the following requirement:

    NFPR2. The system shall lower the wooden arms in 1–3 s.

17. *The system stores the transponder 's identification number (ID), the date, and the time.*
    Tester calls Parking and Transportation on 6th Street and Highland and asks if his car is in the garage. If they confirm that his car is not in the garage, then Tester records that the following requirement has passed its test:

    FR3. The system shall store the identity of every car in the garage.

18. *Tester drives home [exit test procedure].*

Test of Alternate Flows

Precondition: Tester must ensure that (1) Tester 's car is outside of the parking garage, (2) the system believes that Tester 's car is outside of the parking garage (He can verify this by calling Parking and Transportation.), and (3) Tester 's car transponder is valid.

1. *Tester approaches the Entrance Gate.*
2. *The system interrogates the car 's transponder.*
3. *The system determines that the transponder is valid and that the car is not already in the garage.*
4. *The system opens the gate.*
5. *Tester drives through the Entrance Gate.*
6. *The system detects that the car is clear of the Entrance Gate.*
7. *The system closes the gate.*
8. *The system stores the transponder 's identification number (ID), the date, and the time.*

This next step tests the alternate flow "3b. The system determines . . . that the car is already in the garage."

9. *Tester parks the car and carries his transponder back to the entrance gate. He waves it under the sensor. The gate should not open. Tester records the result of this test. This is a second test of the requirement.*

   FR6. The system shall allow a car to enter the garage only if the transponder 's ID is valid and the database shows that the car is outside of the garage.

10. *Tester drives up to the Exit Gate.*
11. *The system interrogates the car 's transponder.*
12. *The system determines that the transponder is valid and that the database says that the car is actually in the garage.*
13. *The system opens the gate.*
14. *Tester drives through the Exit Gate.*
15. *The system detects that the car is clear of the Exit Gate.*
16. *The system closes the gate.*
17. *The system stores the transponder 's identification number (ID), the date, and the time.*
    This next step tests the alternate flow "12b. The system determines . . . that the car is not in the garage."
18. *Tester parks the car and carries his transponder back to the exit gate. He waves it under the sensor. The gate should not open. Tester records the result of this test. This is a second test of the requirement.*

    FR6. The system shall allow a car to exit the garage only if the transponder 's ID is valid and the database shows that the car is inside of the garage.

18. *Tester drives home [exit test procedure].*
    The parking garage system is in compliance with its requirements if and only if it passes all of these tests.

1.11.2.3 *Analysis Classes*

Next, to identify the classes, we underline nouns in the ConOps, the use case, the requirements, and the glossary. Using this technique and stereotypical classes (boundary, entity, and controller), we identified the following analysis classes: Driver, Transponder (for now, we will treat the Driver and the Transponder as one class called Driver), Entrance Gate, EntranceGateReader (boundary class), EntranceGateSensor (boundary class), EntranceGateController (control class), Exit Gate, ExitGateReader (boundary class), ExitGateSensor (boundary class), ExitGateController (control class), GarageDatabase (entity class), and WoodenArm (supporting actor because it is CotS equipment).

Now let us use these classes to make a sequence diagram for the entrance gate portion of the main success scenario of the Park in Garage use case. We can then use the sequence diagram to find the functions of these classes (Fig. 1.32).

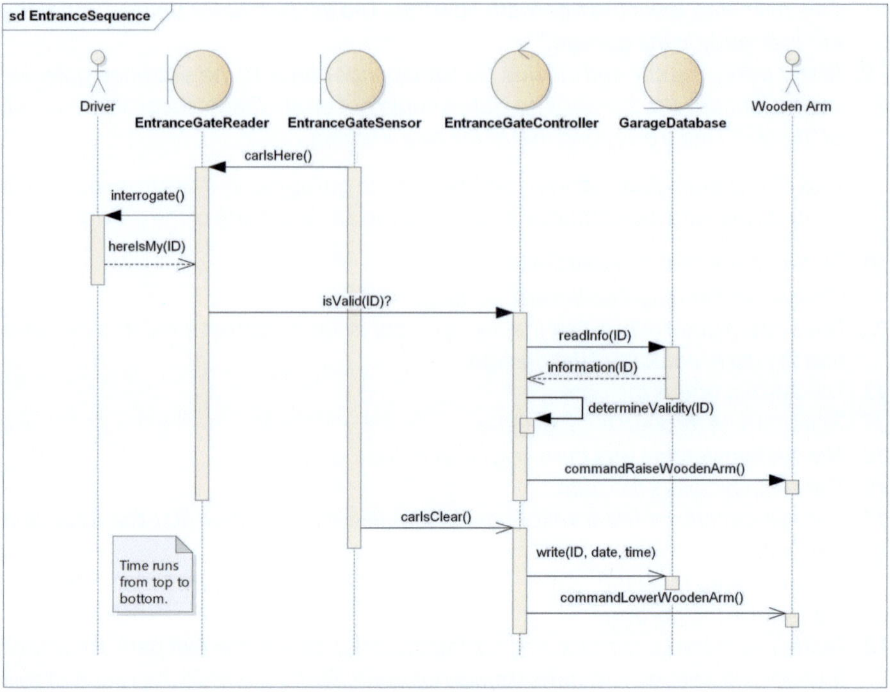

Fig. 1.32 A sequence diagram for the entrance gate portion of the main success scenario of the Park in Garage use case. Time runs from top to bottom

Later, we will discuss a life-cycle model in Fig. 4.8 that has activities labeled as requirements, analysis, design, implementation, and operations. When we are performing the analysis activity, we talk about analysis classes. When we are performing the design activity, we talk about design classes. The only difference between an analysis class and a design class is the amount of detail.

1.11.2.4 *Attributes and Functions of the Classes*

A class diagram has three compartments. The top one contains the class name and its stereotype if it has one. The middle compartment has attributes of the class, their legal (and perhaps default) values. The bottom compartment contains the functions of the class and any parameters that may be passed. Only the top compartment is mandatory.

Class name <<stereotype>>
Attributes: legal values
Functions (passed parameters)

An attribute is a small memory unit. If it is Boolean, it can be set and cleared.

Driver <<actor>>
ID: String
hereIsMy(ID)

In later, more detailed, models of this parking garage system, the Driver will be decomposed into the Human, the Transponder, and the Automobile. For now, we will treat the Driver as an object with an identification (ID) number. The Reader can read this number.

EntranceGateSensor <<boundary class>>
carIsClear: Boolean=yes
carIsHere() *setCarIsClear()*

In the actual physical system, the Sensor is probably a coil of wire, maybe 6 ft by 6 ft, buried in the asphalt. When a car is over it, its inductance is greatly increased. This change signals the presence of a car.

An important observation. At this stage, we do not discuss implementation. However, it often helps to understand the class, if a possible implementation is contemplated and in this case, of course, it is because we are modeling real system. Everyone understands that the implementation can change at any time. The design cannot depend on the example implementation.

EntranceGateReader <<boundary class>>
interrogateTransponder(ID) *sendIsValidQuestion(ID)*

The Reader has an antenna about one foot square about 8 ft off of the ground. It sends an electromagnetic signal to the car 's transponder and reads its identification (ID) number. It sends this number to the Controller to validate the ID.

EntranceGateController <<control class>>
validity: Boolean=no *readInfo(ID)* *determineValidity(ID)* *writeInfo(ID, date, time)* *commandOpenWoodenArm* *commandCloseWoodenArm*
set(validity)

The Controller, which is a computer, communicates with the Sensor, the Reader, the Database, and the WoodenArm.

GarageDatabase <<entity class>>
ID: Integer carInGarage: Boolean= out {in, out} validity: Boolean=invalid date: Date time: Time
readAndWrite(ID, date, time, carInGarage, validity)

The GarageDatabase stores all the valid IDs, the presumed location of all cars, the capacity and number of cars in each garage, and a myriad of other information about parking garages.

WoodenArm <<actor>>
woodenArmState: String {up, down, moving}
raiseWoodenArm() lowerWoodenArm()

A WoodenArm is about 6 in. wide, 8 ft long, and ¾ in. thick. When horizontal, it blocks entry by cars. When it is rotated into the vertical position, it allows cars to pass through. It receives commands from the Controller. The wooden arms are all the same: so there is no reason to specify one as an entrance gate or an exit gate wooden arm. Now they are plastic. This class includes the equipment needed to move the arm, such as a motor, gears, a pivot point, and a stanchion.

Explanations (Figs. 1.33 and 1.34)

Functions	*Explanation*
Functions (or operations) from the sequence and class diagrams	
hereIsMyID()	*The Transponder on the dashboard of the car must be readable by the Entrance and ExitGateReaders. This means it must be perpendicular to the reader and its path must not be obstructed*
raiseWoodenArm	*The system must be able to raise the WoodenArm. Speed of movement is a nonfunctional performance requirement associated with this function*
lowerWoodenArm	*The system must be able to lower the WoodenArm. Speed of movement is a nonfunctional performance requirement associated with this function*
interrogateTransponder(ID)	*The Entrance and ExitGateReaders must interrogate the car 's transponder to read its identification number (ID)*
sendIsValidQuestion(ID)	*The Entrance and ExitGateReaders must send the transponder 's ID in the "isValid(ID)?" message*

(continued)

Functions	*Explanation*
carIsHere()	*The Entrance and ExitGateSensors must sense the position of the car. The WoodenArms should not close on top of a car*
setCarIsClear()	*The Entrance and ExitGateSensors must sense the position of the car. They should set the attribute carIsClear to reflect this condition*
readAndWrite(ID, date, time, carInGarage, validity)	*This is the information that is recorded in the GarageDatabase*
readInfo(ID)	*The Controller must command the database to read information in the GarageDatabase in order to determine if the ID is valid and that the car is outside of the garage*
writeInfo(ID, date, time)	*The Controller must command the database to write the transponder 's ID, the date, and time of entry and exit*
determineValidity(ID)	*The Controller must compare the transponder 's ID with data stored in the GarageDatabase to determine if the ID is valid and that the car is outside of the garage. To determine this it tests the following Boolean expression (carInGarage=out AND validity=valid)*
commandOpenWoodenArm	*The EntranceGateController commands the WoodenArm to open*
commandCloseWoodenArm	*The EntranceGateController commands the WoodenArm to close*
set(validity)	*The EntranceGateController sets the attribute validity to valid or invalid*

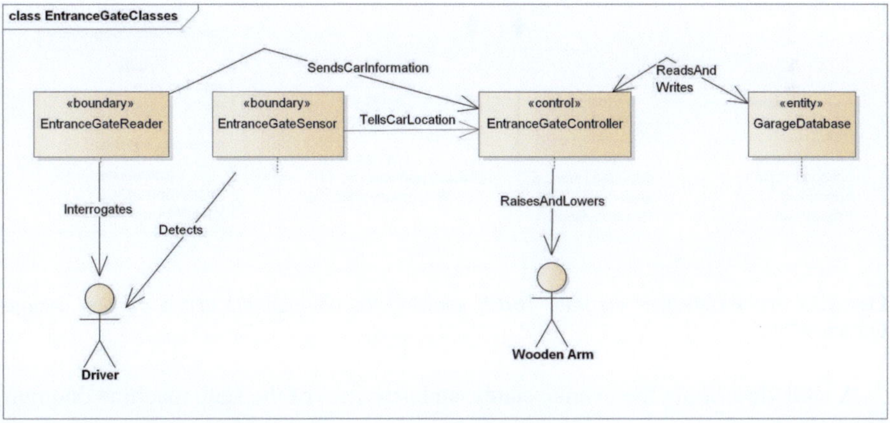

Fig. 1.33 *Entrance Gate class diagram*

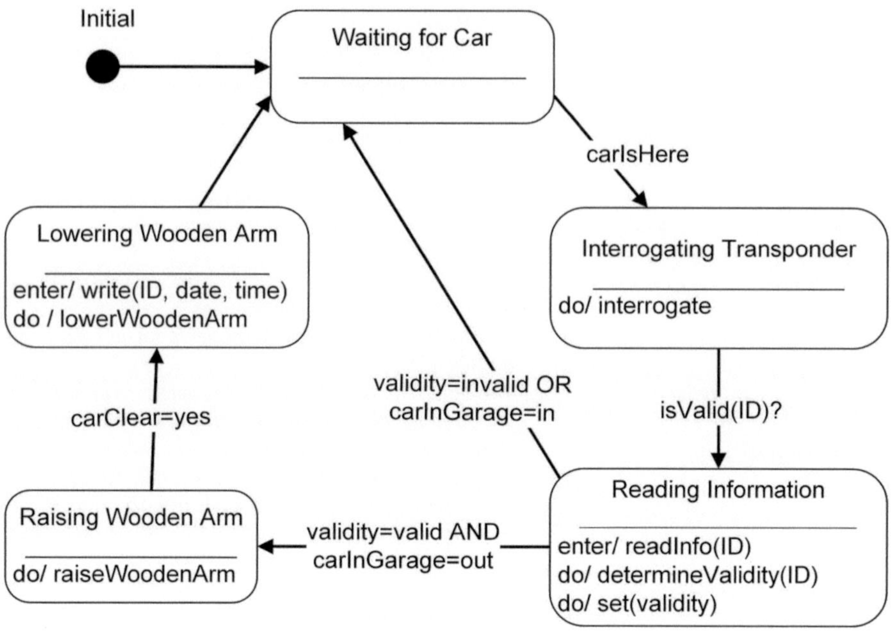

Fig. 1.34 State machine diagram for the EntranceGateController

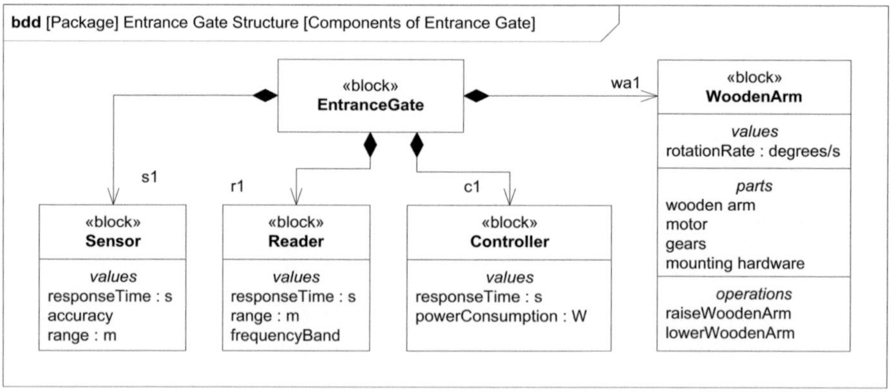

Fig. 1.35 *Block definition diagram (**bdd**) showing the components of the parking garage EntranceGate*

A table describing the events, states, and activities of the state machine diagram would go here.

This parking garage system will have two similar subsystems, one for the Entrance Gate and one for the Exit Gate. They will share a common GarageDatabase.

Explanation of abbreviations and role names used in Figs. 1.35, 1.36, and 1.37.

Usage name or role	*Explanation*
s1	*One particular sensor on an entrance gate*
r1	*One particular reader on an entrance gate*
c1	*One particular controller on an entrance gate*
wa1	*One particular wooden arm on an entrance gate. The term wooden arm includes the motor, gears, mounting hardware, etc The wooden arms are all the same; so there is no reason to differentiate between entrance or exit gate wooden arms*
wam	*Wooden arm motion, the constraint describing the motion of a wooden arm*
sr	*System response, the constraint describing the response time of the system. This includes the response times of the reader, the controller, and the database*
eg	*One particular entrance gate*
Abbreviation	*Meaning*
rr	*Rotation rate in degrees per second*
srt	*Sensor response time in seconds*
crt	*Controller response time in seconds*
rrt	*Reader response time in seconds*
sysrt	*System response time*

The structure of the EntranceGate block was defined with Fig. 1.35. The EntranceGateAnalysis block would be the hardware, software (MatLab, Java, or C

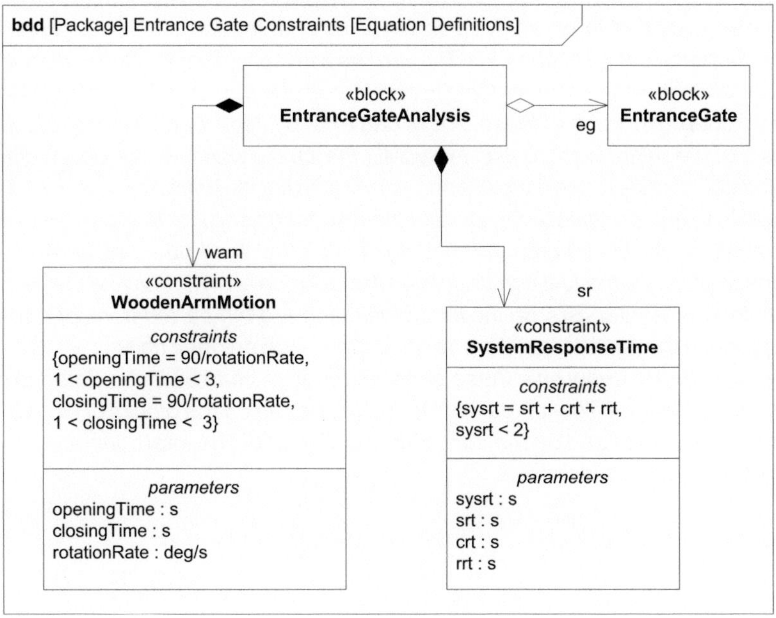

Fig. 1.36 *Block definition diagram (**bdd**) defining constraints for the parking garage EntranceGateAnalysis*

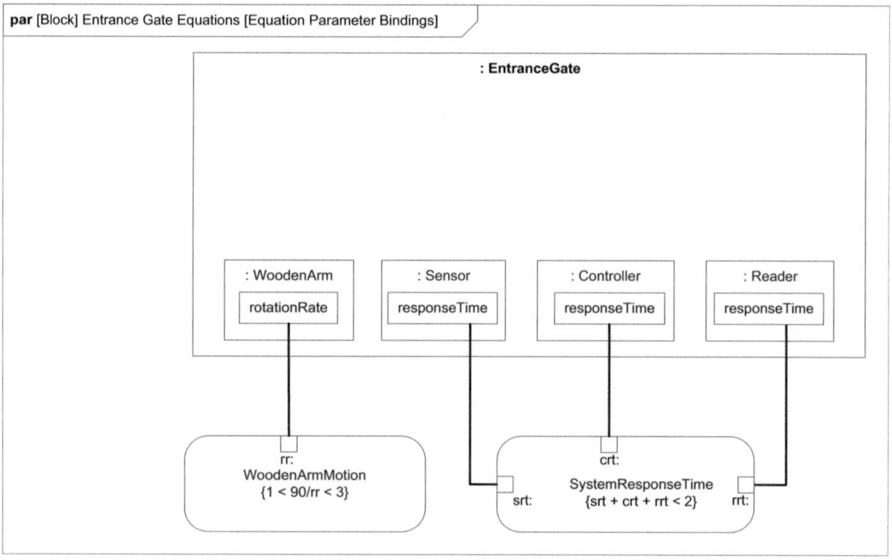

Fig. 1.37 *Parametric diagram (**par**) using nested part notation showing the equation bindings for a particular parking garage EntranceGate*

programs), bioware, and algorithms that are being used to analyze this model. This **bdd** *defines the equations in the <<constraint>> blocks. These equations may have originally been specified in an analysis library. This enables the same equations to be reused for many different contexts. The arrows with solid (black) diamond endings in Fig. 1.35 tell us that the EntranceGate is composed of a Sensor, a Controller, a Reader, and a WoodenArm. Each of these blocks can exist by itself. Each of these blocks has values that we might want to use. In SysML, the arrow with a white (open) diamond head indicates a reference association. This means that the EntranceGate will be associated with a mathematical analysis (like the inspection certificate in an elevator). The arrows with solid (black) diamond endings tell us that the EntranceGateAnalysis is composed of mathematical constraints on the wooden arm motion and the response time. In the parameter lists, on the left of the colon are the parameters, or names as they appear in the equation, and on the right of the colon are their units.*

Now we want to show how these general equations can be instantiated to a particular EntranceGate. The top part of Fig. 1.37 is based on the blocks of Fig. 1.35 and the bottom part is based on the constraints of Fig. 1.36. The interconnections show the bindings of the parameters.

Table 1.10 *Mappings between models*

High-level model of the first iteration of the Parking Garage System	*Analysis model of the second iteration of the Parking Garage System*
Entrance Gate	
[Entrance] gate	*EntranceGateController* *[Entrance Gate] Wooden Arm*
System interrogates transponder	*EntranceGateReader*
System detects car is clear	*EntranceGateSensor*
System stores ID	*GarageDatabase*
There will be similar mappings for the Exit Gate	
Requirements	
Req1	*FR1*
Req2	*FR2*
Req3	*FR3*
Req4	*FR4*
Req5	*FR5*
Req6	*FR6*
Req7	*NFPR1*
Req8	*NFPR2*
Req9	*FR7*
Req10	*NFPR3*

1.11.3 Mapping Between Models

At the start, designs are represented at a high level of abstraction. As the design process continues, levels and details are added. These mappings are documented as the design progresses [22]. For system designs, several models naturally develop. These may result from the process being followed. We start with the business needs, write use cases with requirements, and segue into the customer model. Then, we add details and reduce the level of abstraction to create the analysis model. Much later, we elaborate on the classes, and specify their relationships, to produce the design model. Table 1.10 shows the mapping of the high-level model from the first iteration to the analysis model from the second iteration.

1.11.4 Homework

1.11.4.1 Requirements

Section 10 did not give a requirements diagram. What would be the differences between that diagram and the requirements diagram given in this section?

1.11.4.2 Refine Relationship

Figure 1.31 has a refine relation between FR4 and NFPR2. Which requirement refines the other? Please explain your answer.

1.11.4.3 Sequence Diagram

The sequence diagram of Fig. 1.32 has three classes at the top: EntranceGateReader (boundary class), EntranceGateSensor (boundary class), EntranceGateController (control class), and GarageDatabase (entity class). They have special names and special icons. They are called stereotypes. But haven't we been told since first grade that it is bad to use stereotypes? What are stereotypes and what do these three mean?

1.11.4.4 Class Diagrams

This is the class diagram for the EntranceGateController.

EntranceGateController <<controlclass>>
validity: Boolean=no
readInfo(ID)
determineValidity(ID)
writeInfo(ID, date, time)
commandOpenWoodenArm
commandCloseWoodenArm
set(validity)

What do the double bars on the sides mean?

Some of its operations (functions) pass parameters. For example, the writeInfo(ID, date, time) takes the information ID, date, and time from the EntranceGateController and sends it to the GarageDatabase. Could the command functions be written to pass parameters? For example, command (OpenWoodenArm) and command(CloseWoodenArm).

This is the class diagram for the GarageDatabase.

GarageDatabase <<entity class >>
ID: Integer carInGarage: Boolean = out {in, out} validity: Boolean = invalid date: Date time: Time
readAndWrite(ID, date, time, carInGarage, validity)

Explain each part of this attribute description: carInGarage: Boolean = out {in, out}.

1.11.4.5 State Machine Diagram

The state machine diagram of Fig. 1.34 has the words *entry, do,* and *exit*. What do they mean?

Make a copy of the sequence diagram of Fig. 1.32. Circle and label the messages and commands associated with each of the five states of Fig. 1.34.

1.11.4.6 Block Definition Diagrams (bdd)

Figures 1.35, 1.36, and 1.37 all contain the block *Entrance Gate*. Are they the same thing? What does Fig. 1.35 tell us about the Entrance Gate?

What does Fig. 1.37 tell us about the Entrance Gate?

1.11.4.7 Farmroad–Highway Intersection

There is an intersection between a seldom-used farmroad and a busy highway. A farmer has difficulty getting across the highway, design a system to help him get across. In a previous version of this problem, we did an engineering analysis of the problem including stating the problem and studying alternatives. As a result of that, we decided to install a traffic signal. Therefore, we now want you to design a controller for this traffic light (Fig. 1.38).

Normally, the highway light shall be green and the farmroad light shall be red. When a sensor signals that a tractor is on the farmroad, the highway light shall

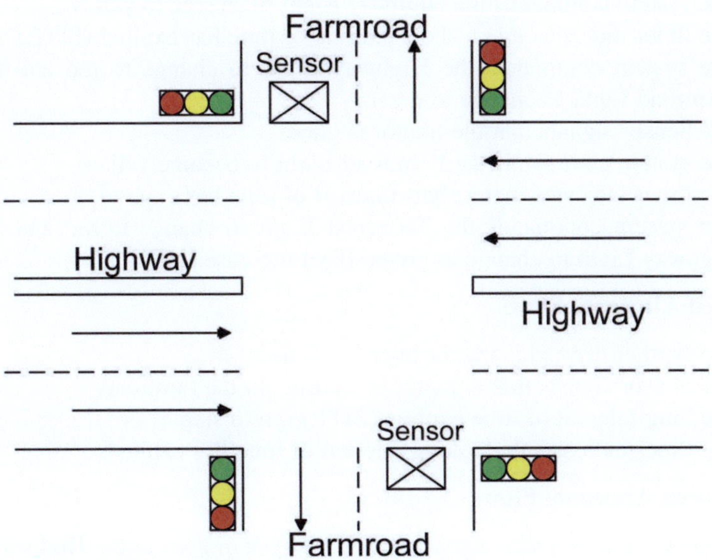

Fig. 1.38 An intersection between a farmroad and a highway

change to yellow. After a *short-time interval* (normally 10 s), the highway light shall change to red and the farmroad light shall change to green. The farmroad light shall stay green until the tractor clears the sensor or after a *long-time interval* (normally 50 s), whichever comes first. Then the farmroad light shall become yellow. After a short-time interval, the farmroad light shall become red and the highway light shall become green. The system shall stay in that state for a least a long-time interval. After a long-time interval, the highway light shall be switched when the sensor detects a tractor. A timer that produces output signals after short-time intervals and long-time intervals will be available. It can be restarted at any time. Here is a use case for this system.

Name: Cross the Highway

Description: A Farmer activates the Sensor on the farmroad. The Controller changes the lights allowing the tractor to cross and then goes back to the normal Highway-Green State. The process cycles through green, yellow, and red lights until stopped by malfunction or maintenance.

Level: Low

Scope: A simple controller for one rural intersection.

Primary Actor: Farmer.

Supporting Actors: Highway and Farmroad Lights.

Precondition: The system is in the Highway Green and Farmroad Red states.

Main Success Scenario

1a. The long interval of time expires (LITE).
2. The Farmer drives his tractor over the Sensor.
3. The Sensor signals that a tractor is waiting on the farmroad.
4. The system commands the Highway Light to change to yellow.
5. The timer indicates that a short interval of time has expired (SITE).
6. The system commands the Highway Light to change to red and then the Farmroad Light to change to green.
7a. The Sensor signals that the tractor is gone.
8. The system commands the Farmroad Light to become yellow.
9. The timer indicates that a short interval of time has expired.
10. The system commands the Farmroad Light to change to red and then the Highway Light to change to green. [Exit use case.]

Anchored Alternate Flows

1b. The Farmer drives his tractor over the Sensor.
1b1. The Sensor signals that a tractor is waiting on the farmroad.
1b2. The long interval of time expires (LITE) [go to step 4].
7b. The timer indicates that a long interval of time has expired.

Unanchored Alternate Flows

A1. Whenever the *initialize* signal occurs, the system goes to the Highway Green and Farmroad Red states.

Postcondition: The system is in the Highway Green and Farmroad Red states.

Rules

Rule 1:. The long-time interval is normally 50 s.
Rule 2:. The short-time interval is normally 10 s.
Rule 3:. While the Farmroad Lights are green or yellow the Highway Lights shall be red.
Rule 4:. While the Highway Lights are green or yellow the Farmroad Lights shall be red.

Frequency: This Cross the Highway use case could be invoked every 2 min, but usually it will be invoked about once every 8 h.
Author/owner: Farmer John
Date: January 31, 2016

We want you to use this use case and construct a use case diagram, a class diagram, a sequence diagram, a state machine diagram, and an activity diagram.

1.12 Section 12. Parking Garage, Third Iteration

Our parking garage has space for 1400 cars. So when there are 1400 cars in the garage, the system should illuminate the "Full" sign at the entrance and allow no more cars to enter until some cars have left. We will now add this additional feature to the Analysis Model of the Second Iteration Design Solution to reflect this new feature. Remember that you have already read the text in the Times New Roman regular black font, so you do not need to scrutinize it this time. *The modifications and additions are in the Italic Sans-Serif font.*

1.12.1 Analysis Model for the Parking Garage

ConOps for a parking garage. A university parking garage should work like this. A person drives up to the entrance gate, the system interrogates the car's transponder, if it is valid, the wooden arm opens, when the system senses that the car has left the gate area, the wooden arm closes. When that person leaves for the day, he or she drives up to the exit gate, the system interrogates the car's transponder, if it is valid, the wooden arm opens. When the system senses that the car has left the gate area, the wooden arm closes. The wooden arm shall not close on top of the car or break the radio antenna. To prevent people from using one transponder for two cars, the entrance gate will not open if the system believes that the car is already inside. The transponder contains only an identification number. All other data are stored in the system. To simplify system design, design it for just one transponder (that is, consider just one car). Also, for simplicity, assume that the system has only two gates: one for entrance and one for exit. Restrict inputs and outputs to artifacts that could actually be in the parking garage. *When the number of cars in the garage reaches 1400, the system shall illuminate the "Full" sign at the entrance and allow no more cars to enter until some cars have left.*

Glossary

Transponder: a passive electronic device (a radio frequency identification tag, RFID) usually mounted in the lower-left corner of a car's windshield. It has the user's identification number stored in it.

Interrogate: the system detects the presence of a transponder (This could be done by continuous scanning or be triggered by the EntranceGateSensor.). Then it sends out electromagnetic signals, in order to read the information stored in the transponder.

Use case (Fig. 1.39)

Name: Park in Garage

Iteration: 3.7

Derived from: Concept of Operations

Brief description: A person drives up to the entrance gate of the parking garage. The system interrogates the car's transponder: if it is valid *and the lot is not full,* then the system opens the gate. When the system senses that the car has left the gate area, the system closes the gate. A similar sequence occurs when the person leaves the garage.

Level: User goal

Scope: Parking Garage

Added value: The Driver can park his car in a safe, shaded environment within a half mile of his or her classroom or office.

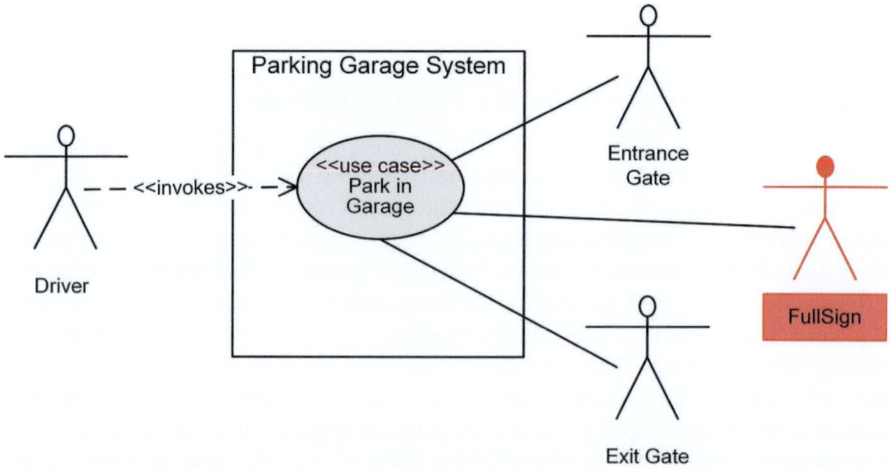

Fig. 1.39 Use case diagram for parking garage system *with a FullSign. We have added the FullSign to this use case diagram*

Primary actor: The student, faculty, or staff member who is The Driver of a car. In this use case, we will treat the Driver and the Transponder as one actor.
Supporting actor: Entrance Gate, Exit Gate, *and FullSign*
Frequency: About 1500 cars use the system each day.
Precondition: The Driver's car is outside of the parking garage, the system believes that the car is outside of the parking garage, *and the garage is not full.*

Main Success Scenario

1. The Driver approaches the Entrance Gate.
2. The system interrogates the car's transponder.
3a. The system determines that the transponder is valid, that the car is not already in the garage, *and that the garage is not already full.*
4. The system raises the wooden arm.
5. The Driver drives through the Entrance Gate.
6. The system detects that the car is clear of the Entrance Gate.
7. The system closes the gate.
8. The system stores the transponder's identification number (ID), the date, and the time.
9. The Driver parks the car and goes to class.
10. The Driver returns to the car and drives up to the Exit Gate.
11. The system interrogates the car's transponder.
12a. The system determines that the transponder is valid and that the database says that the car is actually in the garage.
13. The system opens the gate.
14. The Driver drives through the Exit Gate.
15. The system detects that the car is clear of the Exit Gate.
16. The system closes the gate.
17. The system stores the transponder's identification number (ID), the date, and the time.
18. If it is on, then turn off the FullSign (there should be a dead zone).
19. The Driver drives home [exit use case].

Alternate Flows

3b. The system determines that the transponder is not valid or that the car is already in the garage.
3b1. The system does nothing. (This may be stupid, but it is what the system actually does.) [exit use case].
3c. The system determines that the garage is full.
3c1. The system illuminates the FullSign and does not allow entry [exit use case].
12b. The system determines that the transponder is not valid or that the car is not in the garage.
12b1. The system does nothing (This design flaw must be fixed.) [exit use case].

Postcondition: The Driver's car is outside of the parking garage and the system has recorded that the car is outside of the parking garage. (This does not include the present behavior of step 12b.)

Specific Requirements

Functional Requirements

FR1. The system shall determine that a car is waiting at the Entrance or Exit Gate. This requirement (and other similar compound requirements) will be decomposed later.

FR2. The system shall determine the validity of the transponder's ID.

FR3. The system shall store the identity of every car in the garage (from the ConOps).

FR4. The system shall open and close the Entrance and Exit Gates.

FR5. The system shall determine when a car has cleared the Entrance or Exit Gate areas.

FR7. The system shall illuminate the FullSign.

Derived Requirement:

FR6. The system shall allow a car to enter the garage only if the transponder's ID is valid, the database shows that the car is outside of the garage, *and the number of cars in the garage is less than the garageCapacity.*

Nonfunctional Requirements

NFPR1. When the transponder is valid, the system shall *begin* raising the wooden arm within 2 s of interrogation. This requirement includes the response time of the sensor, the Reader, and the Controller.

NFPR2. The system shall raise and lower the wooden arms in 1–3 s.

NFPR3. The system shall compute the number of cars in the garage and ensure that garage capacity is not exceeded. The parking garage capacity is 1400 cars. Note that this requirement states what not how. There are many ways to implement this requirement. You could (1) increment a counter to count cars, (2) decrement a counter to count empty spaces, (3) use a pointer and an array, (4) add and remove marbles to a tray of a balance scale, (5) use sensors to detect empty and full spaces, (6) measure the inductance of the garage, (7) weight the garage with stain gauges on the earthquake resistant mounts, (8) interrogate the transponders all at one time and measure the strength of the response. In the design model, we will implement this with a class.

Author/owner: Parking and Transportation

Last changed: April 24, 2016

The system requirements *document* will go here. In it, we are becoming more formal: requirements 1–6 map to FR1–FR6 and requirements 7 and 8 become NFPR1 and NFPR2 (Fig. 1.40).

Our design process starts with the use cases. The requirements and everything else follow. Therefore, the dependencies in the above diagram show that, for

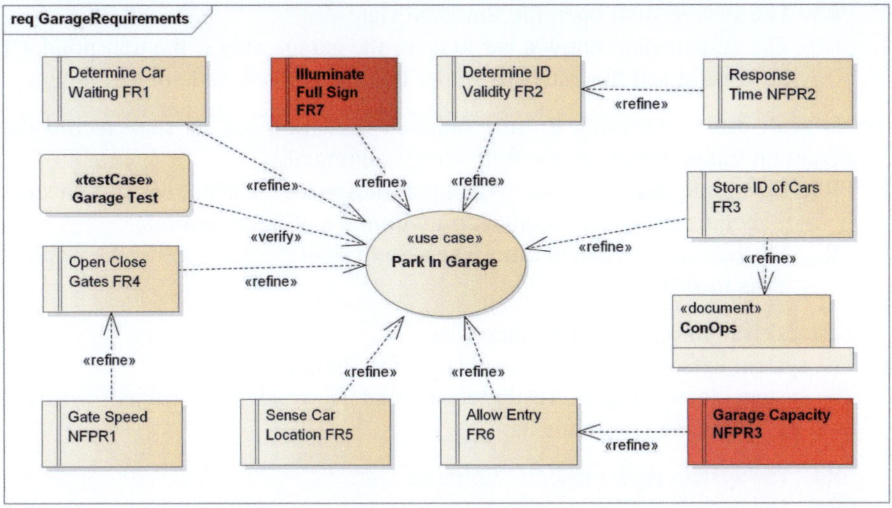

Fig. 1.40 Requirements diagram for parking garage system with a FullSign. FR7 and NFPR3 have been added to this requirements diagram

example, the requirement SenseCarLocation is more detailed and thus refines the use case Park in Garage. If instead, the customer gave us all of the requirements and then we wrote the use cases, then the arrow's direction would show that the use case, being more detailed, refines the requirements.

1.12.2 Parking Garage Test Procedure

1.12.2.1 Original Test of Main Success Scenario

Precondition: Tester must ensure that (1) Tester's car is outside of the parking garage (2) the system believes that Tester's car is outside of the parking garage (He can verify this by calling Parking and Transportation.), and (3) Tester's car transponder is valid.

Main Success Scenario

1. Tester approaches the Entrance Gate.
2. The system interrogates the car's transponder.
3. The system determines that the transponder is valid and that the car is not already in the garage.
4. The system opens the gate. Tester records that the wooden arm went up. This tests the following requirements:

   FR1. The system shall determine that a car is waiting at the Entrance Gate.
   FR2. The system shall determine the validity of the transponder's ID.

FR4. The system shall open the Entrance Gate.
FR6. The system shall allow a car to enter the garage only if the transponder's ID is valid and the database shows that the car is outside of the garage.

Tester shall use a stopwatch to time function execution and he shall record the execution times. This tests the following requirements:
NFPR1. Electronic response time. When the transponder is valid, the system shall begin raising the wooden arm within 2 s of interrogation.
NFPR2. Mechanical movement time. The system shall raise and lower the wooden arms in 1–3 s.

5. Tester drives through the Entrance Gate.
6. The system detects that the car is clear of the Entrance Gate.
7. The system closes the gate. Tester records that the wooden arm went down. This tests the following requirements:

FR4. The system shall close the Entrance Gate.
FR5. The system shall determine when a car has cleared the Entrance Gate area.

Tester shall use a stopwatch to time function execution and he shall record the execution times. This tests the following requirement:
The system shall lower the wooden arms in 1–3 s.

8. The system stores the transponder's identification number (ID), the date and the time. Tester calls Parking and Transportation and asks if his car is in the garage. If they confirm that his car is in the garage, then Tester records that the following requirement has passed its test.

FR3. The system shall store the identity of every car in the garage.

9. Tester parks the car and gets out.
10. Tester returns to the car and drives up to the Exit Gate.
11. The system interrogates the car's transponder.
12. The system determines that the transponder is valid and that the database says that the car is actually in the garage.
13. The system opens the gate. Tester records that the wooden arm went up. This tests the following requirements:

FR1. The system shall determine that a car is waiting at the Exit Gate.
FR2. The system shall determine the validity of the transponder's ID.
FR4. The system shall open the Exit Gate.
FR6. The system shall allow a car to exit the garage only if the transponder's ID is valid and the database shows that the car is inside of the garage.

Tester shall use a stopwatch to time function execution and he shall record the execution times. This tests the following requirements:
NFPR1. Electronic response time. When the transponder is valid, the system shall begin raising the wooden arm within 2 s of interrogation.

NFPR2. Mechanical movement time. The system shall raise and lower the wooden arms in 1–3 s.

14. Tester drives through the Exit Gate.
15. The system detects that the car is clear of the Exit Gate.
16. The system closes the gate. Tester records that the wooden arm went down. This tests the following requirements:

    FR4. The system shall close the Exit Gate.
    FR5. The system shall determine when a car has cleared the Exit Gate area.

Tester shall use a stopwatch to time function execution and he shall record the execution times. This tests the following requirement:
NFPR2. The system shall lower the wooden arms in 1–3 s.

17. The system stores the transponder's identification number (ID), the date, and the time.

Tester calls Parking and Transportation and asks if his car is in the garage. If they confirm that his car is not in the garage, then Tester records that the following requirement has passed its test:
FR3. The system shall store the identity of every car in the garage.

18. Tester drives home [exit test procedure].

1.12.2.2 Original Test of Alternate Flows 3b and 12b

Precondition: Tester must ensure that (1) Tester's car is outside of the parking garage, (2) the system believes that Tester's car is outside of the parking garage (He can verify this by calling Parking and Transportation.), and (3) Tester's car transponder is valid.

1. Tester approaches the Entrance Gate.
2. The system interrogates the car's transponder.
3. The system determines that the transponder is valid and that the car is not already in the garage.
4. The system opens the gate.
5. Tester drives through the Entrance Gate.
6. The system detects that the car is clear of the Entrance Gate.
7. The system closes the gate.
8. The system stores the transponder's identification number (ID), the date, and the time.
   This next step tests the alternate flow "3b. The system determines ... that the car is already in the garage."

9. Tester parks the car and carries his transponder back to the entrance gate. He waves it under the sensor. The gate should not open. Tester records the result of this test. This is a second test of the requirement

   FR6. The system shall allow a car to enter the garage only if the transponder's ID is valid and the database shows that the car is outside of the garage.

10. Tester drives up to the Exit Gate.
11. The system interrogates the car's transponder.
12. The system determines that the transponder is valid and that the database says that the car is actually in the garage.
13. The system opens the gate.
14. Tester drives through the Exit Gate.
15. The system detects that the car is clear of the Exit Gate.
16. The system closes the gate.
17. The system stores the transponder's identification number (ID), the date, and the time.

    This next step tests the alternate flow "12b. The system determines . . . that the car is not in the garage."
18. Tester parks the car and carries his transponder back to the exit gate. He waves it under the sensor. The gate should not open. Tester records the result of this test. This is a second test of the requirement

    FR6. The system shall allow a car to exit the garage only if the transponder's ID is valid and the database shows that the car is inside of the garage.

19. Tester drives home [exit test procedure].

1.12.2.3 *Test of Alternate Flow 3c*

Precondition: Tester must ensure that (1) Tester's car is outside of the parking garage, (2) the system believes that Tester's car is outside of the parking garage (He can verify this by calling Parking and Transportation.), and (3) Tester's car transponder is valid.

1. *Tester calls Parking and Transportation and asks them to set the database parameter numberCarsInGarage to 1399.*
2. *Tester approaches the Entrance Gate.*
3. *The system interrogates the car's transponder.*
4. *The system determines that the transponder is valid and that the car is not already in the garage.*
5. *The system opens the gate.*
6. *The system shall illuminate the FullSign. Tester records the result. This tests the following requirements.*

   FR7. The system shall illuminate the FullSign.

   NFPR3. The parking garage capacity is 1400 cars.
7. *Tester drives through the Entrance Gate.*

8. The system detects that the car is clear of the Entrance Gate.
9. The system closes the gate.
10. The system stores the transponder 's identification number (ID), the date, and the time.
11. Tester approaches the Entrance Gate with a second car with a valid transponder.
12. The system interrogates the car 's transponder.
13. The system determines that the transponder is valid and that the car is not already in the garage.
14. The system shall not raise the wooden arm. Tester records the result. This tests the requirement.
FR6. The system shall allow a car to enter the garage only if the transponder 's ID is valid and the database shows that the car is outside of the garage and the number of cars in the garage is less than the garageCapacity.
15. Tester calls Parking and Transportation and asks them to set the database parameter numberCarsInGarage to the correct number of cars.
A similar procedure is needed to test the exiting of cars.

The parking garage system is in compliance with its requirements if and only if it passes all of these tests.

1.12.2.4 Check for New Classes

Next, to identify the classes, we underline nouns in the ConOps, the use case, the requirements, and the glossary. Using this technique and the stereotypical classes (boundary, entity, and controller), we have identified the following analysis classes: Driver, Transponder (for now, we will treat the Driver and the Transponder as one class called Driver), Entrance Gate, EntranceGateReader (boundary class), EntranceGateSensor (boundary class), EntranceGateController (control class), Exit Gate, ExitGateReader (boundary class), ExitGateSensor (boundary class), ExitGateController (control class), GarageDatabase (entity class), and WoodenArm (supporting actor because it is CotS equipment).

We need an additional class for the FullSign.

Now let us use these classes to make a sequence diagram for the entrance gate portion of the main success scenario of the Park in Garage use case. We can then use the sequence diagram to find the functions of these classes.

The sequence diagram for the entrance gate portion of the main success scenario of the Park in Garage use case is the same as Fig. 1.32 . *No changes are needed for this sequence diagram. Sequence diagrams for other flows would be changed.*

Analysis Classes

Driver <<actor>>
ID: String
hereIsMyID()

In subsequent more detailed models, the Driver will be decomposed into the Human, the Transponder, and the Automobile.

WoodenArm< <actor>>
woodenArmState: String {up, down, moving}
raiseWoodenArm() lowerWoodenArm()

Comment: The wooden arms are all the same: so there is no reason to specify this as an entrance gate or an exit gate wooden arm. They used to be wooden. Now the arms are plastic.

EntranceGateReader <<boundary class>>
interrogateTransponder(ID) sendIsValidQuestion(ID)

EntranceGateSensor <<boundary class>>
carIsClear: Boolean = yes
carIsHere() setCarIsClear()

GarageDatabase <<entity class >>
ID: Integer carInGarage: Boolean = out {in, out} validity: Boolean = invalid date: Date time: Time *numberCarsInGarage: Integer* *garageCapacity: Integer = 1400*
readAndWrite(ID, date, time, carInGarage, validity)

EntranceGateController <<controlclass>>
validity: Boolean=no *garageFull: Boolean* =no
readInfo(ID) determineValidity(ID) writeInfo(ID, date, time) commandRaiseWoodenArm commandLowerWoodenArm set(validity) *set(garageFull)* *displayFull*

FullSign<<actor>>
garageFull: Boolean = false
illuminate

Explanations (Figs. 1.41 and 1.42)

Functions	Explanation
Functions (or operations) from the sequence and class diagrams	
hereIsMyID()	The Transponder on the dashboard of the car must be readable by the Entrance and ExitGateReaders. This means it must be perpendicular to the reader and its path must not be obstructed
raiseWoodenArm	The system must be able to raise the WoodenArm. Speed of movement is a nonfunctional performance requirement associated with this function
lowerWoodenArm	The system must be able to lower the WoodenArm. Speed of movement is a nonfunctional performance requirement associated with this function
interrogateTransponder(ID)	The Entrance and ExitGateReaders must interrogate the car's transponder to read its identification number (ID)
sendIsValidQuestion(ID)	The Entrance and ExitGateReaders must send the transponder's ID in the "isValid(ID)?" message
carIsHere()	The Entrance and ExitGateSensors must sense the position of the car. The WoodenArms should not close on top of a car
setCarIsClear()	The Entrance and ExitGateSensors must sense the position of the car. They should set the attribute carIsClear to reflect this condition
readAndWrite(ID, date, time, carInGarage, validity)	This is the information that is recorded in the GarageDatabase
readInfo(ID)	The Controller must command the database to read information in the GarageDatabase in order to determine if the ID is valid and that the car is outside of the garage
writeInfo(ID, date, time)	The Controller must command the database to write the transponder's ID, the date, and time of entry and exit
determineValidity(ID)	The Controller must compare the transponder's ID with data stored in the GarageDatabase to determine if the ID is valid and that the car is outside of the garage. To determine this, it tests the following Boolean expression (carInGarage = out AND validity = valid)
set(validity)	The EntranceGateController sets the attribute validity to valid or invalid
set(garageFull)	*If garageCapacity – numberCarsInGarage =0, then set (garageFull)=yes*
commandRaiseWoodenArm	The EntranceGateController commands the WoodenArm to rise
commandLowerWoodenArm	The EntranceGateController commands the WoodenArm to lower
numberCarsInGarage: Integer	*The system must keep track of the numbers of cars in the garage*

(continued)

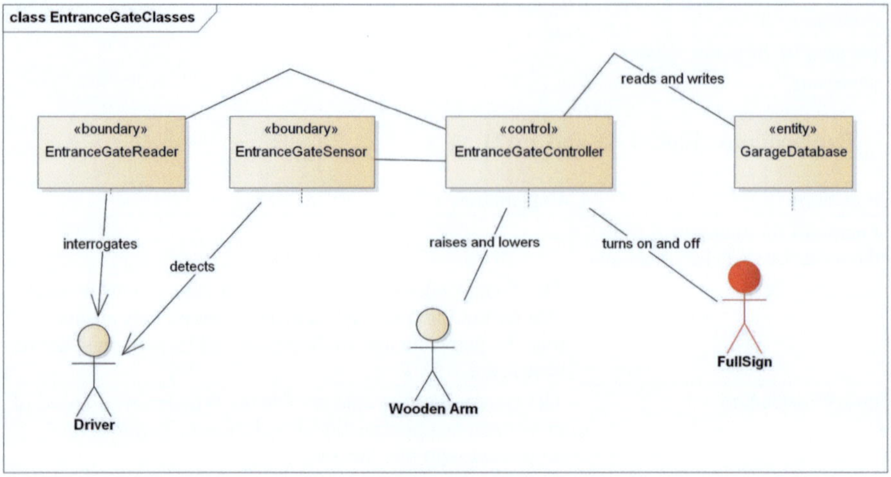

Fig. 1.41 Entrance Gate class diagram. *We have added the FullSign to this class diagram*

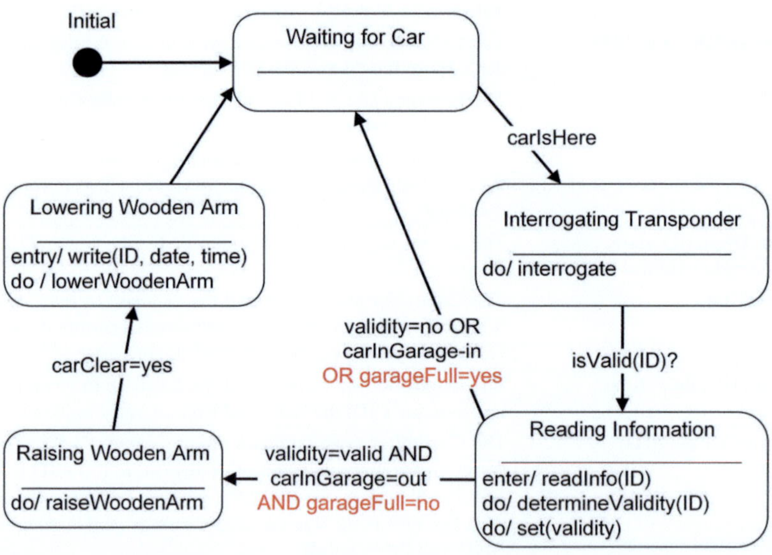

Fig. 1.42 State machine diagram for the EntranceGateController. *The Boolean expressions for the transitions have been changed*

Functions	Explanation
garageCapacity: Integer = 1400	*The capacity of this garage is about 1400 cars*
displayFull	*When the parking garage is full, the system shall illuminate the FullSign*
illuminate	*Turn on the FullSign*

A table describing the events, states, and activities of the state machine diagram will go here.

This parking garage system will have two similar subsystems, one for the Entrance Gate and one for the Exit Gate. They will share a common GarageDatabase.

Explanation of abbreviations and role names used in Figs. 1.43, 1.44, and 1.45.

Usage name or role	Explanation
s1	One particular sensor on an entrance gate
r1	One particular reader on an entrance gate
c1	One particular controller on an entrance gate
db1	The database for the parking garage system
wa1	One particular wooden arm on an entrance gate. The term wooden arm includes the motor, gears, mounting hardware, etc. The wooden arms are all the same; so there is no reason to differentiate between entrance or exit gate wooden arms.
wam	Wooden arm motion, the constraint describing the motion of a wooden arm.
sr	System response, the constraint describing the response time of the system. This includes the response times of the reader, the controller, and the database.
sc	System capacity constraint
eg	One particular entrance gate
Abbreviation	Meaning
rr	Rotation rate in degrees per second
srt	Sensor response time in seconds
crt	Controller response time in seconds
rrt	Reader response time in seconds
ncig	Number of cars in garage
gc	Garage capacity
sysrt	System response time

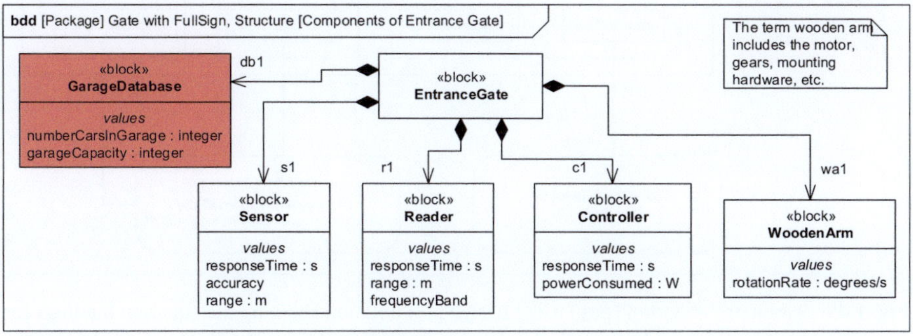

Fig. 1.43 Block definition diagram (**bdd**) showing the components of the parking garage EntranceGate. *The GarageDatabase block has been added. We do not want the FullSign in this bbd, because it does not contribute to the constraints*

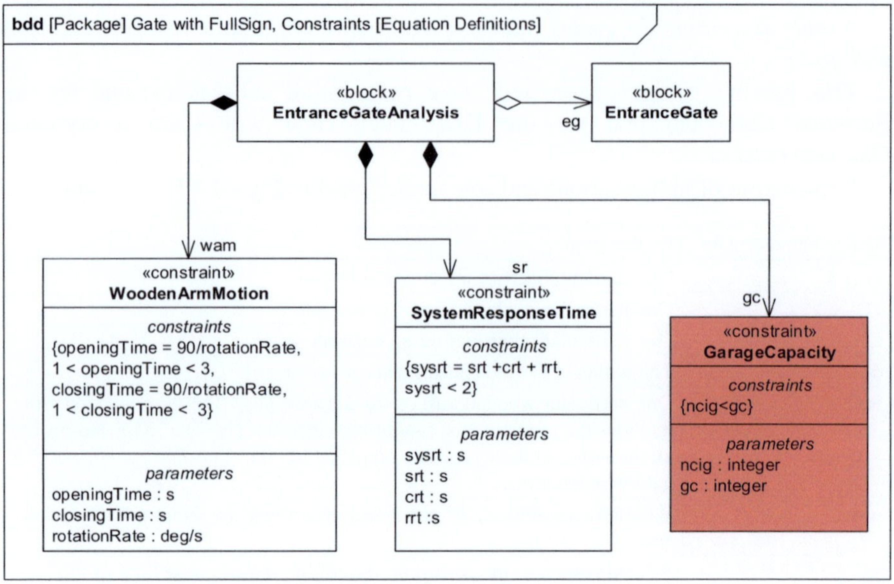

Fig. 1.44 Block definition diagram (**bdd**) defining constraints for the parking garage EntranceGate. *The GarageCapacity constraint block has been added*

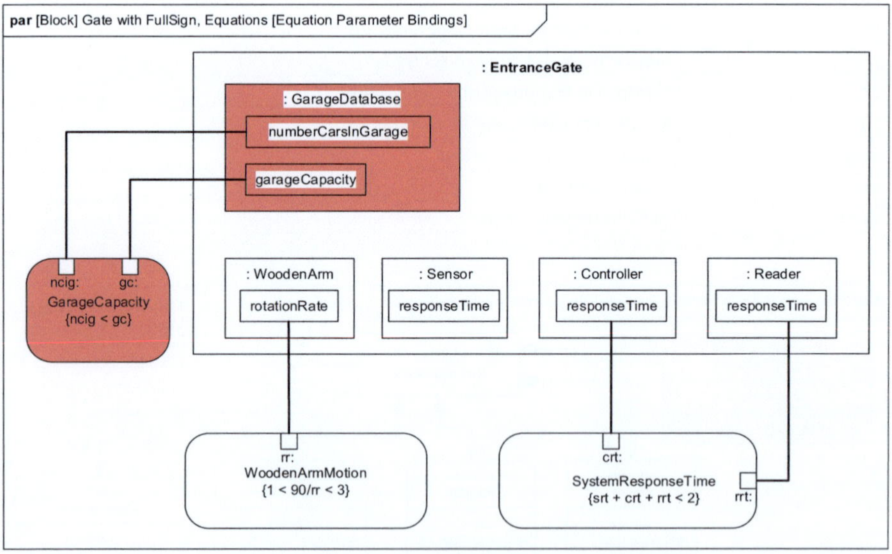

Fig. 1.45 Parametric diagram (**par**) using nested part notation showing the equation bindings for a particular parking garage EntranceGate. *The GarageDatabase and the GarageCapacity blocks have been added*

1.12.3 Homework

1.12.3.1 FullSign

Should the FullSign really be a supporting (secondary) actor?

1.12.3.2 NFPR1

The block definition diagram of Fig. 1.43 shows the response time in three blocks: the sensor, reader, and controller. Which of these would be used for requirement NFPR1, the time delay?

1.12.3.3 NFPR2

This is NFPR2: The system shall raise and lower the wooden arms in 1–3 s. During test, how will this requirement be verified?

During *design* phase, how will we specify the parameters that will ensure satisfaction of this requirement?

1.12.3.4 Wooden Arm

Why is it that sometimes we write that we want to open and close the entrance gate and at other times, we write that we want to raise and lower the wooden arm?

1.13 Section 13. Parking Garage, Fourth Iteration

ConOps for a parking garage. A university parking garage should work like this. A person drives up to the entrance gate, the system interrogates the car's transponder, if it is valid, the wooden arm opens, when the system senses that the car has left the gate area, the wooden arm closes. When that person leaves for the day, he or she drives up to the exit gate, the system interrogates the car's transponder, if it is valid, the wooden arm opens. When the system senses that the car has left the gate area, the wooden arm closes. The wooden arm shall not close on top of the car or break the radio antenna. To prevent people from using one transponder for two cars, the entrance gate will not open if the system thinks that the car is already inside. The transponder only has an identification number on it. All of the other data are stored in the system. To make the system simpler, design it for only one transponder (that is, consider only one car). For simplicity, assume that the system has only two gates: one for entrance and one for exit. Restrict the inputs and outputs to things that could actually be in the parking garage. When the number of cars in the garage reaches 1400, the system should illuminate the "Full" sign at the entrance and allow no more cars to enter until some cars have left.

In UML, activity diagrams describe procedural logic, business process, and workflows. In SysML, activity diagrams also show control inputs and the inputs that are converted into outputs by the functions. Internal block diagrams show the connections between parts of a block. They also show ports and the allocation of functions to physical components: UML could use a deployment diagram for this.

*We will now create an activity diagram (**act**) for this concept of operations and draw an internal block diagram (**ibd**) that assigns functions to hardware for this parking garage system.*

1.13.1 Fourth Iteration Design Solution (Fig. 1.46)

1.13.2 Homework

1.13.2.1 Internal Block Diagrams

This **ibd** of Fig. 1.47 states that the FullSign has a function named illuminate. Where else have we seen this function mentioned?

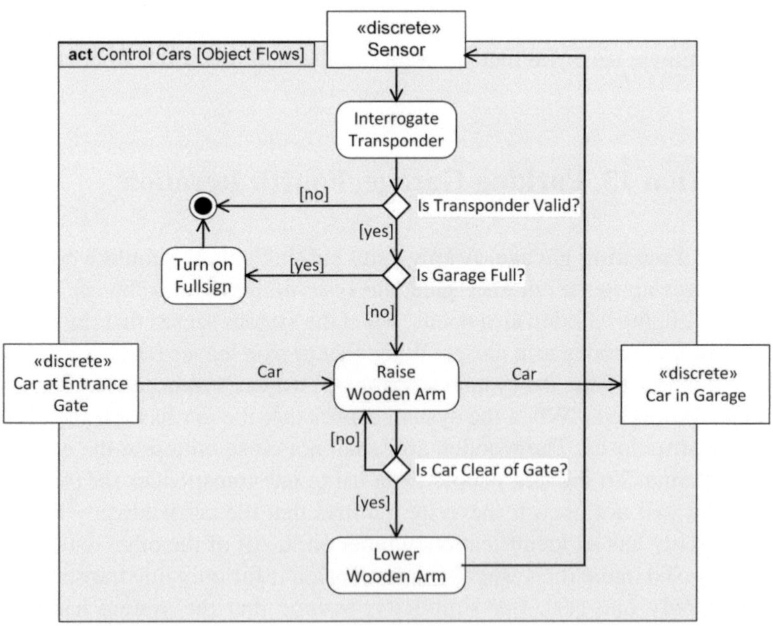

Fig. 1.46 *Activity diagram for the parking garage system. Material flow is horizontal. Control flow is vertical*

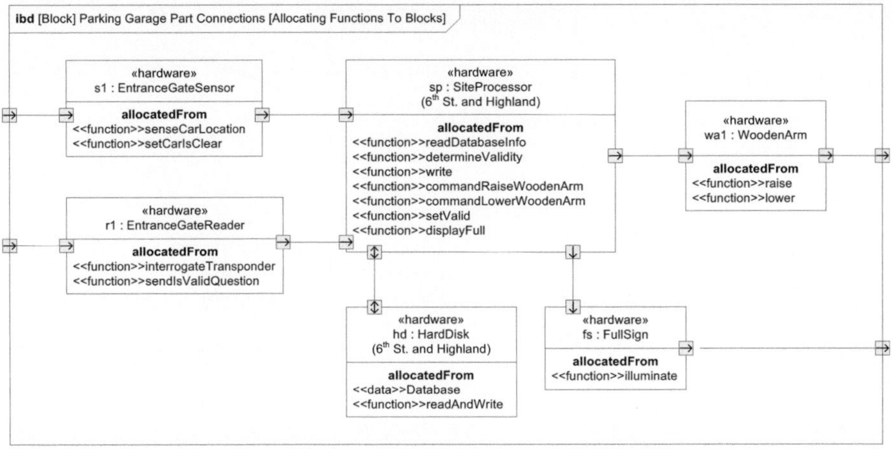

Fig. 1.47 *An internal block diagram (**ibd**) that assigns functions to hardware for this parking garage system. The EntranceGateSensor and the EntranceGateReader are input control ports. The WoodenArm and the FullSign are output control ports*

1.14 Section 14. Parking Garage, Fifth Iteration

The parking garage will also allow people to pay per visit. These people drive up to the entrance gate and push the ticket button. The system gives them a dated ticket and opens the gate for them. When they are ready to leave, they go to the cashier's booth and pay their money. The cashier validates their ticket and gives it back to them. They then drive up to the exit gate and put the ticket in the slot. The exit gate opens for them. We will now modify the Analysis Model for the Parking Garage in the Second Iteration Design Solution to reflect this new feature. Remember text in the Times New Roman regular black font you have read before, so you do not need to scrutinize it. *The modifications and additions are in the Italic Sans-Serif font.*

1.14.1 Fifth Iteration Design Solution

1.14.1.1 Second Analysis Phase Model for the Parking Garage

ConOps for a parking garage. A university parking garage should work like this. A person drives up to the entrance gate, the system interrogates the car's transponder, if it is valid, the wooden arm opens, when the system senses that the car has left the gate area, the wooden arm closes. When that person leaves for the day, he or she drives up to the exit gate, the system interrogates the car's transponder, if it is valid, the wooden arm opens. When the system senses that the car has left the gate area, the wooden arm closes. The wooden arm shall not close on top of the car or break the radio antenna. To prevent people from using one transponder for two cars, the

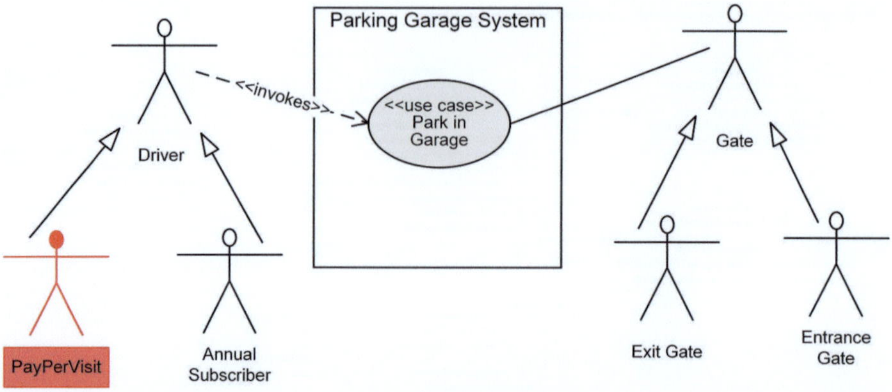

Fig. 1.48 Use case diagram for the Park in Garage use case with pay-per-visit customers. *We have shown specializations of the Driver actor and the Gate actor*

entrance gate will not open if the system thinks that the car is already inside. The transponder contains only an identification number. All of the other data are stored in the system. To make the system simpler, design it for only one transponder (that is, consider only one car). For simplicity, assume that the system has only two gates: one for entrance and one for exit. Restrict the inputs and outputs to things that could actually be in the parking garage. *The system will also allow daily pay-per-visit customers.*

Glossary

Transponder: a passive electronic device (a radio frequency identification tag, RFID) usually mounted in the lower-left corner of a car's windshield. It has the user's identification number stored in it.

Interrogate: the system detects the presence of a transponder (This could be done by continuous scanning or be triggered by the EntranceGateSensor.). Then it sends out electromagnetic signals, in order to read the information stored in the transponder.

Use case (Fig. 1.48)

Name: Park in Garage
Iteration: 3.9
Derived from: Concept of Operations
Brief description: A person drives up to the entrance gate of the parking garage. The system interrogates the car's transponder: if it is valid, the system opens the gate. When the system senses that the car has left the gate area, the system closes the gate. A similar sequence occurs when the person leaves the garage. *The system will also allow daily pay-per-visit customers.*
Level: User goal
Scope: Parking Garage

Added value: The Driver can park his car in a safe, shaded environment within a half mile of his or her classroom or office.

Primary actor: The student, faculty, or staff member who is the Driver of a car. In this use case, we will treat the Driver and the Transponder as one actor. *The Pay-per-visit Customer is also a primary actor.*

Supporting actor: Entrance Gate and Exit Gate

Frequency: About 1500 cars use the system each day.

Precondition: The Driver *or Customer's* car is outside of the parking garage and the system believes that the car is outside of the parking garage.

Main Success Scenario

1. The Driver approaches the Entrance Gate.
2. The system interrogates the car's transponder.
3a. The system determines that the transponder is valid and that the car is not already in the garage.
4. The system opens the gate.
5. The Driver drives through the Entrance Gate.
6. The system detects that the car is clear of the Entrance Gate.
7. The system closes the gate.
8. The system stores the transponder's identification number (ID), the date, and the time.
9. The Driver parks the car and goes to class.
10a. The Driver returns to the car and drives up to the Exit Gate.
11. The system interrogates the car's transponder.
12a. The system determines that (the transponder is valid and that the database says that the car is actually in the garage) *or the ticket is valid.*
13. The system opens the gate.
14. The Driver drives through the Exit Gate.
15. The system detects that the car is clear of the Exit Gate.
16. The system closes the gate.
17. The system stores the (transponder identification number (ID) *or the ticket number*), the date, and the time.
18. The Driver drives home [exit use case].

Alternate Flows:

3b. The system determines that the transponder is not valid or that the car is already in the garage.
3b1. The system does nothing. (This may be stupid, but it is what the system actually does.) [exit use case].
3c. The pay-per-visit Customer pushes the request ticket button.
3c1. The system delivers a dated ticket [return to step 4].
10b. The Customer pays at the cashier, returns to the car, and drives up to the Exit Gate.
10b1. The Customer inserts the ticket into the ticket slot [return to step 13.

12b. The system determines that the transponder is invalid *or the ticket is invalid* or the car is not in the garage.
12b1. The system does nothing (This design flaw must be fixed.) [exit use case].

Postcondition: The Driver or *Customer 's* car is outside of the parking garage and the system has recorded that the car is outside of the parking garage. (This does not include the present behavior of step 12b.)

Specific Requirements

Functional Requirements

Req1. The system shall determine that a car is waiting at the Entrance or Exit Gate. This requirement (and other similar compound requirements) will be decomposed later.
Req2. The system shall determine the validity of the transponder's ID.
Req3. The system shall store the identity of every car in the garage (from the ConOps).
Req4. The system shall open and close the Entrance and Exit Gates.
Req5. The system shall determine when a car has cleared the Entrance or Exit Gates.
Req14. When the request ticket button is pressed, the entrance gate shall produce a parking ticket only if a car is present at the entrance gate. Rationale: This requirement is designed to prevent a pedestrian from coming back from class, requesting a new ticket and paying for only a few minutes of parking time.

Derived Requirement:

Req6. The system shall allow a car to enter the garage only if the transponder's ID is valid and the database shows that the car is outside of the garage. This prevents two cars from using the same transponder.
Req11. When the request ticket button is pressed, the entrance gate shall produce a parking ticket containing date and time of entry and some sort of identification.
Req12. The exit gate shall accept and read paid parking tickets.

Nonfunctional Requirements

Req7. When the transponder is valid, the system shall *begin* opening the wooden arm within 2 s of interrogation. This requirement includes the response time of the Sensor, the Reader, and the Controller.
Req8. The system shall open and lower the wooden arms in 1–3 s.
Req13. The system shall produce a parking ticket within 3 s of the request ticket button being pushed.

Author/owner: Parking and Transportation
Last changed: April 24, 2016

The system requirements document would go here. We are becoming more formal. Requirements 1–6 map to functional requirements FR1–FR6 and

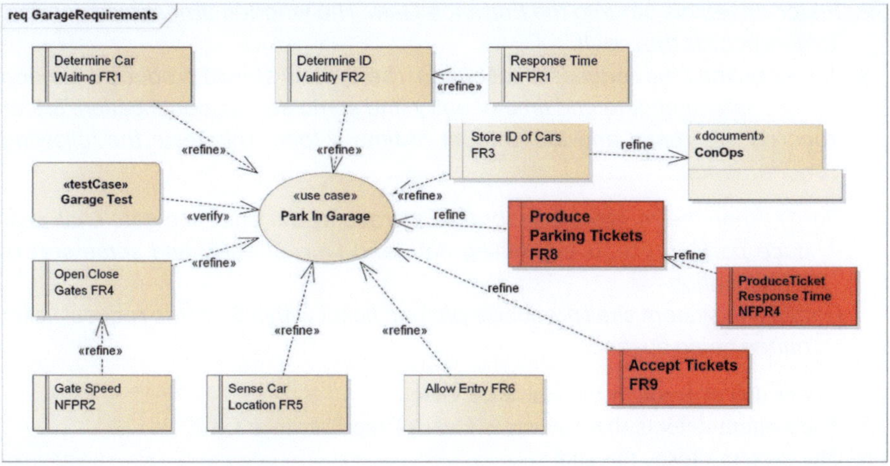

Fig. 1.49 Requirements diagram for parking garage entrance gate with pay-per-visit customers. *Three new requirements, FR8, FR9, and NFPR4, have been added*

requirements 7 and 8 become nonfunctional performance requirements NFPR1 and NFPR2 (Fig. 1.49).

Our design process starts with the use cases. The requirements and everything else follow. Therefore, the dependencies in the above diagram show that, for example, the requirement SenseCarLocation is more detailed and thus refines the use case Park in Garage. If instead, the customer gave us all of the requirements and then we wrote the use cases, then the arrow's direction would show that the use case, being more detailed, refines the requirements.

1.14.1.2 Parking Garage Test Procedure

The old test procedure will remain the same. This new procedure will be added for the pay-per-visit Customers.

Precondition: Tester must ensure that Tester 's car is outside of the parking garage and that it does not contain a valid transponder.

Main Success Scenario

1. *Tester parks his car away from the entrance gate. He walks up and presses the request ticket button. The system should not dispense a parking ticket. Tester records the result. This tests*

   Req14. When the request ticket button is pressed, the entrance gate shall produce a parking ticket only if a car is present at the entrance gate. Rationale: This requirement is designed to prevent a pedestrian from coming back from class, requesting a new ticket and paying for only a few minutes of parking time.

2. *Tester drives his car into the Entrance Gate. The wooden arm should not rise. Tester records the result.*
3. *Tester pushes the request ticket button. The system should produce a parking ticket containing date and time of entry and some sort of identification. Tester records this result and the amount of time it took. This tests the following requirements:*

   Req11. When the request ticket button is pressed, the entrance gate shall produce parking tickets containing date and time of entry and some sort of identification.

   Req13. The system shall produce a parking ticket within 3 s of the request ticket button being pushed.

4. Tester drives through the Entrance Gate.
5. The system detects that the car is clear of the Entrance Gate.
6. The system closes the gate.
7. Tester parks the car.
8. *Tester goes to the Cashier and pays the money.*
9. *Tester returns to the car and drives up to the Exit Gate.*
10. *Tester puts the paid ticket into the ticket slot.*
11. *The system determines that the ticket is valid.*
12. The system opens the gate. Tester records that the wooden arm went up. *This tests the following requirement:*

    Req12. The exit gate shall accept and read paid parking tickets.

14. Tester drives through the Exit Gate.
15. The system detects that the car is clear of the Exit Gate.
16. The system closes the gate.
18. Tester drives home [exit test procedure].

The parking garage system is in compliance with its requirements if and only if it passes all of these tests.

Next, to identify the classes, we underline nouns in the ConOps, the use case, the requirements, and the glossary. Using this technique and the stereotypical classes (boundary, entity, and controller), we have identified the following analysis classes: Driver, Transponder (for now, we will treat the Driver and the Transponder as one class called Driver), Entrance Gate, EntranceGateReader (boundary class), EntranceGateSensor (boundary class), EntranceGateController (control class), Exit Gate, ExitGateReader (boundary class), ExitGateSensor (boundary class), ExitGateController (control class), GarageDatabase (entity class), and WoodenArm (supporting actor because it is CotS equipment), Ticket (entity class), Cashier, and the *TicketMachine that comprises the TicketButton, and the TicketSlot.*

Now let us use these classes to make a sequence diagram for the entrance gate portion of the main success scenario of the Park in Garage use case. We can then use the sequence diagram to find the functions of these classes (Fig. 1.50).

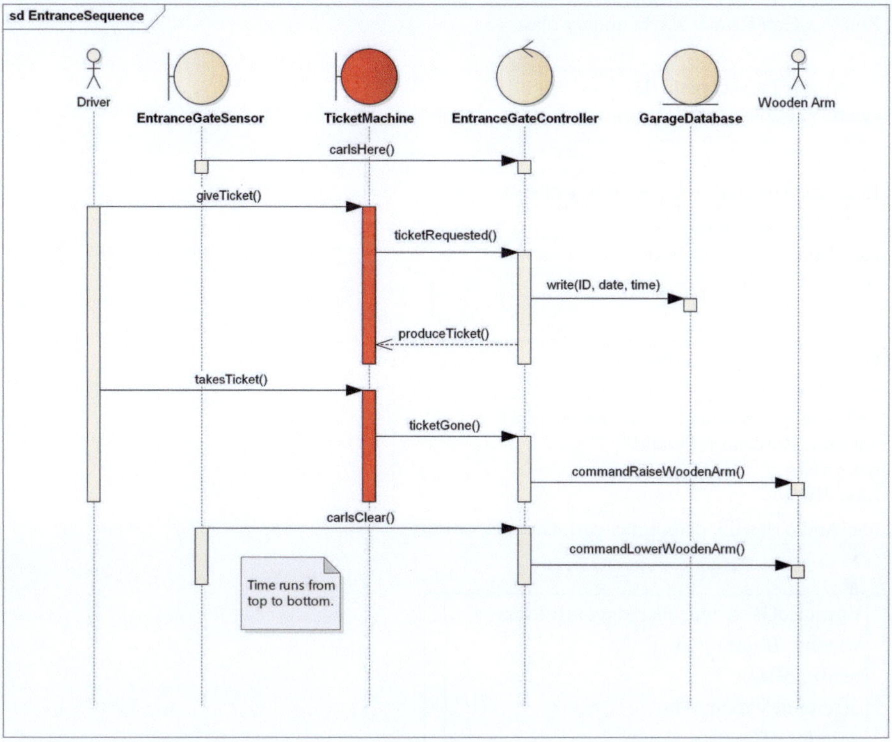

Fig. 1.50 A sequence diagram for the entrance gate portion of the main success scenario of the Park in Garage use case with pay-per-visit customers. *This is a new sequence diagram for the new use case alternative flow at steps 3c and 10b*

1.14.1.3 Analysis Phase Classes

Driver <<actor>>
ID: String
hereIsMyID()

Later, in models with more detail, the Driver will be decomposed into the Human, the Transponder, and the Automobile.

WoodenArm <<actor>>
woodenArmState: String {up, down, moving}
raiseWoodenArm() lowerWoodenArm()

Comment: The wooden arms are all the same: so there is no reason to specify this as an entrance gate or an exit gate wooden arm. Now they are plastic. It should probably be named the EntranceGate. It should not be strong and rigid like steel, because it should break easily in case of emergency and when it breaks, it should protect the rest of the system from harm.

EntranceGateReader <<boundary class>>
interrogateTransponder(ID) sendIsValidQuestion(ID)

EntranceGateSensor <<boundary class>>
carIsClear: Boolean = yes
carIsHere() setCarIsClear()

GarageDatabase <<entity class >>
ID: Integer carInGarage: Boolean = out {in, out} validity: Boolean = invalid date: Date time: Time
readAndWrite(ID, date, time, carInGarage, validity)

EntranceGateController <<controlclass>>
validity: Boolean=no
readInfo(ID)
determineValidity(ID)
writeInfo(ID, date, time)
commandOpenWoodenArm
commandLowerWoodenArm
set(validity)

TicketMachine <<boundary class>>
senseButtonPush() *produceTicket()* *acceptTicket()*

Explanations (Fig. 1.51)

Functions	**Explanation**
Functions (or operations) from the sequence and class diagrams	
hereIsMyID()	The Transponder on the dashboard of the car must be readable by the Entrance and ExitGateReaders. This means it must be perpendicular to the reader and its path must not be obstructed
raiseWoodenArm	The system must be able to open the WoodenArm. Speed of movement is a nonfunctional performance requirement associated with this function

(continued)

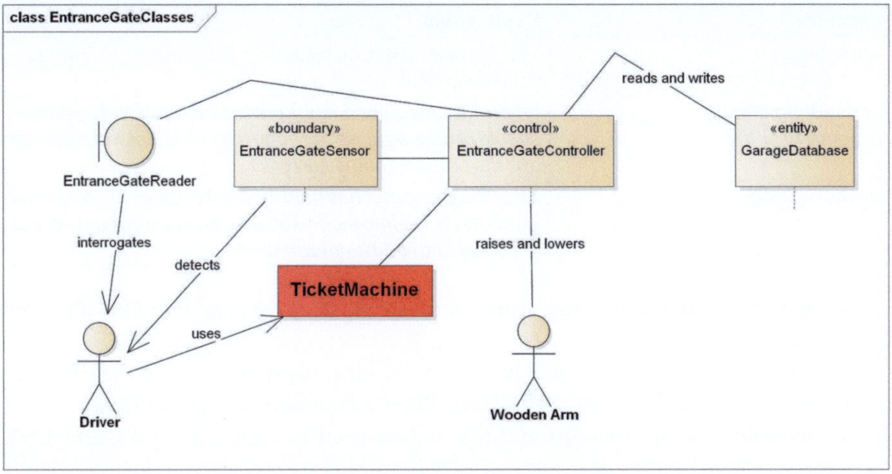

Fig. 1.51 Entrance Gate class diagram with pay-per-visit customers. *We have added the TicketMachine*

Functions	**Explanation**
lowerWoodenArm	The system must be able to lower the WoodenArm. Speed of movement is a nonfunctional performance requirement associated with this function
interrogateTransponder(ID)	The Entrance and ExitGateReaders must interrogate the car's transponder to read its identification number (ID)
sendIsValidQuestion(ID)	The Entrance and ExitGateReaders must send the transponder's ID in the "isValid(ID)?" message
carIsHere()	The Entrance and ExitGateSensors must sense the position of the car. The WoodenArms should not close on top of a car
setCarIsClear()	The Entrance and ExitGateSensors must sense the position of the car. They should set the attribute carIsClear to reflect this condition
readAndWrite(ID, date, time, carInGarage, validity)	This is the information that is recorded in the GarageDatabase
readInfo(ID)	The Controller must command the database to read information in the GarageDatabase in order to determine if the ID is valid and that the car is outside of the garage
writeInfo(ID, date, time)	The Controller must command the database to write the transponder's ID, the date, and time of entry and exit
determineValidity(ID)	The Controller must compare the transponder's ID with data stored in the GarageDatabase to determine if the ID *or the ticket* is valid and that the car is outside of the garage. To determine this it tests the following Boolean expression (carInGarage = out AND *(validity=valid OR ticket=valid))*
commandOpenWoodenArm	The EntranceGateController commands the WoodenArm to open.
commandLowerWoodenArm	The EntranceGateController commands the WoodenArm to close

(continued)

Functions	**Explanation**
set(validity)	The EntranceGateController sets the attribute validity to valid or invalid.
senseButtonPush() *produceTicket()*	*When the Customer pushes the TicketButton, the system should produce a ticket through the TicketSlot with the date and the time and some sort of ID*
acceptTicket()	*After the Customer has paid the Cashier, the Customer will insert the ticket into the TicketSlot. The system must accept the ticket and read its information*

A table describing the events, states, and activities of the state machine diagram will go here.

This parking garage system will have two similar subsystems, one for the Entrance Gate and one for the Exit Gate. They will share a common GarageDatabase.

Explanation of abbreviations and role names used in Figs. 1.52, 1.53, and 1.54.

Usage name or role	Explanation
s1	One particular sensor on an entrance gate
r1	One particular reader on an entrance gate
c1	One particular controller on an entrance gate
wa1	One particular wooden arm on an entrance gate. The term wooden arm includes the motor, gears, mounting hardware, etc. The wooden arms are all the same; so there is no reason to differentiate between entrance or exit gate wooden arms.
wam	Wooden arm motion, the constraint describing the motion of a wooden arm.
sr	System response, the constraint describing the response time of the system. This includes the response times of the reader, the controller, and the database.

(continued)

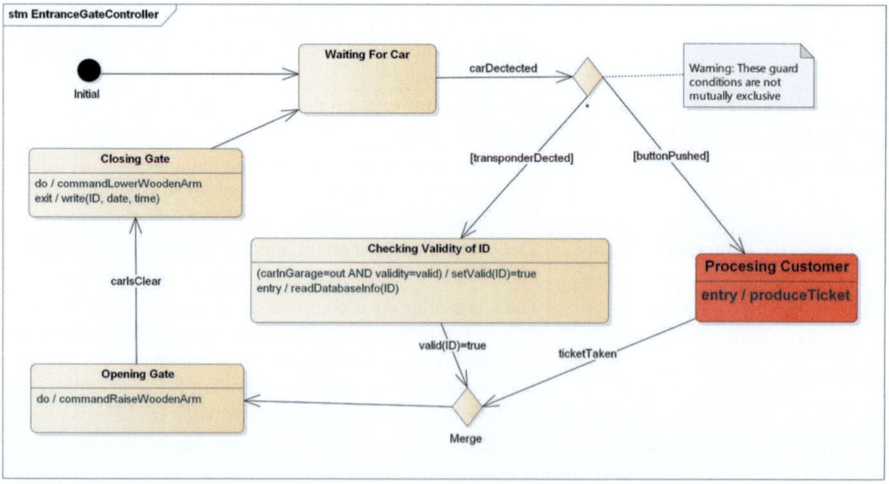

Fig. 1.52 State machine diagram for the EntranceGateController with pay-per-visit customers. *A state and a decision have been added*

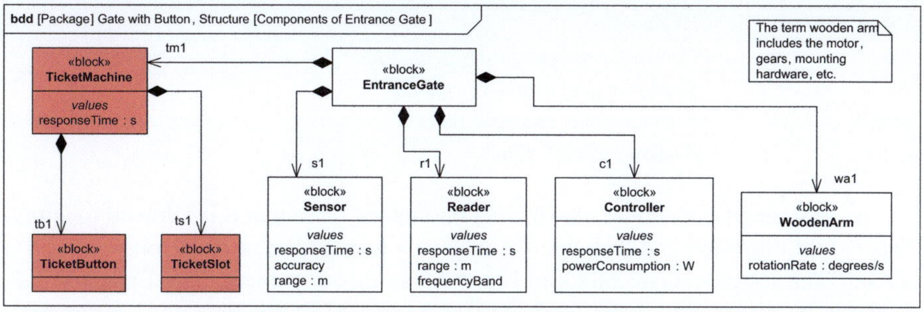

Fig. 1.53 Block definition diagram (**bdd**) showing the components of the parking garage EntranceGate with pay-per-visit customers *with the TicketMachine that comprises the TicketSlot and the TicketButton*

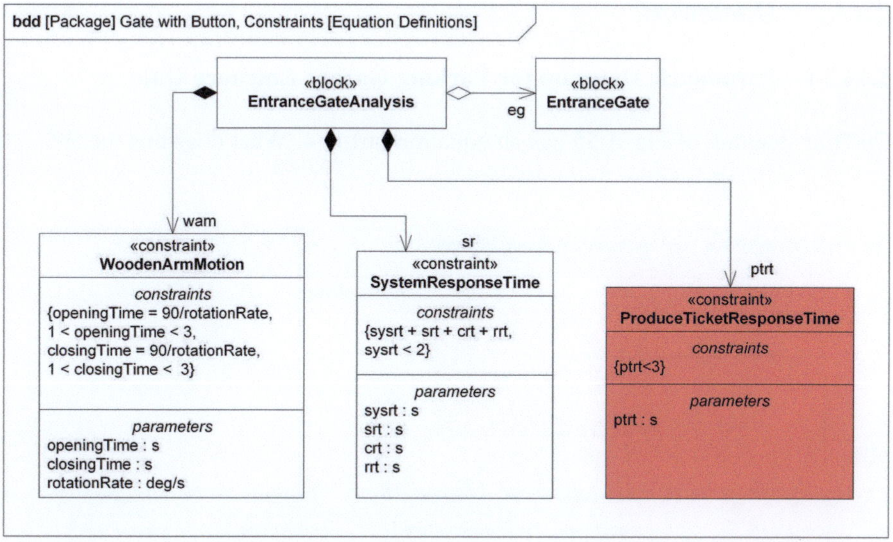

Fig. 1.54 Block definition diagram (**bdd**) defining constraints for the parking garage EntranceGate with pay-per-visit customers *with the ProduceTicket response time constraint*

Usage name or role	Explanation
eg	One particular entrance gate
tm1	Ticket machine 1
tb1	Ticket button 1
ts1	Ticket slot 1
Abbreviation	Meaning
rr	Rotation rate in degrees per second
srt	Sensor response time in seconds

(continued)

Usage name or role	Explanation
crt	Controller response time in seconds
rrt	Reader response time in seconds
ptrt	Produce ticket response time
sysrt	System response time

The Entrance Gate should also have a maximum force that it is allowed to apply. This would minimize the damage if it were to close on top of a car, or worse yet a person. The EntranceGate block is Fig. 1.53 in this document. The EntranceGate Analysis block would be the hardware, software (MatLab, Java or C programs), bioware, and algorithms that are being used to analyze this model.

1.14.2 Homework

1.14.2.1 Parametric Diagram for Parking Garage Entrance Gate

The **par** diagram of Fig. 1.55 has an equation ptrt < 4. What does this mean?

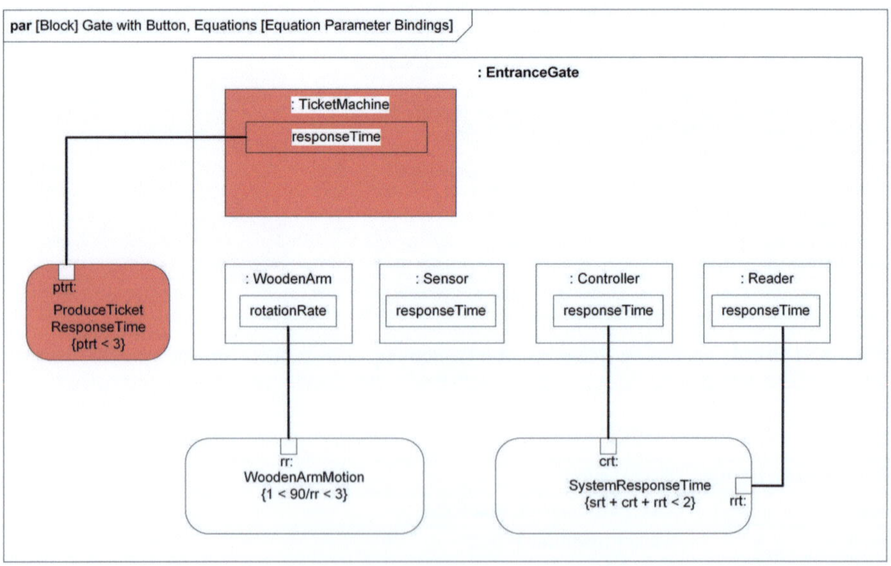

Fig. 1.55 Parametric diagram (**par**) using nested part notation showing the equation bindings for a particular Parking Garage EntranceGate with pay-per-visit customers *with the ProduceTicket response time equation*

1.14.2.2 Abstraction in the Parking Garage Problem

Between Figs. 1.39 and 1.48, the Driver actor was replaced with a PayPerVisit actor and an AnnualSubscriber actor. The Entrance Gate and Exit Gate were replaced with a Gate actor. Is this an example of abstraction or generalization?

Describe details and nonessential components that were not included in Section 10, Design of a Parking Garage System. In other words, what was added in Section 11.

1.15 Section 15. When Is a State Machine Necessary?

In memoryless systems, the output depends only on the present inputs. Whereas, in dynamic systems the output depends on the present inputs and the sequence of previous inputs. Memoryless problems can be modeled and implemented as dynamic systems. But not vice versa. This section is based on Botta, Bahill, and Bahill [81].

This section is very important. If the problem is a memoryless problem then the simplest solution will be a memoryless system. However, a dynamic system solution would also solve the problem, but it would be more complex. This section was not placed immediately after Section 8, because students would not appreciate it until after they have created many dynamic system models on their own.

Consider the household three-way light system of Fig. 1.56 with one light and two switches, one at each end of a hallway. Define the inputs to be the position of the switches {up, down}.

We can create the dynamic model from this information. Figure 1.57 presents a state machine model for the three-way light system. For simplicity, input combinations that do not produce a change of state are not shown; also, simultaneous changes of both switches are not shown. Given the present state and an input trajectory (sequence of inputs), we can compute the state trajectory. Therefore, the three-way light system can be modeled as a dynamic system as in Fig. 1.57, and can be implemented as a flip-flop logic circuit. At this point, one might ask—is a state machine system needed for this problem? Or is there a simpler solution?

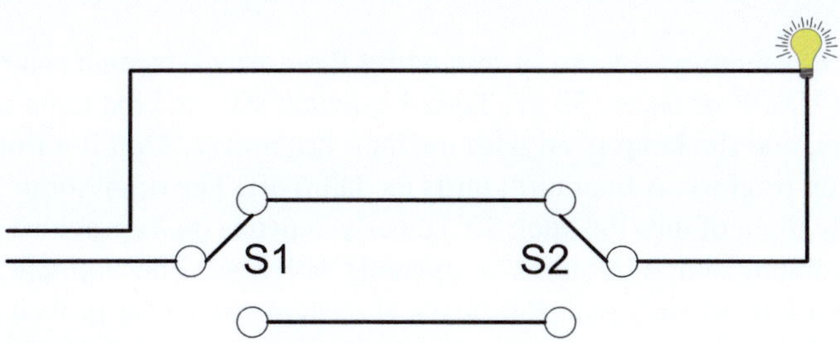

Fig. 1.56 Wiring diagram for a three-way light system

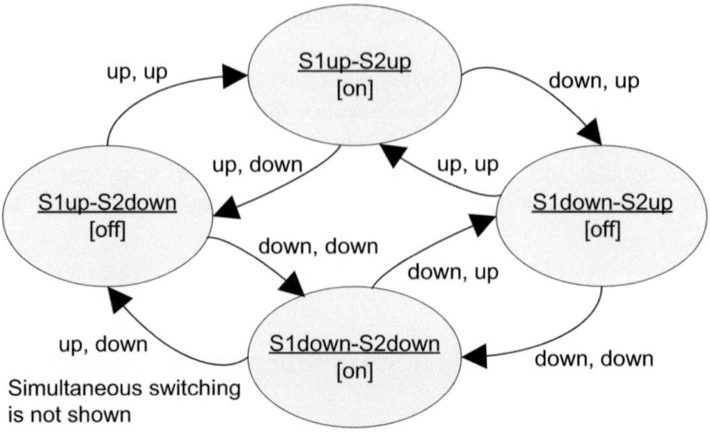

Fig. 1.57 State machine solution for a three-way light system

Table 1.11 A truth table

S2	S1	Light
Down	Down	On
Down	Up	Off
Up	Down	Off
Up	Up	On

In actuality, the three-way light system model can be a *memoryless* system representation and implementation: it does not *need* a dynamic *solution*. The present state depends only on the present positions of the switches. The system can be modeled with the truth table of Table 1.11. From Table 1.11, we can derive the following Boolean equation

$$\text{Light} = \overline{\text{S1}} \cdot \overline{\text{S2}} + \text{S1} \cdot \text{S2}$$

This equation (or model) can be implemented using only AND and OR gates. This memoryless system implementation is simpler than the dynamic system implementation presented in Fig. 1.57. Furthermore, the equivalence of this memoryless model with the physical system can be proven using *only* input/output behavior.

Contrast this three-way-lamp system with a three-way *lamp* that can be off, on at 50 W, on at 100 W or on at 150 W. Turn the switch 90°, and the lamp is on dimly. Turn it again, and the lamp is on with medium brightness. Turn it a third time and the lamp is on brightly. A final turn, turns the lamp off. The behavior of this system (with its only input of turn the knob 90°) clearly depends on its previous state: it is a dynamic problem and it requires a dynamic solution. The equivalence of the dynamic solution (model) with the physical system cannot be proven using only input/output behavior.

Therefore, in the real world, there are memoryless problems that should have memoryless solutions: however, they could also have dynamic solutions. An engineer *can* prove the equivalence of these memoryless problems using only input/output behavior. However, real-world dynamic problems must have dynamic solutions. And the equivalence of these solutions *cannot* be proven using only input/output behavior [41, 81].

1.15.1 What Type of Systems Have States and What Type of Models Need States?

In modeling a system, one of the most important questions that needs to be answered is deciding whether the system is memoryless or dynamic. To elaborate on this concept, let us consider a few state equation systems. An ideal spring is a memoryless system, because the output position depends only on the input force applied. Whereas, an ideal mass is a dynamic system, because the output position depends on the applied input force as well as the initial position and velocity. An ideal resistor is a memoryless system, because the output voltage depends only on the input current. Whereas, a resistor-capacitor system is dynamic, because the output voltage depends on the input current as well as the initial capacitor voltage. Now, let us now consider some well-defined game systems and decide whether they are memoryless or dynamic.

If a baseball game were called due to rain, "What state information would you have to store to restart the game at a later date?" You would need to store total runs scored by the visiting team, total runs scored by the home team, the current inning, whether it is the top or the bottom of the inning, last batter for the visiting team, last batter for the home team, outs in this half of the present inning, balls on present batter, strikes on present batter, name of runner on first base, name of runner on second base, name of runner on third base, and a list of players who have been removed from the game (this is important, because in baseball a removed player cannot return). You would need all this state information to restart a rain-delayed game. For information complexity, this is about 120 bytes. If you wish to go into excruciating detail, you could also consider the batting order and the number of times each manager and coach has gone out to talk to the pitcher in the inning. For Little League games you could list how many innings each pitcher pitched (pitchers are limited to six innings per week) and the children who were in attendance (children who were not there for the original game cannot play in the resumed game).

The game of poker is an extremely complex, dynamic system. Before you make a bet, you need to know all the cards that have been played, all the bets that have been made, and the state of each opponent's mind. To assess their individual states of mind, you must be able to read their body language and vocal patterns, and use

Table 1.12 For which systems are states needed?

Game	Is memory needed?	Needed information
Roulette		
Coin Toss		
Lottery		
Craps		
Chess		
Checkers		
Baseball	Yes	Batter, inning, count, score, etc.
Poker	Yes	Bets, cards in deck, previous behavior, etc.
Blackjack	No	For most people
Blackjack	Yes	If you are a card counter and you vary your bet based on which cards have been played from the deck

this information in combination with their previous betting behavior. Poker requires astute observation and a large memory to store a large number of states.

In contrast, Blackjack is a much simpler dynamic system, because you are only playing against the dealer, who needs to abide by strict rules. So the only state information you need is the cards that have been played. For information complexity, on average 26 cards would have been played and you can record each card with 2 bytes of information. So information complexity is about 50 bytes. Table 1.12 summarizes the state behavior of these games.

1.15.2 Homework

1.15.2.1 When Are States Needed?

Please indicate whether the following well-defined games are memoryless or dynamic. In Roulette, you spin the wheel, look at the input and pay the output. State lotteries draw the numbers and pays the winners. In the card game *Show Low*, each player cuts the deck and shows his card. The player showing the lowest card wins.. The game of Craps requires knowledge of "the point."

Often a question will help you discover the state information that a dynamic system must store in its memory. For instance, if you are playing a game of chess by mail, "What state information does your opponent need in order to respond to your move?" How about the game of checkers? Fill in Table 1.12 for these games.

1.15.2.2 Game of Craps

Draw a state machine diagram for the game of Craps. Make sure that you explain your inputs, outputs, and states. Here is how craps is played. Players stand around the craps table. The game starts when the players place their bets and the stickman pushes the dice to a shooter. The shooter rolls the dice across the table so they hit the wall on the opposite end of the table. This first roll of the dice is called a "come out roll." There are three things that can happen:

1. If the dice come up as 7 or 11, the shooter wins. The same shooter continues to roll the dice.
2. If the dice come up as 2, 3, or 12, the shooter loses. This is called craps. The same shooter still rolls the dice.
3. If the dice come up as anything else (4, 5, 6, 8, 9, or 10) that number is called *the point*. The shooter continues shooting the dice until the point is rolled again, in which case the shooter wins and the game goes back to the start. If a 7 is rolled before the point is rolled, the shooter loses. The game is over, a new shooter is picked (the next person around the table) and a new game begins.

1.15.2.3 Candy Machine-2

In our geedunk machine, candy cost 15 cents. The machine sorts coins into three categories: nickels, dimes, and other. If the customer deposits either three nickels, or a nickel and a dime, the machine will drop candy. If the customer deposits foreign coins or illegal combinations (for example, two dimes), all deposited coins should be returned. Model this system with if-then statements. Describe your inputs and outputs. State all assumptions you make. Compare your solution for this problem to your solution for the Candy Machine homework problem in Section 9. Are they both correct? Which one is more complicated?

1.15.2.4 Gasoline Pump-2

Each gasoline pump contains a computer. It first asks if you want to pay with cash or a credit card. Assume that if you say cash, it will charge you $3.00 per gallon, but if you say credit card it will charge $3.09 per gallon. Then it asks if you want to buy $5, $20, $45, or fill up the tank. (To simplify this problem, let us ignore the possibility of filling up the tank.) When the total amount you owe equals the amount you wanted, it turns off the pump. Design such a system using if-then statements. Describe your input and output ports. State all assumptions you make. Compare your solution for this problem to your solution for the Gasoline Pump homework problem in Section 9. Are they both correct? Which one is more complicated?

1.15.2.5 Three-Way lamp

A three-way lamp has simple house wiring, but a complex three-way bulb. It can be off, on at 50 W, on at 100 W or on at 150 W. Turn the switch 90°, and the lamp is on dimly. Turn it again, and the lamp is on with medium brightness. Turn it a third time and the lamp is on brightly. A final turn, turns the lamp off. Model this system with a state machine diagram. Its only input should be the 90° turn of the knob.

1.16 Summary

In this chapter, we covered system design. In a nutshell, a system design starts with the concept of abstraction. The system designer is given a number of goals, objectives, requirements, desired capabilities, expectations, constraints, assumptions, and a concept of operations. In other words, he is given the whole bull. The system designer then looks at different views and abstracts out the *essence* of the system. These comprise the initial set of abstractions. This abstraction process was compared to the abstraction process of Picasso shown in Fig. 1.58. This abstraction activity is followed by a process of systematic elaboration. The abstraction is elaborated into the problem statement. The problem statement contains a simple story that describes what the system is intended to do and how it is expected to operate. This story evolves into multiple use cases that describe the operation of the system. While writing use cases, functional and nonfunctional requirements are specified. Once the use cases are written, tests are designed. Thereafter, test engineers create test requirements while systems engineers derive technical requirements. These requirements and their relationships are depicted in requirements diagrams. The requirements in conjunction with use cases lead to descriptions of system behavior using formalisms such as state machine diagrams and activity diagrams.

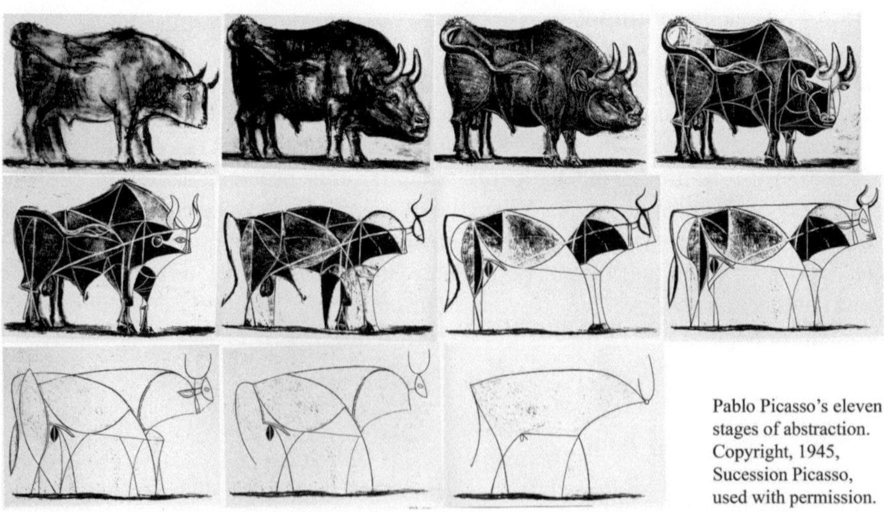

Fig. 1.58 Picasso's 11 stages of abstraction. Copyright ©, 1945, Succession Picasso, used with permission

Against the foregoing backdrop, we organized Chap. 1 of this book. We began by emphasizing the importance of models and processes in understanding and making decisions about system design. We introduced the reader to the general process of design and important concepts such as abstraction, generalization, and design iteration. We employed a simple example to illuminate the key aspects of design and reinforced the concepts through homework problems at the end of each section. We went into progressively greater detail about the system design process and employed UML and SysML diagrams as appropriate to convey the different modeling perspectives that collectively "tell the story" of the evolving system design. We began by introducing the reader to the SIMILAR process, which we employed to solve problems encountered during system design. We noted that the functions involved in the SIMILAR process are performed in parallel, and in iterative fashion.

In Section 1, we described each function in the SIMILAR process in suitable detail, and reinforced the key elements of each function through homework problems. In Section 2, we discussed the concept of abstraction, and gave several "garden-variety" examples to convey the different stages of abstraction. We followed that up with homework problems on abstraction and generalization to reinforce the concepts presented in Section 2.

In Section 3, we synopsized the design process and introduced the reader to model-based systems engineering (MBSE). We identified MBSE as a desirable and much needed shift from document-centric to model-based design with the ability to create smaller, focused documents. We focused on a suitably sized example to convey the various aspects of the design process such as concept exploration, and template-driven use case definition. The illustrative example that we chose to convey key concepts was that of a chocolate chip cookie making system. This example was chosen because it was sufficiently easy to understand and yet rich enough to convey the key modeling concepts associated with system design. We then discussed the differences in writing functional and nonfunctional requirements, and followed that up with a discussion of writing a test plan. Thereafter, we addressed risk, unintended consequences, and BiST using the illustrative example to convey the key concepts. Then we covered system-modeling constructs such as ports, SysML flow ports, inputs and outputs, and state machines within the context of the illustrative example. Subsequently, we covered sensitivity analysis in some depth. At this point, we reminded the reader that this section could be skipped without loss of continuity if the reader was unfamiliar with the modeling constructs and the underlying formalisms. In keeping with our overall philosophy in this book, we reinforced the key concepts with homework problems.

In Section 4, we covered eight different document types that are typically generated during system design: the problem situation document, the customer requirements document, the derived requirements document, the verification and validation document, the concept exploration document, the use case model, the design model, and the mappings and management document. We then gave examples of these documents using a familiar example—the disposal of old financial records. Once again, we reinforced the key concepts with homework problems.

In Section 5, we delved into the fundamental principles of good design. Specifically, we enumerated a set of principles and identified their use in the total life cycle of the system. Once again, we used illustrative examples to convey the key principles of good design.

In Section 6, we explored the principles of functional design. Specifically, we discussed the principles of functional decomposition and their allocation to components using the principles of good design. In other words, the allocation to components is not arbitrary, but abides by a set of principles. We employed appropriate examples to illuminate the key concepts. And, we reinforced the key concepts with appropriate homework problems.

In Section 7, we applied our overall approach to specific use cases through three iterations for a generic, multi-service, tactical missile (i.e., one that can be launched from an aircraft, a ship, or a ground station). The Fire Control System, which communicates with both the missile and the human operator, was used as an illustrative example to convey the key elements of the use cases and design iterations. Once again, homework problems were used to reinforce the key concepts in this section.

In Section 8, we covered model-based systems engineering (MBSE). Specifically, we employed UML and SysML diagrams to convey the different perspectives of MBSE. We discussed the concept of state-based modeling approaches. We provided homework problems to reinforce the key concepts of state-based modeling.

In Section 9, we discussed the design of an automobile seatbelt system using state-based approaches. Specifically, we described the use of statecharts in modeling the Automobile Seatbelt System. We followed up the example with homework problems to reinforce the key points in applying state-based modeling approaches to real-world examples.

In Section 10, we went into great depth with the illustrative example of the Parking Garage System. Parking garage documents were searched for nouns that might indicate potential classes. These classes were described in class diagrams or basic block diagrams. These classes and the actions and activities of the state and activity diagrams were then used to construct sequence diagrams. The system design got progressively larger as existing diagrams were elaborated and new diagrams were added. New basic block diagrams were used to show the constraints and parametric diagrams were used to show the equations.

Like all systems engineering tools, abstraction is not done just once at the beginning of the project. It is done whenever a new concept is introduced. The figures shown in Sections 10 to 14 show the continual interchange of the abstraction process and the elaboration process. For example, Fig. 1.29 is not the behavior of the parking garage: it is an abstraction of the behavior. Figure 1.30 is a further abstraction. Figure 1.34 is a model for a component of the parking garage. It is at a lower level with more detail. It is an abstraction of the behavior of the entrance gate of the parking garage. Figures 1.42 and 1.52 are elaborations of Fig. 1.34: more components are being added. These figures show the continual interchange of the abstraction process and the elaboration process.

Starting with Section 10, we devoted five sections to this illustrative example, with each section devoted to one pass through the design. We felt it was important to clearly demonstrate how design evolves from iteration to iteration, rather than merely state that design is an iterative process. We began the series of iterations with an overarching concept of operations for the Parking Garage. The iterations covered both high-level and lower level requirements. In keeping with our strategy of reinforcement learning, we included homework problems after each section (iteration) to reinforce the concepts highlighted and presented in each of the five sections. Each iteration was described in the form of a textual narrative and supported by appropriate UML/SysML diagrams for activities such as requirements definition and test plan generation. We specifically addressed the mapping from the high-level abstraction to more detailed, lower levels, as well as the documentation of these mappings. We followed the discussion up with homework problems to reinforce the key concepts associated with requirements diagram, sequence diagram, class diagram, state machine diagrams, and block definition diagram using specific examples. We also described the use of block definition diagram to define constraints for the Parking Garage System and parametric diagrams to show nested part notation using the equation bindings for a particular Parking Garage Entrance Gate. We followed this discussion with homework problems to reinforce the key concepts in each design iteration. We then defined the activity diagram for the Parking Garage System, and the internal block diagram that assigns functions to hardware for the Parking Garage System. Once again, we followed up on these discussions with homework problems to reinforce the concepts covered in the preceding discussion. In the last iteration, we developed a use case diagram for the Park in Garage use case with pay-per-visit. We also created a requirements diagram for Parking Garage Entrance Gate with pay-per-visit. We then generated a sequence diagram for the entrance gate portion of the main success scenario of the Park in Garage use case with pay-per-visit customers. We also created Entrance Gate class diagram with pay-per-visit customers, and a state machine diagram for the Entrance Gate Controller with pay-per-visit customers. We finally created a block definition diagram (bdd) showing the components of the Parking Garage Entrance Gate with pay-per-visit customers, a bdd defining constraints for the Parking Garage Entrance gate with pay-per-visit customers, and a parametric diagram using nested part notation showing equation bindings for a particular Parking Garage Entrance Gate with pay-per-visit customers. We concluded the last iteration with homework problems to reinforce the modeling concepts form the last iteration of the Parking Garage example.

In Section 15, we discussed when a state machine is necessary and communicated these concepts through examples. Specifically, we discussed the types of systems that have states and the types of models that require state representation. In keeping with our overall philosophy in this book, we followed up with homework problems to reinforce the concepts associated with the use of state machines (Fig. 1.59).

The parking garage models of Sections 10–14 were a part of the gradual elaboration of a specific system design. The nautilus spiral life cycle model

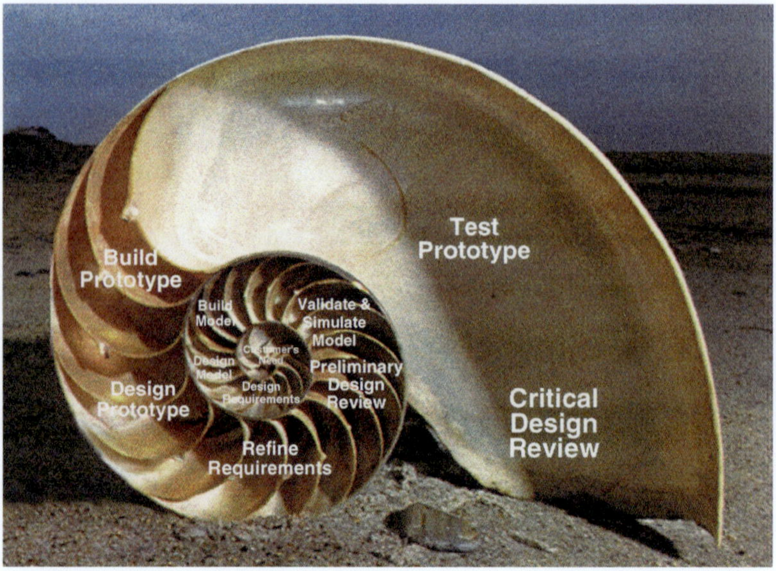

Fig. 1.59 The nautilus life cycle spiral

(Fig. 1.58) generalizes the growth of the design documents for the first two cycles of the design spiral. The system design process is iterative and hierarchical. Therefore, creating documents is necessarily iterative, and not a serial process. With multiple iterations, there are multiple opportunities for parallel processing.

References

Uncategorized References

1. Bahill AT, Gissing B (1998) Re-evaluating systems engineering concepts using systems thinking. IEEE Trans Syst Man Cybern Part C Appl Rev 28(4):516–527
2. Bahill AT, Dean FF (1999) Discovering system requirements. In: Sage AP, Rouse WB (eds) Handbook of systems engineering and management. Wiley, New York, pp 175–220
3. Madni AM (2014) Expanding stakeholder participation in upfront system engineering through storytelling in virtual worlds. Syst Eng 18(1):16–27
4. Madni AM, Brenner M, Costea I, MacGregor D, Meshkinpour F (eds) (1985) Option generation: problems, principles, and computer-based aiding. Proceedings of the 1985 international conference on systems, man, and cybernetics, Tuscon, AZ
5. Madni AM (2013) Generating novel options during systems architecting: psychological principles, systems thinking, and computer-based aiding. Syst Eng 17(1):1–9

6. Madni A, Freedy A, Estrin G, Melkanoff M (eds) (1991) Concurrent engineering workstation for multi-chip module product development process. Invited paper presented at QALS Q QE Washington'91 conference and exposition, Washington, DC
7. Moody JA, Voorhees F, Bahill AT, Chapman WL (1997) Metrics and case studies for evaluating engineering designs. Prentice Hall PTR, Upper Saddle River
8. Santayana G (1905) Flux and constancy in human nature. The life of reason, vol 1, Chapter XII. Charles Scribner's Sons, New York, p 6 and 62. https://www.wikipremed.com/reading/philosophy/The_Life_of_Reason.pdf
9. Bahill AT, Baldwin DG, Ramberg JS (2009) Effects of altitude and atmospheric conditions on the flight of a baseball. Int J Sports Sci Eng 3(2):109–128
10. Rechtin E, Maier M (2000) The art of systems architecting, 2nd edn. CRC, Boca Raton
11. Mourlot F (2015) Le Taureau de Picasso. http://mourlot.free.fr/english/fmtaureau.html Accessed Dec 2015
12. OMG. Documents associated with the unified modeling language (UML) 2015, Version 2.5. http://www.omg.org/spec/UML/2.5/
13. Bahill AT (2010) Design and testing of an illuminance management system. ITEA J 31(1):63–89
14. Bahill AT (2012) Diogenes, a process for identifying unintended consequences. Syst Eng 15(3):287–306
15. Daniels J, Bahill AT (2004) The hybrid process that combines traditional requirements and use cases. Syst Eng 7(4):303–319
16. Friedenthal S, Moore A, Steiner R (2014) A practical guide to SysML: the systems modeling language. Morgan Kaufmann, San Francisco
17. Moore GA (2014) Crossing the chasm, marketing and selling disruptive products to mainstream customers, 3rd edn. Harper Business, New York
18. Madni AM (2012) Elegant systems design: creative fusion of simplicity and power. Syst Eng 15(3):347–354
19. Botta R, Bahill AT (2007) A prioritization process. Eng Manag J 19(4):20–27
20. Wymore AW (1993) Model-based systems engineering. CRC, Boca Raton
21. Bahill AT, Szidarovszky F, Botta R, Smith ED (2008) Valid models require defined levels. Int J of Gen Syst 37(5):553–571
22. Madni AM, Sievers MW (2015) Model based systems engineering: motivation, current status and needed advances. University of Southern California, Systems Architecting and Engineering Program, Technical Report, SAE-TR-006-2015
23. Chapman WL, Bahill AT, Wymore AW (1992) Engineering modeling and design. CRC, Boca Raton
24. Evans E (2004) Domain-driven design: tackling complexity in the heart of software. Addison-Wesley Professional
25. Booch G, Rumbaugh J, Jacobson I (1999) The unified modeling language user guide. Pearson Education India, New Delhi
26. Boehm BW (1988) A spiral model of software development and enhancement. Computer 21(5):61–72
27. Simon HA (2012) The architecture of complexity. The roots of logistics. Springer Science + Business Media, New York, pp 335–361
28. Quintanar GJ (1999) An age of interfaces. Distrib Comp 15–18
29. Schulz AP, Fricke E, Igenbergs E (eds) (2000) Enabling changes in systems throughout the entire life-cycle—key to success? Proceedings of the 10th annual INCOSE conference, Minneapolis, MN
30. Browning TR (2002) Process integration using the design structure matrix. Syst Eng 5(3):180–193
31. Simon HA (1957) Models of man: social and rational. Wiley, New York
32. Bahill AT (2006) Tradeoff studies: a systems engineering skills course. BAE Systems, San Diego, CA

33. Douglass BP (2004) Real time UML: advances in the UML for real-time systems. Addison-Wesley Professional
34. Wymore AW (eds) (2004) The nature of research in systems engineering. Conference on systems engineering research, University of Southern California, Los Angeles, CA. Honourcode
35. Madni AM, Jackson S (2009) Towards a conceptual framework for resilience engineering. IEEE Syst J 3(2):181–191
36. Neches R, Madni AM (2013) Towards affordably adaptable and effective systems. Syst Eng 16 (2):224–234
37. Madni AM, Sievers M (2015) A flexible contract-based design framework for evaluating system resilience approaches and mechanisms. IIE annual conference and expo, 30 May–2 June, Nashville, TN
38. Zachman J (2006) The Zachman framework for enterprise architecture. Zachman Framework Associates
39. Blaha M, Rumbaugh J (2005) Object-oriented modeling and design with UML. Pearson Education, Upper Saddle River
40. Overgraad G, Palmkvist K (2005) Use cases patterns and blueprints. Addison-Wesley Reading
41. Wymore AW, Bahill AT (2000) When can we safely reuse systems, upgrade systems, or use COTS components? Syst Eng 3(2):82–95
42. Botta R, Bahill AT (2006) A prioritization process. INCOSE international symposium, vol 16, no 1, pp 1626–1633
43. Botta R, Bahill AT (2004) 6.5.3 the Zachman framework populated with baseball models. INCOSE international symposium, vol 14, no 1, pp 1333–1350
44. Madni AM, Yee-yeen CA (1988) Design support system for crew system automation. Proceedings of the 1988 I.E. international conference on systems, man, and cybernetics. Institute of Electrical & Electronics Engineers (IEEE)
45. Parnas DL (1972) On the criteria to be used in decomposing systems into modules. Commun ACM 15(12):1053–1058
46. Suh NP (1990) The principles of design. Oxford University Press, New York
47. Fowler M (2004) UML distilled: a brief guide to the standard object modeling language. Addison-Wesley Professional
48. Daniels J, Werner PW, Bahill AT (2001) Quantitative methods for tradeoff analyses. Syst Eng 4(3):190–212
49. Chrissis MB, Konrad M, Shrum S (2011) CMMI for development®: guidelines for process integration and product improvement. Part of the SEI Series in Software Engineering Series. Addison-Wesley Professional
50. Alistair C (2001) Writing effective use cases. Addison-Wesley
51. Kulak D, Guiney E (2001) Use cases. SIGSOFT Soft Eng Notes 26(1):101
52. Gomaa H (2006) Designing concurrent, distributed, and real-time applications with UML. Proceeding of the 28th international conference on software engineering—ICSE'06. Association for Computing Machinery (ACM)
53. O'Connor PD (2001) Test engineering: a concise guide to cost-effective design, development and manufacture. Wiley, Chichester
54. Christian J, Olds J (2005) A quantitative methodology for identifying evolvable space systems. First space exploration conference: continuing the voyage of discovery. American Institute of Aeronautics and Astronautics (AIAA)
55. Sutter JF, Spenser JP (2006) 747: creating the world's first Jumbo Jet and other adventures from a life in aviation. Smithsonian Books, New York
56. Browning TR (2003) On customer value and improvement in product development processes. Syst Eng 6(1):49–61
57. Pugh S (1991) Total design: integrated methods for successful product engineering. Addison-Wesley

58. Smith ED, Piattelli-Palmarini M, Bahill AT (2008) Cognitive biases affect the acceptance of tradeoff studies. Decision modeling and behavior in complex and uncertain environments. Springer, New York, pp 227–249
59. Smith ED, Szidarovszky F, Karnavas WJ, Bahill AT (2008) Sensitivity analysis, a powerful system validation technique. Open Cybernet Syst J 2:39–56
60. Madni AM, Ross A (2016) Chapter 10: Exploring concept trade-offs. In: Parnell GS (ed) Trade-off analytics: creating and exploring the system tradespace. Wiley
61. Bahill AT, Henderson SJ (2005) Requirements development, verification, and validation exhibited in famous failures. Syst Eng 8(1):1–14
62. Fricke E, Schulz AP (2005) Design for changeability (DfC): principles to enable changes in systems throughout their entire lifecycle. Syst Eng 8(4):342–359
63. Coster-Mullen J (2016) Atom bombs: the top secret inside story of Little Boy and Fat Man. Coster-Mullen, Waukesh Wisconsin. OCLC 298514167
64. OMG. http://www.uml.org/
65. Weilkiens T (2007) SysML—the systems modeling language. Systems engineering with SysML/UML. Elsevier BV, pp 223–270
66. Bahill AT, Karnavas WJ (1993) The perceptual illusion of baseball's rising fastball and breaking curveball. J Exp Psychol Hum Percept Perform 19(1):3–14
67. Watts RG, Bahill AT (2000) Keep your eye on the ball: curve balls, knuckleballs, and fallacies of baseball., Potomac Books
68. Long J (2002) Relationships between common graphical representations in systems engineering. Vitech white paper, Vitech Corporation, Vienna, VA, p 70
69. Bock C (2006) SysML and UML 2 support for activity modeling. Syst Eng 9(2):160–186
70. Bahill AT, LaRitz T (1984) Why can't batters keep their eyes on the ball. Am Sci 72 (3):249–253
71. McHugh DE, Bahill AT (1985) Learning to track predictable target waveforms without a time delay. Invest Ophthalmol Vis Sci 26(7):932–937
72. Bahill AT, Karnavas WJ (1992) Inventorsbat selector. U.S. patent number 5,118,102, 2 June 1992.
73. Bahill AT, Baldwin DG (2003) The vertical illusions of batters. Baseb Res J 32:26–30
74. Bahill AT, Baldwin DG (2004) The rising fastball and other perceptual illusions of batters. Bioengineering, mechanics, and materials: principles and applications in sports. Springer Science + Business Media, New York, pp 257–287
75. Bahill AT (2005) Tradeoff study on tradeoff study software. BAE Systems, San Diego
76. Bahill AT, Baldwin DG (2007) Describing baseball pitch movement with right-hand rules. Comput Biol Med 37(7):1001–1008
77. Baldwin D, Bahill AT, Nathan A (2007) Nickel and dime pitches. Baseb Res J 35:25–29
78. Mellor SJ (2004) MDA distilled: principles of model-driven architecture. Addison-Wesley Professional
79. Harel D (1988) On visual formulations. CACM 5:514–530
80. Madni AM, Nance M, Richey M, Hubbard W, Hanneman L (2014) Toward an Experiential Design Language: Augmenting Model-based Systems Engineering with Technical Storytelling in Virtual Worlds. Procedia Comput Sci 28:848–856
81. Botta R, Bahill Z, Bahill AT (2006) When are observable states necessary? Syst Eng 9 (3):228–240

Chapter 2
Decisions and Mistakes

2.1 Section 16. Human Foibles

To understand human decision-making, it is necessary to distinguish between how people make decisions when they behave rationally and when they don't [1]. A rational decision requires a goal, perfect information, and an optimal course of action that maximizes the expected value of the decision. This is the *prescriptive* model of decision-making [2]. This is how people *should* make decisions in an ideal world. In such a utopia, there is a shared single goal, decision makers (DM) have access to all requisite information and they possess the cognitive capacity to use that information efficiently. Furthermore, they know all options that are available to them and the consequences of each option. They must also have the time and resources to make the requisite computations. If these conditions are satisfied, then the optimal course of action can be described as the one that, in the long run, is more valuable (or profitable) than any other course of action [3].

As an illustration of rational decision-making, consider a coin-flipping contest with the following rules. If you call heads and the coin comes up heads, then you win $10. If you call tails and the coin comes up tails, you win $1. Otherwise, you get nothing. Would you call heads or tails? Rational individuals would choose heads, because that would maximize the amount of money that they expect to get. The expected value of choosing heads is 0.5 (the probability of a heads) times $10 (the value of the bet) equaling $5. Whereas, the expected value of calling tails is 50¢.

In contrast, a *descriptive* model describes how people actually do make decisions. This model is discussed in the following pages. Typically when making decisions there is always some uncertainty and insufficient time and resources to fully explore the problem space. Therefore, people are unable to make rational decisions [2]. To explain what people actually do in this situation, Herb Simon [4], the renowned Noble Laureate, coined the term *satisficing*. Simon postulated that people do not even attempt to find an optimal solution. Rather, they search for *good enough* alternatives, that is, alternatives that *satisfice*. Similarly, engineers seek

A.T. Bahill, A.M. Madni, *Tradeoff Decisions in System Design*,
DOI 10.1007/978-3-319-43712-5_2

satisficing designs. Unlike philosophers, engineers tend to be pragmatic. While philosophers might spend endless hours trying to phrase a proposition so that it can have only one interpretation, engineers strive to be unambiguous to the extent feasible, given time and cost constraints.

Humans are not rational. Mark Twain's refrain, "It ain't what you don't know that gets you into trouble. It's what you know for sure that just ain't so," makes this point nicely. Humans can often be very certain of knowledge that is simply false. Let us see how. Consider the following questions. What major American city is directly north of Santiago Chile? If you travel from San Diego, California to Reno, Nevada, in what direction would you travel? Do you think that there are more words that start with the letter r, than there are with r as the third letter. Answers to these questions are likely to provide instances of erroneous information that we carry in our heads. Consider the first question. What American city is directly north of Santiago, Chile? Most Americans would say that either New Orleans or Detroit is north of Santiago, instead of Boston, as shown in Fig 2.1. Now consider the second question. If you were traveling from San Diego to Reno, in what direction would you be going? Most Americans would suggest that Reno is northeast of San Diego, instead of northwest. Furthering this thought train, which end of the Panama Canal is farther west, the Atlantic side or the Pacific side? Most people would say the Pacific. Now the third question. Because it is easier to visualize an r at the beginning of a word, most people think that there are more words that begin with r, than have r as the third letter.

Here is another example of the brain reaching a conclusion that contradicts the facts. Identify an imaginative person, preferably one that knows you. Family and friends work best for this. Ask if he or she wants to participate in an interesting demonstration. Get him to extend his arms in front of him, straight and relaxed (the way one expects a zombie would). Tell him to close his eyes and imagine that he is holding a bucket in his left hand. Not only is the bucket heavy, but it is getting heavier because he is holding it under a tap with running water. Tell him that the bucket is filling up with water and getting *heavier and heavier*. Emphasize the word "*heavier*." Now, tell him to imagine a string tied around his right wrist. The string is attached to a blue helium-filled balloon. The balloon is trying to rise, pulling his hand up in the process. Tell him that as the bucket gets heavier with the inflow of water, the balloon tugs more on his wrist pulling his arm up.

Now, watch his left arm. If the subject has a vivid imagination, and you have performed the "experiment" correctly, you should see his left arm drop from the weight of the water bucket while his right arm gradually rises from the upward pull of the balloon. Now, with his eyes still closed, ask him to tell you where his arms are. If he has not already noticed the effect (most people become aware of the effect while their eyes are still closed), tell him to open his eyes and look at his arms. Surprise!

It is entirely possible that this experiment may not work for you, because it does not work for everyone. It works by the power of suggestion, and works best on people with good imaginations and after you have practiced it many times on your friends. If it does not work for you early on, practice may change that. You can draw encouragement from Dave Baldwin [5], a major league pitcher with a career 3.25 ERA, who famously wrote that if you lose a game, don't blame the umpire or your teammates; just go home and practice harder.

Here are a few more questions that people tend to get wrong.

- A blind beggar had a brother who died. But the dead man's brother is not the blind beggar. What relation was the blind beggar to the brother who died? Think about this for a bit, to make sure you get it right. Brother is not the answer: the answer is his sister. Most people make an erroneous implicit assumption about gender, in this case, that the blind beggar is a man. Making unwarranted assumptions is all too common in engineering design and decision-making in a variety of contexts.
- Here is another interesting question that people are likely to get wrong. How much dirt is there in a hole that measures 2 ft by 3 ft by 4 ft? Take a moment to calculate this number. That's not it! Remember, a hole does not have dirt in it!
- Homicide, diabetes, and suicide are three common causes of death in the United States. List them in order of frequency of occurrence from the most to the least. Jot down your answer on a piece of paper, right now. In actuality, there are more suicides than homicides, but since they do not draw media attention, they do not rank high in your head. The correct order is diabetes, suicides, and then homicides. The point of these examples is that humans carry erroneous information in their minds.

Cognitive illusions can also impair decision-making. We call them illusions because people tend to believe them with as much certitude as they do optical illusions. The following example of a cognitive illusion is from Piattelli-Palmarini [6]. In this study, people were shown a wheel of fortune with numbers from one to hundred (Fig. 2.2). Then the wheel was spun. Sometimes the wheel pointed to a low number such as one, two, three, four, five, or six, and the subjects were then asked a seemingly off the wall question like, "How many African nations are in the United Nations?" They responded and their answers were recorded. How about you? How many African nations do *you* think are in the United Nations? Please write that number on a piece of paper. The cognitive illusion of this example is called anchoring. When estimating numerical values, a person's first impression dominates all further thought. Continuing this example, sometimes the wheel pointed to a high number like 95, 96, 97, 98, 99, or 100. The subjects were then asked, "How many African nations are in the United Nations?" They responded and their answers were recorded. How about you? How many African nations do *you* think are in the United Nations? Please write that number on a piece of paper. Here is what happens. If the wheel showed a small number such as five, then the subjects inevitably underestimated the correct number. On the other hand, if the wheel showed a large number, such as 97, then their subjects inevitably overestimated the correct number. Did you? This shows that a random unrelated event, like spinning a wheel, can predispose answers to questions. The point here is that humans are not always reliable decision makers, because they often carry erroneous information in their heads without actually knowing it [7].

Fig. 2.1 Map showing that Boston is directly north of Santiago Chile

Fig. 2.2 A wheel of fortune with numbers from 1 to 100

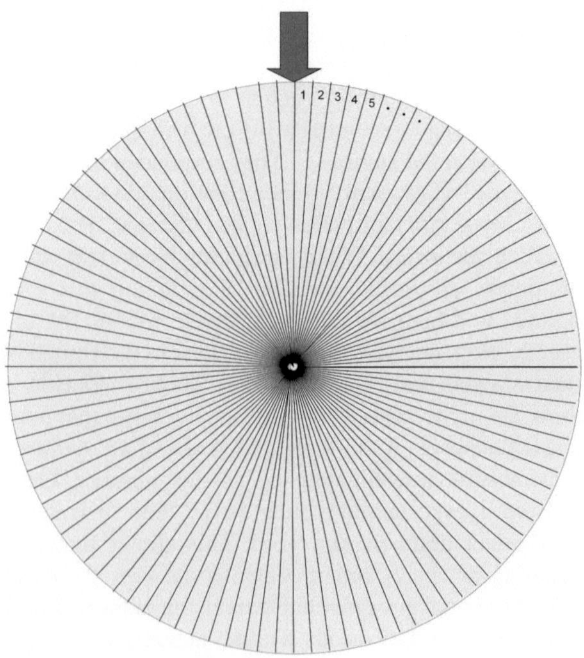

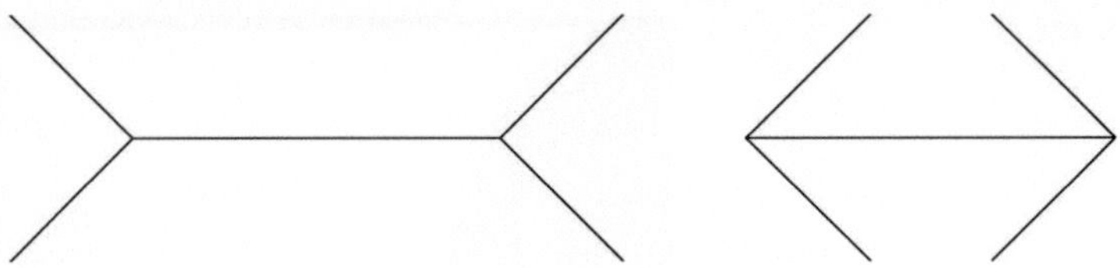

Fig. 2.3 The Müller-Lyer illusion

Fig. 2.4 M. C. Escher's waterfall

To make matters even worse, the information humans gather with their senses is often wrong. In 1641 Descartes [8] wrote that our senses "sometimes misled us; and it is the part of prudence not to place absolute confidence in that by which we have even once been deceived ([8], Meditations I, paragraph 3)." Probably the most famous example of our visual sense misleading us was created by Franz Müller-Lyer in 1889. In Fig. 2.3, which of the two horizontal line segments do you think is longer? Although your visual sensory system tells you that the one on the left is longer, a ruler will confirm that they are of equal length.

Some other examples of our visual sense deceiving us are Escher's waterfall (1961) and Rubin's vase (1915). In Escher's waterfall of Fig. 2.4, let your eyes follow the flow of water. Then explain this apparent perpetual-motion machine. In Rubin's sketch, do you see one white vase in the middle or two black human profiles on the sides? You can see the vase or the profiles, but not both at the same time (Fig. 2.5).

The tactile system sensory system can also deceive us. Imagine this experiment. Put your left hand in a big bowl of cold water and put your right hand in a big bowl of hot water, for around a minute. Then put both of your hands in a big bowl of

Fig. 2.5 Edgar Rubin's vase

lukewarm water. The lukewarm water will simultaneously feel warm to the left hand and cold to the right hand. In 1690, John Locke [9] referred to such an experiment explaining "how water felt as cold by one hand may be warm to the other."

When two sensory systems contradict each other, there will be problems. For example, if a man were below deck on a ship in a troublesome sea, his vestibular system would tell him that he was moving violently, but because his head would be moving with his visual frame of reference, his visual system would tell him that he was stationary. This conflict between sensory systems would cause seasickness. Going above deck, where he could see the horizon, might relieve it.

Two sensory systems can combine to confuse us. In the taste–temperature illusion, putting an ice cube (which has no taste) on the side of the tongue will give rise to a clearly perceptible salty taste sensation. In this case, delivering a sensation via one sensory modality (temperature) gives rise to a sensation in a different sensory modality (taste). Another example of two sensory systems combining to confuse us is when we think a dummy is speaking the words of a ventriloquist. Magicians are also adept at exploiting such foibles of our sensory systems.

In summary, our senses mislead us. This puts false information in our minds. Our minds further muddle this information and then we make decisions based on this knowledge. Good luck!

2.1.1 *Homework*

2.1.1.1 Ethics 1

One of issues that comes up in the business world is the practice of giving and accepting gratuities. When do you accept them and when do you reject them? In some cases, it is a fine line—is it a gift without any thought of receiving anything in return or is it a gift with the idea of "courting" a favor? The following is a typical case.

Scott Bennett is the engineer assigned to deal with vendors who supply needed parts to the Upscale Company. Larry Newman, sales representative from one of Upscale's regular vendors, plays at the same golf club as Scott. Larry and Scott are old friends from grade school.

One morning they go off in the same foursome. Sometime during the round, Scott mentions that he is really looking forward to vacationing in Florida next month. Larry says his uncle owns a three-bedroom condominium, on the water, in Florida that he rents out during the months he and his family are up north. Larry offers to see if the condominium is available next month—assuring Scott that the rental cost would be quite moderate.

The next day Larry calls Scott and tells him that he can rent his uncle's condominium for $200 a week. "My uncle," Larry says, "gets nervous when he rents to total strangers. He likes to have reliable people stay in his condo; the condo is paid for, and my uncle isn't interested in making money on it—he just wants a little help meeting basic operating expenses and the taxes." Scott accepts the offer and begins making plans for his vacation.

1. Would you consider this a gratuity?
2. Would you accept Larry's offer?
3. Would this sway your judgment if, at some point in the future, you were analyzing products offered by Larry's company?

2.1.1.2 Ethics 2

Jack and Jill had a baby while they were Assistant Professors at a university. The department head and the faculty were very supportive, gave them special assignments, and reduced their teaching loads so that they could care for their child. Then, Jill had an affair with Sam, another faculty member, and divorced Jack. At one time Jack and Jill were both assigned to the department committee evaluating the annual faculty reports. Most people expected them to recuse themselves, when it came to evaluating Jack, Jill, and Sam. But surprisingly, Jack said, "We are divorced, but we are on good terms. I will not nix her. I can fairly evaluate her." Jack had a reputation for being honest. So no one objected. Subsequently, Jack evaluated Jill much higher than his average for the other faculty members and higher than the rest of the committee did. Similarly, Jill evaluated Jack much higher than her average for the other faculty members and higher than the rest of the committee did. What ethical problem existed here?

2.1.1.3 Ethics 3

In the future, when you are an expert in some domain, you may become a consultant, either as a member of a consulting company or on your own. Assume that you have had a relationship with a particular company for a long time. It is a good contract; it pays well and is solid income. It comes time for a review with the president of that company. You have been studying their system/process for weeks and you discover that it is seriously flawed. You have written a report that details the mistakes and offers ways to fix most of them. What should you put in your presentation? If you write a hard, accurate presentation, it will embarrass the people who did the original work. The boss will be forced to fire either you or them. In almost all cases, it will be easier to fire you. You could use euphemisms to point out the mistakes without stating how serious they were. But here your suggestions would also have to be weak. You could present your suggestions for improvement, without saying that their system had mistakes in it. You could praise the company and say that they have a very good product, with just a few things that might be improved. Finally, you could say that their product is good and you would keep your consulting job for another year. What would you do?

2.1.1.4 Erroneous Knowledge

Dale Carnegie's book *How to Win Friends and Influence People* stated that humans only use 10 % of their brains. It suggested that reading the book could help people to use more of their brains. Do you think this is true?

Could our brains contain as much as three million miles of neural connections?

Are some people left-brained and others right-brained?

Can subliminal messages influence people to buy products?

Is it reasonable to think that playing Mozart's music to infants or college students will increase their reasoning abilities?

Is adolescences an inevitable time of psychological turmoil?

Is hypnosis useful for retrieving memories of forgotten events.

Do you think that IQ tests are biased against certain groups of people? (Fig. 2.6)

These questions and our answers are from

http://emilkirkegaard.dk/en/wp-content/uploads/50-Great-Myths-of-Popular-Psychology.pdf.

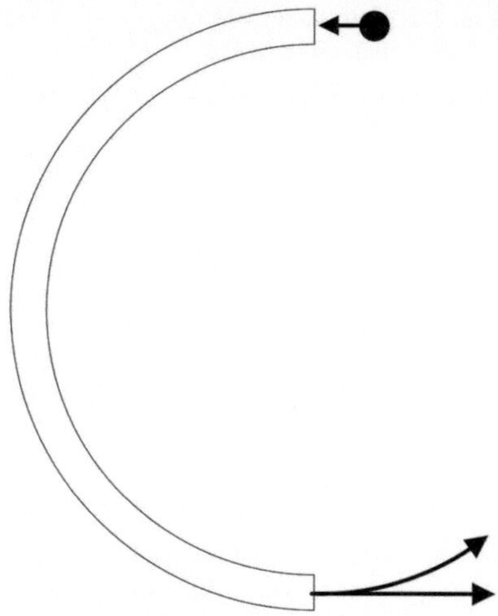

Fig. 2.6 Which path will the ball follow after it leaves the tube?

2.1.1.5 What Color Is the Bear?

A hunter in his camp saw a bear one mile directly south. He stealthily crept to that location. But by the time he got there, the bear had traveled one mile directly east. He shot the bear and walked directly to the dead bear. Then he dragged the bear one mile directly north, back to his camp. What color was the bear?

2.2 Section 17. Prospect Theory

Prospect theory is a widely accepted theory of how humans make decisions . It was developed by the psychologists Tversky and Kahneman[1] [11]. This theory explains the subjective nature of the human decision-making process in terms of heuristics and biases that are present in the assessment of information, and the commonly observed deviations from rational decision-making that result.

At this point, we need to define a few terms. *Value* reflects the *objective* worth of an item. For example, the present market value of an ounce of silver is about $17. *Utility* is a *subjective* measure of the happiness, satisfaction, or reward a person gains (or loses) from receiving a good or service. *Objective probability* means a number derived from observations of and/or experiments on real-world events. *Subjective probability* means a number in your brain that reflects your knowledge

[1]Kahneman won the Nobel Prize in Economics in 2002 "for having integrated insights from psychological research into economic science, especially concerning human judgment and decision-making under uncertainty" [10]. Tversky did not share this prize because Nobel prizes are not awarded posthumously.

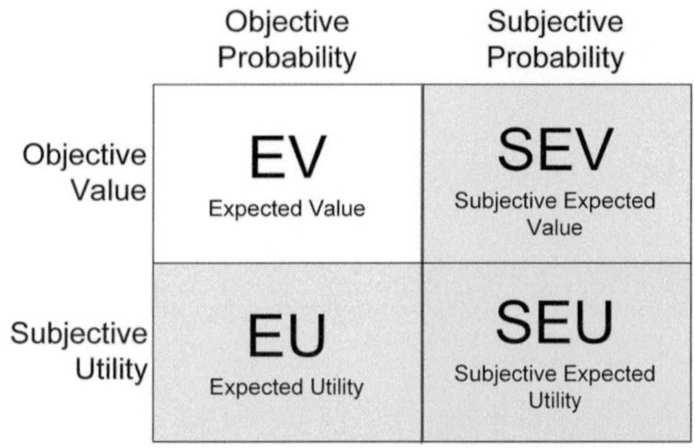

Fig. 2.7 Four variations of the expected value model

of the frequency of an event. *Expected value* is, roughly speaking, the mean or average value of a function. Now we wish to examine combinations of value, utility, objective, and subjective probability.

There are two types of probability: objective and subjective. Objective probability is usually based on experiments and measurements. It was invented by Pascal [12]. Subjective probability is used to describe a person's mental model of the likelihood of a thing being true. It was invented by Locke [13].

Human decision-making is based on probability, value, and utility, Smith [14]. Figure 2.7, based on Edwards [15], differentiates four variations of the expected value model. The upper-left quadrant represents the rational, normative, or prescriptive mathematical theory of expected value (EV) , where objective probabilities and values are considered. This is a rational model, it is independent of the opinion of any individual. The upper-left quadrant is free of heuristics and biases. It is the economist's model for human decision-making, maximizing expected value. The lower-right quadrant of Fig. 2.7 represents the subjective, behavioral, biological, descriptive theory of Subjective Expected Utility (SEU) , where subjective probabilities and subjective utilities are used. The two bottom quadrants involve subjective utility, while the two right quadrants involve subjective probabilities.

Thus, there are two models (or views or theories) of human decision-making: the prescriptive and the descriptive. The *prescriptive* model in the upper-left quadrant prescribes how economists say that humans *should* make decisions in accord with a rational calculus when the objective probabilities are multiplied by corresponding objective values to give the overall expected value. The *descriptive* model in the lower-right quadrant describes what humans actually do.

For 40 years (from around 1950 to around 1990) economic theories used a *prescriptive* model to describe how humans made decisions. The fundamental assumption of their model was that humans are rational. In contrast, the most accepted *descriptive* model of human decision-making is prospect theory, developed by Kahneman and Tversky, [16], which explains the nature of the subjective human decision-making process in terms of the heuristics and biases present in the assessment of information, and the common deviations from rational decision-making that result.

The Subjective Expected Utility model and prospect theory differentiate between actual probability and subjective probability. What they called subjective probability is what John Locke called probability. Locke stated that probability is the degree to which a person believes something to be true. "Probability is likeliness to be true..." Locke [13].

> Our knowledge, as has been shown, being very narrow, and we not happy enough to find certain truth in everything which we have occasion to consider; most of the propositions, we think, reason, discourse—nay, act upon, are such that we cannot have undoubted knowledge of their truth: yet some of them border so near upon certainty, that we make no doubt at all about them; but assent to them as firmly, and act, according to that assent, as resolutely as if they were infallibly demonstrated, and that our knowledge of them were perfect and certain. But there being degrees herein, from the very neighborhood of certainty and demonstration, quite down to improbability and unlikeness, even to the confines of impossibility; and also degrees of assent from full assurance and confidence, quite down to conjecture, doubt, and distrust: I shall come now, (having, as I think, found out the bounds of human knowledge and certainty) in the next place, to consider the several degrees and grounds of probability, and assent or faith. Locke [13].

Thus, Locke thought that humans have a mental model for what Stephen Courbet called *truthiness* [17] of each proposition. This value would run, say, from 0 to 1, representing, respectively, belief that the proposition is completely false and belief that the proposition is completely true. This is subjective probability.

2.2.1 Subjective Utility

Prospect theory divides subjective decision-making into a preliminary screening stage and a secondary evaluation stage. The effect of these two stages is that values are considered not in an absolute sense (from zero), but subjectively from a reference point established by the subject's perspective and wealth before the decision. This psychological process is a part of framing. Figure 2.8 shows how objective values translate into subjective utilities [16].

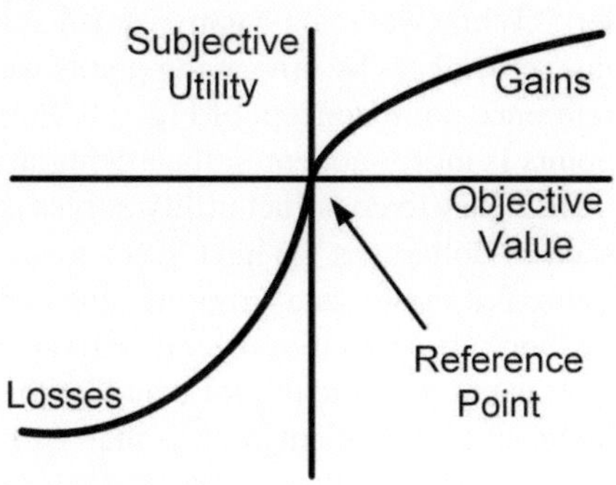

Fig. 2.8 The change in human subjective utility versus numerical value according to prospect theory, Kahneman and Tversky [16]

The three important points shown in Fig. 2.8 are that (1) gains are valued about half as much as losses (the slopes near the reference point are in a 1-to-2 ratio), (2) gains and losses are relative to a reference point, $10 means a lot more to a poor man than to a rich man, and (3) both gains and losses show saturation (slowly approaching upper and lower limits). This graph will go in the lower-left corner of Fig. 2.7.

There are dopamine cells in the ventral tegmental area (VTA) of the human brain that respond like this. Their firing increases if the reward exceeds the reference expectation and decreases if the reward is less than expected.

Even with the requisite knowledge and resources, people tend to not make rational decisions, because they do not evaluate utility rationally. Most people would be more concerned with a large potential loss than with a large potential gain. Losses are felt more strongly than comparable gain, as is shown in Fig. 2.8. Which of the following wagers would you prefer to take?

A *$2 bet*: you win $2 with probability of 0.5, or $0 with probability 0.5. For example, we put down a two-dollar bill and flip a coin to see if you get it or not.

A *$1 bet*: you win $1 with probability of 0.99 or $1,000,000 with probability 0.00000001. For example, we put down a one-dollar bill and a state-lottery ticket. If the lottery ticket is a winner, you keep the $1 million, else you keep the dollar bill.

A *$3 bet*: you win $3 with probability of 0.999999 or you lose $1,999,997 with probability 0.000001. The *$3 bet* has consequences that you might have to give us almost two million dollars. Here is an example of the *$3 bet*. We have a pot containing a million quarters. They are all Arizona quarters, except for one California quarter. We will put a hand into the pot and pull out one-quarter. If it is an Arizona quarter, you win $3. But if it is the California quarter, you owe us almost two million dollars!

Which of these three wagers would you prefer to take? Most engineers prefer the *$2 bet*. Very few people choose the *$3 bet*, because they fear the possibility of a great loss. However, all three of these bets have the same expected value of $1.

A neophyte was watching an NCCA basketball game and his roommate walked in and asked, “Who’s winning?” The TV watcher replied, “The game is nearly over and we are losing 69 to 65.” That is not the way a basketball aficionado would answer. The aficionado would respond, “We are down by four with two minutes left.” While watching a game, an aficionado does not keep the score in his head. He just remembers by how many points we are up or down. This is in keeping with the reference point concept of Fig. 2.8. Furthering the comparison, being down by four points is more saddening than being up by four would be gladdening.

It is easy to construct utility curves as in Fig. 2.8 if the value is in consistent units such as dollars, as we have just shown. But what if the **value** is in different units? Table 2.1 shows data from the renewable energy trade-off study in Chap. 5 of this textbook. It compares several criteria for many alternatives.

It is not possible to plot data like these on a graph like Fig. 2.8. But fortunately, there is a convenient metric that will work, the power-to-price ratio. This ratio

Table 2.1 Rated power and price for several alternative renewable energy systems of Chap. 5

Criteria	Sun electronics PV	DM solar PV	Jacobs Wind turbine	Bergey wind power	Abundant renewable energy	Sky stream	Nordtank 65
Rated power (kW)	10.2	10.4	20	10	10	2.4	65
Purchase price ($)	$34,000	$34,000	$55,000	$29,500	$50,000	$14,000	$62,000

PV means photovoltaic solar panels

Table 2.2 Power-to-price ratio for several alternative renewable energy systems

Criteria	Sun electronics PV	DM solar PV	Jacobs wind turbine	Bergey wind power	Abundant renewable energy	Sky stream	Nordtank 65
Rated power (kW)	10.2	10.4	20	10	10	2.4	65
Purchase price ($)	$34,000	$34,000	$55,000	$29,500	$50,000	$14,000	$62,000
Power-to-price ratio (W/$)	$0.30	$0.31	$0.36	$0.34	$0.20	$0.17	$1.05

shows saturation. The Nordtank system has a high ratio, but it did not generate three times the excitement of the PV systems, because of this saturation (Table 2.2).

Figure 2.8 shows saturation at the left and right, a non-zero reference point and that gains and losses are not weighted equally, as the following example demonstrates.

Situation 1. Assume that you are offered a choice between

1. A sure gain of $100 or
2. A 50 % chance to win $200 and a 50 % chance to win $0

Which would you choose? Most people (75 %) take the $100.

Situation 2. Assume that you are offered a choice between

1. A sure loss of $100 or
2. A 50 % chance to lose $200 and a 50 % chance to lose $0

Which would you choose? Most people (75 %) choose the gamble, because they want to escape a loss.

People prefer to avoid losses more than they prefer to get gains.

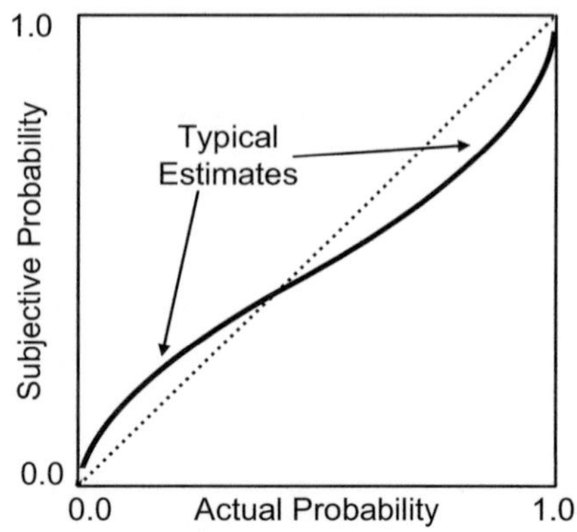

Fig. 2.9 Subjective probability assessment compared to actual probabilities, Tversky and Kahneman [11]

2.2.2 *Subjective Probability*

Prospect theory describes subjective evaluation of probabilities according to the experimentally obtained curve in Fig. 2.9 [11]. Subjectively, small probabilities are overestimated and large probabilities are underestimated.

People tend to overestimate the probabilities of low-likelihood events, such as being killed by a terrorist, or dying in an airplane crash, and underestimate the probabilities of high-likelihood events, such as adults dying of cardiovascular disease. The existence of state lotteries depends on such overestimation of small probabilities. On the right side of this figure, the probability of a brand new car starting every time is very close to 1.0. But people still carry jumper cables in the trunk and routinely buy memberships in AAA. This figure will go in the upper-right corner of Fig. 2.7.

Animals also exhibit tendencies of overestimating low probabilities while not distinguishing much between intermediate probabilities. For example, rats show this pattern [18].

For events where the actual probabilities are between 0.1 and 0.4, humans typically overestimate the probabilities. Moreover, if the event is said to have high severity, then humans overestimate the probability even more [19].

2.2.2.1 Context Dependence (Framing)

People exhibit context dependence: whether A is chosen more often than B can depend on the presence of an irrelevant third choice C, which is dominated and is never chosen [20]. Context dependence means people compare choices within a set rather than assigning separate numerical utilities. Honeybees exhibit the same pattern [21]. Animals also exhibit risk aversion, as will be discussed in Chap. 6.

Next, we consider an example of erroneous estimation of probabilities. Bill is an intelligent, well-organized individual who carries a laptop and dresses conservatively. What is his most likely major in college, Arts and Humanities or

Engineering? A bit of background might help here. In 2009, there were 1,410,000 students majoring in Engineering in American colleges, whereas there were 4,916,000 majoring in Arts and Humanities. Therefore, regardless of what you think of Bill, he is most likely an Arts and Humanities major.

2.2.2.2 Monty Hall Paradox

Humans are not good at estimating probabilities. This fact is amply illustrated by the Monty Hall Paradox, invented by Martin Gardner. This paradox was first published in his Scientific American column in 1959. It is called the Monty Hall paradox because of its resemblance to the TV show *Let's Make a Deal*. This version comes from Massimo Piattelli-Palmarini [6].

Suppose that we are running a game that we can repeat hundreds of times. Resting on a table in front of us are a stack of ten-dollar bills and three identical boxes (Fig. 2.10). You, the reader, are the subject. Here are the rules for each game. You leave the room and while you are out, we place a ten-dollar bill in one of the three boxes. Then we close the lids on all the boxes. We know which box contains the ten-dollar bill, but you do not. Next, we invite you back into the room. You have to guess which box contains the money. If you guess correctly, you get to keep the ten-dollar bill. Each game is divided into two phases. In the first phase, you point to your choice. You cannot open, lift, weigh, shake, or manipulate the boxes in any fashion. The boxes remain resolutely closed. After you make your choice, we open one of the two remaining boxes. Having seen one empty box (the one that we just opened) you now see two closed boxes, one of which contains the ten-dollar bill. Now here is your decision-making problem. Are you better off sticking to your original choice or switching? Many people say it makes no difference. There are two boxes and one contains a ten-dollar bill. Therefore, your chances of winning are 50/50. Almost all people refuse to switch, which is possibly explained by the endowment effect or the status quo bias wherein people prefer to stick with what they have. However, the laws of probability say that you *should* switch. The box you originally chose has, and always will have, a one-third probability of containing the ten-dollar bill. The other two, combined, have a two-thirds probability of containing the ten-dollar bill. But at the moment when we open an empty box

Fig. 2.10 Boxes for the Monty Hall Paradox

(we always open an empty box because we know where the ten-dollar bill is), then the other one alone will have a two-thirds probability of containing the ten-dollar bill. Therefore, your best strategy is to *always* switch! This is a difficult problem, because humans tend to struggle with probabilities.

2.2.3 Subjective Expected Utility

Subjective expected utility, the lower-right quadrant of Fig. 2.7, combines two subjective concepts: utility (subjective value) and subjective probability [22]. Utility is a subjective measure of the worth of an object. Subjective probability is the person's assessment of the frequency or likelihood of the event occurring. The subjective expected utility is the expected value of the product of the utility times the subjective probability. Subjective expected utility theory models human decision-making as *maximizing subjective expected utility*: (1) maximizing, because people choose the set of alternatives with the highest total utility, (2) subjective, because the choice depends on the decision maker's values and preferences, not on reality (e.g., advertising improves subjective perceptions of a product without improving the product), and (3) expected, because the expected value is used. This is called prospect theory: it is a first-order descriptive model of human decision-making.

Now, with a reasonable understanding of the key differences between objective and subjective values and probabilities, we can apply prospect theory to the expected value model of Fig. 2.7 and obtain the descriptive model of Fig. 2.11.

The upper-left corner is rational behavior, represented by the ideal **E**xpected **V**alue (represented by the big capital E and the big capital V). The upper-right

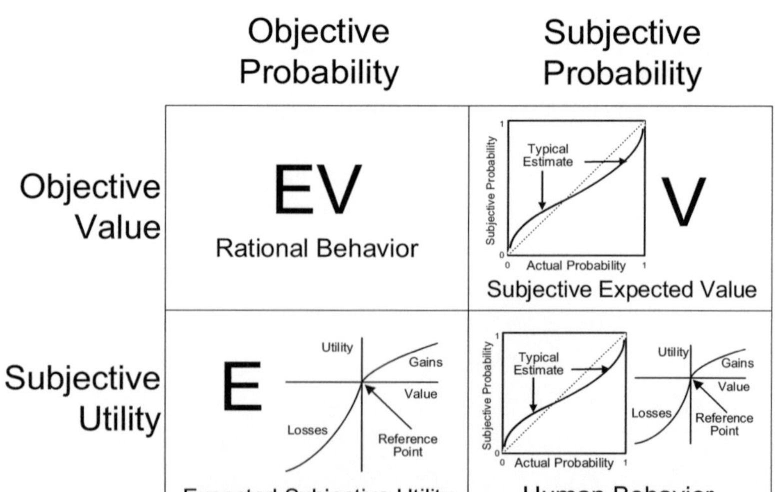

Fig. 2.11 Prospect theory model for human decision-making from Smith [14]

corner illustrates that humans cannot compute expected values, because they are poor at estimating probabilities. Therefore, instead of capital E, for expected value, we insert the human subjective estimate graph, which is then applied to the objective **V**alue (capital V). In the lower-left corner, we apply the **E**xpected value (capital E) to the human's subjective utility, which is represented by the graph of utility versus value. Finally, in the lower-right corner, instead of the expected value, we use the human estimated probability and instead of the objective value, we use the human subjective utility. Thus, the lower-right corner shows actual human behavior whereas the upper-left corner shows ideal rational behavior.

2.2.4 Implications

In the preceding paragraphs, we have reviewed the theoretical foundations of decision-making and human limitations in estimating probabilities. At this point, it is worth asking what conclusions we can draw from these findings. Upon examining human decision-making and expected decision practices, the presence of cognitive biases—even in extensively vetted analyses—can never be ruled out [1, 23]. This is the contention of the economic school of heuristics and biases, which produced prospect theory [24], a theory that describes how people respond to choices under risk and uncertainty. Innate human biases, and external circumstances, such as the framing or context of a question, can compromise decisions. It is important to note that subjects maintain a strong sense that they are acting rationally even when they are exhibiting these biases. Piattelli-Palmarini, in his book *Inevitable Illusions* [6], introduces this subject as follows:

> The current term for these biases is "cognitive" illusions , to indicate that they exist quite apart from any subjective, emotional illusion and from any other such habitual, classical, irrational distortion by a particular subject. The pages that follow provide ample documentation, and contain suggestions as to how we may take urgent and sensible precautions against these illusions.
>
> It never ceases to surprise me that, more or less 20 years after these illusions were first discovered, and after dozens of books and hundreds of articles have been printed on the subject of cognitive illusions, almost no one except for a select circle of specialists seems to have taken this discovery seriously [6, p. x].

> In simple and basic fashion, this book proposes to set out the recent scientific discovery of an unconscious. Not the unconscious or subconscious explored by psychoanalysis, but one that *always and unbeknownst to us* involves the cognitive; that is, the world of reason, of judgment, of the choices to be made among different opportunities, of the difference between what we consider probable and what we consider unlikely [6, p. 1].

Because trade-off studies and risk analyses play such a vital role in decision-making, we reexamine the components of both with the intent of identifying how biases and illusions are introduced in subtle ways in each. In Chaps. 5 and 6, the elements of trade-off studies and risk analyses are examined together with common biases and illusions that may unconsciously and inadvertently affect what is presumed to be rational decision-making.

2.2.5 *What Is Risk?*

The utility of an item includes both subjectively assessed value and the assessor's attitude toward risk. Engineers use the word risk to evaluate and manage bad things that could happen. Risk is quantified by the frequency (or probability) of occurrence times the severity of the consequences. However, in economics and psychology uncertainty risk is defined as the variance of the expected value, which is uncertainty [18].

Table 2.3 explains three bets: A, B, and C. The p's are the probabilities, the x's are the outcomes, μ's are the means, and σ's are the variances. This table shows, for example, that half the time bet C pays $1 and the other half of the time it pays $19. Thus, this bet has an expected value of $10 and a variance of $9. This is a comparatively big variance, so the uncertainty risk is said to be high. Most people prefer bet A, the certain bet.

In choosing between alternatives that are identical with respect to quantity and quality of the reinforcement (expected value, μ), but differ with respect to probability of reinforcement, humans, rats [25], bumblebees [26], honeybees [21], and gray jays [21] prefer the alternative with the lower variance.

To avoid the confusion caused by system engineers and decision theorists using the word *risk* in two different ways, we will refrain from using the word "risk" in this section and instead use ambiguity, uncertainty, and hazards [18].

2.2.5.1 Ambiguity, Uncertainty, and Hazards

A few years ago, a wild fire was heading in the direction of a residential home. The residents packed their car with their valuables, but had no room to save everything. Consequently, they laid their wine bottles in the swimming pool. They put the dog in the car and drove off. When they came back, the house was burned to the ground, but the swimming pool survived. However, the labels were soaked off of the wine bottles. Later, the residents, feeling quite fortunate, invited you to a dinner party to celebrate their safe return. They decided to serve mushrooms that they had picked

Table 2.3 Mean and variance of bets A, B, and C

Name of bet	p_1	x_1	p_2	x_2	μ	σ	Uncertainty, risk
A	1.0	$10			$10	$0	None
B	0.5	$5	0.5	$15	$10	$5	Medium
C	0.5	$1	0.5	$19	$10	$9	High

in the forest, while they were waiting for the fire to subside. There may be some hazard here, because they are not mushroom experts. To accompany dinner, they decided to drink some of their saved wine. There is some uncertainty here, because while you know that none of their wines are cheap, you also know that some are much better than others. Finally, they announce that their sauce for the mushrooms contains saffron and truffles. This produces ambiguity, because you most likely do not know what these ingredients taste like. In light of the foregoing, how would you respond to each of these choices?

Hazard: Would you prefer their forest picked mushrooms or chanterelle mushrooms from the grocery store?
Uncertainty: Would you prefer one of their wines or a Duckhorn Vineyard merlot?
Ambiguity: Would you prefer their saffron and oyster sauce or marinara sauce?

Decisions involving these three concepts are believed to be made in different parts of the brain. The Ellsberg paradox is used to explain the difference between ambiguity and uncertainty [27]. The experimenters gave their subjects a deck of cards and told them it contained 10 red cards and 10 blue cards (the uncertain deck). Another deck had 20 cards, either red or blue, but the percentage of each was unknown (the ambiguous deck). The subjects could take their chances drawing a card from the uncertain deck: if the card were the color that they predicted they won $10, else they got nothing. Or instead of picking a card, they could just take $3 and quit. Most people picked a card. Then their subjects were offered the same bets with the ambiguous deck. Most people took the $3, thus avoiding the ambiguous decision. The experimenters recorded functional magnetic resonance images (fMRI) of their brains while their subjects made these decisions. While contemplating a decision about the uncertain deck, the dorsal striatum showed the most activity and when contemplating a decision about the ambiguous deck the amygdala and the orbitofrontal cortex showed the most activity. Ambiguity, uncertainty, and hazards are three similar but, in fact, different things. Interestingly, people, to different degrees, prefer to avoid all three.

2.2.6 *Confirmation Bias*

Arguably, the most important principle in the fallibility of human decision-making is confirmation bias [1, 23]. Humans hear what they want to hear and reject what they do not want to hear. Humans filter-out information that contradicts their preconceived notions, and remember things that reinforce their beliefs. Confirmation bias causes decision makers to actively seek out and assign more weight to evidence that confirms their hypotheses and ignore or under weigh evidence that could disconfirm their hypotheses. For example, mothers emphasize good deeds of their children and de-emphasize their bad deeds. This is why we often hear the mother of a terrorist crying out, "My boy is innocent. He could never have killed all those people." People who think that they have perfect memory and perfect recall tend to ignore instances when they forgot something and tend to secure in long-term memory instances when they (but not others) correctly recalled events and facts. Senior citizens often believe that they are good drivers despite instances that show that they have poor vision and slow reflexes. Thirty years ago, most smokers were in denial about the hazards of smoking. Some people say, "There must be a storm coming, because my arthritic joints are hurting."

Social media is making this worse. Not only do you filter what you believe, but also Facebook filters what you see. They present to you things from the friends you care about. These friends are probably ideologically similar to you, which accentuates the filtering process.

Nickerson [28] reported many common instances of confirmation bias. In one, the subjects were given a triplet such as (2, 4, 6) and were asked to imagine the rule that was used to generate the triplet and then try to prove or disprove that rule. After each of their guesses, they were told if they were right or wrong. For example, if the subject's mental model for the rule was "successive even numbers," they might guess (10, 12, 14) or (20, 22, 24), triplets that would confirm their mental model, but seldom would they try (1, 3, 5) or (2, 4, 8), triplets that might disprove their mental model. He also presented another example of confirmation bias, witches.

The execution of 40,000 suspected witches in seventeenth-century England is a particularly horrific case of confirmation bias functioning in an extreme way at the societal level. From the perspective of the people of the time, belief in witchcraft was perfectly natural and sorcery was widely viewed as the reason for ills and troubles that could not otherwise be explained. In one test of being a witch, the mob would throw the suspect into a river, if the suspect floated, it was proof that she was a witch and she was executed.

Until the nineteenth century, physicians often did more harm than good because of confirmation bias. Virtually anything that could be dreamt up for the treatment of disease was tried and, once tried, lasted decades or even centuries before being given up. It was, in retrospect, the most frivolous and irresponsible kind of human experimentation. They used blood-letting, purging, infusions of plant extracts and solutions of metals and every conceivable diet including total fasting, most of these were based on no scientific evidence. How could such ineffective measures continue for decades or centuries without their ineffectiveness being discovered? Probably, because sometimes patients got better when they were treated, sometimes they did not and sometimes they got better when they were not treated at all. Peoples' beliefs about the efficacy of specific treatments seem to have been influenced more strongly by those instances in which treatment was followed by recovery or those instances in which recovery occurred spontaneously, than by those instances in which there was no recovery. A tendency to focus on positive cases could explain why the discovery that diseases have a natural history and people often recover from them with or without treatment was not made until much later.

In the middle of the nineteenth century, a Hungarian physician reasoned that so many mothers got infections when delivering babies in a hospital was that the doctors and nurses did not wash their hands before delivering the baby.

Preconceived notions in the medical community allowed doctors to reject his advice for decades.

Most people react to news articles with confirmation bias. If a liberal reads a news story about a scientific study that showed how effective it was to give money to poor people. He might think, "That's interesting. I'll remember that." However, if one of these liberals reads about a new study showing that giving people money when they are unemployed just makes their lives worse. Then he might start looking for flaws in the study. If a person has a long-felt belief that the income gap between the rich and the poor in America is too large and is growing fast, then a new study that challenges this belief might be met with hostility and resistance. However, if that person readily accepts a study that reinforces his beliefs, that is confirmation bias.

People do not think like scientists: they think like lawyers. They form an opinion and then emphasize only evidence that backs up that opinion.

2.2.7 Severity Amplifiers

Interpersonal variability in evaluating the seriousness of a situation depends on the framing. That is, the circumstances surrounding the event will affect how a person responds to it [1, 2, 29–31]. An evaluation may depend on factors such as how the criterion affects that person, whether that person voluntarily exposed himself to the risk, how well that person understands the alternative technologies and the severity of the results. The following are severity amplifiers: lack of control, lack of choice, lack of trust, lack of warning, lack of understanding, manmade, newness, dreadfulness, personalization, ego, recallability, vividness, uncertainty, and immediacy.

Here are some examples of these severity amplifiers: *Lack of control*: a man may be less afraid driving his car up a steep mountain road at 55 mph than having his wife drive him to school at 35 mph. *Lack of choice*: we are more afraid of risks that are imposed on us than those we take by choice. *Lack of trust*: we are less afraid listening to the head of the Center for Disease Control explain anthrax than listening to a politician explain it. *Lack of warning*: people dread earthquakes more than hurricanes, because hurricanes give hours or days of warning. People in California follow strict earthquake regulations in new construction. People in New Orleans seem to ignore the possibility of hurricanes. *Lack of understanding*: we are more afraid of ionizing radiation from a nuclear reactor than of infrared radiation from the sun. In the 1980s engineers invented nuclear magnetic resonance imaging (NMRI). When the medical community adopted it, they renamed it magnetic resonance imaging (MRI). They dropped the adjective nuclear. *Manmade*: we are more afraid of nuclear power accidents than solar radiation. *Newness*: we are more afraid when a new disease (e.g., swine flu, SARs, bird flu, Ebola and Zika) first shows up in our area than after it has been around a few years. *Dreadfulness*: we are more afraid of dying in an airplane crash than of dying in a hospital bed from heart disease. *Fear*: if a friend tells you that, a 6-ft rattlesnake struck at him, how long do you think the snake was? We suspect 3 ft. But, of course, the length of the snake is irrelevant to the harm it could cause. *Personalization*: a risk threatening me, it is worse than that same risk threatening you. *Ego*: a risk threatening my reputation is more serious than one threatening the environment. *Recallability*: we are more afraid of cancer if a friend has recently died of cancer. We are more afraid of traffic accidents if we have just observed one. *Vividness of description*: an Edgar Allen Poe story read by Vincent Price will be scarier, than one that I read. *Ambiguity or uncertainty*: most people would rather hear their ophthalmologist say, "You have a detached retina. We will operate tonight." than, "You might have a detaching vitreous, or it could be a detaching retina, or maybe its cancer. We will let you know the results of the tests." *Immediacy*: a famous astrophysicist was explaining a model for the life cycle of the universe. He said, "In a billion years our sun will run out of fuel and the earth will become a frozen rock." A man who was slightly dozing awoke suddenly, jumped up, and excitedly exclaimed, "What did you just say?" The astrophysicist repeated, "In a billion years our sun will run out of fuel and the earth will become a frozen rock." With a sigh of relief, the disturbed man said, "Oh thank God. I though you said in a *million* years."

2.2.8 Framing

We have mentioned *framing* several times in this section. With Fig. 2.8 we wrote that utility is considered not in an absolute sense (from zero), but rather subjectively from a reference point established by the Decision Maker's (DM) perspective and wealth before the decision, which is his frame of reference. Utility is also affected by whether the event is phrased as a gain or a loss. We wrote that framing (the context of a question) could affect his decision. The section on Severity Amplifiers stated that interpersonal variability in evaluating the seriousness of a situation depends on framing. That is, the circumstances surrounding the event will affect how a DM responds to it. An evaluation may depend on factors such as how the criterion affects that DM, whether that DM voluntarily exposed himself to the risk, how well that DM understands the alternative technologies and the severity of the results. Then we explained over a dozen more severity amplifiers that would affect the framing of a problem.

In contrast to defining framing in passing, as we have done so far, we will now explain framing directly [32]. The DM has a vision, a mission, values, morals, ethics, beliefs, evaluation criteria, and standards for how things should be and how people ought to behave. Collectively these are called *principles*. They are what the DM, the group, or the organization stands for. They limit the goals that are worthy of pursuing and acceptable ways of pursuing these goals. These principles are difficult to articulate, but they powerfully influence the DM's behavior. They are the foundation of the DM's decisions and goals; actions that contradict them will be unacceptable. The utility of the outcomes of decisions derives from the degree to which these decisions conform to and enhance the DM's principles.

Goals are what the DM wants to accomplish. The goals are dictated by the principles, the problem, the problem statement, opportunities, desires, competitive issues, or gaps encountered in the environment. Goals might seed more principles. Goals should be SMART: specific, measurable, achievable, realistic, and time-bound.

The DM has *plans* for implementing the goals. Each goal has an accompanying plan. Each plan has two aspects: (1) tactics are the concrete behavioral aspects that deal with local environmental conditions and (2) forecasts are the anticipation of the future that provide a scenario for forecasting what might result if the tactics are successful. The plans for the various goals must be coordinated so that they do not interfere with each other and so that the DM can maintain an orderly pursuit of the goals. The plans are also fed back to the principles; therefore, they might foment more principles.

Framing means embedding observed events into a context that gives them meaning. Events do not occur in isolation; the DM usually has an idea about what led up to them. This knowledge supplies the context, the ongoing story that gives coherence to experiences, without which things would appear random and unrelated. A frame of reference consists of the principles, goals, and plans that are deemed relevant to the decision at hand and that fixes the set of principles that influence that decision.

The DM uses contextual information to probe his or her memory. If the probe locates a contextual memory that has similar features to the current context, then the current context is said to be recognized. Recognition defines which principles, goals, and plans are relevant to the current context and provides information about the goals and plans that were previously pursued in this context. If a similar goal is being pursued this time, then the plan that was used before may be used again.

In summary, framing means describing all aspects of the problem, the problem statement, and the DM's mind that will affect decisions.

2.2.9 Framing Is the Overarching Process

Confirmation bias, severity amplifiers, conflict of interest, and many other reasoning mistakes may be subsets of bad framing. In all of these, the secret to success is creating a frame of reference with just the right amount of detail.

In previous centuries, physicians often did more harm than good. They used bloodletting, purging, infusions of plant extracts, injections of metal solutions, and numerous other horrible practices. How could such futile medical care have continued for decades or centuries without its ineffectiveness being discovered? Perhaps because the treatment decisions were not framed properly. When patients recovered, that was remembered. But when they got worse, it was forgotten. The frame of reference contained only successes: this is confirmation bias.

The two dozen severity amplifiers given in a section above, all showed one basic decision being made within two frames of reference.

In March 1945, General Patton sent an armored column behind German lines to rescue POWs in OFLAG XIII-B, near Hammelburg. The mission was a failure. General Eisenhower reprimanded Patton for ordering this operation. Patton said that his mistake was sending a force too small to perform the mission: "I can say this, that throughout the campaign in Europe I know of no error I made except that of failing to send a combat command to take Hammelburg." The complete frame of reference shows that Patton may have had a conflict of interest in deciding to initiate that mission, because his son-in-law was a POW in that camp.

Imagine standing beside railroad tracks. You see a train coming down the tracks heading for a group of small children. You are right next to a switch that can reroute the train down a siding, where an old woman has fallen on the tracks. Do you throw the switch? If you don't, many children will die, but you would not have been the cause of their deaths. However, if you throw the switch, the old woman will die and you would have been the *direct* cause, because *your action* caused her death? Do you think that this could be the reason why the pilot of jet fighter does not fire the missile? Instead, he sends a launch signal to Fire Control and it launches the missile. What if the old woman was your mother? Would that change your frame of reference?

Should you always tell the truth? It depends on your frame of reference. Immanuel Kant invented the following dilemma. Suppose an honorable friend says to you, "Quick hide me. Someone is trying to kill me." Then a murderous looking man with a gun shouts at you, "Is that scoundrel hiding in your house? I swear just saw him slither in." Should you lie to protect the innocent man, or should you tell the truth?

All of these reasoning mistakes can be ameliorated by a better description of the frame of reference. We have often said that the best answer most systems engineering questions is, "It depends." Well, it depends on what? It depends on the frame of reference.

2.2.10 Section Summary

In this section, we have discussed salient differences between perfect (rational) behavior and actual human behavior. In Chaps. 5 and 6, we will show how trade-off studies and risk analyses can help move an individual toward making decisions that are more rational (upper-left quarter of Fig. 2.11).

Some day you may find yourself helping a decision maker make informed and defensible decisions. We know that when viewed in formal terms this is a difficult, iterative task. It entails discovering the decision maker's values, principles, importance weights, scoring functions, and preferred combining functions. You need to get into the head of the decision maker to discover his or her preferences and values. Also, different people have different personality types. The Myers-Briggs model is one way of describing these personality types. *Sensory-Thinking-Judging* people are likely to appreciate the decision-making concepts that are in this book. *Intuitive-Feeling* people might not feel that way. Nevertheless, concepts such as those presented in the preceding paragraphs might help you to understand the values and preferences of the decision maker that you are assisting.

In addition to human foibles described by prospect theory, there are several other factors that can adversely affect human decisions. These include corporate culture, framing of the decision, information displayed, wording of questions, context, the decision itself, the effort required to make the decision, the difficulty in making the decision, the time available to make the decision, needed decision accuracy, cost of the decision, likelihood of regret and, finally, the decision maker's values, personality types, risk averseness, biases, illusions, and use of heuristics.

2.2.11 Homework Problems

2.2.11.1 Prospect Theory 1

Tversky and Kahneman ran an experiment concerning a hypothetical outbreak of a new unusual disease that was expected to kill 600 people. Their subjects were divided into two groups and each was given two programs to ponder.

One group of subjects were told that
If Program A were adopted, 200 people would be saved.
If Program B were adopted, there would be a 1/3 probability that 600 people would be saved, and 2/3 probability that no people would be saved.
In this group, 72 % of the subjects choose Program A and 28 % choose Program B.
A second group of subjects were told that
If Program C were adopted, 400 people would die.
If Program D were adopted, there would be a 1/3 probability that nobody would die, and 2/3 probability that 600 people would die.
In this group, 22 % of the subjects choose Program C and 78 % choose Program D.

Note: The difference between 72 % and 78 % is not important. What is important is that one group chose Program A by a 3-to-1 margin and that the other group choose Program D by a 3-to-1 margin.

Compute the expected value for each of these four programs, assuming that value is determined by the number of lives saved.

Try using the prospect theory graph of Fig. 2.8 to compare the various options. Let us assume that the following data were derived for their experimental subjects.

Specific numbers for a curve like Fig. 2.8.

Objective value	Subjective utility
−600	−6
−400	−4.5
$0	$0
+200	+1.5
+600	+3.25

Now calculate the subjective utilities:

Does using subjective utility postulated by prospect theory explain the way people respond to these various options?

2.2.11.2 Prospect Theory 2

When our subject (S) is tasked with a decision, the problem will be set within a frame of reference containing a reference point. This allows S to classify the results of the decision as either gains or losses. S then forms the subjective expected utility, which equals the subjective utility times the subjective probability. The subjective utility is formed using a value function similar to Fig. 2.8 and the subjective probability is formed using a weighting function similar to Fig. 2.9.

Here is an example. S owns a house on the Florida coast valued at one million dollars. He is trying to decide whether or not to buy a hurricane insurance policy costing $500,000 per year. The weather service estimates a 50 % chance that a hurricane will destroy his house this year. Under the expected value criterion, he should feel neutral about buying the policy, because the cost of the insurance is monetarily equal to the potential loss. However, under prospect theory, we first form a frame of reference in which we will consider the decision; we will use his current state as the reference point. The data in the following table are for this frame and Fig. 2.8.

Specific numbers for a curve like Fig. 2.8.

Objective value	Subjective utility
−$100,000	−$500,000
−$500,000	−$400,00
−$10,000	−$30,000
$0	$0
+$500,000	+$300,000
+$1,000,000	+$350,000

This table shows that a $1,000,000 actual loss subjectively *feels* like a $500,000 loss, and a $500,000 actual loss subjectively feels like a $400,000 loss.

The data in the following table show his subjective probabilities versus the actual probabilities. This table shows, for example, that an actual 0.5 probability feels like a 0.4 probability.

Specific numbers for a curve like Fig. 2.9.

Actual probabilities	Subjective probabilities
0.01	0.1
0.5	0.4

To bring it all together, S's choice is a $400,000 subjective loss if he buys the insurance, versus a 0.4 subjective probability of a $500,000 subjective loss if he does not. Buying insurance has a subjective expected utility of −$400,000, not buying insurance has a subjective expected utility 0.4 times $500,000 which is −$200,000. So S rejects the insurance.

Now, let us assume that S owns a one million dollar house in New Hampshire, where there is only a 1 % chance of a hurricane destroying his house. But a hurricane insurance policy only cost $10,000 in New Hampshire. Should he buy insurance?

Suddenly, S receives a stupendously lucky break! He receives an e-mail that tells him that he is the lucky winner of either $500,000 in cash or a 50 % chance on a one million dollar house in New Orleans. Which should he choose? To redeem his prize he should provide by return e-mail his preference, his bank account number (so that they can deposit the money in his account), and his SSN (because they must withhold federal income tax). How would *you* reply?

2.2.11.3 Decision Mistakes

Make a list of the best batters in major league baseball since 1900? What mistakes might you have made in making this decision?

2.2.11.4 Estimating Probabilities and Utilities

For an average major league batter, in a single plate appearance, what do *you think* is the probability of him getting a homerun (HR), a double (2B), a single (1B), a run batted in (RBI), a base on balls (BB), reaching base on error (ROE), a strikeout (SO), and grounding into a double play (GDP)? At first, do not look these up on the Internet, just give your best estimates. Put your estimates in the middle column of Table 2.4. After you see our solutions, record whether you over- or underestimated the probabilities.

Now assume that there are batters on first and second base. Use runs batted in (RBI) as the input objective value. Calculate the utility (subjective worth) of a homerun (HR), a double (2B), a single (1B), a base on balls (BB), reaching base on error (ROE), a strikeout (SO), and grounding into a double play (GDP)? Put your estimates in the right column of Table 2.4. For a reference point, use reaching base safely on an error. For this situation, a HR should produce three RBIs. A stand-up

Table 2.4 Subjective probability and utility estimates

Batting outcomes	Subjective probability estimated by a knowledgeable human, events per at bat (AB)	Subjective worth (utility) assessed by a knowledgeable human
Homerun (HR)		
Double (2B)		
Single (1B)		
Bat in a run (RBI)		
Base on balls (BB)		
Reach base on error (ROE)		
Strikeout (SO)		
Ground into double play (GDP)		

double should produce two RBIs plus leave a man in scoring position. A single should produce one RBI plus leave two men on base. On average, each team scores four runs and suffers 27 outs in a game.

2.2.11.5 Rewrite This Statement

We have a surgical procedure that will cure your problem. Statistically 1 % of the people who undergo this surgery die. Would you like to have this surgery?

2.2.11.6 Confirmation Bias

Suppose that you are given a triplet such as (1, 1, 2) and are asked to hypothesize the rule that generated this triplet. Create some test triplets that will prove or disprove your hypothesis. Assume that after each of your guesses, you will be told if your triplet could have been generated by the rule.

2.3 Section 18. Potential Mental Mistakes in Trade-off Studies

Studies in Judgment and Decision-Making, Cognitive Science and Behavioral Economics can be used to shed light on various aspects of requirements analysis, trade-off studies, and risk assessment. Of course, since many experiments in these fields were designed to reveal truths about human choice at a basic level, they do not address high-level cognitive decisions involved in requirements discovery, trade-off studies, and risk analyses. To shed light on this point, the following sections compare the elements of the experiments at an abstract level. This section deals primarily with trade-off studies and secondarily with requirements discovery and risk analysis. This section is based on Smith, Son, Piattelli-Palmarini, and Bahill [33, 34].

This decision-making literature uses terms such as biases, cognitive illusions, emotions, fallacies, attribute substitution, simplifying heuristics, fear of regret, psychological traps, and paradoxes. In this section, when we refer to specific experiments we use the terminology used in the basic research papers. However, in showing how these factors could adversely affect a trade-off study, we call all of these factors by the collective term "mental mistakes."

Humans often make mental mistakes when conducting requirements analysis, trade-off studies, and risk assessment. Smith [14] extracted seven dozen mental mistakes including heuristics, biases, representativeness, anchoring, base-rate neglect, the conjunction fallacy, cognitive illusions, emotions, fallacies, fear of regret, psychological traps, and paradoxes from the field of *behavioral economics* (Thaler's memoir [32] summarizes this field nicely) and showed how they could induce mental mistakes in trade-off studies. Many of these are discussed in [6, 35–38].

We studied these mental mistakes and determined whether each one principally affects (1) magnitudes, such as evaluation data, weights, and probabilities or (2) textual descriptions, such as stating the problem, writing criteria, or proposing alternatives.

The next section describes these mental mistakes and offers recommendations for ameliorating their effects on the trade-off study components. Our research draws on existing, rigorous experimental results in the cognitive science literature, and provides a cautionary alert about the likelihood of large-scale occurrences of cognitive biases within trade-off studies. To this end, this work offers guidelines for the systems engineering community to consider when developing systems. The positive experimental identification of each bias—specifically within trade-off studies and trade-off study components—is not a goal generally pursued. Therefore, experimentation for all the biases in a trade-off study context can potentially take decades.

2.3.1 Prospect Theory

The seminal paper on cognitive biases and heuristics under uncertainty was authored by Tversky and Kahneman [39]. This work led to the creation of prospect theory [24]. Prospect theory breaks subjective decision-making into a preliminary screening stage and a secondary evaluation stage. In the screening stage, values are considered not in an absolute sense (from zero), but subjectively from a reference point established by the subject's perspective and wealth status prior to making the decision. Please recall Fig. 2.8. In this stage, losses are weighted more heavily than gains. In the evaluation stage, a final value for every prospect (alternative) is calculated. Subsequent refinements to this theory, including subjective probability assessment, were published in Tversky and Kahneman [11], which was reprinted in

Tversky and Kahneman [40]. Kahneman won the Nobel Prize in Economics in 2002 "for having integrated insights from psychological research into economic science, especially concerning human judgment and decision-making under uncertainty" [41, 42].

Several heuristics and biases, notably representativeness, anchoring, base-rate neglect, and the conjunction fallacy are now considered by Kahneman to be instances of a super-heuristic called *attribute substitution*. Judgment is mediated by this heuristic when, without realizing its existence [43]

> an individual assesses a specified target attribute of a judgment object by substituting another property of that object—the heuristic attribute—which comes more readily to mind. Many judgments are made by this process of *attribute substitution*. For an example, consider the well-known study by Strack, Martin, & Schwarz [44], in which college students answered a survey that included these two questions: How happy are you with your life in general? How many dates did you have last month? The correlation between the two questions was negligible when they occurred in the order shown, but it rose to 0.66 when the dating question was asked first [45, p. 53].

As we see clearly in this instance, the target attribute (happiness) is assessed by mapping the value of another attribute (number of dates last month) onto the target scale. This process of attribute substitution

> "will control judgment when these three conditions are satisfied: (1) the target attribute is relatively inaccessible (happiness); (2) a semantically and associatively related candidate attribute is highly accessible (number of dates last month); and (3) the substitution of the heuristic attribute in the judgment and the immediacy of the response is not rejected by the critical operations of" thoughtful thinking [45, p. 54].

Our goal is to recommend actions that will help people avoid making specific mental mistakes when conducting trade-off studies. Later we will consider mistakes that are specific to requirements discovery and risk analyses. Our intention is to help people in defining trade-off studies to choose among alternatives. We want people to have confidence in their decision. We also present arguments to convince people to validate their systems. These recommendations are the principal outputs of this research (Fig. 2.12).

The components of a trade-off study are:

- Problem statement
- Evaluation criteria
- Weights of importance
- Alternative solutions
- Evaluation data
- Scoring functions
- Normalized output scores
- Combining functions
- Preferred alternatives
- Sensitivity analyses

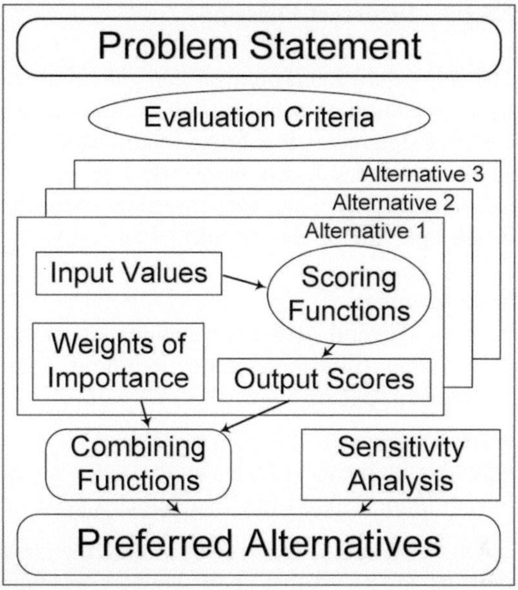

Fig. 2.12 Elements of a trade-off study

2.3.2 Problem Statement Mistakes

Problem statement errors can arise from a variety of mental mistakes such as

- Bad problem stating
- Incorrect phrasing
- Attribute substitution
- Political correctness
- Feeling invincible

Each of these sources is discussed next.

2.3.2.1 Bad Problem Stating

The problem of system design must be stated strictly in terms of functions and requirements, not in terms of a solution, or a class of solutions [46]. It is a serious mistake to state a problem in terms of a solution, instead of customer needs and expectations.

Recommendation: Question the customer to determine his or her values and real needs.

2.3.2.2 Incorrect Phrasing

The way a question is phrased can potentially influence the answer. With respect to the formulation of public policy, Kahneman and Ritov [47] showed their subjects (1) brief statements that looked like headlines and (2) proposed methods of intervention. Some subjects were asked to indicate their willingness to pay for the interventions by voluntary monetary contributions; other subjects were asked which intervention they would rather support.

Issue Mammals
Problem: Several Australian mammal species are nearly wiped out by hunters.
Intervention: Contribute money to a fund to provide a safe breeding area for these endangered species.

Issue Skin Cancer
Problem: Skin cancer from sun exposure is common among farm workers.
Intervention: Support free medical checkups for threatened groups.

On being asked how much money they would be willing to contribute, most subjects indicated that they would contribute more money to provide a safe breeding area for the Australian mammal species than they would to support free medical checkups for the threatened farm workers. However, when the subjects were asked which intervention they would support, they indicated that they would rather support free medical checkups for threatened workers.

Recommendation: Questions designed to elicit a value for a criterion should be tightly coupled to the criterion.

The way a question is phrased will determine, in large part, the answer one gets. When asked whether they would approve surgery in a hypothetical medical emergency, many more people accepted surgery when the chance of survival was given as 99 % than when the chance of death was given as 1 %. In a supermarket, assume you see two packages of ground beef. One is labeled, "80 % fat free." The other is labeled, "Contains 20 % fat." Which one do you buy?

2.3.2.3 Attribute Substitution

Attribute substitution occurs when a subject, while assessing an attribute, substitutes a related attribute that comes more readily to mind. In effect, "people who are confronted with a difficult question sometimes answer an easier one instead" [42, p. 707]. When confronted with a choice among alternatives that should properly be decided by a full trade-off study, there is a strong tendency to substitute a seemingly equivalent yet much simpler decision question in place of the trade-off study process.

Here is one more example. Suppose that you want to measure the salinity of water for drinking or irrigation. You could measure it directly or you could use attribution substitution and instead measure the electrical conductivity. This usually works very well and is a lot cheaper. But as the system progresses we might learn that some ions are more dangerous than others. Then we would have to stop substituting and actually measure the amount of chloride, Mg^{3+}, and HCO_3^- individually and electrical conductivity would not be a good proxy.

Recommendation: Sponsors of trade-off studies should recognize that a premature reduction of the trade-off space invariably leads to a simpler decision question. While this is a common heuristic that is inadvertently applied, it fails to address the original multi-objective decision.

2.3.2.4 Political Correctness

Fear of being politically incorrect causes many top-level decision makers to be reluctant to state the problem clearly and concisely.

2.3.2.5 Feeling Invincible

Teen-agers, especially boys, are notorious for thinking that they are indestructible and immune to adversity. I won't get caught. I can't get hurt. I will avoid car accidents. I won't cause an unwanted pregnancy. I won't get sexually transmitted disease. I don't have to back up my hard drive, because my computer won't crash. And, They can't do that to me. In 1912, the White Star line said that the Titanic was "unsinkable." This mental mistake is particularly insidious when performing risk analyses.

Recommendation: Decision makers must learn to and have the freedom to question all assumptions (no sacred cows!).

It is worth noting that codebreakers have been systematically making and breaking codes for more than 600 years. During WWII, the Allied commanding generals often had copies of Hitler's battle orders before the German generals did. Yet, the Germans were absolutely certain that no one had broken their Enigma codes.

When President Franklin Delano Roosevelt (FDR) was campaigning for his first term as president, he made a memorable speech in Pittsburgh in which he promised that he would slash government spending and balance the budget. Returning to that city in 1936, seeking reelection after 4 years of record government outlays, he asked his top speechwriter, Sam Rosenman, how he should spin this reversal. Rosenman replied, be bold, "Deny you were ever in Pittsburgh" [48].

2.3.3 *Evaluation Criteria Mistakes*

Evaluation criteria mistakes fall into three main categories:

- Dependent criteria
- Relying on personal experience
- Forer Effect

Each category is discussed next.

2.3.3.1 Dependent Criteria

Evaluation criteria should be independent . For example, when evaluating humans, Height and Weight are not independent, but Gender (male versus female) and Intelligence Quotient are.

Recommendation: Dependent subcriteria should be grouped together into a single criterion. In selecting a car, the following dependent subcriteria are not independent: Maximum Horse Power, Peak Torque, Top Speed, Time for the Standing Quarter Mile, Engine Size (in liters), Number of Cylinders, Time to Accelerate 0–60 mph. Therefore, they should be grouped into a single criterion, Power.

The nameplate data for an induction motor will usually include rated voltage, peak current, and maximum power. For an induction motor, these may be independent, because peak-starting current is much larger than steady-state current. However, for most other electrical products, voltage, current, and power are dependent.

Before the Transportation Security Administration came into being, a statistician we know refused to fly on airplanes after reading the alarmingly high probability that there could be a bomb on any given plane. Later he found out that the probability of there being two bombs on any given flight was very low. Now, whenever he flies, he always carries a bomb with him!

2.3.3.2 Relying on Personal Experience

"We are all prisoners of our own experience." Criteria may be chosen based solely on the analyst's experience, with insufficient customer input and environmental confirmation. Uncorrelated to objective knowledge, self-assessed knowledge (what people think they know, as discussed in Section 16) influences their selection of search strategies [49]. Aspects of sole reliance on past personal experience include preconceived notions, false memories, and overconfidence in judgments [50].

Recommendation: It is imperative to engage customers and other stakeholders to acquire balanced and relatively objective views.

Human memory is notoriously fallible. Eyewitness accounts and identification in court trials can be inaccurate and downright harmful [51–53]. Of the criminal defendants who were released on recent DNA evidence, 75 % were convicted based largely on incorrect eyewitness testimony.

Recommendation: Never depend solely on eyewitness testimony; look for corroborating evidence.

2.3.3.3 Forer Effect

Previously existing criteria are likely to be adopted if (1) the analyst believes that the criteria apply to the present problem, (2) the criteria are presented well, and (3) the analyst believes in the *authority* of the previous criteria writer. The analyst might fail to question or rewrite criteria from a legacy trade-off study that originated from a perceived authority and is now seemingly adaptable to the trade-off at hand. This is called the Forer effect [54] The analyst might fail to question or rewrite criteria from a legacy trade-off study that originated from a perceived authority and is now seemingly adaptable to the trade-off at hand. This is another example of confirmation bias.

Recommendations: Spend time to consider and formulate criteria from scratch, before consulting and automatically reusing criteria in the company Process Assets Library (PAL). Practically speaking, generic criteria taken from a company's PAL must be tailored for the project at hand.

In a famous experiment performed at Yale University by Milgram [55] and replicated by Burger [56], subjects were told to administer electric shocks to a learner when he made mistakes on a list of words he was memorizing. As long as the man in the white coat (the authority) nodded and said, "Go on," most subjects administered shocks even though they saw that the learner was in excruciating pain.

2.3.4 Mistakes in Importance Weights

The sources of mistakes in importance weights are:

- Choice versus calculation
- Ignoring severity amplifiers

Each source of mistakes in importance weights is discussed next.

2.3.4.1 Chosen Versus Computed Value

An importance weight can have different values depending on whether the subject chooses the value from a predetermined set, or calculates a value. For example,

Tversky, Sattath, and Slovic [57, pp. 376–377] told their subjects, "About 600 people are killed each year in Israel in traffic accidents. The ministry of transportation investigates various programs to reduce the number of casualties. Consider the following two programs, described in terms of yearly costs (in millions of dollars) and the number of casualties per year that is expected following the implementation of each program. Which program would you prefer?"

Choose a program		
	Expected casualties	Cost
Program X	500	$55 M
Program Y	570	$12 M

Given this "Choose a program" formulation, 67 % of the subjects choose Program X, saying that "human life is priceless." However, when the same set of alternatives was set up as a calculation problem, where the subjects were asked to calculate a cost for the cell with the question marks that would make Program X and Program Y equally attractive, fewer than 4 % of the subjects chose $55 M or more as the worth of Program X.

Calculate a value		
	Expected casualties	Cost
Program X	500	???
Program Y	570	$12 M

In this study, the subjects' mental values were different depending on whether they chose a value from a predetermined set, or made a calculation.

The importance weight of an evaluation criterion can be determined by choice or by calculation. When a legacy trade-off study is used to guide a current study, the weights are determined by choice. When a trade-off study is constructed from scratch, the weights are calculated for the current alternatives. When faced with a choice involving nonnumerical attributes, people will use principles and categories to make a decision; but when faced with a calculation problem, they will shift a great deal of attention to numbers, ratios and differences, sometimes to the point of making narrow-minded decisions based solely on numbers alone.

Recommendation: The use of choice and calculation should be consistent within a trade-off study. The systems analyst might use one of the formal weight derivation tools referenced in Section 20.

2.3.4.2 Ignoring Severity Amplifiers

When a group of people is asked to assign an importance weight to an evaluation criterion, each individual might produce a different value. Different weights arise not only from different preferences, but also from irrational severity amplifiers [58]. These include the factors of lack of control, lack of choice, lack of trust, lack of warning, lack of understanding, being human-made, newness, dreadfulness/direness, personalization, memory limitations, and immediacy. Excessive disparities occur when a person assesses an importance weight after framing the problem differently. An evaluation may depend on factors such as how the criterion affects that person, how well that person understands the alternative technologies, and the direness of the results. As a result, each person might assign a different weight of importance to any criterion.

Recommendation: Interpersonal variability can be reduced through education, peer review of the assigned weights, and group discussions. But you need to be aware of the "lemmings-effect." If you reveal how others are voting, then people are likely to respond with the most popular answers. It is also important to keep a broad view of the whole organization, so that criteria in one area are considered in light of all other areas. A sensitivity analysis can show how important each weight is. For unimportant weights, move on. For important weights, spend more time and money trying to get consensus. This might include showing the recommended alternatives for several different sets of weights.

2.3.5 Alternative Solution Mistakes

Mistakes associated with alternative solutions include:

- Serial consideration of alternatives
- Isolated or juxtaposed alternatives
- Conflicting criteria
- Adding alternatives
- Maintaining the status quo
- Uneven level of detail

Each of these categories of mistake is discussed next.

2.3.5.1 Serial Consideration of Alternatives

When solving a problem, people often exhibit a confirmation bias [28] by seizing on a hypothesis as a solution, and holding on to it until it is disproved. Once the hypothesis is disproved, they proceed to the next hypothesis and hold on to it until it is disproved [59]. This confirmation bias can potentially persist throughout a trade-off study, as the analyst uses the whole study to try to prove that a currently favored alternative is the best.

Recommendation: Alternative solutions should be evaluated in parallel from the beginning of the trade-off study, so that a collective and impartial consideration permits the selection of the best alternative from a complete solution space.

All alternatives should be given substantial consideration [60]. Strategies that require equal consideration for all alternatives are more effective [61, 62].

2.3.5.2 Isolated or Juxtaposed Alternatives

In the following example the two music dictionaries described in this box were evaluated in isolation and jointly [63].

Music dictionary	Number of entries	Condition
A	10,000	Like new
B	20,000	Cover is torn

When the dictionaries were evaluated in isolation, most subjects were willing to pay more for dictionary A than for B. However, when the dictionaries were evaluated jointly, most subjects were willing to pay more for dictionary B.

The importance weights for the number of entries and the book's condition evidentially changed. In isolation, each dictionary was judged more critically according to its condition. However, when the dictionary descriptions were juxtaposed, the number of entries became easier to compare, and the importance attached to the number of entries increased. Features that were hard to assess in separate evaluations were easier to evaluate in a comparative setting.

This phenomenon has implications for the selection of alternative solutions. Specifically, solutions that are evaluated serially (perhaps as they are conceived) may not receive the attention they would if they were evaluated in parallel with the other solutions.

Recommendation: New alternative solutions should be stored and be subject to elimination only after comparison to all alternative solutions.

2.3.5.3 Conflicting Criteria

In the following experiment to analyze the effect of conflicting choices, the subjects were told, "You can either select one of these gambles, or you can pay $1 to add one more gamble to the choice set The added gamble will be selected at random from the list you reviewed" [64].

Conflicting criteria situation			
Gamble	Winning (%)	Payoff	Expected value (not shown to subjects)
X	65	$15	$9.75
Y	30	$35	$10.40
Result: 55 % of subjects chose to pay for the addition of a yet unknown gamble from a pre-reviewed set			

This is an example of conflicting criteria. Gamble X is better on Winning percentage but worse on Payoff and vice versa. Next, the criteria values were changed so that one choice dominated the other and the following result occurred.

Criteria dominance situation			
Gamble	Winning %	Payoff	Expected value (not shown to subjects)
X	65	$15	$9.75
Z	65	$14	$9.10
Result: 30 % of subjects chose to pay for the addition of a yet unknown gamble from a pre-reviewed set			

In this example, Gamble X is equal to or better than Gamble Z on both Winning percentage and Payoff, so it dominates Gamble Z. It is seen that a choice among alternatives with conflicting features will cause a continued search for alternatives, while the presence of clear dominance among the alternatives will increase the chance of the decider finalizing his or her decision.

Recommendation: Assure that the alternative solutions represent the global solution space. Do not curtail the search for alternatives when you perceive criteria dominance by one alternative.

2.3.5.4 Adding Alternatives

A similar study in medical decision-making [65] investigated distinctiveness by adding alternatives. The objective was to determine whether situations involving multiple options could paradoxically influence people to choose an option that would have been declined if fewer options were available. The following problem statement was given to the subjects:

The following patients have been scheduled for carotid endarterectomy [cleaning of plaque from the carotid arteries that supply the brain with blood], but only one operating room slot is available (more slots will not be available for 2 weeks). Of these patients, who should have a higher priority?

Patient M. S. is a 52-year-old employed journalist with TIA's (Transient Ischemic Attack: a mini-stroke caused by temporary interruption of blood supply to an area of the brain) experienced as transient aphasia. She has had one such episode occurring 10 days ago which lasted about 12 h. Angiography shows a 70 % stenosis of the left carotid. Past medical history is remarkable (noteworthy) for past alcoholism (no liver cirrhosis) and mild diabetes (diet controlled)

Patient A. R. is a 72-year-old retired police officer with TIA's experienced as left hand paralysis. He has had two such episodes over the last 3 months with the last occurring 1 month ago. Angiography shows a 90 % stenosis of the right carotid. He has no concurrent medical problems and is in generally good health.

One group of decision makers was given just these two choices: patients M. S. and A. R and were asked on which patient they should operate first. 38 % chose Patient A. R.

Another group of decision makers was given an additional patient.

Patient P. K. is a 55-year-old employed bartender with TIA's experienced as transient monocular blindness. She has had one such episode a week ago, that lasted less than 6 h. Angiography shows a 75 % stenosis of the ipsilateral carotid. Past medical history is remarkable for ongoing cigarette smoking (since age 15 at a rate of one pack per day).

In the group of decision makers that was given all three patients, 58 % chose Patient A. R. This is counter intuitive, because when you add an alternative you would expect all of the existing alternatives to get fewer votes. Phrased another way, when there were two patients, M. S. got 62 % of the votes and when there were three patients, she only got 33 % of the votes.

Adding an additional alternative can increase decision difficulty and thus the tendency to choose a distinctive alternative. Note that the distinctiveness of the alternative was rather unnoticeable before the additional alternative was added.

Recommendation: All alternative solutions should be evaluated in parallel from the beginning of the trade-off study. If an alternative needs to be added in the middle of a study, then the most similar alternative will lose support.

Here is a different example of adding alternatives. People at a movie theater were offered a choice of a small box of popcorn for $3 or a large bucket for $7. Most, 80 % chose the small size. Then another group was offered a choice of a small box of popcorn for $3, and an *addition* medium sized box for $6.50 or a large bucket for $7. This time most chose the large size.

Here is a third example of the effect of adding alternatives. A group of study participants was offered the opportunity to buy a bottle wine. They were offered three merlots from the Napa Valley. They were priced at $8, $26, and $31. Most chose the $26 bottle. Some said that the $8 bottle seemed too cheap. Some said that they seldom buy the most expensive item on a list. Then a $52 bottle of wine was added to the display and it was shown to a new group of people. This time a lot of people bought the $31 bottle that few had bought before. Marketers know all about this. If they put out such a display, they never expect to sell the $52 bottle. It is only there to coax more people into buying the $31 dollar bottle. But capitalizing on such biases is risky business.

2.3.5.5 Maintaining the Status Quo

In another experiment, students were given $1.50 for filling out a questionnaire [64]. They were then offered either a metal zebra pen for their $1.50, or they were offered a choice of a metal zebra pen or two plastic pilot pens for their $1.50. The probability of keeping the $1.50 was lower when they were offered only the zebra pen (25 %) than when they were offered the choice of pens (53 %). An increase in the conflict of the choice thus increased a decision to stay with the status quo. It seems that the students thought the increase in conflicts within the subset of pen options increased the chance of making a bad choice within the subset.

Thus to change behavior, do not denigrate an existing concept. Rather extol the virtues of a new concept, because people are more willing to accept a new concept than to reject an old one.

Recommendation: The more alternatives that exist and the more complicated the decision, the more the status quo is going to be favored. Do not needlessly increase the number of alternatives in a trade-off study. Having more alternatives increases the difficulty of the decision. However, in the very beginning of a project it is good to have many alternatives to better understand the problem and requirements.

In another similar scenario involving a patient with osteoarthritis, family physicians were less likely to prescribe a medication (53 %) when deciding between two medications than when deciding about only one medication (72 %). The difficulty in deciding between the two medications led some physicians to recommend neither.

This mental mistake has a very big part in the Monty Hall paradox of Section 16. We did not tell you at that time, but over 90 % of the contestants refused to switch.

2.3.5.6 Uneven Level of Detail

Uneven level of detail in the description of the alternatives is likely to confuse a naive reader. If alternatives are abstracted at a different level of detail, it will be difficult to assign scores to the alternatives.

2.3.6 Evaluation Data Mistakes

Evaluation data mistakes fall into the following categories:

- Relying on personal experience
- Magnitude and reliability
- Judging probabilities poorly

Each category is discussed next.

2.3.6.1 Relying on Personal Experience

Estimates for evaluation data may faultily come from personal experiences. People may be completely oblivious to things they have not experienced, or they may think that their limited experience is complete. What people think they know may be different from what they actually know (remember Section 16) [66].

Recommendation: The source of evaluation data for public projects must be subject to peer and public review. Decision analysts must be willing to yield absolute control over choosing evaluation data.

2.3.6.2 Magnitude and Reliability

Numbers have two components: value and variance. People tend to judge the validity of data first on its magnitude or strength (expected value) and then according to its reliability or variance [67]. Therefore, data with outstanding magnitudes but poor reliability are likely to be chosen and used in trade-off study. Remember that when our bumblebees of Table 2.3 were choosing between alternatives that were identical with respect to quantity and quality of reinforcement (expected value), but differed with respect to the probability of reinforcement they preferred the alternative with the lower variance.

Recommendation: Either data with uniform reliability should be used, or the speciousness of data should be taken into account.

2.3.6.3 Judging Probabilities Poorly

Humans are poor at judging probabilities. Now it is worth recalling (Fig. 2.9) that probabilistic evaluation data usually come from guesses by domain experts. Later, these guesses are replaced by quantitative estimates, then with simulations, and finally by prototypes. So, initial data often come from human evaluation and humans are terrible at estimating probabilities, especially for events with probabilities near 0 or 1 [11]. People overestimate events with low probabilities, like being killed by a terrorist and they underestimate high probability events, such as adults ultimately dying of cardiovascular disease. The existence of state lotteries depends upon such overestimation of small probabilities. On the right side of Fig. 2.9, when you are about to depart on an airplane flight, your probability of a safe landing is almost 1.0. But while waiting in the terminal, a lot of people buy flight life insurance.

Recommendation: Distrust human probability estimates, especially for improbable and very probable events. Use frequency of occurrence or likelihood of occurrence instead of probability.

Typically, risk is defined as the probability of occurrence times the severity of consequences. Therefore, half of a risk analysis is based on a task that humans do poorly.

Subjective Probability

In general, people tend to overestimate small probabilities—perhaps gambling on the unlikely, but possibly important appearance of some phenomenon. Please recall

Fig. 2.9. People also tend to underestimate large probabilities—perhaps because a more commonly occurring phenomenon loses psychological impact [11, 68, 69].

Surprisingly, this curve seems to be reversed for children [70], who tend to underweight small probabilities and overweight high probabilities.

Certainty Effect

People prefer reducing chances of something bad happening from something to nothing, more than reducing chances of something bad happening by the same amount but not to zero. Plous [71, p. 99] writes, "Zeckhauser observed that most people would pay more to remove the only bullet from a gun in Russian roulette than they would to remove one of four bullets." Notice that the reduction in probability of being shot is the same in both cases. (In the first case from 1/6 to 0 and in the second from 4/6 to 3/6.) The certainty effect can be seen to affect scoring functions. Since scoring functions by definition give scores between 0 and 1, they always approach small numbers (or their difference) at their endpoints.

Sub-additivity

This morning, Mr. Smith's car engine did not start. He had to call a mechanic. Rate the following probabilities for the cause of the malfunctioning.

No gas in the tank	P1
Insufficient battery charge	P2
Faulty carburetion	P3
Other causes	P4

A definite wrong answer is that: $P1 + P2 + P3 + P4 < 1$

However, mechanics with over 15 years of experience are almost invariably prone to considerable sub-additivity effects.

Super-Additivity

An instance of super-additivity occurs when complementary, yet mutually exclusive, probabilities are judged to add up to greater than a 100 % chance of occurring.

P(a): What is the probability that the freezing point of gasoline is lower than the freezing point of alcohol?

P(b): What is the probability that the freezing point of alcohol is lower than the freezing point of gasoline?

- Wrong answer: $P(a) + P(b) > 1$

The ease and the speed with which we can mentally summon seemingly relevant instances are of decisive importance in producing this error.

2.3.6.4 Ignoring the First Measurement

Often when a measurement (test) reveals an unexpected result, the physician and/or the patient will ask for a second measurement. If the second measurement is satisfactory, then the first measurement is discarded and only the result of the last measurement is recorded. This is probably related to confirmation bias. The decision maker rejects the results of all tests until one confirms the decision maker's preconceived notions.

Recommendation: If there is no evidence showing why the first measurement was in error, then it should not be discarded. A reasonable strategy would be to record the average of the two measurements. For example, if you take your blood pressure and the result is abnormally high, then you might measure it again. If the second measurement indicates that blood pressure is near the normal range, and you do not have proof that the first reading was a mistake, then do not record only the second reading, either record both measurements or record the average of the two readings.

2.3.7 Scoring Function Mistakes

Scoring function mistakes fall into three categories:

- Mixing gains and losses
- Not using scoring functions
- Anchoring

2.3.7.1 Mixing Gains and Losses

With scoring functions (such as the utility curve of Fig. 2.8), objective value is translated to subjective utility. Input values become normalized output scores. Scoring functions must be elicited from the customer (Fig. 2.13).

2.3.7.2 Treating Gains and Losses Equally

Loss/Gain Utility Disparity

Loss Aversion

It is important to realize that people do not treat gains and losses equally. Figure 2.8 shows how values are translated into utilities according to prospect theory. People

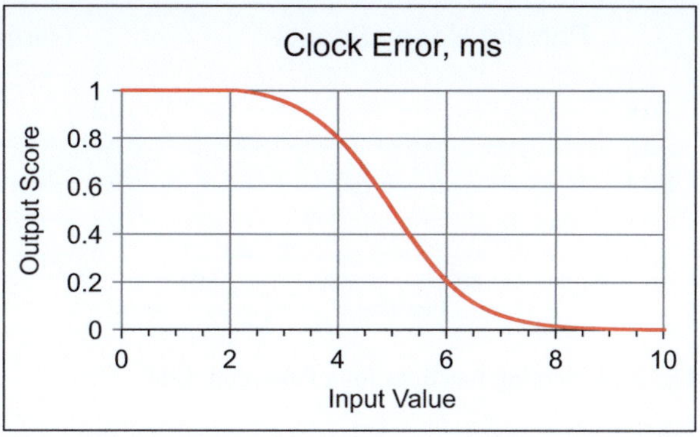

Fig. 2.13 A scoring function where less is better

prefer to avoid losses more than to acquire gains. Prospect theory [16] suggests that psychologically losses are twice as powerful as gains. Would you rather get a 5 % discount, or avoid a 5 % penalty? Most people would rather avoid the penalty. In a trade-off study, you will get a different result if the scoring function expresses losses rather than gains.

The same change in price, framed differently, has a significant effect on consumer behavior. Though traditional economists consider this to be completely irrational, it is important to the fields of marketing and behavioral economics. The implicit assumption of conventional economics is that the only relevant metric is the magnitude of the absolute change in expenditure. In the above example, *saving* 5 % is considered equivalent to *avoiding* paying 5 % extra.

The **Pinewood Derby Trade-off Study** is a real-world trade-off study that also serves as a reference model. The original study was published in Chapman, Bahill, and Wymore [72, Chapter 5] http://sysengr.engr.arizona.edu/pinewood.pdf. It was subsequently implemented in Excel with a complete sensitivity analysis.

> Over 80 million people have participated in Cub Scout Pinewood Derbies. Pinewood is a case study of the design of a Cub Scout Pinewood Derby for one particular scout pack. The system helps manage the entire race from initial entry through final results. Many alternatives for race format, scoring, and judging are presented [72, p. 83].

The Pinewood Derby trade-off study had these criteria

Percent Happy Scouts
Number of Irate Parents

Because people evaluate losses and gains differently, the preferred alternatives would have been different if they had used

Percent Unhappy Scouts
Number of Ecstatic Parents

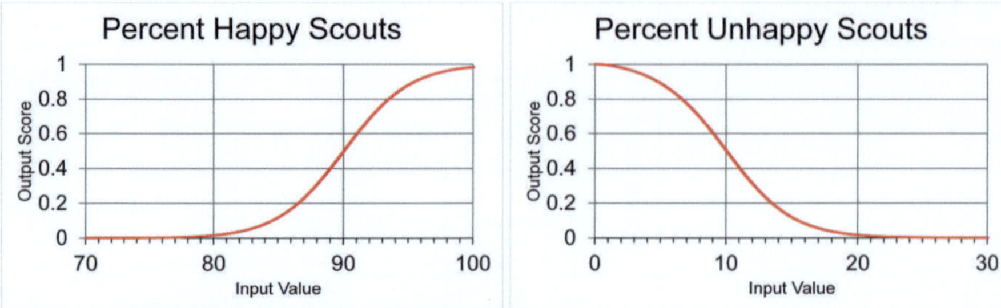

Fig. 2.14 Scoring functions for a Pinewood derby

Recommendation: Scoring functions in a trade-off study should express gains rather than losses. When the Percent Happy Scouts scoring function of Fig. 2.14 (top) is shown to people, and they are asked, "How would you feel about an alternative that gave 90 % happy scouts?" they typically say, "It's pretty good." In contrast, when the Percent Unhappy Scouts scoring function of Fig. 2.14 (bottom) is shown to people and they are asked, "How would you feel about an alternative that gave 10 % unhappy scouts?" they typically say, "It's not very good." When people are allowed to change the parameters, they typically push the baseline for the Percent Unhappy Scouts of Fig. 2.14 scoring function to the left.

Human unequal treatment of gains and losses suggests that scoring functions in a trade-off study should uniformly express either gains or losses. Principles of linguistic comprehensibility suggest that criteria should always be phrased in a positive manner, for example, use Uptime rather than Downtime, use Mean Time Between Failures rather than Failure Rate, use Probability of Success rather than Probability of Failure, and use Clock Accuracy rather than Clock Error. Therefore, when using scoring functions make sure more output is better, not like Fig. 2.13.

Recommendation: Scoring functions in a trade-off study should uniformly express gains rather than losses.

2.3.7.3 Not Using Scoring Functions

In *infrastructure bias*, the location and availability of preexisting infrastructure such as roads and telecommunication facilities influences future economic and social development. In science, an availability bias is at work when existing scientific work influences future scientific observations. For example, when sampling pollutants, most samples may be taken in towns or near roads, as they are the easiest places to get to. Other examples occur in astronomy and particle physics, where the availability of particular kinds of telescopes or particle accelerators acts as constraints on the types of experiments performed.

Most trade-off studies that we have observed in industry did not use scoring functions. In some cases, scoring functions were explained in the company engineering process, but for the sake of convenience, they were not used.

Recommendation: Standard scoring functions [46] (or similar functions, or fuzzy sets or utility functions) should be used in trade-off studies. Easy to use scoring functions should be referenced in company systems engineering processes.

2.3.7.4 Anchoring

When estimating numerical values, a person's first impression dominates all further thought. In this example from Piattelli-Palmarini [6] that was shown in Section 16, people were shown a wheel of fortune with numbers from one to hundred. The wheel was spun and then the subjects were asked to estimate the number of African nations in the United Nations. If the wheel showed a small number, like 12, the subjects inevitably underestimated the correct number. If the number on the wheel were large, like 92, the subjects overestimated the correct number.

Recommendation: View the problem from different perspectives. Use different starting points. When estimating values for parameters of scoring functions, think about the whole range of expected values for the parameters. For example, when estimating the baseline for the Percent Happy Scouts of Fig. 2.14 (top), the systems engineer could show the expert two scoring functions, one with a baseline of 80 and another with a baseline of 98. Then allow the expert to express his or her preference.

Recommendation: The order in which the alternatives are listed has a big effect on the values that humans assign to the evaluation data. Therefore, a trade-off study matrix should be constructed showing the evaluation criteria in the rows and the alternatives in the columns. Then the trade-off study matrix should be filled out row by row with the status quo as the first alternative. This makes the evaluation data for the status quo the anchors for estimating the evaluation data for other alternatives. This strategy implies that the anchoring alternative is known, is consistent and as importantly you have control over it. Make the status quo the first alternative. In the first iteration, examine the scores left to right, and in the next iteration examine them right to left.

2.3.8 Output Score Mistakes

The key category of output score mistake is:

- False precision

This is discussed next.

The most common mistake in trade-off studies is implying false precision. For example, an analyst might ask a subject matter expert to estimate values for two criteria. The expert might say something like, "The first criterion is about 2 and the second is around 3." The analyst inserts these numbers into a calculator and computes the ratio as 0.666666667. This degree of precision is totally unwarranted, but these nine significant digits may be used throughout the trade-off study. The Forer Effect might explain this. The analyst believes that the calculator is an impeccable authority in calculating numbers. Therefore, the calculation is not questioned.

Recommendation: Use significant digit methodology. Furthermore, in numerical tables, print only a sufficient number of digits after the decimal place as is necessary to show a difference between the preferred alternatives and others.

2.3.9 Combining Function Mistakes

Combining function mistakes fall into the following two categories:

- Lack of knowledge
- Lack of availability

Each category is discussed next.

2.3.9.1 Lack of Knowledge

The typical engineer is not familiar with the nuances of combining functions and their impact on trade-off studies. This is a judgment derived from a 20-year search for effective trade-off studies in industry [73].

Recommendation: Training in the use of combining functions is necessary. Discussions of combining functions are found in Section 22 and in [74, 75].

2.3.9.2 Lack of Availability

Decision support software is invariably equipped with limited types of combining functions. For example, one of the best commercial tools has only the Sum and the Product combining functions. Most others have only the Sum.

Recommendation: Spreadsheet-formulated trade-off studies have the greatest potential to achieve variety in combining functions.

2.3.10 Preferred Alternative Mistakes

The three categories of preferred alternative mistakes are:

- Overconfidence
- Ignoring the need for expert opinion
- Failure to talk with the customer

Each of these categories is discussed next.

2.3.10.1 Overconfidence

Trade-off studies often begin with overconfidence on the part of the analyst. Thereafter, the analyst continues to maintain a state of overconfidence without examining details. Griffin and Tversky [67, p. 411] state,

> People are often more confident in their judgments than is warranted by the facts. Overconfidence is not limited to lay judgment or laboratory experiments. The well-publicized observation that more than two-thirds of small businesses fail within 4 years [76] suggests that many entrepreneurs overestimate their probability of success [77].

Recommendation: To compensate for this bias, there is no better teacher than performing trade-off studies and then presenting the results at reviews that demand high-quality work in all trade-off study components.

Consider the following example. A ball and a baseball bat together cost $110. The bat cost $100 more than the ball. How much does the ball cost? Kahneman models the human brain as two systems, cleverly named System 1 and System 2. System 1 relies on frugal heuristics yielding intuitive responses. It is quick, but sometimes wrong. It says that the ball cost $10. System 2 relies on deliberate analytical reasoning. It is slower, but often more accurate. It says that the ball costs $5. This is the right answer.

2.3.10.2 Ignoring the Need for Expert Opinion

The analyst holds a circular belief that expert opinion or review is not necessary because no evidence exists that suggests the need for expert opinion. This is especially true if no expert has ever been asked to comment on the trade-off study.

Humans store about seven units or "chunks" of information at a time [78], in short-term memory, irrespective of skill level. However, chess masters' chunks are larger and richer in information than amateurs' chunks [79]. Thus, a novice often "can't see the forest for the trees," and cannot perceive the refinement that occurs in an expert's thought.

Recommendation: Experts should be sought formally or informally to evaluate trade-off studies.

2.3.10.3 Cannibals

Five cannibals manage to get hired as engineers in an aerospace defense company. As part of welcoming them, the boss says, "You're all part of our team now. You can earn good money here. Our cafeteria is conveniently located for you to run over and get something to eat when you want. So try not to trouble other employees with trivial questions." The cannibals promised. Four weeks later, the boss returns and says, "You're all working hard. I'm satisfied with all of you. However, one of our janitors is missing. Would any of you know what happened to him?" The cannibals shake their heads indicating that they did not. After the boss leaves, the leader of the cannibals admonishes the others, "Which of you idiots ate the janitor?" A hand raises hesitantly. The leader scolds, "You fool! For four weeks, we've been eating experts and no one noticed anything. And now you had to go and eat a janitor!"

2.3.10.4 Failure to Talk with the Customer

The second most common trade-off study mistake is failing to talk to the customer. Both students and engineers seem to feel that it is a sign of weakness to ask for help.

Recommendation: Talk with your customer.

2.3.11 Sensitivity Analysis Mistakes

The categories that go with a mistake in sensitivity analysis are:

- Lack of training
- Hawthorne effect

Each is discussed next.

2.3.11.1 Lack of Training

Most personnel are not well trained in the use of methods and tools for sensitivity analysis. They often fail to find interactions by computing second-order partial derivatives. When estimating partial derivatives they often use too large a step size. When estimating partial derivatives of functions of two parameters, they often use the wrong formula. Smith et al. [80] has showed that interactions among parameters in trade-off studies can be very important, step sizes for the approximation of effects should be very small, and second-order derivatives should be calculated accurately. It is expected that only the highly trained personnel will know of such results, illustrating the gap between reality and training.

Recommendation: Investments in sensitivity analysis training is worthwhile. Perhaps enabling software can substitute for much sensitivity analysis knowledge. Section 26 describes the use of sensitivity analyses.

2.3.11.2 Hawthorne Effect

Individuals may alter their behavior simply because they know they are being studied. In a research project from 1927 to 1932 in the Hawthorne Plant of the Western Electric Company in Cicero, Illinois, workers improved in performance for no other reason than that human attention was given to them. This research was initiated by Mayo [81]. They studied the physical and environmental influences of the workplace (for example, brightness of lights, temperature, and humidity) on human performance and later moved into the psychological aspects (for example, coffee breaks, group pressure, working hours, and managerial leadership) [82].

These studies showed that regardless of the experimental manipulation employed, the productivity of the workers improved because of the attention from the experimenters. Other findings of the study were that (1) the workplace is a social system, (2) the informal organization of work groups and their internal norms for behavior affect performance, and (3) the measured aptitude of individuals is often a poor predictor of their performance.

Studies in the 1960s [83] produced similar behavior changes that the author called the Pygmalion Effect. When teachers were told that a certain group of randomly selected students was expected to perform particularly well in that school year, the teachers' expectations for those students increased. With increased expectations, the students performed better. This effect was repeated in military training centers and in business. According to these studies, when managers had high expectations for their employees, the workers became more productive. When military instructors believed that their trainees had superior skills, the trainees performed better.

There is a similar effect in physics, the Heisenberg Uncertainty Principle [84], which describes how precisely we can simultaneously measure the position and momentum of a particle. If we increase the precision in measuring one quantity, we are forced to lose precision in measuring the other.

In other words, interacting with a person or a system will change its behavior. Giving added attention to workers will increase their productivity. Having higher expectations for students will increase their performance. Increasing the precision in measuring position or momentum of a small particle will decrease the precision in measuring the other. You will be studying alternatives and you will be measuring system performance using evaluation criteria. Be sure that you understand how your activities affect the behavior of the system. You will be interacting with decision makers. They will change their minds. Understand how your interactions affect them. Your decision makers will change their preferences; however they should not change their values. They also should change their propensity for making mental mistakes.

Your job is to help a decision maker to make valid and effective decisions that he or she (and other stakeholders) can have confidence in. This is a difficult and iterative task. It entails discovering the decision makers preferred weights, scoring functions, and combining functions. You must also discover the decision maker's mental mistakes and ameliorate them or mitigate their impact. You must get into the head of the decision maker to discover his or her values.

It is important to recognize that different people have different personality types. The Myers-Briggs model [85] is one way of describing personality types. In the language of this model, Sensing-Thinking, Thinking-Judging people are likely to appreciate the trade-off study techniques we will present. Intuitive-Feeling-Perceiving people most likely will not.

2.3.12 Section Summary

Humans usually consider alternatives sequentially. They often hastily choose one alternative after having their attention anchored to just one or a few criteria. To make good, rational choices among alternatives, a decision maker's awareness of cognitive biases must increase. Limited awareness and possible consequential irrationality can cause erroneous estimation or misuse of trade-off study components. Therefore, trade-off study components should be examined individually. The higher level decision then becomes a calculation resting on a broad base of rationally considered assumptions, components, and calculation methods. Decision makers should understand the mathematical methods that allow the parallelization of human decision processes through trade-off studies.

2.3.13 Homework

2.3.13.1 Accelerated Life-Cycle Testing

On one episode of the TV series MythBusters, they investigated myths concerning pirates. In particular, they wanted to know if pirates could have removed stains from blood, sweat, and tears from their clothing with urine. The pirates were said to have one barrel that all the pirates urinated in. After a few months, they would use that stale urine to wash their clothes and take stains out. The MythBusters did not want to wait a few months for the urine to get stale, so they boiled it down to 50 % of its volume. What mental mistake does this illustrate.

2.3.13.2 Mental Mistakes in the Spin Coach

The following is from the Spin Coach documents. Point out mental mistakes that you find.

Document 1: The Problem Situation

When a spinning object (like a baseball) is put in a moving fluid (like air) it will experience a force that pushes it sideways. Some highly successful baseball players have written that they see this spin of the ball and use this information to predict its future trajectory. But at present, there is no system that can teach high school and college baseball and softball players to predict this spin-induced deflection of the pitch. Our customer needs such a system. The goal of this project is to design a web site that will display images of spinning baseballs, allowing the subject to predict the spin induced deflection and providing feedback to facilitate learning.

Document 5: Concept Exploration

Alternative architectures for the Spin Coach

3. System on CD-ROM. Balls spinning on drills can be photographed and their images stored on CD-ROM disks. Such videos are on my web site. http://www.sie.arizona.edu/sysengr/baseball/index.html under the tag line "here is a ten-second video of the four-seam (left) and two-seam (right) fastballs. These images along with the software program (application) will be transferred from the CD-ROM to the user's hard disk using a license key provided by BICS. Thereafter the user runs the Spin Coach from his or her hard disk. The user must login for each session. The information gathered at login is used to track user performance history. System upgrades will be provided with new CD-ROMs.

3b. Application Sold Online. A slight variant of the CD-ROM alternative is to sell the application through an online store with distribution via Internet downloads.

Evaluation Criteria

Fidelity of Images. How realistic are the images? Are they two or three dimensional? What is the resolution? What is the color depth? What is the update rate? Will the presentation vary depending on the processor speed or the communications bandwidth? This traces to the Operational Concept Description (OCD) and is also related to Validation. Importance weight is 6.

Varying image size. When the pitcher releases a baseball, the ball is 54 ft from the batter's eye and it subtends 0.26° of arc. When it is two-thirds of the way to the contact point, when the swing must begin, the ball subtends 0.77° of arc. Perhaps our displays should have a changing image size. This traces to the Operational Concept Description (OCD) and is also related to validation. Importance weight is 4.

Distortion. Are the images distorted? Is the image of the ball circular? Are the images in focus? This traces to the Operational Concept Description (OCD) and is also related to Validation. Importance weight is 7.

Ease of Verification. Verification means proving that our system works correctly. We could help verification by asking, "Can we gather data that shows the player's progress while using the Spin Coach?" Importance weight is 9.

Ease of Validation. Validity indicates the fidelity of transferring learned knowledge from our product to the real world. We can help validation by asking, "Can we gather data that shows improvements in the player's performance in the real game?" Importance weight is 9.

Autonomy. Do we control the needed assets, or must we depend on outside sources? Negotiating a contract could take months. Finding and buying equipment might take months. This traces to SR1. Importance weight is 10.

2.3.13.3 What Mistake Was Made Here?

A Modification of SIERRA

	Input/output figure of merit	Weight	Normalized weight
1	Number of collisions	8	0.258065
2	Trips on circular track	7	0.225806
3	Trips on elliptical track	7	0.225806
4	Spurious stops on circular track	3	0.096774
5	Spurious stops on elliptical track	3	0.096774
6	Availability	2	0.064516
7	Reliability	1	0.032258

2.3.14 Summary of Section 18

This section, which is based on an article by John G. Cramer, http://www.npl.washington.edu/AV/altvw146.html provides a good summary for this section. Human beings have evolved with brains that have amazing capabilities for rational thought, pattern recognition, judgment, creativity, and imagination, none of which can be readily duplicated by the most powerful computers. However, there is one area in which the human brain is sadly lacking: the ability to accurately assess probabilities and act on these assessments. The successes of state lotteries and casinos in Las Vegas and on Indian reservations provide ample evidence that when it comes to estimating probabilities and acting accordingly, humans are deficient. Humans erroneously believe that winning streaks are real, that slot machines are overdue to pay off jackpots, and that past patterns of random events somehow influence the odds of occurrence of the next event. Consider the irony in this next story. Long ago, when humans began to emerge from the woods and started to think as humans, there was a wide river separating two hostile tribes that constituted most of humanity. On one side of the river lived the Probabilist Tribe, a group that had the ability to understand probabilities and act accordingly. On the other side of the river lived the Numbskull Tribe, a group more like modern humans who believed in luck and winning streaks, and other such fantasies. The Probabilists and the Numbskulls hated each other and wanted to get rid of each other. On one fateful night, the Probabilists wanted to rush across the river and kill the Numbskulls. So, they did a risk analysis, computing probabilities and consequences. They concluded that it was too risky to cross the river and engage in combat with the Numbskulls. The Numbskulls made no such calculation and felt no such reluctance. On a night when the moon was no more than a crescent, they quietly made their way across the river, and launched a surprise attack on the Probabilists killing them all. Therefore, today we are all descendants of the Numbskulls tribe.

2.4 Section 19. Heuristics: Cost-Effective Shortcuts to Desired Outcomes

Heuristics are mental shortcuts that humans employ when making rapid decision and judgments without spending a lot of time investigating, analyzing, and evaluating problem situations. Heuristics are quick, informal, and intuitive algorithms the brain uses to generate approximate answers to questions that require reasoning. Heuristics have been effective in system architecting, systems engineering, management, and systems integration. Heuristics have been the cornerstone of methods to cope with complexity in systems. The term "heuristics" is short for heuristic reasoning [86, 87] . Heuristics range from declarations (i.e., warnings, cautions) and true statements (i.e., descriptive text without obvious implications) to prescriptions (i.e., effective responses in particular situation)

Effective heuristics tend to exploit contextual and domain knowledge, analogies and analogs, and exemplars and archetypes. Unlike scientific laws, they tend to be qualitative, suggestive, and less amenable to replicable measurements. They tend to be "generalization from specific examples, not conclusions reached from the application of general principles." The power of a heuristic stems from coverage (breadth) with minimal exceptions (conditions). Heuristics embrace the principle of satisficing instead of optimizing as a means to conquer complexity. Consequently, heuristics do not guarantee a solution, much less an optimal one. Therefore, it is prudent to question heuristics and periodically revisit them in light of evidence. It is entirely possible that a heuristic in use may be discarded in favor of another, based on evidence.

It is important to recognize that heuristics can lead to errors in judgment. When heuristics fail to produce correct judgments, they can result in cognitive biases, i.e., tendencies to draw incorrect conclusions in specific circumstances based on cognitive factors. This is an important concern in that judgements are routinely made in trade-off studies and risk analyses. Khalil Walker states that the difference between a heuristic and a bias is normative judgment. Heuristics are helpful biases, whereas biases are hurtful heuristics. In this regard, there are three heuristics that can have a negative impact on decision-making in general and trade-off studies in particular:

Availability Heuristic: This is a mental shortcut that helps us make a decision based on how easy it is to bring something to mind. In other words, humans tend to think of examples when making a decision or judgement. For example, if one were to ask what percentage of crime involved violence in 2001, most would guess high because of the nonstop news coverage of 9/11 yet the Federal Bureau of Investigation (FBI) reported that violent crimes were less than 12 % of all crimes in the USA in 2001. The main problem with the availability heuristic is that humans tend to assume that if there are several readily available examples, then the event or topic is commonplace. This is not the case in many circumstances.

Representativeness Heuristics: This is a mental shortcut that helps us make a decision by comparing available information to our mental stereotypes. For example, motor cyclists are troublemakers, skateboarders are slackers, and politicians are corrupt. And, of course, we all know how wrong stereotypes can be!

Base Rate Heuristic: This is mental shortcut that helps us make decisions based on probabilities associated with prevailing context. For example, in an unsafe neighborhood, an automobile back firing might be concluded as a gut shot.

In psychology, heuristics are simple, efficient rules, learned or hard-coded by evolutionary processes that have been proposed to explain how people make decisions, arrive at judgements, and solve problems when facing complex problems or incomplete information [88, 89].

Heuristics can vary in power or potency. Table 2.5 presents the characteristics of a powerful heuristic.

Heuristics have been used successfully in architecting, engineering, design, maintenance, and test [86, 87]. Heuristics are the basis for shortcuts, sanity checks, guidance and hints, alternatives processing, and lessons learned. Heuristics cannot be tested in the traditional sense because they are nonquantitative. Therefore, they need to be applied in particular contexts and their efficacy evaluated statistically. Ideally, heuristics should have prescriptive power. Some heuristics written as declarative statements can be rephrased in prescriptive terms.

Table 2.5 Characteristics of a powerful heuristic

• The opposite of a heuristic does not make sense or leads to failure [86]
• Coverage regime with no/few exceptions
• Prescriptive, implies/suggests appropriate action
• Actionable, can be implemented in context

Table 2.6 Defining heuristics (adapted from [92])

• Heuristics do not guarantee a solution
• Two heuristics may contradict/give different answers to the same question and still be useful
• Heuristics permit the solving of unsolvable problems and reducing search time to a satisficing solution
• A heuristic depends on the immediate context instead of absolute truth as a validity standard

For example, consider the popular refrain: "Keep It Simple, Silly (KISS)." This heuristic can be rewritten as a prescriptive heuristic:

> The simplest solution is usually the correct one (Occam's razor) [90].

Similarly, consider the cautionary (descriptive) heuristic:

> Don't assume that the original problem statement is necessarily the best, or even right

This heuristic can be rewritten in prescriptive form as:

> Probe the original problem statement in terms of objectives, assumptions and constraints; doing so, may result in reformulation/revision of the original problem statement

Heuristics have become an essential aspect of modeling [91]. Heuristics are easy to define using the characteristics offered by Koen [92] (Table 2.6).

A heuristic is anything that provides a plausible aid or direction in the solution of a problem but, in the final analysis, does not guarantee a solution. Its real value is in its ability to guide, discern, and reveal [89, 93].

Heuristics are a key part of Koen's definition of the engineering method [92]:

> The engineering method is the use of heuristics to cause the best change in a poorly understood situation within the available resources

Typical engineering heuristics include:

- Rules of thumb and orders of magnitude
- Safety factors
- Heuristics that determine the engineer's attitude toward his or her work
- Heuristics that engineers employ to keep risk within acceptable bounds
- Rules of thumbs that are the key to resource allocation

Heuristics that pertain to human decision-making and trade-off analysis [86, 94] are

- It is easier to match the system to the human rather than the other way round
- The probability of implementing new ideas depends on the number of people in the chain leading to their implementation and on the probability that each person understands and retransmits that idea
- If social cooperation is required, the way in which a system is implemented and introduced must be an integral part of its architecture
- The choice between architectures may well depend upon which set of drawbacks the client can handle best
- Build in and maintain options as long as possible in the design and implementation of complex systems. You will need them

- Sometimes, but not always, the way to solve a difficult problem is to expand it

More recently, with the emphasis on adaptive systems, newer heuristics have become important:

- Systems need to be architected for change and extension beyond currently recognized needs
- It is likely that there will be more stakeholders and more needs per stakeholder at the end rather than at the start
- Not all stakeholders can be satisfied equally; thus, trade-off analysis is needed to engage in reasoned compromise to achieve a balance among competing objectives/quality attributes
- Align incentives with desired outcomes to ensure quality and performance don't suffer
- Introduce flexibility in architecture, design, and program to cope with uncertainty and risk.

In the light of the foregoing, Madni [93, 95] compiled a set of prescriptive and actionable heuristics pertaining to decision-making and trade-off analysis for systems architecting and engineering (Table 2.7). These heuristics are classified as modeling, problem solving, system development, and stakeholder management.

Each heuristic in the four categories of Table 2.7 are discussed in the following paragraphs.

1. A model should be only as complicated as needed to answer questions specific to a particular engineering phase in the system life cycle. It is important to realize that no model is complete. The question is whether or not the model development is useful in answering the questions at hand. If the answer is "yes," then the model is deemed useful. What if the model can answer some of the questions but not all? In this case, the model can be refined, elaborated, and disaggregated along dimensions that help answer the remaining questions.
2. Scale up techniques to address the problem; resist at all cost the tendency to dumb down the problem to fit known techniques. Continuing to simplify a problem until it begins to look familiar or resemble something previously known is a common habit that people inadvertently fall into or consciously make to make the problem amenable to being solved using known techniques. This tendency toward oversimplification invariably results in solving a different wrong problem, than the one originally intended. Instead, the problem solver needs to explore other disciplines (e.g., biology, sociology, economics) and new metaphors and paradigms that can help frame the problem in a new light and thereby pave the way for solving the problem of the right scope and at the right scale and complexity.
3. A model needs to be interpreted within the context of use, and not taken as an absolute. All models make assumptions in the absence of required information, or as a means to reduce complexity. When needed information becomes available, the assumptions need to be revisited and revised, and the model

Table 2.7 Prescriptive heuristics [95, 96]

Modeling
1 A model should be only as complicated as needed to answer questions associated with a particular engineering phase or stage in the system life cycle
2 A model needs to be interpreted within its context, and not taken as an absolute
3 It is best not to employ a model for a purpose other than originally intended unless prior provision is made to repurpose the model. Even then, repurposing a model will require adding details (to abstractions), new perspectives, and new visualizations formats
4 Understand models to be an aid, not a replacement for common sense; act accordingly. Model to increase understanding; analyze to increase insight
Problem solving
1 Scale up techniques to address the problem; resist dumbing down problems to fit known techniques
2 Exploit query reformulation (about a problem) to elicit a more complete set of requirements
3 Probe for high level objectives; don't confuse objectives with options
4 Keep the option space open as late as possible to assure ability to select the best option
5 Cross-check results from heuristics with actual testing against high fidelity models
6 Exploit apt analogies from other disciplines such as biology to increase understanding and develop effective solutions, e.g., an ant colony or the human immune system
7 Ensure that the trade space spans the problem domain before undertaking trade space analysis
8 Periodically revisit assumptions and constraints in light of new evidence/data; revise as needed
9 Supplement heuristics with "patterns" and "cases" as appropriate
System development
1 Build in flexibility in the product and process at specific points based on cost-benefit and risk-benefit analysis to ensure the ability to adapt to new circumstances and exploit new opportunities
2 Architect systems to accommodate change and support future extensions that go beyond the current needs of the stakeholders
3 Separation of concerns is a way to manage complexity; it is about striking an effective balance between parts of a system problem/solution while keeping the whole in mind
Stakeholder management
1 (Re)balance people and process perspectives with increasing system complexity
2 Perform process decomposition to a level where the tasks are directly assignable to people, or where automation handles all lower levels
3 Exploit storytelling especially in upfront engineering to ensure participation of all stakeholders
4 Iterate with stakeholders during the requirements engineering phase, because the first pass will not yield a complete set of stakeholders or their needs
5 Incentives and inhibitors shape behaviors; reward value delivered, not milestones met
6 Focus on the decision-making behavior of stakeholders, not just their roles and skillsets
7 Aid human decision-making by recognizing that individuals will often make decisions that satisfice, rather than optimize. This results, in part, because humans lack the necessary bandwidth to assimilate all relevant information and weigh alternatives
8 Groups tend to make bolder decisions than individuals acting alone, but group norms and pressures produce conformity in decision-making in crucial situations

modified accordingly in light of the new context defined by the new assumptions and models.

4. It is best not to employ a model for a purpose other than what was originally intended. However, under special circumstances, if there is a compelling need to do so, then it should be recognized that the model might require elaboration, abstraction, aggregation, or disaggregation before being put to use. There are two examples that stand out in this regard. First, the attempt to use a training simulator network SIMNET [97] as an engineering tool. It was quickly discovered that the approximate models (based on the principle of selective fidelity) that were used for training were inadequate for engineering and had to be redone in terms of computational accuracy. Second, the attempt to use MYCIN (an expert system for medical diagnosis) as a medical teaching system [98]. It was quickly discovered that the models used in MYCIN (an aiding expert system) were not sufficiently detailed for training purposes.
5. Employ problem-related query reformulation to elicit a more complete set of requirements. It has been widely observed that query reformulation about a problem statement can illuminate new perspectives which, in turn, can facilitate the identification of a more complete set of requirements. Restating the problem can also enhance stakeholder participation, because stakeholders may relate to the reformulated problem statement more than the original problem statement.
6. Probe for higher level objectives to ensure that lower level objectives or options have not been inadvertently accepted as objectives. It is important to realize that people often confuse options with objectives. Probing for higher level objectives will avoid this problem.
7. Keep the option space open as late as possible to assure selection of the best possible option when additional information is available. To achieve this objective requires building in sufficient flexibility and not prematurely pruning options before needed information becomes available and uncertainty is reduced.
8. Build in flexibility in the product and process at appropriate points to combat the uncertainty that might exist at that point. This capability is essential to adapt to new circumstances and exploit new opportunities. Real options are an engineering analog of financial options. They provide the ability to introduce flexibility in a cost-effective fashion. Financial options offer the right but not the obligation to exercise an option at a particular price by a particular date. Real options provide targeted flexibility to systems architects and engineers to exploit an emergent opportunity, or cope with change in the face of uncertainty.
9. Models are an aid, not a replacement for common sense; so, act accordingly. It is important to remember that models can be fallible. When they produce results that "fly in the face" of common sense, they should be examined closely. Two possibilities exist here. Either the models are simply wrong, or some unexpected, counter-intuitive phenomenon or behavior has been discovered. Therefore, it is important to ascertain which of the two options hold, and act accordingly.

10. Cross-check results from heuristics with actual testing against high fidelity models and real-world operational tests data. While heuristics provide a cost-effective, shortcut in general, they are fallible. In other words, they don't guarantee a solution, much less an optimal one. Therefore, they need to be validated against higher fidelity, models, and/or real-world testing.
11. Model with the intent of increasing understanding, and analyze (the model) with the intent of increasing insights. The very act of modeling systems increases understanding of the system structure, components, and their relationships. The analysis of the model from different perspectives and over time illuminates system behavior and uncovers dynamic interactions between the system and the environment.
12. Exploit apt analogies from biology, economics, sociology, and other disciplines to increase understanding of the system and develop effective solutions. The ant colony, human immune system, swarming behavior, and real options are all being exploited in multi-UAV missions, cybersecurity, social sciences, and design flexibility. Several analogies of this ilk can be exploited to not only enhance understanding but also discover effective solutions.
13. Ensure that the trade space is sufficiently representative of the problem domain before undertaking trade space analysis. For trade-off analyses to be of value, the trade space should span the problem domain. Various methods exist to expand the trade space, including constraint relaxation, query reformulation, role-playing, question answering, and objective probing.
14. Periodically revisit assumptions, constraints, and beliefs in light of new evidence/data. This practice will allow revision of beliefs and models, and development of new insights. With complex systems, this is even more important.
15. You need to balance people and process emphasis with increasing system complexity. With increasing system complexity, a process description needs to be replaced by human judgment and trade-off analyses. This finding is becoming increasingly more apparent in software development where process-driven development is being replaced by people-driven, process-enabled development.
16. Perform process decomposition to a level where tasks are directly assignable to people or automation. This heuristic is the key to designing systems with different levels of automation. The "done" condition occurs when the tasks at the leaf nodes are directly performed by humans or handed off to automation.
17. Supplement heuristics with "patterns" and "cases" as appropriate to simplify complex problem solving. Patterns enable the transformation of problem-solving activities into pattern-matching activities. Similarly, cases simplify reasoning by providing historical precedents to draw on. The combination of patterns and cases can dramatically enhance systems architecting and engineering when used in conjunction with heuristics.
18. Exploit storytelling to ensure participation of all stakeholders especially in upfront requirements engineering. Storytelling is also compelling in terms of its persuasive powers. In particular, interactive storytelling (for example, in virtual

worlds) can provide insights to all stakeholders by allowing them to explore alternate futures with different assumptions, constraints, and technologies.

19. Architect systems to accommodate anticipated change and support potential extensions that go beyond current stakeholder needs. This is especially necessary for DOD systems that are typically long-lived. Part of this requirement is addressed by Pre-planned Product Improvement (P3I), but part of it needs to be handled "on the fly" when the system encounters unexpected operational environments.
20. Iterate with stakeholders during requirements engineering phase because it is highly unlikely that a complete set of stakeholders or their needs will be identified on the first pass. Therefore, it will invariably be the case that system developers will have to iterate with the stakeholders a few times as needed to define a stable set of requirements. Complicating matters is the fact that systems engineers employ engineering notation to capture system models. This practice leaves out several stakeholders (mostly nonengineers) unfamiliar with engineering notation. To overcome this deficiency several new approaches rooted in interactive technical storytelling in virtual worlds can be employed.
21. Since incentives and inhibitors shape behaviors, reward value delivered, not milestones met. The behavior of developers and program managers, in large part, is shaped by the incentives employed. Tying the reward structure to milestones met or tasks completed says little about the quality of the product being developed. Instead, the reward structure should be tied to the value delivered through the product or service. This strategy ensures that the project team is not being rewarded for going through the motions of satisfying milestones without regard to the quality of the product/service being created.
22. Separation of concerns is a way to manage complexity; it is about striking an effective balance between attending to parts of a system (problem/solution) while keeping the whole in mind. This is a form of the well-known "divide and conquer" strategy in which by looking at the system from separate viewpoints (concerns) allows the stakeholders to get their arms around a complex system. Addressing each concern individually enhances comprehension, communication, and issue resolution.
23. Focus on the decision-making behavior of developers in addition to their roles and skillsets. Traditionally, the focus has been on the roles and skillsets of the development team members. This focus is necessary, but not sufficient. Focus should also be on the information seeking and decision-making behavior of the development team members.
24. Aid human decision-making by recognizing that individuals will often make decisions that satisfice, rather than optimize. This behavior results in part from the lack of cognitive bandwidth to assimilate all relevant information and weigh alternatives. Humans can conduct limited search and are poor at aggregating information. This circumstance calls for decision aiding/automation to overcome human cognitive limitations.
25. Resist "group think" but employ teams whenever possible to address different aspects of a problem. Groups tend to make bolder decisions than individuals

acting alone, but group norms and pressures produce conformity in decision-making in crucial situations. Thus, it is also important to encourage divergent thinking.

2.4.1 Homework for Section 19

2.4.1.1 Powerful Heuristics

Table 2.5 states that for a powerful heuristic the opposite of the heuristic does not make sense or leads to failure. The following are some heuristic from Rechtin and Maier [86, 87]. Rewrite them so that they say the opposite. Do they make sense? Which do you follow?

"In architecting a new software program, all the serious mistakes are made in the first day."
"The most dangerous assumptions are the unstated ones."
"Do the hard parts first."
"If you can't explain it in five minutes, either you don't understand it or it doesn't work."
In a medical "triage: let the dying die. Ignore those who will recover on their own. And treat only those who would die without help."
It is easier to match the support system to the human it supports than the reverse.
The number of defects remaining in a software system after a given level of test or review is proportional to the number found during that test or review.
"High quality, reliable systems are produced by high quality architecture, engineering, design and manufacture, not by inspection, test and rework."
"Before ordering a test decide what you will do if it is (1) positive or if (2) it is negative. If both answers are the same, then don't do the test."

2.4.1.2 How Do Heuristics Help

Give an example of how heuristics can allow one to attack a complex problem successfully.

Dumb Things

What kinds of biases can cause competent people to do "dumb" things?

2.4.1.3 Trade Space Exploration

How do heuristics help with trade space exploration?

2.4.1.4 Domain-Specific Heuristic

Give an example of a domain-specific heuristic and explain how it helps with problem solving.

2.4.1.5 Upsides and Downsides

What are the upsides and downsides of using heuristics?

2.4.1.6 Expert Knowledge

In what ways can heuristics codify expert knowledge?

2.5 Summary

This chapter covered common mental mistakes that people make. Mistakes that can adversely affect trade-off studies and risk analysis. We began by providing instances of erroneous knowledge that people carry in their heads. We then reviewed prospect theory, a Nobel Laureate's collection of erroneous human thinking. A dozen examples of mental mistakes that people tend to make when conducting trade-off studies and risk analysis were presented. We then addressed heuristics and cognitive biases that come into play when conducting trade-off studies.

Imagine that you find yourself in a position to help a decision maker make informed trade-offs and defensible decisions. We know that when viewed in formal terms these are difficult, iterative tasks. They entail discovering the decision maker's importance weights, scoring functions, and preferred combining functions. To accomplish this discovery, you need to get into the head of the decision maker to discover his or her preferences and values. As importantly, different people have different personality types. The Myers-Briggs model is one way of describing these personality types. *Sensory-Thinking-Judging* people are likely to appreciate the decision-making concepts that we have presented. *Intuitive-Feeling* people might not appreciate them. Nonetheless, having an understanding of the foregoing concepts can potentially help you in understanding the values and preferences of the decision maker.

In Section 16, we discussed human foibles that can adversely impact trade-off analysis. Specifically, we discussed the salient differences between rational and irrational human behavior. We defined a rational decision as one that requires a goal, perfect information, and an optimal course of action that maximizes the expected value of the decision. We noted that this is the *prescriptive* model of decision-making. In particular, we defined an optimal course of action as one that, in the long run, is more valuable (or profitable) than any other available course of action. We then provided several examples of rational decision-making. Thereafter, we discussed how people actually make decisions, the *descriptive* model. We introduced the concept of *satisficing* attributed to Nobel Laureate, Herb Simon. We then gave several examples of humans not being rational decision makers, and making decisions and reaching conclusions not supported by facts. We discussed cognitive illusions and how they can impair judgment and adversely impact decision-making. In keeping with our overall philosophy of the structure of each section, we provided homework problems to reinforce the concepts in Section 16. In Parts V and VI, we will show how trade-off studies and risk analyses can help move an individual toward making decisions that are more rational.

In Section 17, we covered prospect theory developed by Tversky and Kahneman. We noted that this theory explained the subjective nature of decision-making in terms of heuristics and biases that are ever-present in human assessment of information, and the commonly observed deviations from rational decision-making that results as a consequence. We defined key terms such as Expected Value Model of human decision-making, and clarified the distinction between Value and Utility, and between Objective Probability and Subjective Probability. We presented Edward's four-quadrant representation of the four variations on the Expected Value Model. We then applied prospect theory to the Expected Value model to realize a descriptive model of human decision-making.

After reviewing the theoretical foundations of decision-making and human fallibility in estimating probabilities, we drew specific implications. We cautioned the reader about the existence of cognitive biases even in extensively vetted analysis, a contention of prospect theory. We discussed how biases and illusions can influence trade-off studies and risk analysis, and cause them to produce irrational decisions.

We then focused on *risk* and noted that systems engineers and decision theorists use the word differently. To avoid confusion, we chose to focus on terms such as ambiguity, uncertainty, and hazards that have common interpretations in both the systems engineering and decision sciences communities. We discussed the existence of human filters that are responsible for biases. We noted that confirmation bias is the most prevalent and important in that humans hear what they want to hear and tune out the rest. Said differently, humans filter information that contradicts their preconceived notions (biases), and remember things that reinforce their beliefs.

We discussed the concept of *severity amplifiers* and gave specific examples, including lack of control, lack of choice, lack of trust, ambiguity, uncertainty, and immediacy. We concluded this section by stressing the significance of the Myer–

Briggs model as an effective instrument to describe personality types and explain how personality types influence decision-making. In addition to human foibles described by prospect theory, we noted that there are several other factors that affect human decisions. These include corporate culture, information displayed, wording of questions, context, the decision itself, the effort required to make the decision, the difficulty in making the decision, the time available to make the decision, needed decision accuracy, cost of the decision, likelihood of regret and, finally, the decision maker's values, personality types, risk averseness, biases, illusions, and use of heuristics.

In Section 18, we covered potential mental mistakes made in trade-off studies. We cautioned that humans are susceptible to mental mistakes, and identified several mental mistakes that people make when conducting trade-off analyses. We reviewed prospect theory, which breaks down subjective decision-making into a preliminary screening stage and a secondary evaluation stage. We defined the seven key components of a trade-off study and identified issues in each component that can potentially compromise the results of a trade-off study. We illuminated some of the key points using the Pinewood Derby Trade-off Study as a real-world example.

We noted that humans tend to consider alternatives sequentially during decision-making. They often hastily choose one alternative after having their attention anchored in just one or a few criteria. We cautioned that to make good, rational choices among alternatives, a decision maker needs to become aware of cognitive biases. Limited awareness and possible consequential irrationality can cause erroneous estimation or misuse of trade-off study components. Therefore, trade-off study components need to be examined individually. The higher level decision then becomes a calculation that rests on a broad base of rationally considered assumptions, components, and calculation methods. We recommended that decision makers should understand the mathematical methods that allow the parallelization of human decision processes through trade-off studies. We followed up with a set of homework problems that reinforced the key points made in Section 18.

In Section 19, we discussed the body of knowledge associated with heuristics and explained how heuristics can potentially provide effective shortcuts to complex problems. In particular, we identified three heuristics that could negatively impact trade-off studies: the availability heuristic, the representativeness heuristic, and the base rate heuristic. We defined the characteristics of a powerful heuristic and showed how a descriptive heuristic can be transformed into a prescriptive heuristic that is actionable. We provided examples of this transformation. We followed that up with specific criteria for defining heuristics, typical engineering heuristics and heuristics that pertain to human decision-making and trade-off analyses. We then focused on heuristics that specifically pertain to adaptive systems. Thereafter, we presented Madni's prescriptive heuristics that span modeling, problem solving, system development, and stakeholder management. Finally, we described each heuristic under these four categories in greater detail, and followed up the discussion with homework problems to reinforce these concepts.

References

1. Madni AM, Sage AP, Madni CC (2005) Infusion of cognitive engineering into systems engineering processes and practices. In: 2005 I.E. international conference on systems, man and cybernetics: Institute of Electrical & Electronics Engineers (IEEE)
2. Madni AM, Yee-yeen C (1988) A design support system for crew system automation. In: Proceedings of the 1988 I.E. international conference on systems, man, and cybernetics: Institute of Electrical & Electronics Engineers (IEEE)
3. Von Neumann J, Morgenstern O (2007) Theory of games and economic behavior. Princeton University Press, Princeton
4. Simon HA (1955) A behavioral model of rational choice. Q J Econ 69(1):99
5. Baldwin D, Bahill AT (2008) Mechanics of baseball pitching and batting. Applied biomedical engineering mechanics. Informa UK Limited. pp 445–488
6. Piattelli-Palmarini M (1994) Inevitable illusions: how mistakes of reason rule our minds. Wiley, New York
7. Madni AM (2011) Towards a generalizable aiding-training continuum for human performance enhancement. Syst Eng 14(2):129–140
8. Descartes R. Descartes' meditations, 1641 [cited 2016 January]. http://www.wright.edu/~charles.taylor/descartes/
9. Locke J. An essay concerning human understanding, Book II: Ideas 1690. [cited 2016 January]. http://www.earlymoderntexts.com/assets/pdfs/locke1690book2.pdf
10. Nobel (2002) The Bank of Sweden Prize in Economic Sciences in memory of Alfred Nobel 2002 Stockholm, Sweden: Nobel Foundation. [March 2006]. http://nobelprize.org/economics/laureates/2002/index.html
11. Tversky A, Kahneman D (1992) Advances in prospect theory: cumulative representation of uncertainty. J Risk Uncertain 5(4):297–323
12. Pascal B (1654) Fermat and Pascal on probability. www.york.ac.uk/depts/maths/histstat/pascal.pdf. Accessed Jan 2016
13. Locke J (2004) An essay concerning human understanding. In: Fraser AC (ed). Barnes & Nobel 1689, New York
14. Smith ED (2006) Tradeoff studies and cognitive biases. Ph.D. dissertation, University of Arizona, Tucson, AZ
15. Edwards W (1955) An attempt to predict gambling decisions. In: Dunlap JW (ed) Mathematical models of human behavior. Dunlap and Associates, Stamford, pp 12–32
16. Kahneman D, Tversky A (1979) Prospect theory: an analysis of decision under risk. Econometrica 47(2):263
17. Colbert S (2005) The Colbert Report: comedy central home video
18. Kagel JH, Battalio RC, Green L (1995) Economic choice theory: an experimental analysis of animal behavior. Cambridge University Press, New York
19. Harris AJ, Corner A, Hahn U (2009) Estimating the probability of negative events. Cognition 110(1):51–64
20. Madni AM (2009) Integrating humans with software and systems: technical challenges and a research agenda. Syst Eng 13(3):232–245
21. Shafir S, Waite T, Smith B (2002) Context-dependent violations of rational choice in honey-bees (Apis mellifera) and gray jays (Perisoreus canadensis). Behav Ecol Sociobiol 51 (2):180–187
22. Savage LJ (1954) The foundations of statistics. Wiley, New York
23. Madni A, Madni C, Garcia S (2005) Cognitive model-enabled simulation-based training of aerospace operations center operators. Infotech@Aerospace. American Institute of Aeronautics and Astronautics (AIAA)
24. Kahneman D, Tversky A (1979) Prospect theory: an analysis of decision under risk. Econometrica 46(2):171–185

25. Battalio RC, Kagel JH, MacDonald DN (1985) Animals' choices over uncertain outcomes: some initial experimental results. Am Econ Rev 75(4):597–613
26. Real LA (1981) Uncertainty and pollinator-plant interactions: the foraging behavior of bees and wasps on artificial flowers. Ecology 62:20–26
27. Hsu M, Bhatt M, Adolphs R, Tranel D, Camerer CP (2005) Neural networks responding to degrees of uncertainty in human decision-making. Science 310:1680–1683
28. Nickerson RS (1998) Confirmation bias: a ubiquitous phenomenon in many guises. Rev Gen Psychol 2(2):175–220
29. Madni AM, Samet MG, Freedy A (1982) A trainable on-line model of the human operator in information acquisition tasks. IEEE Trans Syst Man Cybern 12(4):504–511
30. Madni A, Moini A, Madni C (2010) Cross-cultural decision making training using behavioral game-theoretic framework. Advances in cross-cultural decision making. Informa UK Limited. pp 385–394
31. Orellana D, Madni A (2012) Extending model based systems engineering for human machine interaction analysis and fault tolerant design. Infotech@Aerospace 2012. American Institute of Aeronautics and Astronautics (AIAA)
32. Thaler RH (2015) Misbehaving: the making of behavioral economics. WW Norton & Company, New York
33. Smith ED, Son YJ, Piattelli-Palmarini M, Bahill AT (2007) Ameliorating mental mistakes in tradeoff studies. Syst Eng 10(3):222–240
34. Smith ED, Piattelli-Palmarini M, Bahill AT (2008) Cognitive biases affect the acceptance of tradeoff studies Decision modeling and behavior in complex and uncertain environments. Springer, New York, pp 227–249
35. Beach LR, Connolly T (2005) The psychology of decision making: people in organizations, 2nd edn. Sage, Thousand Oaks
36. Hammond JS, Keeney RL, Raiffa H (2006) The hidden traps in decision making. Harv Bus Rev 84(1):118
37. Shafir E (ed) (2004) Preferences, belief and similarity: selected writings by Amos Tversky. The MIT Press, Cambridge
38. Kahneman D, Slovic P, Tversky A (eds) (1982) Judgment under uncertainty: heuristics and biases. Cambridge University Press, Cambridge
39. Tversky A, Kahneman D (1974) Judgment under uncertainty: heuristics and biases. Science 185(4157):1124–1131
40. Tversky A, Kahneman D (eds) (2000) Choices, values and frames. Cambridge University Press, New York
41. RSAS (2002) Press release: Nobel Prize in Economics 2002: The Royal Swedish Academy of Sciences. [cited 2006 September]. http://nobelprize.org/nobel_prizes/economics/laureates/2002/press.html
42. Kahneman D (2003) A perspective on judgment and choice: mapping bounded rationality. Am Psychol 58(9):697–720
43. Smith ED, Bahill AT (2010) Attribute substitution in systems engineering. Syst Eng 13 (2):130–148
44. Strack F, Martin LL, Schwarz N (1988) Priming and communication: the social determinants of information use in judgements of life-satisfaction. Eur J Soc Psychol 18:429–442
45. Kahneman D, Frederick S (2002) Representativeness revisited: attribute substitution in intuitive judgment. In: Gilovich T, Griffin D, Kahneman D (eds) Heuristics and biases. Cambridge University Press, New York, pp 49–81
46. Wymore AW (1993) Model-based systems engineering. CRC, Boca Raton
47. Kahneman D, Ritov I (1994) Determinants of stated willingness to pay for public goods: a study in the headline method. J Risk Uncertain 9:5–38
48. Shogan R (1998) The fate of the union: America's Rocky Road to Political Stalemate. Basic Books, New York

49. Park CW, Mothersbaugh DL, Feick L (1994) Consumer knowledge assessment. J Consum Res 21:71–82
50. Arkes HR (1981) Impediments to accurate clinical judgment and possible ways to minimize their impact. J Consult Soc Psychol 49:323–330
51. Costandi M (2013) Evidence-based justice: corrupted memory. Nature 500:268–270
52. Clifasefi SL, Bernstein DM, Mantonakis A, Loftus EF (2013) "Queasy does it": false alcohol beliefs and memories may lead to diminished alcohol preferences. Acta Psychol (Amst) 143 (1):14–19
53. Loftus EF (1996) Eyewitness testimony. Harvard University Press, Cambridge, MA
54. Forer BR (1949) The fallacy of personal validation: a classroom demonstration of gullibility. J Abnorm Psychol 44:118–121
55. Milgram S (1969) Obedience to authority. Harper & Row, New York
56. Burger JM (2009) Replicating milgram: would people still obey today? Am Psychol 64(1):1
57. Tversky A, Sattath S, Slovic P (1988) Contingent weighting in judgment and choice. Psychol Rev 95(3):371–384
58. Bahill AT, Karnavas WJ (2000) Risk analysis of a pinewood derby: a case study. Syst Eng 3 (3):143–155
59. Petroski H (2003) Framing hypotheses: a cautionary tale. Am Sci (Jan–Feb):18–22
60. Wickelgren WA (1974) How to solve problems: elements of a theory of problems and problem solving. Freeman, San Francisco
61. Slovic P, Fischhoff B (1977) On the psychology of experimental surprises. J Exp Psychol Hum Percept Perform 3:544–551
62. Koriat RM, Lichtenstein S, Fischhoff B (1980) Reasons for confidence. J Exp Psychol Hum Learn Mem 6:107–118
63. Hsee CK, Loewenstein G, Blount S, Bazerman M (1999) Preference reversals between joint and separate evaluations of options: a theoretical analysis. Psychol Bull 125(5):576–590
64. Shafir E, Tversky A (1992) Thinking through uncertainty: nonconsequential reasoning and choice. Cogn Psychol 24(4):449–474
65. Redelmeier D, Shafir E (1995) Medical decision making in situations that offer multiple alternatives. JAMA 273(4):302–305
66. Schacter DL (1983) Feeling of knowing in episodic memory. J Exp Psychol Learn Mem Cogn 9(1):39–54
67. Griffin D, Tversky A (1992) The weighting of evidence and the determinants of confidence. Cogn Psychol 24:411–435
68. Preston MG, Baratta P (1947) An experimental study of the auction-value of an uncertain outcome. Am J Psychol 61:183–193
69. Karmakar US (1978) Subjectively weighted utility: a descriptive extension of the expected utility model. Organ Behav Hum Perform 21:61–72
70. Harbaugh WT, Krause K, Vesterlund L (2002) Risk attitudes of children and adults: choices over small and large probability gains and losses. Exp Econ 5:53–84
71. Plous S (1993) The psychology of judgment and decision making. McGraw-Hill, New York
72. Chapman WL, Bahill AT, Wymore AW (1992) Engineering modeling and design. CRC, Boca Raton
73. Bohlman J, Bahill AT (2013) Examples of mental mistakes made by systems engineers while creating tradeoff studies. Stud Eng Technol 1(1):22–43
74. Keeney RL, Raiffa H (1976) Decisions with multiple objectives: preferences and value tradeoffs. Wiley, New York
75. Daniels J, Werner PW, Bahill AT (2001) Quantitative methods for tradeoff analyses. Syst Eng 4(3):190–212
76. Bradstreet D (1967) Patterns of success in managing a business. Dun & Bradstreet, New York
77. Cooper AC, Woo CY, Dunkelberg WC (1988) Entrepreneurs' perceived chances for success. J Bus Ventur 3:97–108

78. Miller GA (1956) The magic number seven, plus or minus two: some limits on our capacity for processing information. Psychol Rev 63:81–87
79. Chase WG, Simon HA (1973) Perception in chess. Cogn Psychol 4:55–81
80. Smith ED, Szidarovszky F, Karnavas WJ, Bahill AT (2008) Sensitivity analysis, a powerful system validation technique. Open Cybern Syst J 2:39–56
81. Mayo E (1933) The human problems of an industrial civilization. Macmillan, New York
82. Roethlisberger FJ, Dickson WJ (1939) Management and the worker. Harvard University Press, Cambridge, MA
83. Rosenthal R, Jacobson L (1968) Pygmalion in the classroom. Urban Rev 3(1):16–20
84. Heisenberg W (1927) Über den anschaulichen Inhalt der quantentheoretischen Kinematik und Mechanik. Zeitschrift für Physik 43(3-4):172–198
85. Keirsey D (1998) Please understand me II. Prometheus Nemesis Book Company, Del Mar
86. Rechtin E (1991) Systems architecting: creating & building complex systems. Prentice Hall, Englewood Cliffs
87. Rechtin E, Maier M (2000) The art of systems architecting, 2nd edn. CRC, Boca Raton
88. Madni AM (2014) Expanding stakeholder participation in upfront system engineering through storytelling in virtual worlds. Syst Eng 18(1):16–27
89. Madni AM, Brenner M, Costea I, MacGregor D, Meshkinpour F (1985) Option generation: problems, principles, and computer-based aiding. In: Proceedings of the 1985 international conference on systems, man, and cybernetics, Tuscon, AZ
90. Jefferys WH, Berger JO (1992) Ockham's razor and Bayesian analysis. Am Sci 80:64–72
91. Starfield AM, Smith K, Bleloch AL (1993) How to model it: problem solving for the computer age. McGraw-Hill, New York
92. Koen BV (1985) Definition of the engineering method., ERIC
93. Madni AM (2013) Generating novel options during systems architecting: psychological principles, systems thinking, and computer-based aiding. Syst Eng 17(1):1–9
94. Rechtin E (1992) The art of systems architecting. IEEE Spectr 29(10):66–69
95. Madni AM (2012) Elegant systems design: creative fusion of simplicity and power. Syst Eng 15(3):347–354
96. Madni AM (2012) Game-based simulation for cross-cultural decision making training. Hum Factors Ergon Manuf Serv Ind 23(2):85–94
97. Thorpe J (1995) Trends in modeling, simulation, & gaming: personal observations about the past thirty years and speculation about the next ten. Proc IEEE 1995:83(8)
98. Buchanan BG, Shortliffe EH (1984) Rule-based expert systems. Addison-Wesley, Reading

Chapter 3
Common Methods and Processes

3.1 Section 20. Evaluation Criteria

Three activities are fundamental in deriving evaluation criteria: stating the problem, investigating alternative concepts, and validation.

3.1.1 *The Problem Statement*

The problem statement starts with a description of the purpose of the system, the top-level functions that the system is intended to perform, or a deficiency that needs to be ameliorated. The system's requirements should be traceable to this problem statement [1]. The problem statement should be in terms of *what* must be done, not *how* to do it. It may be stated in words and/or in the form of a model. Inputs are supplied by stakeholders consisting of end users, operators, maintainers, suppliers, acquirers, owners, bill payers, regulatory agencies, victims, sponsors, manufacturers, and others. Stating the problem correctly is one of the engineer's most important tasks. An elegant solution to an incorrectly formulated problem or the wrong problem is worthless. Problem framing (defined in Section 2.2.8), using the right problem statement, is more important than problem solving. A problem statement describes the customer's needs, states the goals of the project, delineates the scope of the problem, reports the concept of operations, describes the stakeholders, lists the deliverables, and presents key decisions that need to be made.

The problem statement shows that you understand the problem and its context [2, 3]. This problem statement document is rewritten periodically as more information becomes available.

A.T. Bahill, A.M. Madni, *Tradeoff Decisions in System Design*,
DOI 10.1007/978-3-319-43712-5_3

3.1.2 Concept Exploration

Concept exploration helps refine the problem statement and define the evaluation criteria. Concept Exploration is used to study alternative system designs through brainstorming, modeling, simulation, and virtual and physical prototyping [2, 3]. In early iterations, this exploration helps formulate the incipient system architecture and produce evaluation criteria. Evaluation criteria are used in later design iterations, in which trade-off studies are conducted leading to the preferred alternatives being suggested by the available data.

3.1.3 Validation

Validation means proving that the object of study conforms to law, logic, and facts, and that the object is sound, effective, and cogent. Its sister term verification means proving through testing that the object of study is correct. Validating a system means building the *right system*, making sure that the system does what it is intended to do. Verifying a system means building the *system right*, ensuring that the system complies with its requirements and conforms to its design specification. Another key point is that validation ensures that the system will satisfy the actual needs of the customer. Validation starts at the beginning of system design and continues throughout the design process. System validation artifacts can be collected at the following discrete events: trade-off studies, phase reviews, life cycle reviews, red team reviews, system requirements review (SRR), preliminary design review (PDR), critical design review (CDR), and field test. System validation artifacts that can also be collected continuously throughout the life cycle include results of modeling, simulation, and a count of the number of operational scenarios (use cases) modeled. Detectable system validation defects include mismatches between the model/simulation and the real system. A sensitivity analysis can reveal validation errors. If a system is particularly sensitive to parameters over which the customer has no control, then it may be the wrong system for that customer. If sensitivity analysis reveals the most important parameter and that is a surprise, then it may be the wrong system. If a system is more sensitive to its parameters than to its inputs, then it may be the wrong system or the wrong operating point. If the sensitivities of the model are different from the sensitivities of the physical system, then it may be the wrong model. Validation defects can be detected at inspections: the role of Tester should include an additional responsibility, validation. Testers should read the vision/mission, problem statement, and the concept of operations, and specifically look for such system validation problems and proofs.

3.1.4 Designing Evaluation Criteria

When evaluating systems and investigating trade-offs, the analyst and the customer must jointly select measures that encompass the customers' preferences and values regarding system designs with respect to the problem at hand. Simple, objective, and quantitative techniques are needed to analyze alternative system designs and compare attributes such as performance, cost, schedule, and risk. Selecting a technique for conducting these analyses is typically handled with approaches such as multiattribute utility theory. The general tasks involved in multiattribute utility measurement are structuring objectives, choosing evaluation criteria, eliciting single-attribute scoring functions, eliciting weights, and selecting combining functions. This section deals with creating evaluation criteria.

When designing systems, the analyst and the customer must identify the customers' preferences and values. With these metrics, one can infer the overall desired quality or performance of the system as judged by the customer. As noted earlier in this section, we call such system measures or properties evaluation criteria. Evaluation criteria are specific items that need to be quantified to determine how well the system under study satisfies the design requirements from the stakeholders' point of view. Consider the simple example of evaluating personal computer systems. When comparing computer systems, one may use evaluation criteria such as cost, processor speed, amount of RAM, and hard disk size to determine how well each computer system meets the requirements associated with these metrics. Evaluation criteria are most commonly derived for trade-off requirements, or system requirements that exhibit compensation and can be traded off. Occasionally one is faced with the situation where the customer states mandatory requirements in such a way that they end up being in conflict with each other. In these unusual instances, it may be necessary to conduct a trade-off among mandatory requirements.

We need to make a distinction between *mandatory requirements* and *trade-off requirements*. Mandatory requirements state specific conditions that must hold in order for the system to be considered acceptable. In general, mandatory requirements are stated as *hard* limits and show little or no compensation. In this example, a mandatory requirement could be, `The processor speed shall be above 3 GHz`, Fig. 3.1. Consequently, any system demonstrating a processor speed below 3 GHz constitutes an infeasible alternative in the eyes of the customer. We will see later that mandatory requirements can play an important role when considering evaluation criteria for trade-off requirements.

In summary, evaluation criteria need to be established and maintained. Each criterion must have a weight of importance. Each criterion should link to a trade-off requirement, that is, a requirement whose acceptable value can be more or less depending on quantitative values of other requirements. It is recommended that companies have a repository of generic, hierarchically arranged, evaluation criteria. The top-level criteria might be performance, cost, schedule, and risk. Management should prioritize these four criteria at the beginning of the project and then make sure that everyone is made aware of these priorities.

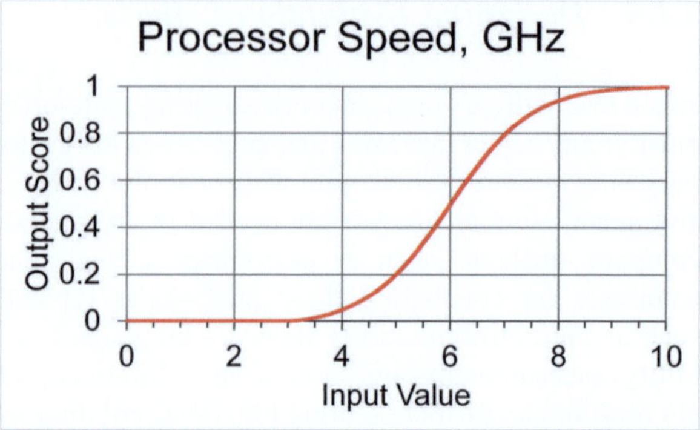

Fig. 3.1 A simple scoring function . Processor speed is the input and the score is the output

3.1.4.1 Hierarchy

Requirements analyses, trade-off studies, and risk analyses are commonly structured as hierarchical decompositions of evaluation criteria. While in principle hierarchies can be arbitrarily deep, Kirkwood [4] advises that smaller hierarchies are desirable, because they can be communicated more easily, and the analysis requires fewer resources. When actually performing the analysis, each evaluation criterion is evaluated using its scoring function and the resulting scores are combined or "rolled up" at each level in the hierarchy to achieve an overall numerical evaluation of the system under study. The overall numerical value is used for comparative evaluation of system designs (Table 3.1).

In designing the hierarchical evaluation structure for a system, evaluation criteria that have some natural relationship with one another are usually grouped [5]. Sometimes while designing evaluation criteria, it is beneficial to first list possible categories of concern to the stakeholder and then derive appropriate evaluation criteria that belong to each category identified. Common top-level categories in the hierarchy are performance, cost, schedule, and risk. Each top-level category is decomposed into finer subcategories as needed to capture the full spectrum of stakeholder criteria.

The hierarchical structuring of evaluation criteria enables contributions from the bottom-level evaluation criteria to the top-level categories [6]. The hierarchical structure also helps in dealing with dependence issues between evaluation criteria. Since evaluation criteria within any given category should be inherently related, independence among them assures that no specific category will contaminate the overall evaluation.

3.1.4.2 Properties of Evaluation Criteria

When selecting evaluation criteria, combining methods and verification processes, it is important to ensure that they satisfy certain properties. Some of the properties mentioned here pertain to the evaluation criteria themselves while some may

Table 3.1 Requirements hierarchy for the Cookie acquisition system

Top-level requirement: satisfy student's craving for cookies	High-level requirements	What	When is it applicable	Function	Detail
	Performance				
		Built-in Self-Test	At start of use case	Perform BiST	
		Heat the oven	Before baking	Raise oven temperature	Raise oven temp from room temp to 350 °F in less than 5 min
		Baking Time (BT)			
			Beginning of baking	Accept desired BT input	Accept times between 1 min and 10 h
			ditto	Display desired BT and remaining BT	BT display shall be visible to a person with 20/20 vision standing 5 ft away with room illuminance between 0 and 1000 lx
			During baking	Update BT display	Update remaining BT display every minute
			End of baking	Indicate visually when BT is over	BT over display shall be visible from 5 ft. away
			ditto	Emit audible signal when BT is over	Intensity level 80 ± 2 dB at 30 cm
					Frequency 440 Hz
		Nutrition	While eating cookies	Four cookies shall contain no more than	520 cal
					24 g of fat
					72 g of carbohydrates
	Cost				
	Schedule				
	Risk				
		Cookie quality	Incorrect ingredients		
			Product is out of date		
			Oven temperature is too high		
			Baking time is too long		
		House danger	Failure to turn off oven		
		Student risk	Student gains weight		
			Student burns his or her hand		

concern the combining method or verification process for the evaluation criterion. However, problems that arise when these properties are violated can usually be ameliorated by reengineering the evaluation criteria.

3.1.4.3 Quantitative

Evaluation criteria should be *quantitative*. In other words, it must be possible to build scoring functions (like Fig. 3.1) for the evaluation criteria. We desire quantifiable metrics so that we are able to measure and collect data that can be fed directly into the scoring function for evaluation. "Soft" or "fuzzy" metrics such as "safety culture" are not directly measurable as stated, and violate the quantitative property requirement of evaluation criteria. Such soft metrics are better candidates as higher level categories containing quantifiable evaluation criteria. Alternatively, the analyst should seek out measurable quantities that convey the required information. Customers usually think in terms of "soft" metrics, so it is our job as engineers to derive appropriate, measurable metrics that can then be used to objectively evaluate the system with respect to the information sought by the customer. For the *safety culture* evaluation criterion stated earlier, one possible measure might be *the number of safety training courses taught per year*.

3.1.4.4 Independent

Evaluation criteria should be *independent*. This implies that information about one evaluation criterion should not lead to inferences about another, which would cause complications, because in combining the data to extract results we would not be obtaining orthogonal or independent combinations. This can result in "double counting" some criteria. This would make it difficult or impossible to ascertain which of the evaluation criteria is in fact contributing to the analysis and which are confounded due to interactions with other evaluation criteria [7].

Here are some examples. Human Gender and Intelligence Quotient (IQ) are independent. However, Human Height and Weight are dependent. To understand the importance of independence, consider one couple buying a new car. For the wife, safety is very important. For the husband, horsepower is most important. Now consider a second couple buying a new car. The wife says that safety is most important. The husband says that the most important considerations are horsepower, peak torque, top speed, time for the standing quarter mile, engine size (in liters), number of cylinders, and time to accelerate 0–60 mph. What kind of a car do you think they are going to buy? One ad hoc method that is used to reduce the effects of dependence is to bundle the husband's dependent evaluation criteria into a higher level category called *power*. This way if there are any adverse outcomes due to interdependence, they will remain localized in that category and will minimize contamination of subsequent analysis. The baseball homework problem has a heuristic for assessing independence.

3.1.4.5 Transitive

Results of combining evaluation criteria should exhibit *transitivity*. Transitivity means that if A is preferred to B, and B is preferred to C, then A should be preferred to C. Violations of this property show inconsistencies that can produce invalid conclusions. The fundamental axioms of decision analysis state that this is an essential ingredient to sound analysis. The consequences of violating this property can be illustrated by the "money pump" argument of Buede [7]; an impartial party may lure us into offering an infinite amount of money by providing us with a sequence of trades among three alternatives.

I would like to have a rich, intransitive uncle. Assume that I have an Alpha Romero Spider and a BMW Z6. And my Uncle has a Corvette Stingray.

I would love to hear him say, "I prefer your BMW to my Corvette, therefore I will give you $2000 and my Corvette for your BMW."

Next he might say, "I prefer your Alpha Romero to my BMW, therefore I will give you $2000 and my BMW for your Alpha Romero."

And finally, I would wait with baited breath for him to say, "I prefer your Corvette to my Alpha Romero, therefore I will give you $2000 and my Alpha Romero for your Corvette."

We would now have our original cars, but I would be $6000 richer. I would call him Uncle Money Pump.

3.1.4.6 Objective

Evaluation criteria should be *objective*, or observer independent. Data that depends heavily on "engineering judgment" tends to be susceptible to bias and error. Experimental design, simulation, or similar techniques should be employed to generate data whenever possible to avoid such difficulties. For example, one should avoid evaluation criteria such as "prettiness" or "niceness" for selecting crewmembers in a mission to Mars. In sports, most valuable player (MVP) selections often tend to be controversial. Deriving a consensus for the best football player of the century would be impossible, because it violates the objectivity property of good evaluation criterion.

3.1.4.7 Correctly Worded

Evaluation criteria should be designed and worded such that a higher output score implies a more desirable outcome. This design aids in interpreting results, that is, when comparing two alternatives, the alternative with the higher score is preferred to the other. If scoring functions are used, then the shape of the scoring function determines how scores are allocated; this feature allows more flexibility in the statement of the evaluation criterion. However, care should be exercised in the wording of the evaluation criterion. For example, the use of double negatives or complex phraseology should be avoided. For example, don't use no double negatives. If scoring functions are not used, then wording the evaluation criterion in a way that assures that a higher score will be more desirable will satisfy this property.

Evaluation criteria should be worded using positive terms, so that more of anything implies better. Thus, uptime is preferred to downtime, and mean time between failures is preferred to failure rate. Similarly, replacement rate is preferable to failure rate, and likelihood of success is preferred to likelihood of failure. When using scoring functions, it is important to make sure that more output is better. However, sometimes there will be an optimum or a pessimum value. Wording can make this clear. For example, there is a preferred human body temperature, namely 98.6 °F.

3.1.4.8 Independent of Temporal Order

The *temporal order* of verifying or combining evaluation criteria should not be important, for example, the clothing for our astronauts should be flame proof and water resistant, but verification should not depend on which we test first. Questionnaires are often written to lead the reader down a primrose path. Whether a question is asked early or late in a list might affect the given answer. Let us consider a combining function (CF) that adds two numbers truncating the fraction: (0.2 CF 0.6) CF 0.9 = 0, however, (0.9 CF 0.6) CF 0.2 = 1. The answer depends on the order in which the inputs were presented and therefore it would not be a good combining function for a trade-off study or a risk analysis. In the field of Boolean Algebra, it is said that such functions do not obey the Law of Association, for example, as with the Boolean NAND function (↑): $(0 \uparrow 1) \uparrow 1 = 0$ however, $(1 \uparrow 1) \uparrow 0 = 1$. If temporal order is important, then an expert system might be more appropriate than a trade-off study or a risk analysis.

3.1.4.9 Show Compensation

Pairs of evaluation criteria for trade-off requirements should show *compensation* meaning they can be traded off. For an example of perfect compensation, consider our astronauts growing food on a trip to Mars. Assume we have two evaluation criteria: amount of rice grown and amount of beans grown with the single objective of maximizing yield. A lot of rice and a few beans are preferred equally to a lot of beans and little rice. This shows that we are able to trade off beans for rice and vice versa. Now consider another scenario that demonstrates *little* compensation: a system that produces oxygen and water for our astronauts. A system that produced a huge amount of water, but no oxygen might get the highest score, but clearly, it would not support life for long. Care must be exercised in stating trade-offs between evaluation criteria for requirements that do not show compensation to reveal situations as described above. Evaluation criteria of this type often fall into the mandatory requirements class. The baseball homework problem has a heuristic for judging compensation.

3.1.4.10 Synonyms

Even though the following terms have different nuances, people tend to treat them as synonyms: evaluation criteria, attributes, objectives, metrics, measures, quality characteristics, figures of merit (FoM), acceptance criteria, indicators, factors, scales, measures of effectiveness, and measures of performance [8].

3.1.4.11 Template for Evaluation Criteria

Evaluation criteria are used to help select amongst alternatives in trade-off studies, quantify requirement satisfaction, and assess performance of the system. Criteria should be created with the intent of reuse. To this end, a company should have a library of generic criteria.

Name of criterion: Criterion names should be short, but should not be abbreviations.

Description: This is an expansion of the criterion name. It explains what the criterion is supposed to assess. It should describe how this criterion will determine how well the alternative, requirement, or system satisfies an attribute.

Weight of importance: This is usually a number between 0 and 10 (or maybe between 0 and 1) that states how important this criterion is compared to other criteria. Weights of importance will be derived in discussions with stakeholders.

Basic measure: The basic measure is the value of what will be measured or calculated. A measure indicates the degree to which an entity possesses and exhibits a quality or an attribute. The basic measure's name is often similar to the name of the criterion.

Measurement method: This describes the method that will be used to measure and/or calculate the basic measure. This will vary with the phase in the system life cycle.

Units: SI units are preferred, except when interfacing with a human. For example, when communicating with a human, we use baseball units: miles per hour, inches, feet, and ounces. However, inside the system, all calculations are done with SI units.

Scoring functions will not be discussed until Section 21.

Scoring function inputs: Input values are constrained by the domain of definition (legal values) and expected values.

Scoring function type and parameters: Specify the scoring function and its parameters.

Scoring function output: This will probably be 0–1. More output should be better than less output.

Properties: Is this evaluation criterion quantitative, objective, and worded correctly? What other criteria might this criterion depend on or be dependent on? What other criteria might this criterion be traded off with (exhibit compensation with)?

Trace to a function, a trade-off requirement, the concept of operations, or a customer statement.
Owner is responsible for creating, maintaining, and approving changes to this evaluation criterion.
Date of last change or some other configuration management item

Example Evaluation Criterion

This example is from the Pinewood Derby documentation [20].

Name of criterion: Total Event Time
Description: The total event time will be calculated by subtracting the start time from the end time.
Weight of importance: 8
Basic measure: Duration of the derby from start to finish.
Units: Hours
Measurement method: Observation, recording, and calculation by the Pinewood Derby Marshall.
Scoring functions will not be discussed until Section 21.
Scoring function inputs: The domain is 0–8 h. The expected values are 1–6 h.
Scoring function type and parameters: SSF5(1, 3, 4, 5, 6, 1, −2, RLS[0–8])
Scoring function Output: 0–1
Properties: This evaluation criterion will interact with other criteria that measure the number of races. This evaluation criterion has an optimal value at 4 h.
Trace to concept of operations
Owner: Bill Karnavas
Date of last change: November 14, 2016

Second Example Evaluation Criterion

Many of our homework problems investigate the SpinCoach. For the baseball batter who needs to predict the trajectory of the pitch, the SpinCoach is a training system that helps him or her to recognize the spin on the pitch and predict the ball's spin-induced movement; unlike present coaches and books, the SpinCoach shows the batter how each pitch spins and helps him to recognize this spin. We want to know, "Does training with the SpinCoach teach batters to predict the spin-induced deflection of the baseball?" But this is too hard to measure; therefore, we substitute the player's batting average as a measure of success. This is attribute substitution that we warned you about in Sections 2.3.1 and 2.3.2.3. Here is an evaluation criterion for the SpinCoach.

Name of criterion: Ability to predict the spin-induced deflection of a ball.
Description: We need to know how well the batter learned to predict the spin-induced deflection of the baseball or softball.

Weight of importance: 10

Basic measure: We will compute each batter's batting average for 20 at-bats before and after the SpinCoach training. The basic measure is the *difference* between the before and after batting averages.

Measurement method: We will compute the batting averages from the league's official score book.

Units: Batting averages are unit less quantities.

Scoring functions will not be discussed until Section 21.

Scoring function input: The domain shall be real numbers between ±0.05. The expected values shall be real numbers between ±0.01.

Scoring function type and parameters: SSF3(−0.01, 0.01, 0.1, 50, RLS[minus 1 to plus 1])

Output: 0–1.

Properties: Note that the basic measure is different from the criterion name.

Trace to the concept of operations.

Owner: Coach Candera

Date of last change: January 7, 2016

3.1.4.12 Verboten Criteria

Availability might be a poor criterion, because it cannot be traded off. Assume oranges are available 6 months out of the year. Would it make sense to do a trade-off study selecting between apples and oranges and give oranges an availability expected value of 0.5? Suppose your trade-off study leads to the selection of oranges, but it happens to be the month of September, and oranges are not available! In this case, the trade-off study makes little sense!

3.1.4.13 Deriving Values for the Criteria

The prioritization process consists of first deriving values for all criteria for all features and then combining the data to reveal the priorities. Common criteria scales include:

(a) Low, Medium, and High
(b) Optional, Conditional, and Essential
(c) Nice to Have, Highly Desired, and Must Achieve Goal
(d) Numeric (for example, 0–10)

However, it is *imperative* to use the same range for all criteria. For example, you should not use a range of 1–3 for one criterion and 0–10 for another, see Bahill and Karnavas [9] and Table 3.7. Obtaining a consensus on criteria values might require a group decision support technique such as voting, Delphi, the analytic hierarchy process, or some specialized facilities and software.

In an ideal world, to get the right criteria values, you would first talk to the customer. In the real world, the following sequence is more realistic. The engineer assigns strawman values to all the criteria. These values are typically integers in the range of 0–10, where 10 is the most important. The next step is to meet with specialty engineers and domain experts. The systems engineer leads a discussion of each criterion and tries to get a consensus value. In the first pass, the engineers might evaluate each criterion and its context and then take the average value. After in-house evaluation is complete, the values are taken to the customer (however many people that might be). The Chief Engineer leads a discussion of each criterion and tries to reach consensus for all the criteria values. If the customer says that all identified criteria are very important, then the process is continued, until later evaluation data help prioritize the features.

Of course, as with all engineering processes, prioritization is not a waterfall process. It is highly iterative with multiple tasks that can and should be done in parallel [10]. In the beginning of a program, no one generally has a good understanding of the complexity, dependencies, and reuse potential. As knowledge about the system is developed, the prioritization process is refined. Prioritization is a communication tool: the numbers that are derived are not as important as the understanding that results from the process.

There are alternatives to the above procedure. (1) Instead of assigning a number between 0 and 10, the systems engineer, in conjunction with the customer, could rank order all the features. Sometimes this technique works in spite of being methodologically flawed. It is flawed because we are adding the weighted scores; therefore we need cardinal numbers (for example, if feature-A gets a score of 6 and feature-B gets a score of 3, then feature-A should have twice as much worth or utility as feature-B), not ordinal (as in rank order) numbers. (2) The systems engineer can help the customer make pair-wise comparisons of all the features and then use the analytic hierarchy process to derive the values [11]. This is not a practical approach without a commercial software tool.

Tools that implement the analytic hierarchy process add value by also producing a consistency index. For example, if the domain expert said that A was preferred to B, and B was preferred to C, then we would expect him or her to say that A is preferred to C. The consistency index indicates how consistent the comparisons were throughout the entire matrix.

3.1.5 *Weights of Importance*

Another critical element of a trade-off study or a risk analysis is selecting the weights through which priorities among evaluation criteria and levels in the evaluation criterion hierarchy are established. That is, evaluation criteria or categories with higher importance are given more weight in the overall evaluation of the system. Weights of importance are assigned in accord with the five principles of Von Neumann and Morgenstern [12] namely that fundamental decision analysis is based on probability, order, equivalence, substitution, and choice. As a first approximation for the desired weights, the customer assigns a number between 1 and 10. The weights are then normalized in each category. While this method is ad hoc and without axiomatic basis, it has been proven to be useful as a rough indicator of stakeholder preferences despite inconsistencies that theoretically might occur with this method. The key point here is to get the decision maker to think about these issues. The exact numbers are less important. After all, decision makers tend to be inconsistent, and they do change their minds.

3.1.5.1 Creating Weights of Importance

There are several issues that should be brought to the attention of a decision maker to help him or her in assigning weights. These include: organization commitment, critically to mission success, architecture, business value, priority of scenarios (use cases), frequency of use, benefit, cost, benefit-to-cost ratio, and (when needed) time required, risk, safety, complexity, implementation difficulty, stability, dependencies, and reuse potential.

3.1.5.2 Deriving Values for Weights of Importance

A dozen methods exist for deriving numerical values for the weights of importance for the evaluation criteria [13]. These methods can be used by individuals or teams. The following is the most common method for deriving weights of importance. It is similar to the method for deriving values for the criteria that was presented in Section 3.1.4.13. The engineer derives strawman weights of importance for all the evaluation criteria. These weights of importance are numbers (usually integers) in the range of 0–10, where 10 is the most important. He presents these weights of importance to the decision maker and helps him adjust them. Then he meets with the customer (however many people that might be). For each criterion, the systems engineer leads a discussion of the criteria and tries to reach a consensus on the priority of each criterion. On the first pass, each stakeholder might be asked to evaluate each criterion and then the engineer computes the average value. Ultimately, the customer's opinion prevails. If the customer only considers one (or two) criterion and declares the criterion to be a 10, then it's a 10. While we would not expect a domain expert to assign a weight of 0, a weight of 0 can be given to criteria that have no effect on the output, but whose consideration should be made prominent. When the output values are used for numerical comparisons in complex high-risk situations, then additional quantitative methods might be useful. When the weights are to be assigned using both the decision makers' relative importance and the expected range of input values, then the method of swing weights (explained in the following paragraphs) can be used.

The Method of Swing Weights

The swing weight method is explained using a new car selection example. For evaluation criteria, we use Five-year Life Cycle Cost, Horsepower, and Safety. The Five-year Life Cycle Cost (in US dollars) includes purchase price, taxes, licenses, loan interest, insurance, gasoline, and maintenance. The Horsepower is the peak SAE net horsepower. (However, the Horsepower-to-Weight Ratio might have been a better criterion.) The Safety rating is 0–5 stars based on the National Highway Traffic Safety Administration. Table 3.2 has values for some typical cars.

Next, we determine the range of each criterion. As noted earlier, there are several choices for the range. For this example, we will use the real data from Table 3.2. For the three cars that we are examining the maximum and minimum Five-year Life Cycle Costs are $52,000 and $22,000 (Table 3.3).

Next, we consider Horsepower. The three cars that we are examining have a minimum Horsepower of 170 and a maximum of 290 (Table 3.4).

Our third criterion is Safety. The three cars have minimum and maximum values of three and five stars (Table 3.5).

We now have definitions and ranges, measured from worst to best, for each of the three criteria that matter the most to the customer in this new-car selection example. Other characteristics, such as color or type of transmission, may also be important considerations in the choice of a car. However, in our example we are assuming that on all these other criteria the differences between the cars are unimportant. This does not mean that these other characteristics do not matter,

Table 3.2 Evaluation criteria and values for three automobiles

Evaluation criteria	Car A	Car B	Car C
Five-year Life Cycle Cost (US$)	$52,000	$34,000	$22,000
Horsepower (hp)	290	240	170
Safety (stars)	4	5	3

Table 3.3 Range of values for 5-year Life Cycle Cost

	Value for the worst alternative	Value for the best alternative
Five-year Life Cycle Cost	$52,000	$22,000

Table 3.4 Range of values for horsepower

	Value for the worst alternative	Value for the best alternative
Horsepower	170	290

Table 3.5 Range of values for safety

	Value for the worst alternative	Value for the best alternative
Safety	3	5

Table 3.6 Addition of weights of importance for selecting a new car

Evaluation criteria	Weight of importance	Car A	Car B	Car C
Five-year Life Cycle Cost (US$)	0.55	$52,000	$34,000	$22,000
Horsepower (hp)	0.27	290	240	170
Safety (stars)	0.18	4	5	3

but only that, in the context of this choice, they are unlikely to vary sufficiently to warrant making explicit trade-offs among them.

Now imagine a *hypothetical* car that is the worst it can be on all three criteria. In other words, its Five-year Life Cycle Cost is $52,000, its Horsepower is 170, and its Safety rating is three stars. Suppose that you can change the value of one (and only one) of these criteria on this hypothetical car from the worst to the best. This means that you can change only one of the following:

- Five-year Life Cycle Cost, from $52,000 (worst) to $22,000 (best)
- Horsepower, from 170 hp (worst) to 290 hp (best)
- Safety, from 3 (worst) to 5 (best).

Which one would you want to change? Suppose you say Five-year Life Cycle Cost. That means that you value a $30,000 drop in price (a change from $52,000 to $22,000) more than you do either an increase of 120 hp or an increase of two stars of Safety. This criterion, the one that you most want to change from worst to best, is the one you weight most highly in the context of this problem. Assign it 100 points.

Now, which criterion do you value second? Let us say it is Horsepower. Ask yourself, "How much less do I value the 120 hp change compared to the $30,000 drop in price?" Suppose that you value it half as much. Then you would assign it 50 points, or half the weight you gave to the most important criterion.

Now let us consider the last criterion, Safety. Because this criterion is ranked below Horsepower, it should get fewer points. For example, if you value it two-thirds as much as Horsepower, then give it 33 points. (Note that this also means that Safety, with its 33 points, is only one-third as important for this decision as the Five-year Life Cycle Cost.) This process produced values of 100, 50, and 33. All that remains is to normalize the weights so that they sum to 1, as shown in Table 3.6.

Other Methods for Deriving Weights

In another method for deriving weights of importance, the decision maker rank orders the criteria. Rank ordering gives ordinal numbers, not cardinal numbers. However, often this technique works, because rank ordering is easier for humans than assigning weights of importance. The analytic hierarchy process (AHP) can also be used to derive weights. The engineer helps the customer make pair-wise comparisons of all the criteria to derive the weights of importance. This is a feasible approach when using a commercial tool. Prioritization, which is the topic of Section 24, is an obvious method that can be used to derive weights of importance.

Many other methods exist for deriving weights of importance, including the ratio method [14], trade-off method [15], swing weights method [4], rank-order centroid techniques [7], and paired comparison techniques discussed in Buede [7] such as the analytic hierarchy process [11], trade-offs [16], balance beam [16], and judgments and lottery questions [15]. These methods are more formal and some have an axiomatic basis.

3.1.5.3 Cardinal Versus Ordinal

Ideally, weights of importance should be cardinal numbers, not ordinal numbers. Cardinal numbers indicate size or quantity. Ordinal numbers merely indicate rank ordering. (This mnemonic may be useful, ordinal is ordering, as in rank ordering.) Cardinal numbers do not just tell us that one criteria is more important than another is, they tell us how much more important. If one criteria has a weight of 6 and another has a weight of 3, then the first is twice as important as the second is. However, in practice this is often violated.

3.1.5.4 Units

Weights of importance for evaluation criteria can be pure numbers or they may have units. They should be chosen according to preferences of the intended end user. We have found that most end users prefer weights between 0 and 10, so that is what we let them use . We choose the appropriate units for our end user and normalize everything inside our system. For a trade-off study or a risk analysis for the game of baseball, our interaction with the end user would be in baseball units, for example, miles per hour, feet, and pounds, but internally everything would be in SI units.

3.1.6 Homework

3.1.6.1 Validation

Names are important. A good name for a system will entice more people to pay attention to your system and that will get more people to support it. Sometimes your system might not match its model. When that happens, you should change your model, perhaps by adding an extra term in your equation, by adding new particles to your collection, or by changing the name of your model or system.

In the early 1900s, Albert Einstein applied the equations of general relativity to the universe and saw that there was a mismatch. So he added a *kludge* factor to his equation. He did not like the name or the idea, but that is what he did. Later, when Eddington showed that the universe was not static, but was expanding, the kludge factor became unnecessary.

In the early 1980s, when astronomers found that their equations could not predict the motions of the stars and galaxies, they invented new particles with cute names like *dark matter* and WIMPs.

In 1949, Fred Hoyle invented the term *Big Bang*. He meant the term to be pejorative and perhaps insulting to his scientific opponents. But the public loved the term, so it stuck.

In 2003, archeologists found fossils of nine individuals of an extinct species of the genus *Homo*. The adults would have been only 3.5 ft tall. Most people paid little attention to *Homo floresiensis* until they called it a *hobbit*.

In the early 2000s, Al Gore invented *global warming*. In subsequent years, when scientific data failed to back up the claims, politicians changed the name to global climate change. This name is a tautology, because the globe has changed its temperature every day for the last 4.5 billion years.

In the late 1970s, the Alvarez team discovered a layer of clay that contained an unusually high concentration of the rare-earth element iridium—30 times the level typically found in the Earth's crust. Although iridium is rare in the earth's crust, it is abundant in many meteorites and asteroids. Therefore, they hypothesized that a large asteroid must have struck the Earth right at the Cretaceous–Tertiary (K–T) boundary, 65 million years ago. But the Alvarez hypothesis did not draw a lot of attention until they posited that this event caused the mass *extinction of the dinosaurs*.

Your homework assignment is to create cute, compelling names for the Cookie acquisition system and the university parking garage system. E-mail your answers to terry@sie.arizona.edu.

3.1.6.2 Cookie Acquisition System

Fill in an evaluation criterion template, like the one in Section 3.1.4.11, for an evaluation criterion that you think would be appropriate for the Cookie acquisition system of Section 3. Ignore the section for scoring functions.

3.1.6.3 Parking Garage System

Fill in an evaluation criteria template, like the one in Section 3.1.4.11, for at least three evaluation criteria that you think would be appropriate for the parking garage system of Section 10. Ignore the section for scoring functions.

3.1.6.4 Springer Site

Go to one of the websites for this textbook and download the file baseball StatisticsforStudents.xlsx.

http://www.springer.com/gp/book/9783319437101#otherversion=9783319437125
http://sysengr.engr.arizona.edu/forStudents/index.html

3.1.6.5 Baseball Evaluation Criteria

In Section 16, you estimated the probabilities of the following baseball batting outcomes: a homerun (HR), a double (2B), a single (1B), a run batted in (RBI), a base on balls (BB), reaching base on error (ROE), a strikeout (SO), and grounding into a double play (GDP). Would these outcomes make good evaluation criteria for judging the batting performance of a baseball player?

First, let us examine the properties that affect only one of the above outcomes at a time.

Are each of these outcomes quantitative?
Are each of these outcomes objective?
Are each of these outcomes worded correctly, so that more is better?
Now, let us examine the properties that affect two or more outcomes at the same time.
Are these outcomes independent?
Are homeruns, doubles, and bases on balls transitive?
Are these outcomes unaffected by the temporal order of evaluation?
Do singles and bases on balls show compensation?

3.2 Section 21. Scoring Functions

A trade-off study or a risk analysis is performed by assessing each evaluation criterion and combining the resulting data (usually on multiple levels) to give a general numerical evaluation of the system. For complex systems with multiple evaluation criteria, it may be difficult or impossible to directly evaluate each criterion in raw units and combine them to elicit useful data. This difficulty is due to the lack of a common basis for combining and comparing (for example, comparing hard disk space in Terabytes to processor speed in Gigahertz). Scoring functions provide a simple technique through which this issue can be resolved. This method is similar to using utility, value, or fuzzy-set functions.

Each bottom-level evaluation criterion is given a scoring function. Intermediate or top-level evaluation criteria may also be assigned scoring functions as prescribed by the stakeholder, although Buede [7] and Wymore [17] advise that scoring functions are not necessary beyond the bottom level, because the data have already been normalized. The scoring function is a mathematical mapping between the evaluation criterion's "natural" or measured values to a "coded" or normalized range of values common to all evaluation criteria. It is formalized as a mathematical function that accepts numerical values that are specific to the evaluation criterion and assigns as output a real number, commonly between 0 and 1. However, it can employ any other consistent scaling. With this technique, an evaluation criterion, no matter what raw units or scale it may be defined in, can be transformed by its scoring function to have a set of possible values common to all other evaluation criteria whose raw values have been transformed by their own scoring functions. Through this mathematical transformation, a common basis for aggregation and comparison can be achieved.

A classical procedure for creating a scoring (or utility) function for a given evaluation criterion is to survey the decision maker concerning his or her preferences and judgments as input values of the evaluation criterion are considered [18]. This information guides the researcher in defining the shape of the scoring function curve. The shape of the curve is important as it explicitly illustrates the degree of variation in the customer's judgment with each incremental change in evaluation criterion input values. With classical methods, it is essential that sufficient data be acquired so that the shape of the curve reflects an accurate portrayal of the customer's assessments. As one might imagine, this method can be tedious and time consuming. Consequently, to speed up this process a finite set of "standard" scoring functions have been developed that possess the properties previously mentioned [17]. Standard scoring functions are characterized by a few parameters and have the flexibility to take on many shapes. This allows the researcher to conveniently capture a customer's preferences by only specifying a few parameters for a predefined curve.

The need for scoring functions is further illustrated by these examples. First, could you add values for something that costs a billion dollars and lasts a nanosecond? Second, assume that you are to choose the alternative with the largest sum.

Alt-1 costs a hundred dollars and lasts one millisecond, Sum = 100.001.
Alt-2 only costs 99 dollars but it lasts two millisecond, Sum = 99.002.

Does the duration have any effect on the decision? No.

3.2.1 *Linear Normalization*

We can create linear functions that normalize all criteria so that they range between 0 and 1. The general procedure is given with the following equations. For criteria with a positive slope, the

$$\text{Normalized score} = \frac{\text{value} - \text{minimum}}{\text{maximum} - \text{minimum}}$$

where value is the input and score is the output. With this equation, the maximum value of each normalized criterion is 1.0 and the minimum is zero. For criteria with a negative slope, the

$$\text{Normalized score} = \frac{\text{maximum} - \text{value}}{\text{maximum} - \text{minimum}}$$

These equations assure that the ranges are from 0 to 1 and the data are linearized.

For example, suppose that a person is interested in buying one of three cars: car-A produces 290 hp, car-B produces 240 hp, and car-C produces 170 hp. We want to normalize these values so that they can be combined with other dissimilar data. First we must choose the input range over which we will normalize. The range could be legal values, the highest to the lowest values that would ever be expected or maximum and minimum constraints given by the customer. If data are available for typical alternatives and software is available to update the weights when new alternatives are introduced, then real data could be used. In this example, let us choose the highest to the lowest values that would ever be expected for *our* customer. Let us choose 100–400 hp.

The general formula for linear normalization is

$$\text{score}_j = \frac{\text{value}_j - \text{value}_j^{\min}}{\text{value}_j^{\max} - \text{value}_j^{\min}} \in [0, 1]$$

where the index *j* indicates the particular feature being evaluated. Let the index *j* be horsepower (hp), then $\text{value}_{\text{hp}}^{\min} = 100$ and $\text{value}_{\text{hp}}^{\max} = 400$ and we get

$$\text{score}_{\text{hp}} = \frac{\text{value}_{\text{hp}} - 100}{400 - 100}$$

$$\text{score}_{\text{hp}}(\text{car-A}) = \frac{290 - 100}{400 - 100} = \frac{190}{300} = 0.63$$

$$\text{score}_{\text{hp}}(\text{car-B}) = \frac{240 - 100}{400 - 100} = 0.47$$

$$\text{score}_{\text{hp}}(\text{car-C}) = \frac{170 - 100}{400 - 100} = 0.23$$

This linearly normalized function is shown in Fig. 3.2.

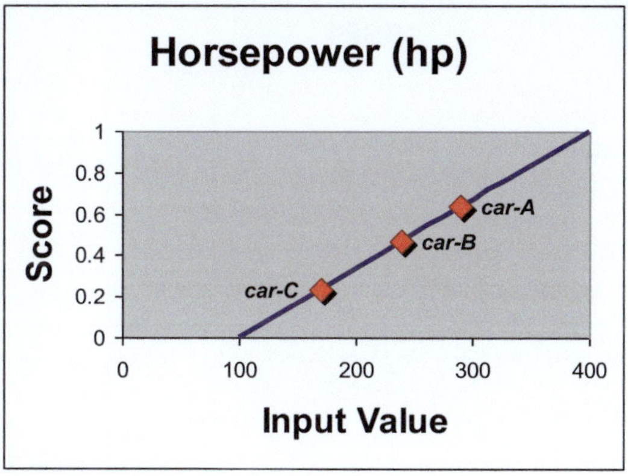

Fig. 3.2 Linear normalization of the horsepower criterion

3.2.2 Wymore's Scoring Functions

There are hundreds of different types of scoring functions, which are also called utility functions. In this section, we consider a set of standard scoring functions developed by Wymore [17]. Students without adequate math background can ignore the equations, because they are embedded in a software program that is available for free on our web sites.

This set has 12 families of scoring functions representing 12 fundamental shapes. Figure 3.3 and the following list describe the general shape of the first six standard scoring functions. Note that these shapes limit their output scores to a zero to one scale.

SSF1:. A shape whose output score is 0 as the input goes from $-\infty$ to a lower threshold point L and then increases asymptotically to 1.

SSF2:. A shape whose output score is asymptotic to 0 as the input approaches $-\infty$, as the input increases the output increases to 1 at a upper threshold point U and gives an output score of 1 from there on.

SSF3:. A shape whose output score is 0 as the input goes from $-\infty$ to a point L and then increases to 1 at a point U and gives an output score of 1 from that point to ∞.

SSF4:. A shape whose output score is asymptotic to 0 as the input approaches $-\infty$, gives an output score that is asymptotic to 1 as the input approaches ∞ and is always increasing.

SSF5:. A hill shape whose output score is 0 until the input reaches a point L, then increases to 1 at the optimal point O, then gives an output that decreases to 0 at a point U and is 0 for larger input values .

SSF6:. A hill shape whose output score is asymptotic to 0 at both ends, but increases to 1 at an input point O.

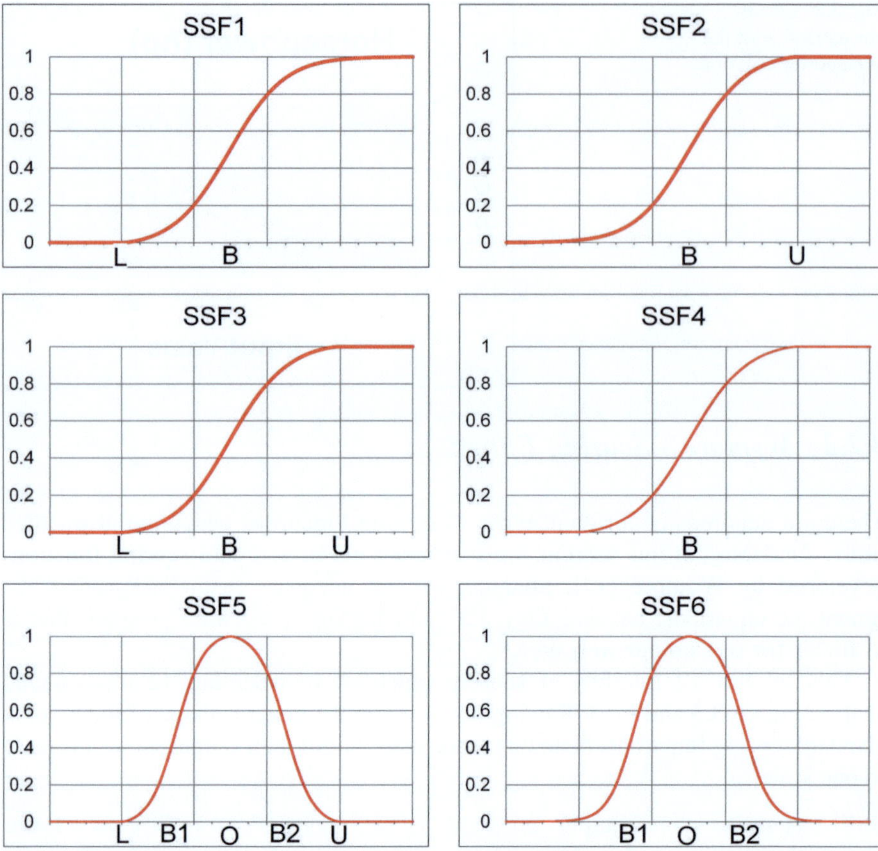

Fig. 3.3 The first six of Wymore's standard scoring functions. The abscissa is the input, the ordinate is the output

For all these functions, S is the slope at the baseline B.

The other six fundamental shapes are obtained by interchanging the roles of 0 and 1 and by replacing "increasing" with "decreasing." The notation SSF* refers to Standard Scoring Function*, where * can be a number between 1 and 18 (for our software tool).

Each of the 12 fundamental scoring functions is defined using the first scoring function called SSF1. This is convenient since if any of the 12 scoring functions needs to be evaluated, only one calculation involving SSF1 needs to be done. Figure 3.4 shows the SSF1 shape and its mathematical equation is given below.

L is the lower threshold

B is the baseline value where the output score is 0.5

S is the slope of the tangent to the curve at B

v is the value of the evaluation criterion, the input

If $v < L$ then Score $= 0$

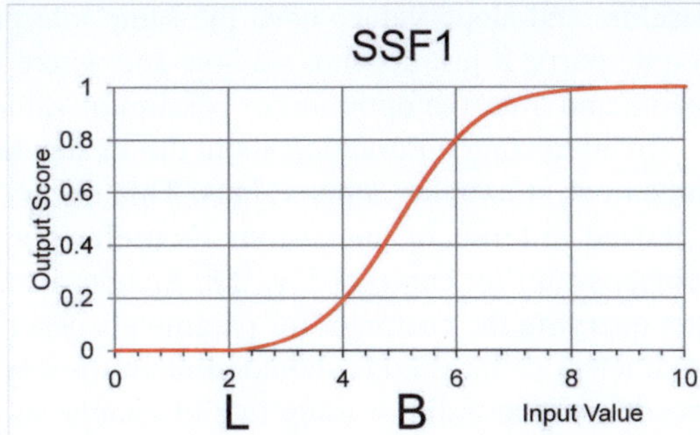

Fig. 3.4 The first standard scoring function, SSF1(L, B, S)

Else, if $v \geq L$ then Score $= \frac{1}{1+\left(\frac{B-L}{v-L}\right)^{2\times S(B+v-2\times L)}}$

Dear reader, do not be intimidated by the equation, because our software computes the results for you. http://sysengr.engr.arizona.edu/forStudents/ScoringFunctionToolbox/index.html

3.2.2.1 Definition of Variables and Parameters

v: is the value for the input variable for the evaluation criterion.
Score: is the output variable of the scoring function.
L: The lower threshold of performance for the evaluation criterion below which the value to the customer is undesirable (but not necessarily unacceptable) and is assigned a zero score.
B: The parameter B is called the Baseline value for the evaluation criterion and can be chosen as the design goal or the status quo for this or similar systems. By *definition*, baseline values are always assigned a score of 0.5.
S: The parameter S determines the behavior of the scoring function in the neighborhood of the baseline value B. Mathematically, S is the slope of the tangent to the scoring function at the baseline value B. Practically speaking, the slope represents the maximum incremental change in the customer's quantitative judgment with each incremental change in input.
D: The parameter D represents the domain of definition of the scoring function. D states the range of input values that are possible from a buildability viewpoint or legal due to mandatory requirements. Values outside this range constitute impossible or unacceptable inputs.

For the non-monotonic "hill" and "valley" shaped scoring functions (SSF5, SSF6, SSF11, and SSF12), we need to define additional parameters such as multiple baselines (B1 and B2), multiple slopes (S1 and S2) values and an optimum (O) or pessimum (P) value. The optimum and pessimum values indicate the maximum or minimum, respectively, of the customer's preferences relative to the input values. For example, SSF5 is written as SSF5(L, B1, O, B2, U, S1, S2, D). The multiple baseline and slope values have the same interpretation as the single parametered counterparts; it just depends on how and where the customer's preferences change before and after the optimum or pessimum values.

An interesting observation about the 12 standard scoring functions is that they all flatten out at extreme input values. This occurs because these functions are characterized in terms of the customer's preferences that are described by the human subjective utility curve of Fig. 2.8. Standard scoring functions are constructed by first querying the customer for parameters describing his or her preferences. These parameters or thresholds should contain the key performance range and sufficient breadth so that a diminishing rate of change in preference is observed at both the upper and lower limits. However, the full range of input values delimited by the customer's preferences may not be attainable from a realizability or requirements perspective. With standard scoring functions, we allow the domain to implicitly truncate the curve at these hard limits.

3.2.2.2 Complete Family of Wymorian Scoring Functions

We now present guidelines for choosing an appropriate standard scoring function along with a short example for each.

SSF1 **Guideline**: If more is better than less, the customer can provide a finite lower bound but no finite upper bound, and the domain can support this, then choose SSF1(L, B, S, D).

Example: Let the units be the capable density in dots per inch of a printer. The higher the better and there is no finite upper bound to this measure.

SSF2 **Guideline**: If more is better than less, the customer can provide a finite upper bound but no finite lower bound, and the domain can support this, then choose SSF2(B, U, S, D).

Example: Depth below the surface of the earth, with greater depth being better. There is a fixed upper limit (3951 miles), but as we go out into space (the left side of our function) the value goes to $-\infty$.

SSF3 **Guideline**: If more is better than less, the customer can provide both a finite upper bound and a finite lower bound, then choose SSF3(L, B, U, S, D). This is by far the most common scoring function used when dealing with an evaluation criterion where more is better.

Example: Let the units be the number of widgets sold. The customer can provide a lower bound below which we can assign a zero score and an upper bound where anything above that value is assigned a score of one.

SSF4 **Guideline**: If more is better than less, the customer can provide neither a finite upper bound nor a finite lower bound on his or her preferences, and the domain can support this, then choose SSF4(B, S, D).

Example: Let the upper range denote happiness and the lower range represent unhappiness. There are no upper or lower bounds, but more happiness is better.

SSF5 **Guideline**: If the customer's preferences can be described as a hill shaped curve where there is one optimum point and values to the left and right of this point are less desirable, and the customer can provide a finite upper bound, a finite lower bound, and the optimum value, then choose SSF5(*L*, B1, *O*, B2, S1, S2, *D*).

Example: Let the units be a measurement specification for a particular piece of material, for example, the width of a printed circuit board trace or the diameter of a hole. The optimum would be the target value and values greater or less than the target value would result in a lower score.

SSF6 **Guideline**: If the customer's preferences can be described as a hill shaped curve where there is one optimum point and values to the left and right of this point are less desirable, and the customer can furnish the optimum value, but not finite upper or lower bounds, then choose SSF6(B1, *O*, B2, S1, S2, *D*).

Example: Let the units be the error in measurement for a particular quantity and the customer cannot say how low or how high the error would have to be in order to assign a score of zero or one. The optimum value would obviously be zero error.

SSF7 **Guideline**: If less is better than more, and the stakeholder can provide a finite lower bound but no finite upper bound (as in the case of a cost), then choose SSF7(*L*, *B*, *S*, *D*).

Example: Number of deaths in airplane crashes. The lower bound would be zero deaths and we assume the customer cannot give an upper limit where if the number of deaths is greater than this limit it is assigned a zero score. However, the domain may restrict this use of the scoring function, because the population of the world is finite.

SSF8 **Guideline**: If less is better than more, and the stakeholder can provide a finite upper bound but no finite lower bound, then choose SSF8(*B*, *U*, *S*, *D*).

Example: Let the abscissa be the logarithm of the size of a RAM cell. The smaller the better, but the customer might not be able to specify how small the cell should be in order to receive a score of one.

SSF9 **Guideline**: If less is better than more, and the stakeholder can provide both a finite upper bound and a finite lower bound, then choose SSF9(*L*, *B*, *U*, *S*, *D*). This scoring function is by far the most frequently used with evaluation criteria where less is better.

Example: Costs. It is usually the case that the customer can provide upper and lower limits on expenditures. For simplicity, we assume the higher the expenditure, the lower the score.

SSF10 **Guideline**: If less is better than more, the stakeholder can provide neither a finite upper bound nor a finite lower bound, and the domain can support this, then choose SSF10(*B*, *S*, *D*).

Example: Assume the customer is trying to build a powerful microscope. Let the units be the size of the smallest viewable object. Here, the smaller the object, the happier the customer.

SSF11 **Guideline**: If the stakeholder's preferences can be characterized by a valley shaped curve, there is one pessimum value where inputs greater or less than this value give more desirable results, and the stakeholder can provide a finite upper bound, a finite lower bound, and the pessimum value, then choose SSF11(L, B1, P, B2, U, S1, S2, D)

Example: Consider a vehicle with unbalanced tires. At a certain speed the car will resonate or shimmy, which creates an undesirable condition. If the driver maintains speeds far away (either faster or slower) from this pessimum speed, then the result is a smoother, more desirable ride.

SSF12 **Guideline**: If the stakeholder's preferences can be characterized by a valley shaped curve, there is one pessimum value where inputs greater or less than this value give more desirable results, and the stakeholder can provide the pessimum value but can provide neither a finite upper bound nor a finite lower bound, then choose SSF12(B1, P, B2, S1, S2, D).

Example: Consider trying to eliminate bacterial growth by varying temperature. There will be a most undesirable temperature where the bacteria grow very fast, and temperatures greater or less than this value will impede bacterial growth.

We mentioned earlier that all 12 fundamental Wymorian scoring functions are defined in terms of SSF1. Next, we will expand this notion by illustrating the explicit manner in which the SSF3 and SSF9 are constructed. We will start with SSF3.

SSF3 has the following four ranges:

1. For input values of the evaluation criterion less than the lower threshold, L, the Score = 0
2. For input values of the evaluation criterion between the lower threshold, L, and the baseline value, B, the Score = SSF1(L, B, S) evaluated at v = input value, or

   Score = SSF1(L, B, S)(input value) with domain D.

3. For input values of the evaluation criterion between the baseline value, B, and the upper threshold, U, the Score = 1 − SSF1($2 \times B - U$, B, S) evaluated at v = ($2 \times B$ − input value), or

   Score = 1 − SSF1($2 \times B - U, B, S$)($2 \times B$ − input value) with domain D. Graphically, take the left part of SSF1 and flip it horizontally then flip it vertically.

4. For input values of the evaluation criterion greater than the upper threshold, U, the Score = 1

SSF9 has similar ranges as SSF3, but is constructed by taking one minus SSF3($L, B, U, -S, D$) and evaluating at v = input value, or Score = 1 − SSF3($L, B, U, -S, D$) (input value). For example, say we have an SSF9 scoring function for an evaluation criterion defined with the following parameters: $L = 1$, $B = 2$, $U = 3$, $S = -4$, D = real numbers between 0 and 10 abbreviated as RLS[0–10], and say the input value v is 2.7. First, note that 2.7 is a valid input since it is within the domain, D. Then evaluate the score (denoted "Score" below) for SSF9 through a series of calculations as follows:

$$\text{Score} = 1 - \text{SSF3}(1,\ 2,\ 3,\ -(-4))(2.7)$$

Where SSF3(1, 2, 3, −(−4))(2.7) = 1 − SSF1(2 × 2 − 3, 2, −(−4))(2 × 2 − 2.7) since the input value, 2.7, is between the baseline value, 2, and the upper limit, 3.

Where SSF1(2 × 2 − 3, 2, − (−4))(2 × 2 − 2.7) =

$$\frac{1}{1+\left(\frac{2-(2\times 2-3)}{(2\times 2-2.7)-(2\times 2-3)}\right)^{2\times(-(-4))(2+(2\times 2-2.7)-2\times(2\times 2-3))}}$$

We have left the numerical values unsimplified so that it can be seen how each value traces back to the original SSF9 problem.

The non-monotonic standard scoring functions ("hill" and "valley" shaped curves) are constructed from the appropriate monotonic functions. The upper limit for the first curve must match the lower limit for the second to obtain a smooth transition from curve to curve. Non-monotonic scoring functions such as hill and valley shapes are often used when trying to balance two or more objectives, or anticipating the actions of others [19].

3.2.2.3 Mathematical Properties of Wymorian Scoring Functions

Wymore's base scoring function SSF1 was chosen in part because it has the desirable mathematical properties listed below. Because these standard scoring functions are differentiable, it is possible to conduct *analytical* sensitivity analyses when evaluating alternative systems. We will now abbreviate SSF1(L, B, S, D) with f_1 where $f_1(v)$ denotes SSF1(L, B, S, D)(v), or SSF1 evaluated at v given L, B, S, and D. The properties are as follows (taken from Wymore, [17], pp. 389–390):

1. The output of $f_1(v)$ evaluated at the baseline value, B, is 0.5
2. The limit of $f_1(v)$ as v approaches infinity is 1
3. The limit of $f_1(v)$ as v approaches L from the right is 0
4. Under the assumption that the slope, S, is greater than $\frac{1}{B-L}$,
   (a) The limit of the derivative of $f_1(v)$ with respect to v as v approaches L from the right is 0
   (b) The value of the derivative of $f_1(v)$ with respect to v is positive for every v that is greater than L, and therefore $f_1(v)$ is strictly increasing over the interval (L, ∞)
   (c) The value of the derivative of $f_1(v)$ with respect to v at the point $v = B$ is S
   (d) The value of the second derivative of $f_1(v)$ with respect to v at the point $v = B$ is 0
   (e) The limit of the second derivative of $f_1(v)$ with respect to v as v approaches L from the right is 0
5. From these facts and the form of the function f_1, it can be concluded that the first and second derivatives of $f_1(v)$ with respect to v are continuous everywhere in D.

Because none of the mathematical manipulations performed on f_1 to create the other 11 fundamental scoring functions affect the derivatives, all the standard scoring functions as well as their first and second derivatives are continuous.

However, the flipping and pasting of f_1 added new parameters and therefore more restrictions on the slopes. Here is how these parameters affect the limits on the slopes.

For SSF1, $\frac{1}{B-L} < S$.
For SSF2, $\frac{1}{U-B} < S$.
For SSF3, $\max\left\{\frac{1}{U-B}, \frac{1}{B-L}\right\} < S$.
For SSF4, $\frac{3}{\text{Domain Max}-\text{Domain Min}} < S$.
For SSF5, $S_2 < \min\left\{\frac{-1}{\text{B2}-O}, \frac{-1}{U-\text{B2}}\right\}$ and $\max\left\{\frac{1}{\text{B1}-L}, \frac{1}{O-\text{B1}}\right\} < S_1$
For SSF6, $S_2 < \frac{-1}{\text{B2}-O}$ and $\frac{1}{O-\text{B1}} < S_1$.

3.2.2.4 How Many Scoring Functions Are Necessary?

Over the last three decades of using scoring functions (see, for example, Chapman, Bahill, and Wymore, [20]) we have found that eight scoring functions are sufficient for all applications that we have encountered. We use two Boolean or step functions, two linear functions, and the four function shapes shown in Fig. 3.5. The step

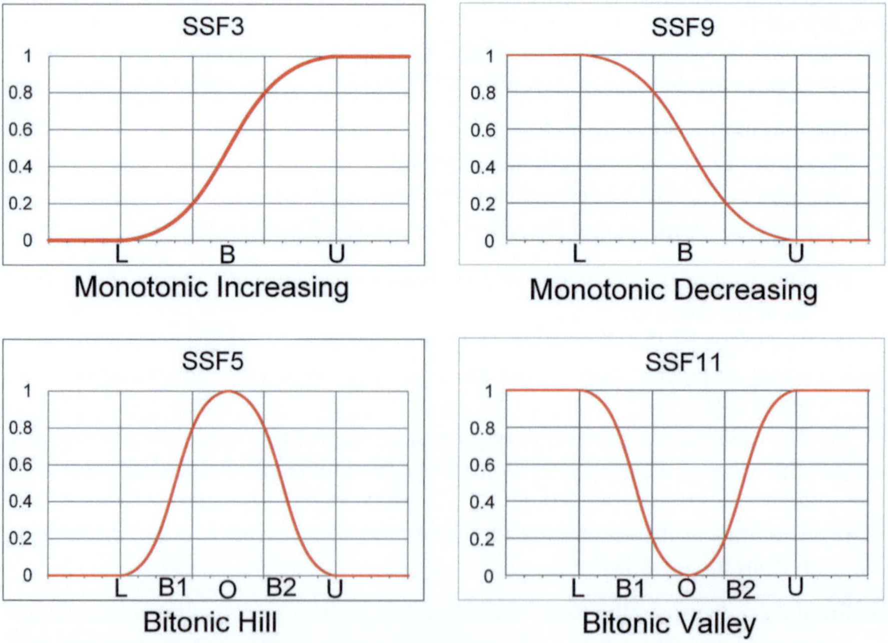

Fig. 3.5 The four essential standard scoring functions. The abscissa is the input, the ordinate is the output

functions are used when a yes/no type input response is required and there is no continuum of possible inputs. We have also found that infinity end points (as with SSF1) are not necessary. It is adequate to use a large number and adjust the slope accordingly. Of these eight scoring functions, we have found the linear scoring functions to be the least frequently used. A program that implements the Wymorian scoring functions of Fig. 3.3 and a dozen others is available at http://sysengr.engr.arizona.edu/forStudents/ScoringFunctionToolbox/index.html

3.2.3 Homework

3.2.3.1 Cookie Acquisition System

In the homework for Section 20, you derived an evaluation criterion for the Cookie acquisition system of Section 3. Now we want you to create a scoring function that is appropriate for this evaluation criterion. You may use your evaluation criterion or the one provided in our solutions.

3.2.3.2 Parking Garage System

In the homework for Section 20, you derived three evaluation criteria for the Parking Garage system of Section 10. Now we want you to create scoring functions that are appropriate for those evaluation criteria. You may use your evaluation criteria or the ones provided in our solutions.

3.2.3.3 Cost Function

In the Pinewood Derby documentation, the hill-shaped scoring function SSF5 was used for the criterion related to the cost of the new components to be purchased. Why do you think this was so? After all, isn't the cheapest alternative always the best? (Fig. 3.6).

Fig. 3.6 SSF5, the hill shaped scoring function

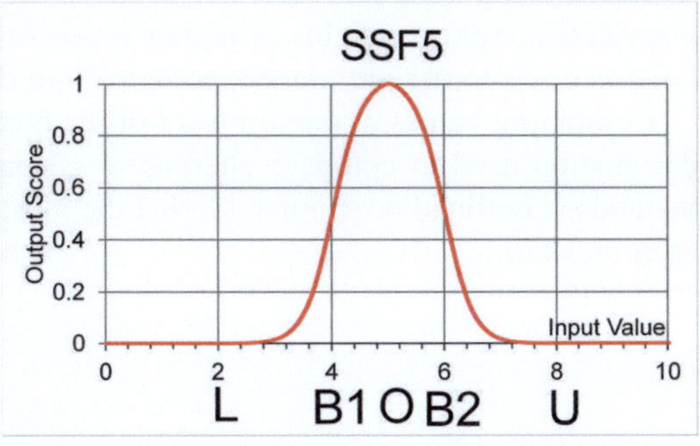

SSF5(L, B1, O, B2, U, S1, S2, D) = SSF5(2, 4, 5, 6, 8, 1, −1, RLS[0–10]) where the notation RLS[0–10] means all real numbers between 0 and 10.

3.2.3.4 Scheduled Meeting

You have a meeting with another person. It is scheduled for noon. At what time should you arrive if it is your boss's boss? At what time should you arrive if it is a student that you are tutoring? Show plots of your scoring function with appropriate numerical values.

3.2.3.5 Length of Sports Games

The average length of a MLB baseball game is 2 h and 50 min. The average length of an NBA basketball game is 2 h and 20 min. The average length of an NFL football game is 3 h and 10 min. Presently, there is much discussion and pressure to shorten all of these games. Assume that you have just been appointed Sports Czar. What scoring function type and parameters would you dictate for the length of these professional sports games? Why?

3.3 Section 22. Combining Functions

Different methods exist for combining scores, or values in a trade-off study or a risk analysis to calculate a numerical measure that can be used to compare alternatives. These methods are discussed in this section. The combining methods described here are used to combine data at all levels of the evaluation criterion hierarchy.

At the lowest level, when we are dealing with individual evaluation criteria, the scores are given as outputs of the scoring functions associated with each of the evaluation criteria, and the weights are based on expert opinion or customer preference. When we move to higher levels in the hierarchy, the scores are no longer derived from scoring functions, but are derived by combining scores at lower levels. Scoring functions are not needed, because these scores are all already normalized. Again, weights at higher levels in the hierarchy are based on expert opinion or customer preference, perhaps from a different category of stakeholder.

Combining values at the top level of the hierarchy yields the overall numerical designation used to compare alternative systems. A selection of data-combining methods is outlined next. First, we list the most popular four, and then we discuss each in detail.

For two evaluation criteria, y and z, we define the *sum combining function* as

$$f = w_y y + w_z z$$

where w_y and w_z represent the weights of importance for y and z, respectively. This is often called the *sum of weighted scores* combining function.

The *product combining function* is

$$f = y \times z.$$

The *sum minus product combining function* is

$$f = w_y y + w_z z - w_{yz} yz$$

where w_{yz} represents a weight of importance for the product of the two evaluation criteria.

The *compromise combining function* is

$$f = \left[w_y^p \cdot y^p + w_z^p \cdot z^p + \right]^{1/p}$$

The parameter p represents a scaling factor that will be discussed in detail.

3.3.1 Common Combining Functions

The following functions combine data from one or many evaluation criteria. In these equations, n is the number of evaluation criteria to be combined, x_i is the output score of the scoring function for the ith evaluation criterion, and w_i is the normalized weight of importance for the ith evaluation criterion. Weights of importance are expected to vary from zero to one. If the weights vary from 0 to 100, then the equations would have to be adjusted. Here are four common combining functions. In the equations of this section, math aversive students should ignore the first-general equation and only pay attention to the second two-dimensional equation.

3.3.1.1 Sum Combining Function

The sum combining function is the simplest and most common method for combining data. It is ideal when the evaluation criteria show perfect compensation, that is, when both criteria contribute to the result and when more of y and less of z is just as good as less of y and more of z. To describe this data combining process, first suppose there are n reasonably independent evaluation criteria to be combined (perhaps they are in the same layer in the evaluation criterion hierarchy). We assign a qualitative weight to each of the n evaluation criteria and then normalize the weights so that they add up to 1. Data is collected for the evaluation criterion, each evaluation criterion is then evaluated with its scoring function and the resulting scores (valued from 0 to 1) are then multiplied by their corresponding weights. The final result is the summation of the weight-times-score for each evaluation criterion. This process is commonly used, for example, when computing the grade point average for a student.

The *sum combining function* is

$$f = \sum_{i=1}^{n} w_i \cdot x_i, \text{ if } n = 2 \text{ then } f = w_y y + w_z z.$$

Most studies use a combination of combining functions.

The sum combining function is appropriate when the decision makers' preferences satisfy additive independence [15], which is the case for most industry applications that we have seen.

The sum combining function, which is the most common method used in engineering, also seems to have great public acceptance. It was the technique initially used to rate the National Football League (NFL) professional quarterbacks.

The National Football League used these four evaluation criteria:

evaluation criterion1 = (Completed Passes)/(Pass Attempts)
evaluation criterion2 = (Passing Yards)/(Pass Attempts)
evaluation criterion3 = (Touchdown Passes)/(Pass Attempts)
evaluation criterion4 = Interceptions/(Pass Attempts)
Rating = {5(evaluation criterion1 − 0.3) + 0.25(evaluation criterion2 − 3) + 20 (evaluation criterion3) + 25(−evaluation criterion4 + 0.095)} × 100/6.

Presently the NFL Total Quarterback Rating is much more complicated. It is based on a computer program comprising one thousand lines of code!

College football also started simple in selecting the college football teams that would play for the national championship in the Bowl Championship Series. They

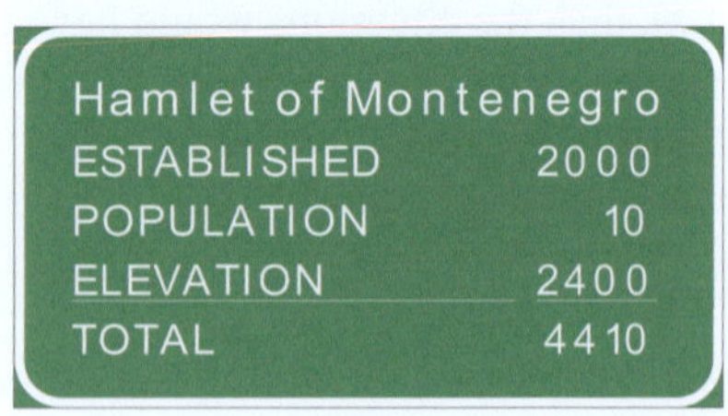

Fig. 3.7 Although the sum combining function is very popular, it is not always the best combining function

used the sum combining function and these four evaluation criteria: subjective *polls* of the writers and coaches, *computer rankings*, *team record*, and *strength of schedule*. Now, their algorithm is also considerably more complicated (Fig. 3.7).

So we will now examine some other combining functions.

3.3.1.2 Product Combining Function

The *product combining function* is

$$f = \prod_{i=1}^{n} x_i, \text{ if } n = 2 \text{ then } f = y \times z.$$

When using the product combining function, it is important that the range of the two evaluation criteria be the same. The reason for this is explained with Table 3.7.

Need for Normalization

Values for the evaluation criteria can come in a variety of formats, for example, {low, medium, high}, {0–10} or natural units that might run, for example, from one thousand to one million dollars. In order to combine apples and oranges like these, the values must be normalized. Table 3.7 shows why.

In this table, the risk (R) has been defined as the frequency of occurrence (F) multiplied by the severity of the consequences (S). The examples in the left and right halves of Table 3.7 have the same frequency of failure, but the severity column in the right half has been turned upside down. The risk columns are different, but the rank order columns are identical. Severity has no effect! In general, if two items are being multiplied and they have different ranges, the one with the bigger range has more weight.

We can generalize this by letting F_i, S_i and R_i and F_j, S_j and R_j be, respectively, the frequency, severity, and risk data for any two rows of Table 3.7. Clearly $R_i > R_j$ if and only if $F_i \times S_i > F_j \times S_j$, that is, when ${}^{F_i}/_{F_j} > {}^{S_j}/_{S_i}$. For the data of Table 3.7, the smallest possible value of ${}^{F_i}/_{F_j} = 10$, for $F_i > F_j$. The largest possible value of

Table 3.7 The problem with different ranges

Example 1				Example 2			
Frequency	Severity	Risk	Rank order	Frequency	Severity	Risk	Rank order
10^{-1}	1	1×10^{-1}	1	10^{-1}	6	6×10^{-1}	1
10^{-2}	2	2×10^{-2}	2	10^{-2}	5	5×10^{-2}	2
10^{-3}	3	3×10^{-3}	3	10^{-3}	4	4×10^{-3}	3
10^{-4}	4	4×10^{-4}	4	10^{-4}	3	3×10^{-4}	4
10^{-5}	5	5×10^{-5}	5	10^{-5}	2	2×10^{-5}	5
10^{-6}	6	6×10^{-6}	6	10^{-6}	1	1×10^{-6}	6

$^{S_j}/_{S_i} = 6$. Since $10 > 6$, the ratio $^{F_i}/_{F_j}$ will always dominate S, regardless of the order of the rows. In general, if two items are being multiplied and they have different ranges, the one with the bigger range has more weight, perhaps secretly. There are three solutions for this problem: giving equal ranges for the two criteria, using a geometric mean, and normalization as discussed in Section 21.

We could adjust the severity scale so that it has the same range as the frequency scale. Then frequency and severity would have the same weight in the quantification of risk, and bias would be eliminated. For example, if the frequency scale goes from 10^{-6} to 10^{-1}, then the range has five orders of magnitude and thus the severity scale must also have a range of five orders of magnitude, for example, from 1 to 10^5.

This problem of different ranges can be insidious if the severity values were based solely on physical terms. For example, if the frequency of occurrence for the events ran from one to ten events per year and the economic losses ranged from a thousand dollars to a million dollars, then the rank order of the products would be determined strictly by the dollar amounts; the frequency of occurrence would have no effect.

We have already discussed two methods for making the ranges the same: linear normalization and the use of standard scoring functions. Another method is discussed here. Assume that the F data extend from 1 to 100, but the S data only extend from 1 to 10. We would then need a function that would transform the S data so that they would have the same range as the F data. First, let us try an exponential function like $f(S) = S^{w_S}$, where the weight of importance for S, $w_S \geq 1$. If $w_S = 1$ then the S data will extend from 1 to 10. If $w_S = 2$ then the S data will extend from 1 to 100. Generalizing, if we need the S data to extend from 1 to α, then we need $f(S) = S^{w_S} = \alpha$. The original maximum value of S was 10 and now it needs to be α. So we need $f(10) = 10^{w_S} = \alpha, \quad \therefore w_S \approx 0.43 \ln\alpha$. In contrast, for a different set of F and S data, we might need to *compress* the data range of S, which we can do using a logarithmic transform, $f(S) = \ln S$. Obviously, both of these transforms are nonlinear.

The need for normalization is mandatory for the product combining function and it is preferred for all other combining functions.

Weight of Importance and the Product Combining Function

By this time, we know that when using the product combining function the evaluation criteria must have the same range. The most common way to achieve this is normalization (either linear or Wymorian). Let us now assume that we have done this with our data, so that $0 \leq y \leq 1$ and $0 \leq z \leq 1$. Now we want to give more weight to one criterion (for example, y) than the other (for example, z). We could make $w_y \geq w_z$ and use

$$f = \prod_{i=1} x_i^{w_i}, \text{ if } n = 2 \text{ then } f = y^{w_y} \times z^{w_z}$$

or

$$f = \prod_{i=1}^{n} x_i^{(1-w_i)}, \text{ if } n = 2 \text{ then} f = y^{(1-w_y)} \times z^{(1-w_z)}$$

However, these combining functions do not have the intended consequences of making y more important than z over the whole range. Let us try one more function. $f = y^{-\alpha} \times z^{-(1-\alpha)}$. This is the well-studied Cobb–Douglas equation from economics. It has the desired property that as α increases so does $y^{-\alpha}$ and $z^{-(1-\alpha)}$decreases. So, it does make y more important than z. But this only holds for certain values of y and z. To show this, we take the partial derivative of f with respect to y.

$$f = y^{-\alpha} \times z^{-(1-\alpha)}$$
$$\frac{\partial f}{\partial y} = -\alpha y^{-\alpha-1} z^{-(1-\alpha)}$$
$$\frac{\partial f}{\partial y} = -\alpha y^{-\alpha-1} \frac{y}{y} z^{-(1-\alpha)}$$
$$\frac{\partial f}{\partial y} = \frac{-\alpha f}{y}$$

The value of this partial derivative depends on the value of y. Likewise, we would find that the value of the partial derivative of f with respect to z depends on the value of z. So the most important parameter depends on the values of y and z. So this is not what we wanted, but it works over a certain range of values for y and z. Because of the complexity, the technique of using weights in the product combining function is not done when discovering requirements or doing trade-off studies.

The product combining function is commonly used, for example, in computing benefit-to-cost ratios and in doing risk analyses. This method favors alternatives where all evaluation criteria have moderate values. It rejects alternatives that have an evaluation criterion value of zero. This method is often used for mission critical functions, where setting one function to zero, or a very low number, is not likely to be overcome by extra effort in the other functions.

3.3.1.3 Sum Minus Product Combining Function

The sum minus product combining function for two evaluation criteria, x and y, is given as:

$$f = w_y y + w_z z - w_{yz} yz$$

where w_{xy} is either uniquely evaluated or is defined as $w_{yz} = w_y \times w_z$

The sum minus product combining function has its origins in probability theory; it is appropriate for computing probabilities of unions for independent events. It is also used in fuzzy logic systems. It also is the function used in Mycin style decision support systems for computing certainty factors when two or more rules with the same conclusion succeed. Certainty factors (CFs) have been used in expert systems for over 30 years [21]. The underlying method surrounding CFs, which is based on probability theory, has stood up to mathematical scrutiny. Thus, a vast knowledge base exists for CFs, and a great deal is known about their properties and uses.

Suppose, the certainty factor for the first premise of a rule is $\mathrm{CF}_1 = w_1 x_1$, where w_1 is the weight (between 0 and 1) and x_1 is the score (output of the scoring function, also between 0 and 1), and the certainty factor for the second premise is $\mathrm{CF}_2 = w_2 x_2$. The equation for determining the certainty factor after both rules are evaluated is

$$\mathrm{CF}_{\mathrm{both}} = \mathrm{CF}_1 + (1 - \mathrm{CF}_1)\mathrm{CF}_2$$

In general, the certainty factor combining equation is

$$\mathrm{CF}_{\mathrm{total}} = \mathrm{CF}_{\mathrm{old}} + (1 - \mathrm{CF}_{\mathrm{old}})\mathrm{CF}_{\mathrm{new}}$$

or

$$\mathrm{CF}_{\mathrm{total}} = w_y y + w_z z - w_{yz} yz$$

which of course is the sum minus product combining function. The nature of the rule restricts CFs to the range $\{0, 1\}$.

The CFs for the remaining evaluation criteria are combined using equation $\mathrm{CF}_{\mathrm{total}} = \mathrm{CF}_{\mathrm{old}} + (1 - \mathrm{CF}_{\mathrm{old}})\mathrm{CF}_{\mathrm{new}}$ to create an aggregate score for their respective subcategories. When we move to the next level up, the x_i's and w_i's become the weights and scores for the subcategories just calculated, and so on. At the highest level of the evaluation criterion structure, $\mathrm{CF}_{\mathrm{total}}$ becomes the overall evaluation from which alternatives can be compared.

An advantage of using CFs as a combining method is that the weights (w_i) do not have to be normalized within any rule or section. This means each time the objectives hierarchy is modified, be it the addition or subtraction of a new evaluation criterion, or a new layer in the hierarchy, it is not necessary to re-normalize the weights as with sum combining function. This feature simplifies computations; however, the weights must still be in the range $\{0, 1\}$. A disadvantage of the certain factor calculus is that it can only be applied to two rules at a time. Furthermore, the sum minus the product combining function is only valid if the evaluation criteria are normalized with a range of zero to one. The most convenient expert system certainty factors are those between 0 and 100, but we must normalize those from 0 to 1.0.

Please note that the sum minus product combining function is different from the difference between the sum combining function and the product combining function.

3.3.1.4 Compromise Combining Function

The compromise combining function is given as:

$$f = \left[\sum_{i=1}^{n} (w_i x_i)^p\right]^{\frac{1}{p}} \text{ or for } n=2\, f = \left[(w_y y)^p + (w_z z)^p\right]^{1/p},$$

Here, y and z represent the outputs of the scoring functions for evaluation criteria y and z, respectively, and the parameter p represents a scaling factor. For the compromise combining function, if $p=1$, it is merely the sum combining function mentioned above. This function shows perfect compensation, because a unit reduction in one alternative can be perfectly compensated by a similar increase in another. This type of compensation is typified by optimization techniques such as linear programming. If $p=2$, it is the Euclidean distance measure, used, for example, to obtain root-mean square measurements. This type of compensation is used in problems with quadratic objective functions, such as certain optimal control problems. If $p=\infty$ then the preferred alternative will be the one that has the largest evaluation criterion. There is no compensation in this case, because only one of the evaluation criteria (the largest) is considered. Consider the Compromise Output $= [y^p + z^p]^{1/p}$. If $y > z$ and $p >> 1$, then $y^p >> z^p$ and the Compromise Output $= [y^p]^{1/p} = y$. This type of compensation is typical in H_∞ control or in nuclear reactors where peak centerline temperature has to be limited. This is called the Chebyshev norm [22]. The compromise combining function with $p=\infty$ might be most appropriate when choosing a hero and the evaluation criteria show no compensation, for example, when selecting the greatest athlete of the century using evaluation criteria based on Number of National Championship Rings and Peak Salary, when picking the baseball player of the week using home runs and pitching strikeouts, when a couple is choosing a movie using evaluation criteria based on romance, action, and comedy, when the NBA teams are drafting basketball players using height and assists, or when a pregnant woman is choosing between pickles and ice cream. In some municipalities, sewerage charges are based on winter water usage only; this prevents people from protesting that a lot of their summer water did not go down the sewer, it went into plants, grass, and swimming pools. This is another example of using the compromise combining function with $p=\infty$.

When given the same inputs most of the combining functions presented in this section produced similar results. When fed random data, the combining functions did not diverge, but produced results that differed only by multiplicative constants [23]. The exceptions were the product combining function with certain values and the compromise combining function with $p=\infty$. The product combining function rejected any design with a zero for any criterion. And the compromise combining function with $p=\infty$ only considered the most important criterion.

3.3.2 *Mathematical Summary*

This section is based on the work of Ferenc Szidarovszky, Emeritus Professor of Systems Engineering at the University of Arizona. Students who are not conversant in mathematics may skip this section without loss of continuity.

Throughout history, mathematics has been used to codify and summarize scientific endeavors. Accordingly, this section provides a mathematical summary of five multicriterion decision-making methods: (1) sequential optimization, (2) ε-constraint, (3) summing weighted scores, (4) ideal point, and (5) anti-ideal point. The mathematical representation allows direct implementation of the rules for these methods, that is, code can be written directly from the rules.

3.3.2.1 Classical Optimization

In classical optimization, there is one decision maker and one objective function. The decision maker chooses the value of the function (**x** in this case) so that it will maximize (or minimize) $f(\mathbf{x})$ subject to $\mathbf{x} \in X$. Which means find a value of **x** that is an element of its allowed values X that will maximize $f(\mathbf{x})$. In this section, vectors will be set in lowercase bold letters. The optimal design, at the end of the following line,

———————————————————● $f(\mathbf{x})$

(1) is better or equivalent to any other feasible design, (2) no better feasible design can be found, and (3) all optimal designs give the same objective function value; therefore, they are equivalent. Methods for implementing this technique include linear programming, nonlinear programming, and discrete programming.

3.3.2.2 Multiobjective Multicriterion Optimization

In multiobjective, multicriterion optimization, there is a single decision maker but multiple evaluation criteria, For example, we could have criteria x_1 and x_2 and objective functions f_1 and f_2. To assure the generality of the mathematical analysis of this section, we allow the objective functions to be mathematical combinations of evaluation criteria, e.g., $f_1 = x_1 + x_2$. Usually, we want to maximize $f_k(\mathbf{x})$ for $k = 1, 2, 3, \ldots, k$ subject to $\mathbf{x} \in X$. Figure 3.8 illustrates this graphically for two-dimensions ($k = 2$).

3.3.2.3 Pareto-optimal Designs

Invariably, there are multiple inferior designs in the feasibility space (the gray area) of Fig. 3.8. There will be no designs at or near the ideal point. Some designs will

Fig. 3.8 Feasibility space

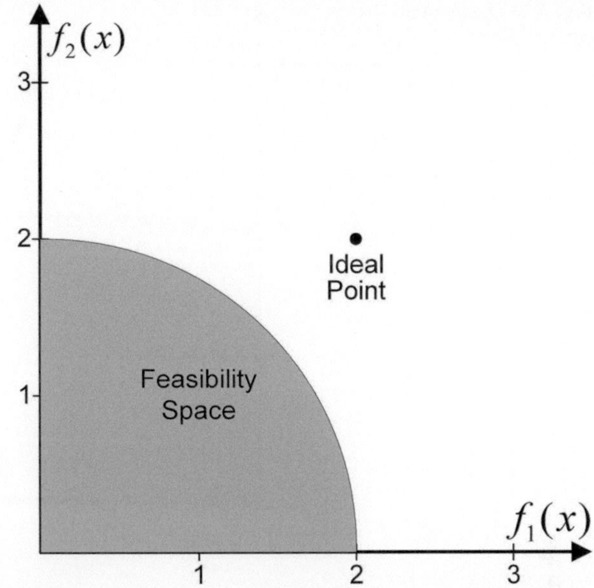

Fig. 3.9 Pareto-optimal designs

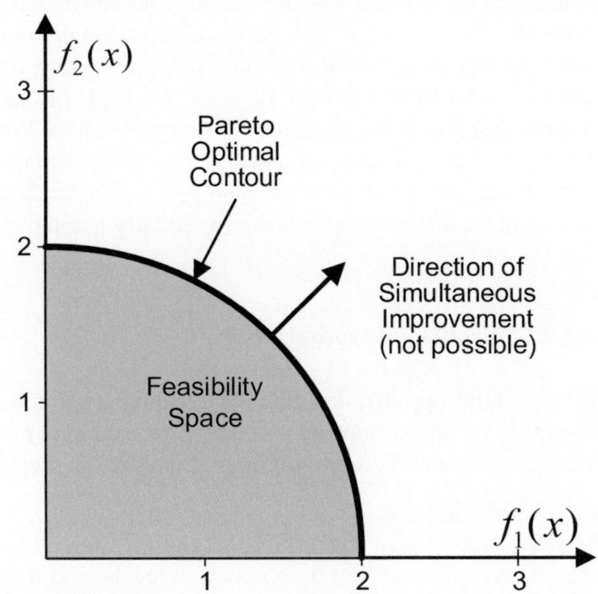

exist along the quarter circle bounding the feasibility space. All designs on this contour are called Pareto-optimal designs. For these designs, not all objective functions can be improved at the same time [24]. In other words, improving one objective will degrade the other (Fig. 3.9).

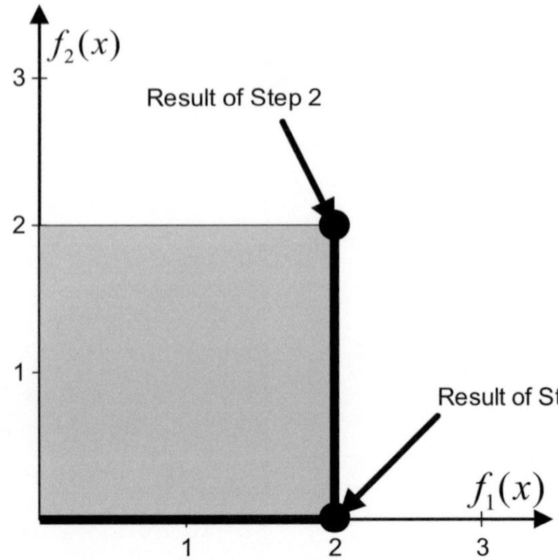

Fig. 3.10 Sequential optimization

It is important to recognize that Pareto-optimal designs (1) are not necessarily better than all feasible designs, (2) have no better feasible designs, and (3) are not equivalent to each other; therefore, an improvement of one objective function must result in worsening of at least one other objective function. There is a possible problem with Pareto-optimal designs. We typically have multiple Pareto-optimal designs that give different objective function values. Which one should be selected? To answer this question, we need additional information from the decision maker concerning his or her priorities and values. Depending on the type of preference information, different designs are potentially available.

3.3.2.4 Sequential Optimization

In sequential optimization (the lexicographic method), strict preference order of objective functions must be available, for example, $f_1 \succ f_2 \succ \cdots \succ f_k$, where the symbol $\succ$ means "is more important than" or "is preferred to." The method is

Step 1. Maximize f_1.
Step 2. Keep f_1 at its maximal level and then maximize f_2.
Step 3. Keep f_1 and f_2 at their maximal levels, and then maximize f_3.

$\vdots$

Step k. Keep $f_1, f_2, \ldots, f_{k-1}$ at maximal levels, and then maximize f_k.

Here is a graph for the two-dimensional case (Fig. 3.10).

A possible problem is that for certain shapes of the feasibility space the result might not depend on all of the criteria as shown in Fig. 3.11.

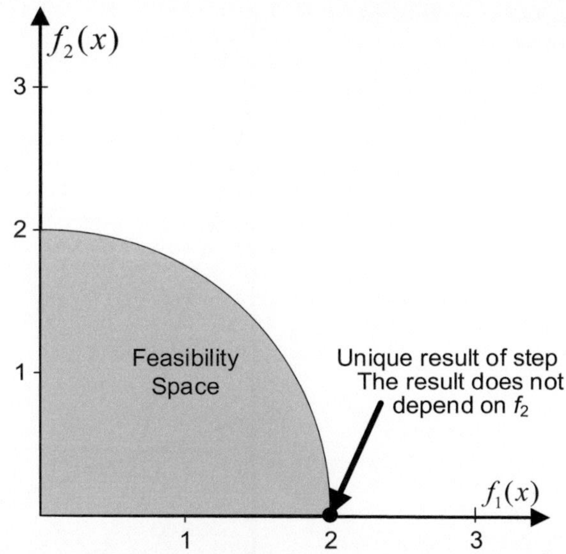

Fig. 3.11 A problem for sequential optimization

Table 3.8 A water treatment plant

	Plant 1	Plant 2
f_1 = profit ($)	1000	1100
f_2 = water quality improvement (g/m^3)	1.2	1

Case 1. If $f_1 \succ f_2$, then Plant 2 is better
Case 2. If $f_2 \succ f_1$, then Plant 1 is better

For some feasibility spaces, the design might not depend on all of the objective functions. As previously stated, the distribution of alternatives in the feasibility space is important. We now show two examples of using this method: designing a water treatment plant and solving an algebraic problem. We use these same two examples for the next three methods.

A Water Treatment Plant Example

Suppose that we are selecting the design for a new water treatment plant. We have the following suggestions (Table 3.8).

The Algebraic Example

We need to maximize the following two objective functions, the sum and the difference of the two evaluation criteria.

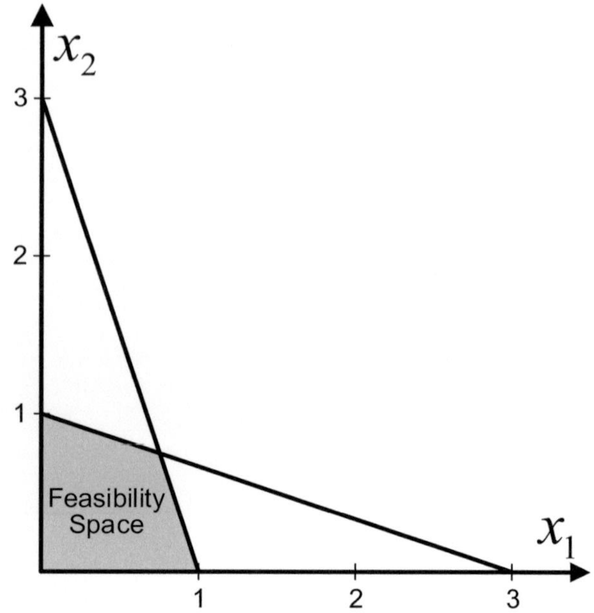

Fig. 3.12 A graph of the constraints

Maximize
$f_1 = x_1 + x_2$ and $f_2 = x_1 - x_2$
subject to these constraints
$x_1, \ x_2 \geq 0$
$3x_1 + x_2 \leq 3$
$x_1 + 3x_2 \leq 3$

Figure 3.12 contains a graph of these constraints.

This $x_1 - x_2$ plane is called the problem domain. It contains the feasibility space, which is the region contain all feasible designs. Figure 3.13 adds lines for the two objective functions.

The double line arrows show the direction of improvement for the objective functions. The optima are at the points where the objective functions terminate at the corners of the feasibility space.

Case 1. If $f_1 \succ f_2$, then the optimum is at $\left(\frac{3}{4}, \frac{3}{4}\right)$.

Case 2. If $f_2 \succ f_1$, then the optimum is at (1, 0).

3.3.2.5 The ε-Constraint Method

In the ε-constraint method, one of the objective functions is the most important and the others have minimal acceptable values.

Using it, we need to maximize $f_1(\mathbf{x})$ subject to $\mathbf{x} \in X$

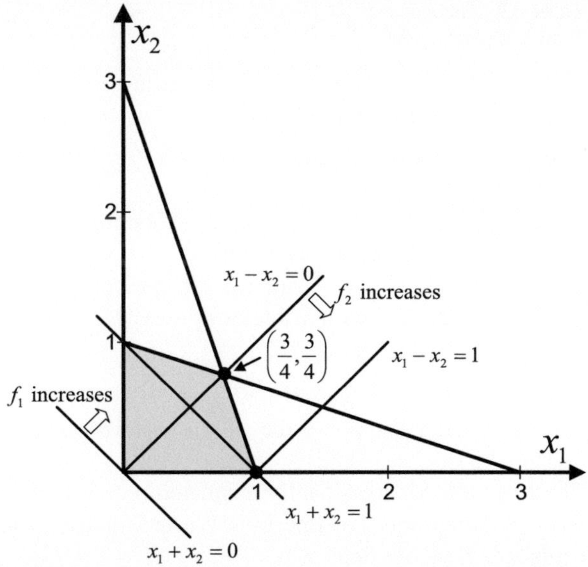

Fig. 3.13 The constraints along with the objective functions

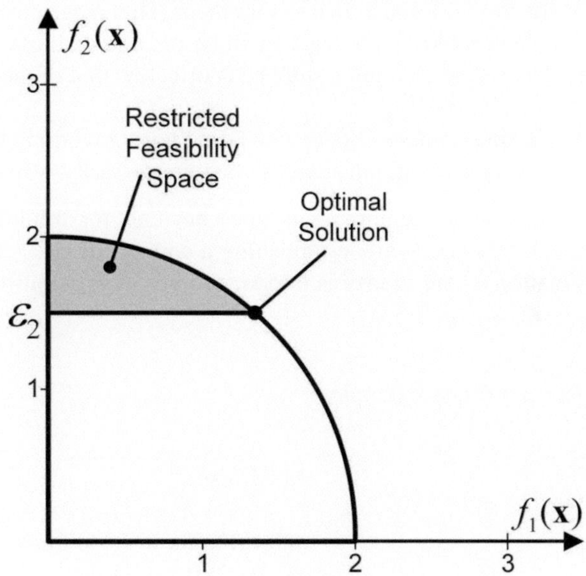

Fig. 3.14 The ε-constraint method

$$
\begin{aligned}
f_2(\mathbf{x}) &\geq \varepsilon_2 \\
f_3(\mathbf{x}) &\geq \varepsilon_3 \\
&\vdots \\
f_K(\mathbf{x}) &\geq \varepsilon_K
\end{aligned}
$$

Figure 3.14 contains a graph for two-dimensions, with $\varepsilon_2 = 1.5$.

Table 3.9 The water treatment plant, again

	Plant 1	Plant 2
f_1 = profit ($)	1000	1100
f_2 = water quality improvement (g/m^3)	1.2	1

The Water Treatment Plant Example (Table 3.9)

Assume f_1 is more important than f_2. Then maximize f_1 subject to $f_2 \geq \varepsilon_2$.
If $\varepsilon_2 \leq 1$, then Plant 2 is better.
If $1 < \varepsilon_2 \leq 1.2$, then Plant 1 is the choice.
If $\varepsilon_2 > 1.2$, then no feasible design exists.

A Real-World Example of the ε-Constraint Method

These are the congressionally mandated requirements for moving nuclear waste to the U.S. Department of Energy's Waste Isolation Pilot Plant (WIPP) located near Carlsbad, New Mexico:

- The cost of the system shall be minimized. (Note: It is usually a mistake to use the term minimize in a requirement. This requirement would be better stated as, "The preferred alternative will be the one with the lowest cost.")
- The risk to the public shall be at or below that of the existing system. (This is the ε-constraint)
- Existing Federal, State, and Tribal laws governing this subsystem must defer to this legislation, but shall not be unnecessarily waived.

This set of requirements does not call for minimization of cost and risk. It minimizes cost while establishing a ceiling for risk. This avoids the questionable situation where money is traded for lives at equal priority.

The Algebraic Example

Maximize
$f_1 = x_1 + x_2$ and $f_2 = x_1 - x_2$
subject to these constraints
$x_1, \ x_2 \geq 0$
$3x_1 + x_2 \leq 3$
$x_1 + 3x_2 \leq 3$

This is the same algebraic example that we studied in Fig. 3.13. Assume that f_1 is more important than f_2. Put a constraint on f_2, $\varepsilon_2 = \frac{1}{2}$, then

Fig. 3.15 The algebraic example with the ε-constraint = 0.5

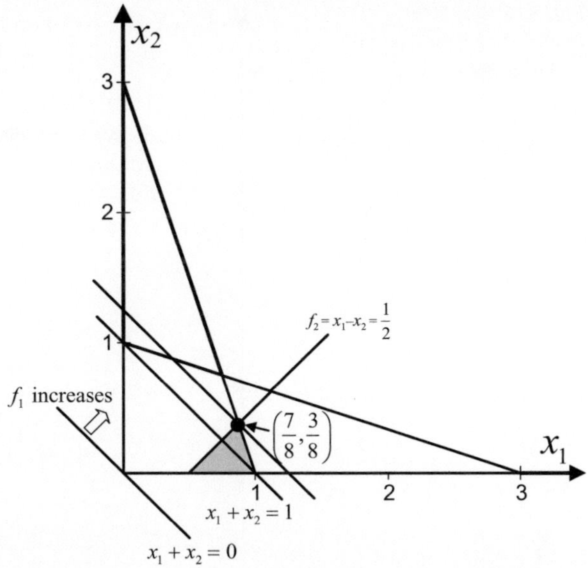

Maximize
$f_1 = x_1 + x_2$
subject to these constraints
$x_1,\ x_2 \geq 0$
$3x_1 + x_2 \leq 3$
$x_1 + 3x_2 \leq 3$
$f_2 = x_1 - x_2 \geq \frac{1}{2}$

Figure 3.15 shows a graph of the algebraic example with the ε-constraint = 0.5.

A possible problem with this method is shown in Fig. 3.16. If the constraints are too high, there might not be a feasible design in the specified region.

3.3.2.6 The Sum Combining Function

In the Sum of Weighted Scores method (used in most trade-off studies), the importance is assigned to the objectives using weights.

$\alpha_1,\ \alpha_2,\ \ \ldots \alpha_k \geq 0$
$\alpha_1 + \alpha_2 + \ \ \ldots\ \ + \alpha_k = 1$
We need to maximize
$\alpha_1 f_1(\mathbf{x}) + \alpha_2 f_2(\mathbf{x}) + \cdots + \alpha_k f_k(\mathbf{x})$
subject to
$\mathbf{x} \in X$

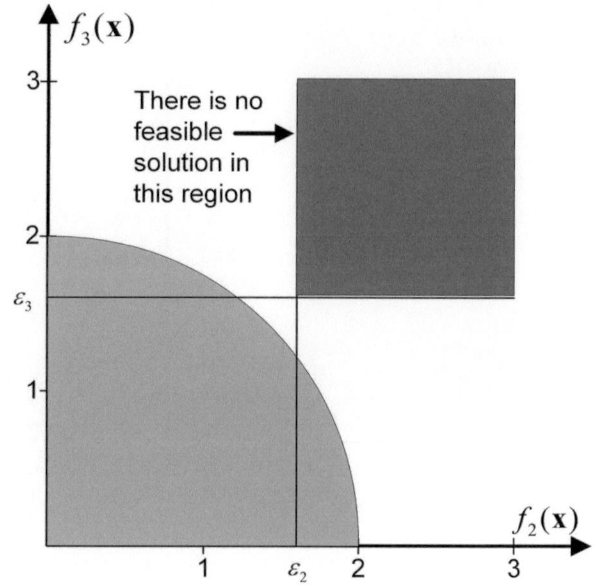

Fig. 3.16 Possible problem: no solution in the specified region

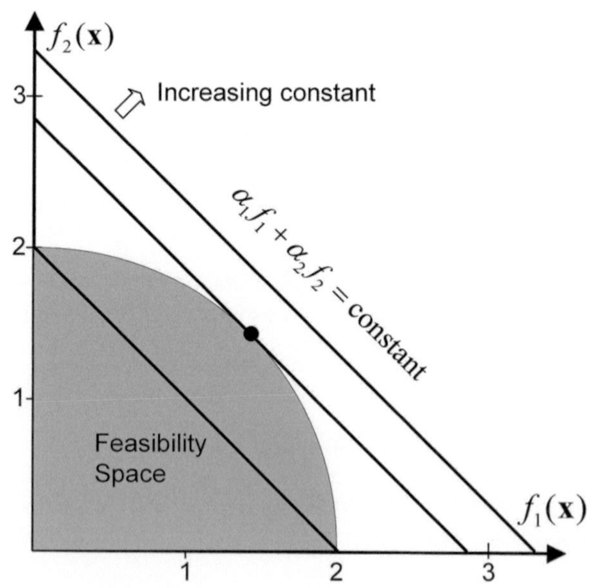

Fig. 3.17 Possible problem: no solution in the specified region

In other words, we are trying to find the design that will produce the maximum value of $\alpha_1 f_1(\mathbf{x}) + \alpha_2 f_2(\mathbf{x}) + \cdots + \alpha_k f_k(\mathbf{x})$.

There two possible problems with this method. First, the new objective function, $\alpha_1 f_1(\mathbf{x}) + \alpha_2 f_2(\mathbf{x}) + \cdots + \alpha_k f_k(\mathbf{x})$ usually has no meaning (apples are added to oranges), therefore, changing units might change the recommended alternative. Second, there might not be a solution in the feasibility space (Fig. 3.17).

Table 3.10 The water treatment plant with weights

	Plant 1	Plant 2	α_k
$f_1 =$ profit ($)	1000	1100	0.5
$f_2 =$ water quality improvement (g/m^3)	1.2	1	0.5
$\alpha_1 f_1 + \alpha_2 f_2$	500 + 0.6	550 + 0.5	
$\alpha_1 f_1 + \alpha_2 f_2$	500.6	550.5	Plant 2 is best

Table 3.11 The water treatment plant with different units

	Plant 1	Plant 2	α_k
$f_1 =$ profit ($1000)	1	1.1	0.5
$f_2 =$ water quality improvement (mg/m^3)	1200	1000	0.5
$\alpha_1 f_1 + \alpha_2 f_2$	600 + 0.5	500 + 0.55	
$\alpha_1 f_1 + \alpha_2 f_2$	600.5	500.55	Plant 1 is best

Clearly, this is a problem

We will now give two examples of solving for the sum of two weighted objective functions.

The Water Treatment Plant Example (Tables 3.10 and 3.11)

Normalizing the Objective Functions

To overcome this problem, we must normalize the objective functions. The following employs linear normalization. (Wymorian scoring functions could also be used).

$$\overline{f}_i = \frac{f_i - f_i^{\min}}{f_i^{\max} - f_i^{\min}} \in [0, 1]$$

The overbar indicates a normalized function. Profit goes from $500 to $2000. Water quality goes from 0 to 2 g/m^3 (Fig. 3.18).

$$\overline{f}_1(\text{for}\, f_1 = 1000) = \frac{1000 - 500}{2000 - 500} = \frac{1}{3}$$

$$\overline{f}_1(\text{for}\, f_1 = 1100) = \frac{1100 - 500}{2000 - 500} = \frac{2}{5}$$

Normalizing f_2, the water quality (Fig. 3.19).

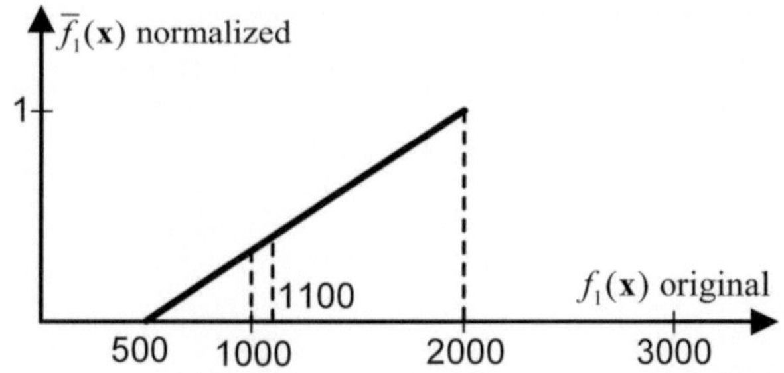

Fig. 3.18 Normalizing the profit in dollars

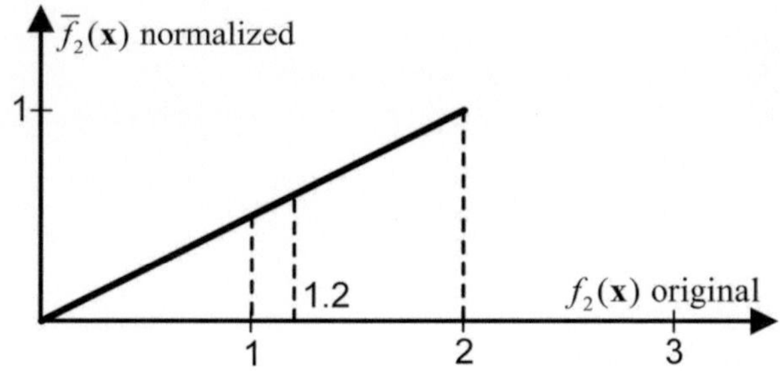

Fig. 3.19 Normalizing the water quality in 2 g/m^3

Table 3.12 The water treatment plant with normalized values

	Plant 1	Plant 2	α_k
$\bar{f}_1$	1/3	2/5	0.5
$\bar{f}_2$	3/5	1/2	0.5
$\alpha_1\bar{f}_1 + \alpha_2\bar{f}_2$	1/6 + 3/10	1/5 + 1/4	
$\alpha_1\bar{f}_1 + \alpha_2\bar{f}_2$	7/15	9/20	Plant 1 is best

$$\bar{f}_2(\text{for}\, f_2 = 1) = \frac{1-0}{2-0} = \frac{1}{2}$$
$$\bar{f}_2(\text{for}\, f_2 = 1.2) = \frac{1.2-0}{2-0} = \frac{3}{5}$$

The Normalized Water Treatment Plant Example (Table 3.12)

The Normalized Algebraic Example

Maximize

$f_1 = x_1 + x_2$ and $f_2 = x_1 - x_2$

subject to these constraints

$x_1,\; x_2 \geq 0$

$3x_1 + x_2 \leq 3$

$x_1 + 3x_2 \leq 3$

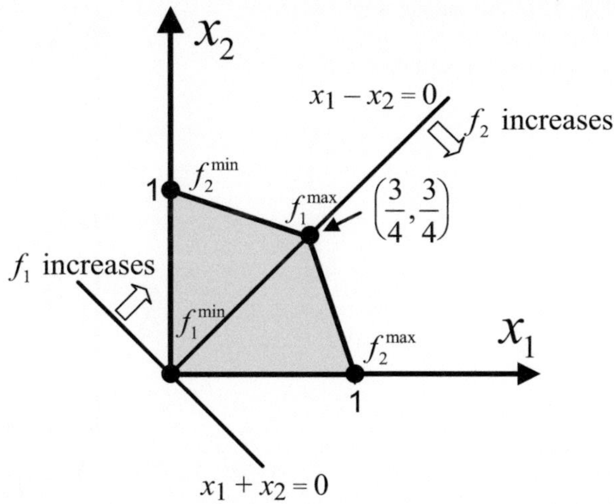

Fig. 3.20 Finding f_1^{max}, f_1^{min}, f_2^{max} and f_2^{min} from Fig. 3.15

This is the same algebraic problem that we studied before in Figs. 3.13 and 3.15. In order to normalize the objective functions we need values for f_1^{max}, f_1^{min}, f_2^{max} and f_2^{min}. We can get them from Fig. 3.15, as is shown in Fig. 3.20.

$$f_1^{max} = \max\{x_1 + x_2\} = 3/2, \qquad f_1^{min} = \min\{x_1 + x_2\} = 0$$
$$f_2^{max} = \max\{x_1 - x_2\} = 1, \qquad f_2^{min} = \min\{x_1 - x_2\} = -1$$

Now we can normalize the objective functions.

$$\bar{f}_1 = \frac{f_1 - f_1^{min}}{f_1^{max} - f_1^{min}} = \frac{(x_1 + x_2) - 0}{3/2} = \frac{2}{3}(x_1 + x_2)$$

$$\bar{f}_2 = \frac{f_2 - f_2^{min}}{f_2^{max} - f_2^{min}} = \frac{(x_1 - x_2) + 1}{2} = \frac{1}{2}(x_1 - x_2)$$

Let us arbitrarily make α_1 twice α_2. So that
$2\alpha_2 + \alpha_2 = 1 \Rightarrow \alpha_1 = \frac{2}{3}, \quad \alpha_2 = \frac{1}{3}$.
Putting these values into the above equations produces

$$\alpha_1\bar{f}_1 + \alpha_2\bar{f}_2 = \underbrace{\frac{2}{3}\cdot\frac{2}{3}}_{\frac{8}{18}}(x_1 + x_2) + \underbrace{\frac{1}{3}\cdot\frac{1}{2}}_{\frac{3}{18}}(x_1 - x_2) = \frac{11}{18}x_1 + \frac{5}{18}x_2$$

Now if you can remember back to the beginning of this section, we were trying to find the design that would produce the maximum value of $\alpha_1 f_1(\mathbf{x}) + \alpha_2 f_2(\mathbf{x}) + \cdots + \alpha_k f_k(\mathbf{x})$. In this example that means finding values for x_1 and x_2 that will maximize $\alpha_1 f_1(\mathbf{x}) + \alpha_2 f_2(\mathbf{x})$ within the feasible region.

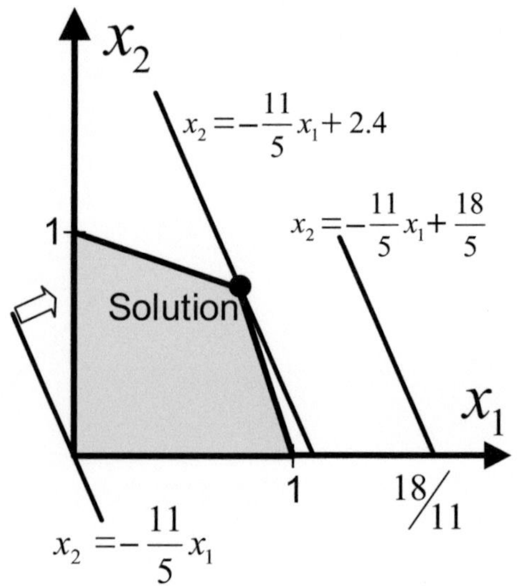

Fig. 3.21 The optimal sum of weighted scores design for the algebraic example

To get some examples, take this equation $\alpha_1\overline{f}_1 + \alpha_2\overline{f}_2 = \frac{11}{18}x_1 + \frac{5}{18}x_2$, set it equal to 0 and to 1, and then solve for x_1 and x_2. First, let us set it equal to 0.

$$\frac{11}{18}x_1 + \frac{5}{18}x_2 = 0$$
$$11x_1 + 5x_2 = 0$$
$$x_2 = -\frac{11}{5}x_1$$

Now let us set it equal to 1.

$$\frac{11}{18}x_1 + \frac{5}{18}x_2 = 1$$
$$11x_1 + 5x_2 = 18$$
$$x_2 = -\frac{11}{5}x_1 + \frac{18}{5}$$

These equations will be the extremities of the problem space. So let us plot these lines on Fig. 3.21.

As we are sweeping the line $\alpha_1\overline{f}_1 + \alpha_2\overline{f}_2$ between 0 and 1, we encounter the line $x_2 = -\frac{11}{5}x_1 + 2.4$. It is the last line that contacts the feasible space. So it is the best design. It has coordinates of (¾, ¾).

3.3.2.7 The Ideal Point

The ideal point is defined to be the point where each objective function takes on its maximum value, and at other times, its definition tends to be subjective, as in this example. The best design is one that is the *closest* to the ideal point. But what does "the closest" mean? To clarify the concept of closeness, we offer the following example which shows three instances of closeness all with the same ideal point, but with different results. Each of the three metrics results in a different best design. The top graph in Fig. 3.22 shows the Euclidian distance metric, ρ_2. It defines the best design as the one that has the smallest hypotenuse to the ideal point. Next, the ρ_1 summation distance defines the best design as the one that has the smallest sum of the two triangle sides. Assume that you are walking on city streets that run horizontally and vertically (city block). Your expense is the sum of your travel distance, so use the ρ_1 summation distance. Finally, we use ρ_∞ and find the design in feasibility space that minimizes the biggest of the triangles sides. Imagine that you and your fellow detective are following a suspect. To lessen the chances that he will see you, *you* only follow when he is going horizontally and your *partner* only follows when he is going vertically. Your exposure risk is the longest leg. Therefore, the best design will be the one with the smallest ρ_∞ distance.

A construction technique for these figures is to pretend that you are blowing up balloons. In the top figure, you blow up the balloon and when it just touches the feasibility space, that is the best design. In the middle figure, your balloon is diamond shaped. In the bottom figure, your balloon is square with its sides parallel to the f_1 and f_2 axes.

Now consider a set of alternatives $\underline{f} = (f_1, \ldots, f_k)$. An underbar indicates a vector, or a collection of points. Where $\underline{f}' = \left(\underline{f}'_1, \ldots, \underline{f}'_k\right)$ represents any general point. Then

$\rho_\infty = \left(\underline{f}, \underline{f}'\right) = \max\{\alpha_1|f_1 - f'_1|, \ldots, \alpha_k|f_k - f'_k|\}$. That is, ρ_∞ is looking for the largest value of $\alpha_i|f_i - f_i'|$. Next, let us consider the ρ_1 sum of weighted scores combining function.

$$\rho_1\left(\underline{f, f'}\right) = \alpha_1|f_1 - f'_1| + \alpha_2|f_2 - f'_2| + \cdots + \alpha_K|f_K - f'_K|$$

In going from one design to the next, compensation among objective functions is possible. In general, $\rho_P\left(\underline{f}, \underline{f}'\right) = \alpha_1|f_1 - f'_1|^P + \alpha_2|f_2 - f'_2|^P + \cdots + \alpha_K|f_K - f'_K|^P$ and we need to minimize $\rho\left(\underline{f}(\underline{x}), \underline{f}^*\right)$ subject to $\underline{x} \in X$. We want the alternative that is the "closest" to the ideal point. We have the same problem with units as with the weighting method, therefore we will use normalized objective functions.

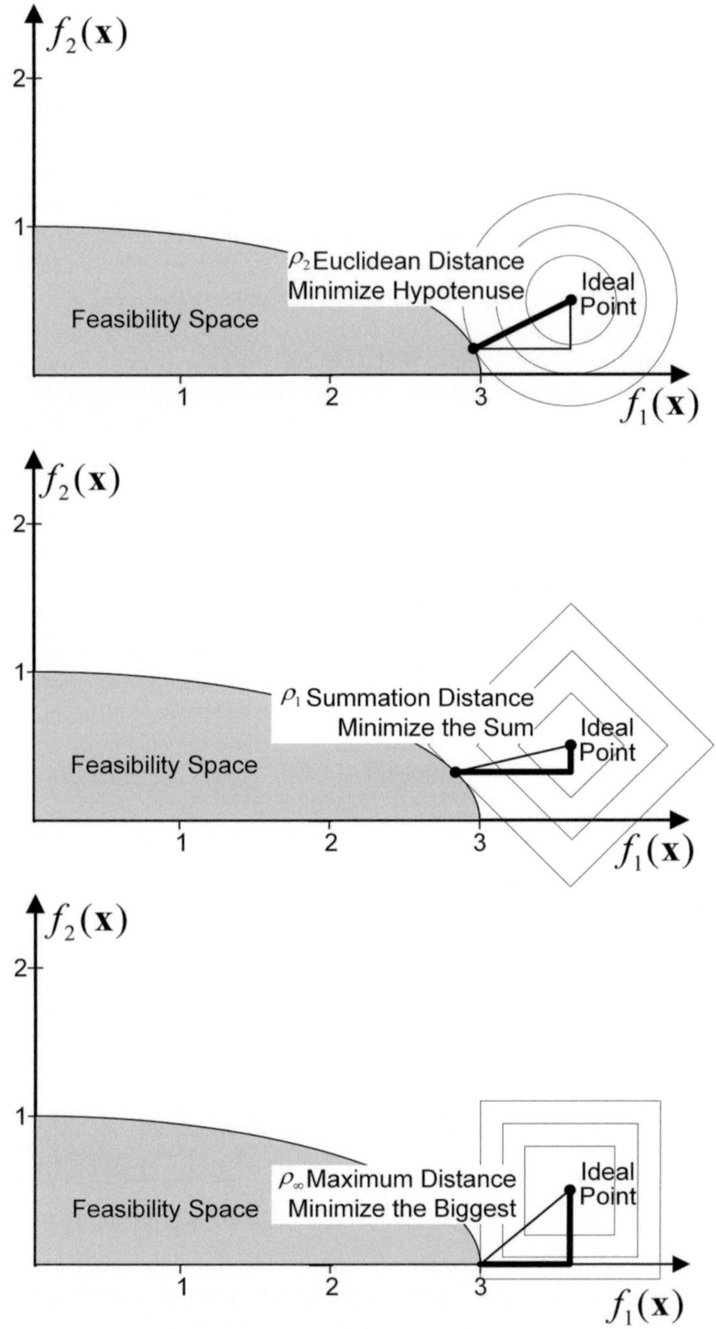

Fig. 3.22 Three definitions of closeness to the ideal point. The *circles* and *squares* are like balloons that blow up until they touch the feasibility region

Table 3.13 The water treatment plant for the ideal point

	Plant 1	Plant 2	α_k
$\overline{f}_1$	1/3	2/5	0.5
$\overline{f}_2$	3/5	1/2	0.5

The Water Treatment Plant Example (Table 3.13)

We define the ideal point to be (1, 1). $\overline{f}_1^* = \overline{f}_2^* = 1$

Now let us find the maximum deviation from the ideal point for Plant 1

$$\underbrace{\rho_\infty\left(\underline{\overline{f}}, \underline{\overline{f}}^*\right)}_{\text{for Plant 1}} = \max\left\{ \underbrace{0.5\left|\frac{1}{3} - 1\right|}_{\frac{1}{3}}, \underbrace{0.5\left|\frac{3}{5} - 1\right|}_{\frac{1}{5}} \right\} = \frac{1}{3}$$

Now find the maximum deviation for Plant 2

$$\underbrace{\rho_\infty\left(\underline{\overline{f}}, \underline{\overline{f}}^*\right)}_{\text{for Plant 2}} = \max\left\{ \underbrace{0.5\left|\frac{2}{5} - 1\right|}_{\frac{3}{10}}, \underbrace{0.5\left|\frac{1}{2} - 1\right|}_{\frac{1}{4}} \right\} = \frac{3}{10}$$

Finally, choose the smallest maximum deviation

Since $\underbrace{\frac{3}{10}}_{0.3} < \underbrace{\frac{1}{3}}_{0.333}$, Plant 2 is the better choice because it is closer to the ideal point.

However, for $\rho = 1$,

$$\underbrace{\rho_\infty\left(\underline{\overline{f}}, \underline{\overline{f}}^*\right)}_{\text{for Plant 1}} = \underbrace{0.5\left|\frac{1}{3} - 1\right|}_{\frac{1}{3}} + \underbrace{0.5\left|\frac{3}{5} - 1\right|}_{\frac{1}{5}} = \frac{8}{15}$$

$$\underbrace{\rho_\infty\left(\underline{\overline{f}}, \underline{\overline{f}}^*\right)}_{\text{for Plant 2}} = \underbrace{0.5\left|\frac{2}{5} - 1\right|}_{\frac{3}{10}} + \underbrace{0.5\left|\frac{1}{2} - 1\right|}_{\frac{1}{4}} = \frac{11}{20}$$

Since

$\frac{8}{15} < \frac{11}{20}$, this time Plant 1 is the better choice

Once again, we are showing that the preferred alternative depends on the input data and the combining function.

Table 3.14 The water treatment plant for the anti-ideal point

	Plant 1	Plant 2	α_k
$\bar{f}_1$	1/3	2/5	0.5
$\bar{f}_2$	3/5	1/2	0.5

3.3.2.8 Anti-ideal Point

In the anti-ideal point method, the anti-ideal point (the nadir) and the distance between objective vectors are given. This ideally worst point might be subjective; in fact, it is often the status quo. We need to find the design that maximizes $\rho\left(\underline{f}(\mathbf{x}),\underline{f}_*\right)$ subject to $\mathbf{x}\in X$ (Table 3.14).

We define the anti-ideal point to be (0, 0).$\bar{f}_{1*}=\bar{f}_{2*}=0$. We want the alternative that is the "farthest" away from the anti-ideal point (0, 0).

For $\rho=\infty$

$$\underbrace{\rho_\infty\left(\underline{\bar{f}},\underline{\bar{f}}_*\right)}_{\text{for Plant 1}}=\max\left\{\underbrace{0.5\left|\frac{1}{3}-0\right|}_{\frac{1}{6}},\ \underbrace{0.5\left|\frac{3}{5}-0\right|}_{\frac{3}{10}}\right\}=\frac{3}{10}$$

$$\underbrace{\rho_\infty\left(\underline{\bar{f}},\underline{\bar{f}}_*\right)}_{\text{for Plant 2}}=\max\left\{\underbrace{0.5\left|\frac{2}{5}-0\right|}_{\frac{1}{5}},\ \underbrace{0.5\left|\frac{1}{2}-0\right|}_{\frac{1}{4}}\right\}=\frac{1}{4}$$

Since $\frac{3}{10}>\frac{1}{4}$, Plant 1 is the better choice

And for $\rho=1$

$$\underbrace{\rho_1\left(\underline{\bar{f}},\underline{\bar{f}}_*\right)}_{\text{for Plant 1}}=\underbrace{0.5\left|\frac{1}{3}-0\right|}_{\frac{1}{6}}+\underbrace{0.5\left|\frac{3}{5}-0\right|}_{\frac{3}{10}}=\frac{7}{15}$$

$$\underbrace{\rho_1\left(\underline{\bar{f}},\underline{\bar{f}}_*\right)}_{\text{for Plant 2}}=\underbrace{0.5\left|\frac{2}{5}-0\right|}_{\frac{1}{5}}+\underbrace{0.5\left|\frac{1}{2}-0\right|}_{\frac{1}{4}}=\frac{9}{20}$$

Since $\frac{7}{15}>\frac{9}{20}$, Plant 1 is again the better choice

3.3.2.9 Section Summary

In light of these combining methods and examples, the best design depends on the objective functions, their importance, their weights, and the constraints. The best design also depended on the units for the objective functions. Therefore, we always normalize the objective functions.

3.3.2.10 Some Nontraditional Combining Functions

Students who are not conversant with the mathematics used in this section may skip this section without loss of continuity. This section presents some novel combining functions. Many of them have not been published in the trade-offs or risk literature. They are presented here to show the large variety of possible combining functions.

The Certainty Equivalent Combining Function

This section is based on the work of Ferenc Szidarovszky, Emeritus Professor of Systems Engineering at the University of Arizona.

For some situations, the variance of the evaluation criteria might be more important than the mean of the evaluation criteria. For example, assume that regulations require an electric utility company to produce 120 ± 5 V rms. In Fig. 3.23, the top example (red line) has an acceptable mean, but a big variance. It might be a more serious violation than the bottom example (blue line) that has an unacceptable mean but a small variance.

In light of the foregoing, we propose a new combining function to deal with this situation. If power supplied fails to meet the power demanded, then there is a

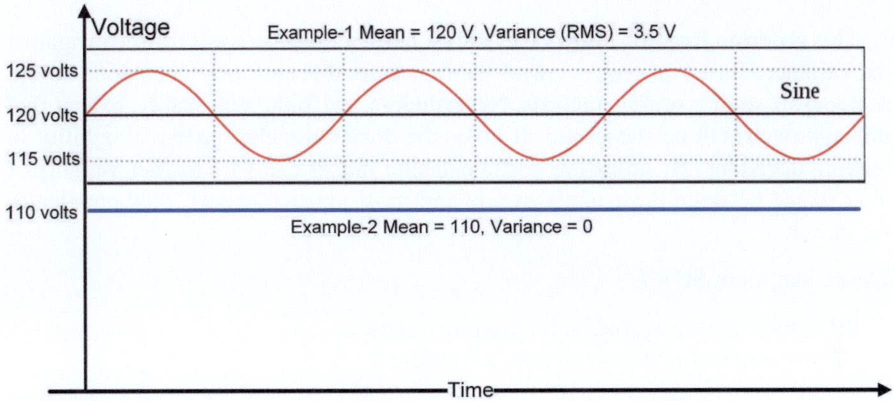

Fig. 3.23 Two possible electric voltage profiles

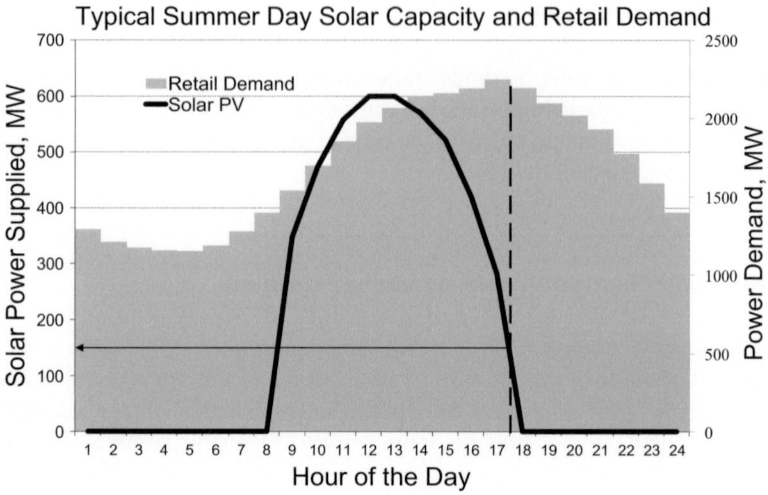

Fig. 3.24 Solar power supplied and electric power demanded

shortage and the utility has a loss due to penalties, etc. Our shortage function is defined as (Fig. 3.24)

$$\text{Shortage} = -\text{Power Supplied (both traditional and solar)} + \text{Power Demanded}$$

Traditional Power Supplied will be treated as a constant. We now have two stochastic variables, Solar Power Supply and Power Demand. We propose a general treatment of these statistics. Specifically, we propose constructing a criterion function that is a trade-off between the mean *and the variance*. (For the time being, think of the *mean* and the *expected value* to be the same.)

$$\text{Criterion Function} = \alpha\text{Expected value of the loss} + (1-\alpha)\text{Variance of the loss}$$

This criterion function allows the analyst to explicitly trade-off the mean against the variance. Based on our review of the literature, this has never been done before in trade-off studies or risk analysis. Nevertheless, we, quite reasonably, expect that this approach will be successful. It gives the decision maker greater flexibility in making decisions. By adjusting α, the decision maker can vary his/her weights of importance between most likely events and most unusual events. Mathematically we have:

Power Supplied: **SP** + **T**

SP = solar power available, a random variable
T = traditional power available, a constant

Power Demanded: **D**, a random variable

Shortage: $s = \mathbf{D} - (\mathbf{SP} + \mathbf{T})$ (for $s > 0$)
the utility's Loss is a function of the shortage: $l(s)$
Expectation of loss:

$$E[l(s)] = \int\int l(\mathbf{D} - \mathbf{SP} - \mathbf{T}) f_1(\mathbf{SP}) f_2(\mathbf{D}) d\mathbf{SP} d\mathbf{D}$$

Where f_1 and f_2 are probability density functions of **SP** and **D**, respectively. We also need the second moment of the Variance of the loss.

$$E\left[\left(l(s)^2\right)\right] = \int\int l(\mathbf{D} - \mathbf{SP} - \mathbf{T})^2 f_1(\mathbf{SP}) f_2(\mathbf{D}) dSPdD$$

and so

$$\mathrm{Var}[l(s)] = E\left[l(s)^2\right] - E[l(s)]^2$$

The function $l(s)$ depends on the strategy being used to deal with the shortage, for example, fast-starting natural-gas motor-generator sets, agreements for purchase of power via the grid, storing energy with batteries, hydropower, load shedding, brownouts, reducing the voltage of the grid, and planning reserves.

The certainty equivalent is then

$$\text{Certainty Equivalent} = \alpha E[l(s)] + (1 - \alpha)\mathrm{Var}[l(s)]$$

The decision maker then chooses alpha and computes the certainty equivalent for all feasible strategies and the smallest value gives the best decision, showing the best way to deal with the potential shortage.

Exclusive OR Combining Function

If the data seem to need a rule like this

If condition-1 is true, then use criterion-1, else use criterion-2,

then the use of the exclusive OR combining function (XOR) may be quite appropriate.

For example,

If the vehicle has a gasoline engine, then use
Criterion-1: Fuel consumption in highway driving, miles per gallon of gasoline.
Baseline = 23 mpg.
Else, if the vehicle has a diesel engine, then use
Criterion-2: Fuel consumption in highway driving, miles per gallon of diesel fuel.
Baseline = 26 mpg.

Table 3.15 Four types of second-generation flowers

		Sperm from purple (Pp) plant	
		P	p
Eggs from purple (Pp) plant	P	Purple flowers (PP)	Purple flowers (Pp)
	p	Purple flowers (Pp)	White flowers (pp)

Genetics

In the latter half of the nineteenth century, Gregor Mendel, an Augustinian monk, founded the field of genetics [25]. He showed that some genes were dominant and others were recessive. For example, one particular gene coded for the flower color of the pea plant. The dominant allele (represented with the upper case P) coded for purple and the recessive allele (represented with the lower case p) coded for white. The following table illustrates Mendel's model for the inheritance of the alleles of this gene. Each plant has two alleles for the gene controlling color, one allele inherited from each parent. First, Mendel produced plants with purple flowers with one dominant gene, P, and one recessive gene, p. Then, as shown in Table 3.15, these plants produced sperm with (Pp) genes and eggs with (Pp) genes. These combined to produce four types of new flowers.

Because the purple allele P is dominant, the flowers will be purple if either gene is P. The flowers will only be white if both of its genes are recessive (pp).

Bayesian Statistics

When substantial statistical data exist and you have to evaluate rules one at a time, then Bayes' theorem and the Bayesian method offer an effective combining function.

The Boolean AND Function

In a requirements database, there is an implied AND between all of the requirements. For example

ReqF1:. Cookie shall provide a mechanism … to enter desired baking time AND
ReqF2:. Cookie shall display the desired baking time… AND
ReqF3:. Cookie shall … cook the food AND
ReqF4:. Cookie shall … display the remaining baking time AND
ReqF5:. Cookie shall emit an audible signal… AND
ReqF6:. Cookie shall visually indicate … when the baking time is over AND
ReqF7:. Cookie shall execute … BiST.

More Combining Functions

There are several other methods for combining evidence. It is important to note that different methods may be used in different parts of the same hierarchy. It should also be noted that in some cases, the choice of a combining method is more important than acquiring accurate data about the system [26, 27]. Therefore, it is at least as important to choose the appropriate combining method as it is to collect precise data. This is particularly true when the distribution of alternatives in the evaluation criteria space is non-convex [27].

3.3.3 Section Summary (Fig. 3.25)

So far in this part, we have exploited the hierarchical nature of systems design by approaching the discussion of trade-off analyses and ultimately scoring functions through selective decomposition of the system design process. The SIMILAR process suggests that effective system design is achieved in a hierarchical and iterative fashion. A critical subfunction of the system design process is the

Fig. 3.25 The phrase "To a man with a hammer everything looks like a nail," is attributed to Mark Twain. We teach students how to use a tool. Then they use that tool for everything. We should teach them to study each problem and then select the best tool for that problem

investigation of alternative designs. Investigating alternative designs is achieved by first deriving an appropriate evaluation criteria architecture based on customer requirements and preferences. Individual evaluation criteria are evaluated by running models or acquiring measurements on the actual system. The evaluation data are then transformed into normalized numerical scores using standard scoring functions that capture the customers' relative preferences. Scoring functions allow comparisons to be carried out across the many diverse evaluation criteria used for evaluating a given alternative. Comparisons of alternative system designs involve combining scoring function outputs in such a way that decisions regarding preferred system alternatives can be made and defended while concurrently preserving customer preferences and requirements.

3.3.4 Homework

3.3.4.1 Flipping Coins

Assume that we are going to flip a coin in the air and either catch it in the air or let it fall on sand, clay, or some other substance that will not allow it to bounce. Is it possible to alter the coin (for example, by shifting the center of mass, filing an edge, bending the coin, attaching putty) or the flip (for example, by painstakingly perfecting a perfectly repetitive toss) in order to make an unfair coin toss, that is, one that does not have 50–50 outcomes? Why or why not?

If you were flipping two coins, what combining function would you use to determine the probability of getting two heads? This is an example of one of our answers. To determine the probability of flipping two heads, we multiply the probability of each coin turning up heads, like this $p_{\text{Two Heads}} = p_{\text{H1}}$ times p_{H2}. This is the product combining function.

If you are flipping two coins what combining function would you use to determine the probability of getting at least one head?

In rule-based expert systems, each rule has a certainty factor that suggests how certain we are of the result depending on that rule being true. If two rules are true then we should be more certain than if only one rule were true. So we want to have a formula for combining certainty factors so that if two rules are both true then we are more certain of the result than if only one rule is true. Assume, certain factors are between 0 and 100. What combining function would you use for an expert system? Why?

Rule 1 CF = 70
Rule 2 CF = 60
After both rules evaluate to true, CF = 88

What combining function would you use to calculate a cost-to-benefit ratio?

A water treatment plant wants to reduce the amount of mercury, lead, and arsenic in the water. Trace amounts are not of concern, because currently, all three of these concentrations are below EPA thresholds. What combining function would you use to choose the best alternative water treatment process? Why?

What combining function would you use to determine the best National Football League (NFL) quarterback using these criteria?

BM1 = (Completed Passes)/(Pass Attempts)
BM2 = (Passing Yards)/(Pass Attempts)
BM3 = (Touchdown Passes)/(Pass Attempts)
BM4 = Interceptions/(Pass Attempts)

During a planned manned mission to Mars, the astronauts will grow beans and rice. Lots of beans and a little rice are just as good as lots of rice and a few beans. What combining function would you use for these criteria? These astronauts will also need a system that produces oxygen and water. What combining function would you use for these criteria?

What combining function would you use to select the greatest athlete of the century using the criteria of Number of National Championship Rings and Peak Salary?

What combining function would you use to select the NCAA softball player of the week using the criteria of home runs and pitching strikeouts?

What combining function would you use to select a movie for your date using the criteria of amount of romance, action, and comedy?

What combining function do teams in the National Basketball Association (NBA) seem to use when drafting basketball players? Assume that their only criteria are height and number of assists.

3.3.4.2 Baseball Combining Functions

In Section 20, you estimated appropriateness of the following baseball batting outcomes for evaluation criteria: a homerun (HR), a double (2B), a single (1B), a run batted in (RBI), a base on balls (BB), reaching base on error (ROE), a strikeout (SO), and grounding into a double play (GDP). Runs batted in (RBI), strikeouts (SO), and grounding into a double play (GDP) had some problems, so we will leave them out. So, suppose we want to evaluate two batters using the criteria of homeruns (HR), doubles (2B), and singles (1B). Weights of importance are 0.4 for homeruns, 0.2 for doubles, and 0.1 for singles. (This is almost the same as slugging percentage.) Fill in Table 3.16.

Table 3.16 Different combining functions for baseball batters

	Weights of importance	Homerun hitter, score	Singles Hitter, score
Normalized homeruns, HR	0.4	0.35	0.1
Normalized doubles, 2B	0.2	0.1	0.3
Normalized singles, 1B	0.1	0.1	0.9
Sum combining function alternative rating	–	–	–
Product combining function with no weights, alternative rating	–	–	–
Product combining function with weights in exponents, alternative rating	–	–	–
Sum minus product combining function alternative rating	–	–	–
Compromise combining function with $p = 2$ alternative rating	–	–	–
Compromise combining function with $p = 10$ alternative rating	–	–	–

Who do you think is the best batter?

3.4 Section 23. Quantitative Assessments

3.4.1 Measures

Evaluation criteria, measures, and technical performance measures are all used to quantify system performance. While these terms are often used interchangeably, a useful distinction can be made. Evaluation criteria are used in requirements discovery, trade-off studies, and risk analyses. Measures are used to help manage a company's processes. TPMs are used to mitigate risk during design and manufacturing.

Performance, schedule, and cost **evaluation criteria** show how well the system satisfies its requirements (for example, the car accelerated from 0 to 60 in 6.2 s). Evaluation criteria are often called measures of performance, measures of effectiveness, or figures of merit. Such measurements are made throughout the evolution of the system, based first on estimates by the design engineers, then on models, simulations, prototypes, and finally on the real system. Evaluation criteria are used to quantify system requirements. Evaluation criteria are also used to help select amongst alternative designs in trade-off studies where criteria are traded off; that is, switching from one alternative to another increases the value of one criterion, while decreasing others. Evaluation criteria describe the failures and their consequences in risk analyses.

Measures (used by the metrics analysis group) usually relate to the process, not the product. Therefore, they do not always relate to specific system requirements. Rather some measures relate to the company's mission statement and subsequent goals. For example, a useful measure is the percentage of requirements that change after the System Requirements Review (SRR).

Measure. A measure indicates the degree to which an entity possesses and exhibits a quality or an attribute. A measure has a specified method, which, when executed, produces values (or metrics) for the measure.

Metric. A measured, calculated, or derived value (or number) used by a measure to quantify the degree to which an entity possesses and exhibits a quality or an attribute.

Measurement. A value obtained by measuring, which makes it a type of metric.

Technical performance measures (TPMs) are tools that show how well a system satisfies critical requirements or avoids risks. TPMs provide assessments of the product and the process through design, modeling, breadboards, prototypes, implementation, and test. If the system primarily employs commercial off-the-shelf (COTS) components, and an evaluation criterion is being poorly met by all alternatives, then that criterion could be made a TPM. Then, if a supplier ends up providing a new product that satisfies that criteria, this fact would be discovered immediately.

3.4.2 Technical Performance Measures

Technical performance measures (TPMs) are tools used to show how well a system satisfies its requirements, meets its goals, or avoids risks. This section is based on Oakes, Botta, and Bahill [28]. TPMs provide assessments of the product and the process through design, implementation, and test. While they assess the product and its associated process, they are primarily employed in the context of the product. TPMs are used to (1) forecast values to be achieved through planned technical effort, (2) measure differences between achieved values and those allocated to the product, (3) determine the impact of these differences, and (4) trigger optional design reviews.

TPMs are critical measures that a project monitors to ensure that the technical objectives (requirements) of a product are going to be realized. Typically, TPMs have planned values at defined time increments, against which the actual values are plotted. See Fig. 3.26. Monitoring TPMs allows trend detection and correction, and helps identify possible performance problems prior to incurring significant cost or schedule overruns.

The purpose of TPMs is to (1) provide visibility of actual versus planned performance, (2) provide early detection or prediction of problems requiring management attention, (3) support assessment of the impact of proposed changes, and (4) help manage risk.

The method for measuring the data for the TPM varies with life-cycle phase. In the beginning, data are used from legacy systems, blue-sky guesses, and approximations. Then data are derived from models and simulations. Later data are collected from prototypes. Finally, data are acquired from measurements on rudiments of the real system. Even the planned values might not be known at the

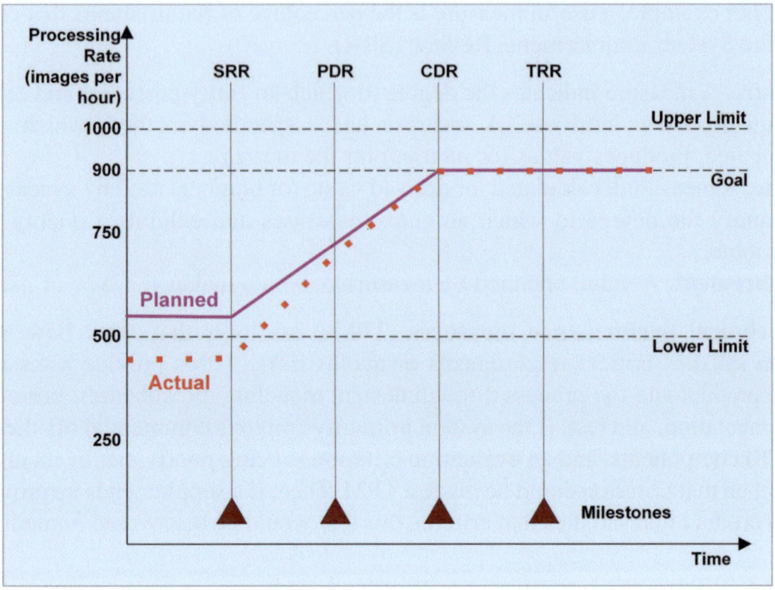

Fig. 3.26 TPM for image processing rate

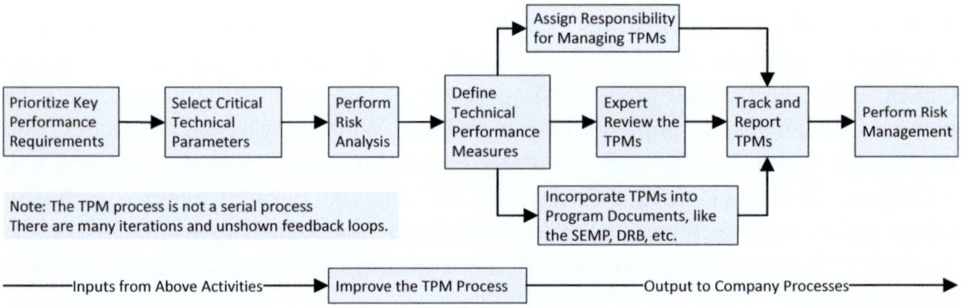

Fig. 3.27 TPM process

beginning of the project. The original estimates are refined through modeling and simulation.

Exemplar attributes, that could serve as TPMs for some systems, include reliability, maintainability, power required, weight, throughput, human factors, response time, complexity, availability, accuracy, image processing rate, achieved temperature, requirements volatility, speed, setup time, change over time, and calibration time.

The TPM process is shown in Fig. 3.27.

3.4.2.1 Characteristics of TPMs

TPMs should be created for requirements and risks that satisfy *all* of the following criteria:

1. High priority requirements that impact mission accomplishment, customer satisfaction, cost, or system usefulness.
2. High risk requirements that have a high probability of not being met. And if they are not met, there is a high probability of project/system failure and great consequences of failure.
3. Requirements where the desired performance is not currently being met. The causes may include the use of new or unproven technology, imposition of additional constraints (for example, a drastic increase in the number of users), or an increase in the performance target.
4. Requirements where the performance is expected to improve with time, where progress toward a goal is expected.
5. Requirements where the performance can be controlled.
6. Requirements where the program manager is able to trade off cost, schedule, and performance. If a TPM exceeds its threshold and indicates imminent cost overruns or fails to meet performance goals, then the associated requirement should be renegotiated with the customer. Occasionally a TPM indicates that performance needs are being greatly exceeded, often because of innovative approaches, and/or advances in technology or materials. This is important to recognize, because resources and trade-offs can be redirected to other critical areas or requirements.

TPMs require quantitative data to determine how well the system under development is satisfying its requirements. Gathering such data can be expensive. Because of the expense, not all requirements have TPMs, only the high priority requirements. As a rule of thumb, less than 1 % of requirements should have TPMs: each program might have half-dozen TPMs.

Because TPMs use quantitative data, they may use evaluation criteria and scoring functions.

TPMs

- Should reveal the project's state of health
- Should be important
- Should be relevant
- Should be relatively easy to measure, analyze, interpret, and communicate with upper management
- Performance should be expected to improve with time
- If the measure crosses its threshold, corrective action should be known and documented
- The measured variable should be controllable. If the project team cannot change the measured variable, then do not measure it. For example, in the design of a

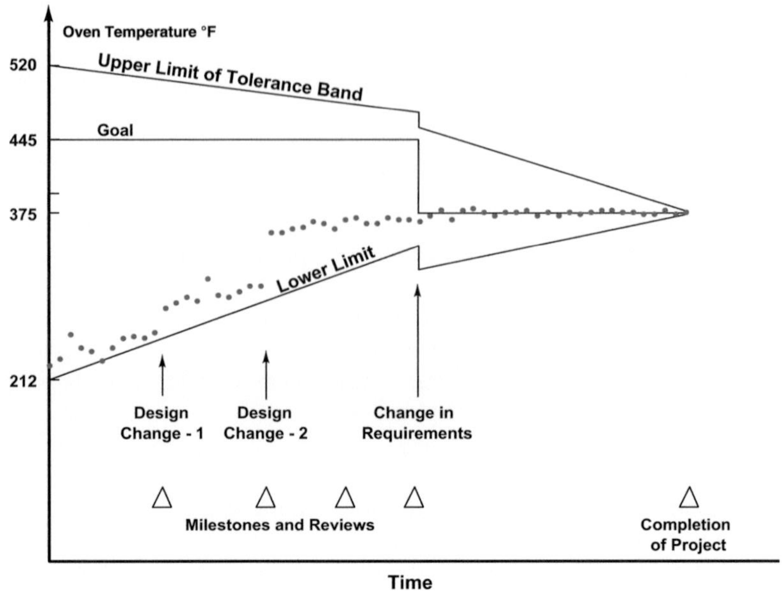

Fig. 3.28 Predicted oven temperature TPM progress

home air conditioning system, the outside air temperature would be an important design input, but a terrible TPM.

- Management should be able to trade off cost, schedule, and performance
- Should be documented
- Should be tailorable for the project

The TPM measures can be displayed using graphs, charts, diagrams, figures, or frames, for example, Statistical Process Control Charts, Run Charts, Flow Charts, Histograms, Pareto Diagrams, Scatter Diagrams, Check Sheets, PERT Charts, Gantt Charts, Line Graphs, Process Capability Charts, and Pie Charts.

For example, consider the design and manufacture of solar ovens. In many societies, particularly in Africa, women tend to spend as much as 50 % of their time gathering wood for their cooking fires. To obviate this squandering of human resources, solar ovens are being designed and built. We will now examine the solar oven design and manufacturing process.

We begin by defining a TPM for our design and manufacturing process. When a loaf of bread is done baking, its internal temperature should be 95 °C (203 °F). To reach this internal temperature, commercial bakeries bake the loaf at 230 °C (446 °F). As initial values for our oven temperature TPM, we chose a lower limit of 100 °C (212 °F), a goal of 230 °C (446 °F), and an upper limit of 270 °C (520 °F). The tolerance band shrinks with time as shown in Fig. 3.28.

At the beginning of the design and manufacturing process, our daily measurements of this metric increased because we were able to find better insulators and better glazing materials (for example, glass and Mylar), while also being able to seal the cardboard box better and aim at the sun better.

At the time labeled "Design Change-1," there was a jump in performance caused by adding a second layer of glazing to the window in the top of the oven. This was followed by another period of gradual improvement as we learned to stabilize two distinct glazing materials.

At the time labeled "Design Change-2," there was another jump in performance caused by a design change that incorporated reflectors to reflect more sunlight onto the window in the oven top. This was followed by another period of gradual improvement as we discovered better shapes and positions for the reflectors.

However, in this case, it seemed that we might not attain our goal. Therefore, we reevaluated the process and the requirements. Bread baking is a complex biochemical process that has been studied extensively: Millions of loaves have been baked each day for centuries. These experiments revealed the following consequences of insufficient oven temperature:

1. Enzymes are not deactivated soon enough, and excessive gas expansion causes coarse grain and harsh texture.
2. The crust is too thick, because of drying caused by the longer baking duration.
3. The bread becomes dry, because prolonged baking causes evaporation of moisture and volatile substances.
4. Temperatures below 160 °C (320 °F) do not activate the Maillard reaction that describes the interaction of amino acids with sugars, which produces the crust flavor, color, and appealing bloom.

After consulting some bakers, our managers decided that 190 °C (374 °F) would be sufficient to avoid the above problems. Therefore, the requirements were changed at the indicated spot, and our system design was then able to meet our intended goal. Of course, this change in requirements forced a review of *all* other requirements and a change in many other facets of the design. For example, the duration versus weight tables had to be recomputed.

If sugar, eggs, butter, and milk were added to the dough, we could get away with temperatures as low as 175 °C (347 °F). But we decided to design our ovens to match the needs of our customers, rather than try to change our customers to match our ovens.

We have stated earlier that when it appears that a TPM cannot be met, the program manager must readjust performance requirements, cost, and schedule. In this example, we renegotiated the requirements.

If the TPM measurement is being displayed with a graph, then the graph might contain the goal, upper and lower thresholds, planned values, and actual values. In Fig. 3.28, the planned values were coincident with the lower limit and were not called out separately.

TPMs are used to identify and track performance requirements that are program critical. TPMs are used to establish the appropriate design emphasis, design criteria and identify levels of technical risk. TPM measurements are collected and tracked against project design objectives in the project's risk register.

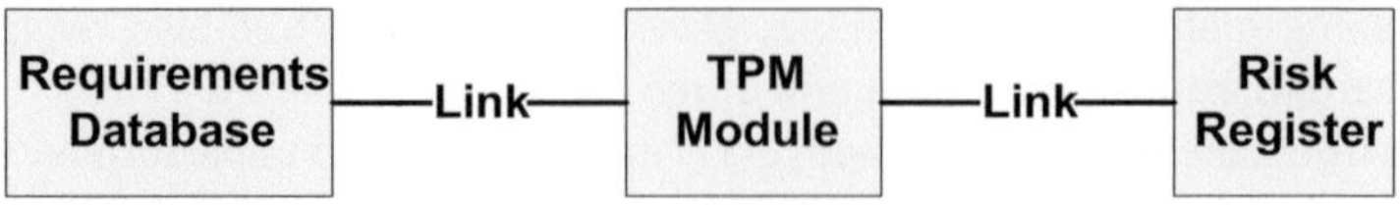

Fig. 3.29 TPM links

3.4.2.2 Technical Performance Measure Template

Name of TPM:
Purpose of TPM: i.e., how will it be used?
Source requirement: a TPM must be tied to a requirement (or a WBS element)
Type of the source requirement: I = contract requirement, II = derived requirement, III = goal in contract, IV = required by reference
Risk level: resolve, manage, monitor, or accept
What should be measured? include units of measurement
How should it be measured? inspection, modeling, simulation, analysis, test, software scripts
How often should it be measured? this could be time-based or tied to milestones
During which phase of the project should it be measured? when do you stop collecting data?
How should it be displayed? a figure may be best
To whom should it be presented?
Thresholds above or below which action is necessary: these may be on the figure
What action should be performed if thresholds are exceeded?
Who should perform this action?

3.4.2.3 TPM Links

The requirements database should have an attribute named TPM. The name of each TPM should be entered in the attribute field of the appropriate requirement and this should be linked to the TPM module as shown in Fig. 3.29. Each TPM should also be put in the project's Risk Register and be evaluated monthly.

Each TPM must be linked to a requirement (or a work breakdown structure element), have quantitative values, have a risk level, and have a priority. TPMs should only be created for requirements when action can be taken to cause a change in the quantity or result being measured.

In summary, TPMs are used to identify and track performance requirements that are program critical. TPMs are used to establish the appropriate design emphasis, design criteria and identify levels of technical risk. TPM measurements are collected and tracked against project design objectives in the project's risk register. TPMs should be created for high-priority requirements that impact mission accomplishment, customer satisfaction, system usefulness, and where performance improves with time, where performance can be controlled, and where management can trade off cost, schedule, and performance.

3.4.3 *Homework*

3.4.3.1 Quantitative Measures Selection Exercise

The following questions pertain to how you run ***your*** daily life. In the following table, for each suggested requirement, state whether you think the requirement should have a Technical Performance Measure (TPM), a measure (metric), an evaluation criterion, or nothing. Remember, criteria are used to quantify requirements and to help select amongst alternative designs in trade-off studies. Measures (which are sometimes called metrics) are used to help manage a company's processes. TPMs are used to mitigate risk during design and manufacturing. You may give a short explanation if you wish.

Suggestion of a requirement	TPM, measure, criterion, or nothing	Reason
The value of your stock portfolio (dollars)	–	–
The Dow Jones Industrial Average	–	–
College grade point average	–	–
Dollars spent per week on groceries	–	–
Dollars spent per week on entertainment	–	–
Your body weight (pounds)	–	–
Your dog's body weight (pounds)	–	–
Daily calorie intake	–	–
Daily fat intake (grams)	–	–
High density lipoprotein (HDL) (mg/dL)	–	–
LDL (lousy cholesterol) (mg/dL)	–	–
Ration total cholesterol over HDL	–	–
Automobile gas mileage (mpg)	–	–
Diameter of the piston rings on your car	–	–
Daily cell-phone long-distance minutes used	–	–
Price being offered for cell-phone service	–	–
Number of pushups you can do	–	–
Monthly cost of health insurance (dollars)	–	–
The high card/low card ratio in Blackjack	–	–
Cumulative annual dollars contributed to charity	–	–
Diastolic blood pressure (mmHg)	–	–
Systolic blood pressure (mmHg)	–	–
Number of visitors to your web site per week	–	–

3.4.3.2 Blood Pressure

Scenario: You have just visited your physician. She says that you have high blood pressure, 155/95. She prescribes Nadolol, Atenolol, and Hydrochlorothiazide. She tells you to exercise for a half-hour a day. She tells you to watch your diet: take vitamins, eat lots of fruits, vegetables, and foods with omega-3 fatty acids; avoid sodium and fats, particularly hydrogenated fats. She says, "Thou shall reduce your systolic blood pressure to below 140 mmHg and your diastolic blood pressure to below 90 within 3 months."

Create a TPM for your blood pressure reduction system.

3.5 Section 24. Prioritization

"Do the hard parts first (Eb Rechtin, 2000)."

To prioritize means to arrange (or deal with) in the order of importance. In practice, everything should be prioritized.

3.5.1 Why Is Prioritization Important?

Assume that you want to select a new car using evaluation criteria of Five-year Life Cycle Cost, Horsepower, and Safety. The Five-year Life Cycle Cost (in US dollars) includes purchase price, taxes, licenses, loan interest, insurance, gasoline, tires, and maintenance. The Horsepower is the peak SAE net horsepower. The Safety rating is 0–5 stars based on the National Highway Traffic Safety Administration's front and side crash test and rollover ratings. Assume that you have narrowed the field to three cars. One is best on Five-year Life Cycle Cost, the second car is best on Horsepower, and the third car is best on Safety. Which car should you buy? Your answer obviously depends on which criteria are most important to you. To help you make this decision, a prioritization process is desirable. This section, which is based on Botta and Bahill [13], describes such a process.

The development of this prioritization process was stimulated by a need to prioritize requirements, but the resulting prioritization process is not limited to requirements. We have also used it to derive weights of importance for trade-off studies and to prioritize risks. In this section, we first discuss some of the reasons to prioritize requirements. Then we discuss prioritization of other items and finally we present the prioritization process.

Prioritization is applicable to discovering requirements, deriving weights of importance for the evaluation criteria in trade-off studies and to prioritizing risks, goals, customer needs, capabilities, directives, initiatives, issues, activities, use cases, technical performance measures, features, functions, and value engineering activities. For these other tasks, certain aspects can be expected to be different. For example, in prioritizing design activities, external constraints or changes in the intended operational environment may render some activities infeasible. Therefore, these activities can be removed from further consideration. This situation is possible, but not likely with requirements. In deriving weights of importance for the criteria in trade-off studies, cost might be treated as an independent variable and hence not be in the criteria set for the trade-off study. An important purpose in prioritizing requirements and design activities is scheduling work. This is not the case for prioritizing technical performance measures where all TPMs are managed throughout. One purpose of prioritization is to identify those features that should be candidates for elimination. This situation is much more common for goals, customer needs, and capabilities than it is for requirements.

When a project is nearing the end of a development phase (or within a development spiral, iteration, or time box, depending on the life cycle model you are using) and it appears that not all of the features that were scheduled during that iteration can be produced, do not slip schedule and delay the review until all the features are finished. Instead, push some low-priority features into the next iteration. Prioritization of features helps determine which features can be delayed. This approach obviously depends on the contracted life cycle model. Furthermore, low-priority features should be renegotiated with the customer to see if they are really necessary. Changes to customer funding profiles may force low-priority features to be deleted.

It is important to realize that in the real world not all features will be implemented equally. Some are likely to receive extra attention, others less. Companies often seek advice from outside consultants to identify or confirm the highest priority features. Also, they often assign their best people to the high-priority features and contract out the low-priority ones. As a noncommercial example, consider Little League baseball. The most important positions are pitcher and catcher. So where do coaches put their best athletes? Pitcher and catcher. Furthermore, the highest priority features might be subjected to more reviews and more thorough testing. Even if the extra testing is not planned, it can be expected to happen because of the nature of regression testing. In regression testing, at the end of a life cycle phase all completed features are tested. In the next phase, more features and tests are added. Then at the end of this second phase, all completed features are tested, that is, those completed in the second phase as well as those completed in the first phase. Thus, features that are completed first will be tested in every iteration.

Priorities tend to change with discussions with the customer, with better understanding of customer's needs, with changes in the environment, with changes in stakeholders' needs and perceptions, as various features are implemented, as the system matures, as its architecture develops, and as uncertainty is resolved. Therefore, the priorities of all features can be expected to change with time [29].

3.5.2 Criteria Used for Prioritization

Table 3.17 lists criteria that are useful for prioritization. Cost is obviously an important criterion in most decisions. Cost should include monetary as well as other resources such as time, labor, overhead, infrastructure, and shipping. In purchasing a new car, total life cycle cost would include purchase price, taxes, loan interest, insurance, gasoline, tires, and maintenance. Benefit is a measure of the good things that accrue due to acquiring a feature. This would include performance measures such as speed, mean time between failures, requests served per minute, market percentage, quality, convenience, testability, and accuracy. In purchasing a new car, horsepower is an important performance measure. Some criteria naturally go together (like peanut butter and jelly or raisins and walnuts). For example, cost and benefit are often combined into a benefit-to-cost ratio. Clearly, if you choose to use the benefit-to-cost ratio, then you should not also include cost and benefit as separate criteria.

In an ideal world, criteria used for prioritization would be completely independent. However, for modern complex systems this is hardly possible. The criteria in Table 3.17 are meant to be orthogonal and as independent as possible. However, some dependencies will creep in. Therefore, it is important that you do not look for derived effects. Here are two examples of what **not** to do. (1) Features that affect system architecture should be given a high priority, because, in general, features

Table 3.17 Criteria that are useful in establishing priorities

Criteria	Rationale
Risk	Work on high-risk features first in order to reduce risk; in addition, high-risk features are likely to change thereby producing changes in other features. So working the high-risk features first will reduce the rework due to changing features. Furthermore, if it were impossible to satisfy the high-risk features and the project were canceled, you would have saved the money that otherwise would have been squandered satisfying low-risk features. High-risk features should have high priorities [30]. Risk should be quantified as frequency of occurrence multiplied by the severity of consequences
Criticality to mission success	Satisfy your customer's critical needs first. These might be stated in the vision or mission statements
Customer satisfaction	Features that will increase customer satisfaction should get high priorities
Commitment	If you told your boss (or your customer) that you *would* do something, then it should have high priority. If the organization said that it wanted a task done, then the task should have high priority
Architecture	Give high priorities to features that will have a large impact on system architecture, because these will cause changes in many other entities
Business value	Different features will have different value to the business: some features have tactical usefulness, whereas others have long-term strategic value. If you are presenting your features to the President of your

(continued)

Table 3.17 (continued)

Criteria	Rationale
	business unit, and you can only present three features, which would they be? These features should have high values for this criterion
Priority of scenarios	Give high priorities to features involved in scenarios (use cases) that are important for business goals
Benefit	Give high priorities to features that increase performance measures such as speed, mean time between failures, and requests served per minute
Cost	A basic business goal is to produce return on investment (RoI). Therefore, high cost features should have low priorities. The cost for each feature would be the summation in dollar values of labor, travel, and material. This then may have to be converted into the range of values being used for the other criteria. Other measures that could be used for cost include internal rate of return, net present value, and payback period
Benefit-to-cost ratio	Benefit and cost are often combined into a ratio. Putting cost in the denominator gives low priorities to high-cost, low-value features
Implementation time	Features that will take a long time to implement should have high priorities; to make sure they get scheduled early
When it is needed	Features that should be scheduled early, because of desire or constraints, should have high priorities
Frequency of use	Things that will be used often should have high priorities. For example, in software, code contained inside nested loops will be executed very often; therefore, it should have high priority and should be optimized
Safety	Safety critical features should have high priorities
Complexity	Complex features should have high priorities. You should assign your best people to complex features
Implementation difficulty	Features that will be hard to implement should have high priorities. Attributes that can make implementation difficult include large size, tedium, uncertainty, novelty, number of people involved, and organizational constraints. Complexity and implementation difficulty are independent. For example, diagnosing illness is complex, but if the diagnosis is correct, implementing the cure is simple. Whereas, describing a program to find 128-bit prime numbers is simple, but implementing the computation is difficult
Stability	Implement stable features first. Identify features that are likely to change. If the changes have an external cause, give the feature a low priority, thereby giving the changes a chance to occur before you work on the feature. But if the changes are likely to be caused by work done on that feature, then give it a high priority to force the changes early
Dependencies	If item B depends on item A, then A's priority should be at least as high as B's. If many other functions depend on a particular function, then give it a high priority. On the other hand, activities, such as optimization, that will have to be redone when other parts of the system change, should be given low priorities
Infrastructure	If a system is a critical part of a bigger system, then it should have a high priority. On the other hand, if it depends critically on other systems, then perhaps its implementation should be delayed until the other systems are stabilized
Reuse Potential	If an item is highly reusable, then give it a high priority

like architecture that cause changes in a lot of other systems should be given a high priorities. For example, we notice that architecture affects customer satisfaction and customer satisfaction is important. Therefore, should we increase the priority for architecture even more? No. (2) We want to reduce cost as much as reasonably possible. We notice that benefit, complexity, and design for reuse all affect cost. So should we reduce the priorities of these three criteria, in order to reduce cost? No. If you continually look for interactions, then you will never get to the end of the prioritization process.

While the criteria of Table 3.17 are not listed in any particular order, the top of the list has criteria that are more general. Of course, the criteria to be used must be tailored for the particular company and for the type of business. Not all criteria in Table 3.17 would be useful in all industries and other criteria would have to be added for some industries. Some customers may allow no flexibility in requirements or schedule. This section is written for people who have flexibility in when and how well various features can be implemented. However, a list should not be prioritized if the cost of prioritizing is not significantly less than the cost of doing the tasks. For example, for most people, it would not make sense to prioritize a grocery-shopping list.

3.5.3 Combining Information

Once we have values and weights of importance for the criteria, we can combine them, as discussed in Section 22. The most common combining function is the sum of weighted scores. This method is implemented with the following equation to compute the priority of the *i*th feature.

$$\begin{aligned}\text{Priority}_i = {} & w_{\text{Risk}} \times \text{Risk}_i + w_{\text{Complexity}} \times \text{Complexity}_i \\ & - w_{\text{Cost}} \times \text{Cost}_i + w_{\text{Reuse}} \times \text{ReusePotential}_i + \cdots\end{aligned}$$

For now, there is a negative sign in front of the cost term. In this equation, the weights are the priorities that we assigned to the criteria of Table 3.17. This simple technique uses very simple mathematics. The weights and the criteria values may be integers between 0 and 10, or if normalized then between 0 and 1. Usually this technique produces good results.

The criteria to be included *must* be tailored for individual decisions. Criteria that do not differentiate between alternatives can be omitted, or given a zero weight. Other criteria, such as security and resources (in addition to the already included cost and time resources), might be added. IBM has a criterion of "Can it be implemented in the future as easily as it can be added today?"

The following alternative equation gives more importance to the cost criterion.

$$\text{Priority}_i = \frac{\left[\sum_{j=1}^{n} w_j \text{Criterion}_j\right]_i}{\text{Cost}_i}$$

where n is the number of criteria you have chosen to include. If you use this equation, you should remove both the cost and the benefit-to-cost ratio from the criteria set being summed in the numerator. Using this equation, cost would no longer need to have the same range as the other criteria, dollar values could be used directly.

Of course, this prioritization process is iterative. The first pass will show the most important features. These features might then be scheduled to be implemented first, which would change their values in the *When it is Needed* criterion.

This simple technique usually works, because the purpose of prioritization is communication. The numbers themselves should not be used in numerical calculations.

If the priority values are to be used in calculations, then the weight of importance must be normalized so that the sum of the weights is 1.0. Also, the values of the criteria must be normalized with scoring functions as shown in Section 21 and [31] so that the scores for the criteria are all between 0 and 1.

Earlier we noted that high cost features, features that are likely to be changed by outside forces, and features that need to be redone as the design progresses should be given low priorities. What does low priority mean? Does it mean a small weight of importance? No, because these criteria may be important. Does it mean a negative weight of importance? No, because the weights must be positive numbers that when normalized sum to one. It means that a large value for one of these criteria should reduce the priority score of the feature under consideration. So, how can we do this? One of the reasons for using scoring functions (even if they are just simple linear transformations) is to ensure that *more is better* holds for all scores. For example, we might consider less cost to be better than more cost. Scoring function SSF9 inverts the cost relationship so that a larger output score is better. Our combining equation now becomes

$$\begin{aligned}\text{Priority}_i = \; & w_{\text{Risk}} \times \text{RiskScore}_i + w_{\text{Complexity}} \\ & \times \text{ComplexityScore}_i + w_{\text{Cost}} \times \text{CostScore}_i + w_{\text{Reuse}} \\ & \times \text{ReusePotentialScore}_i + \cdots\end{aligned}$$

You will notice that now there is a plus sign in front of the cost term. If your customer wants to use the benefit-to-cost ratio instead of cost, the following term could be added to the equation

$$\text{Priority}_i = w_{\text{BenefitCostRatio}} \times \frac{\text{BenefitScore}_i}{\text{CostScore}_i} + \cdots$$

Other methods that have been used for prioritization include Quality Function Deployment [32, 33] and the analytic hierarchy process [11].

3.5.3.1 Completeness of Coverage

At this point, we can ask if the criteria of Table 3.17 would provide even coverage of an organization's aspirations and needs. To answer this question, let us look at frameworks.

Frameworks help people organize models of their enterprises. This organization helps ensure interoperability of systems and helps control development cost. The Zachman framework [34] provides a general schema to organize and assess completeness of descriptive representations for a complex enterprise. To ensure a complete and holistic understanding of the enterprise, it is necessary to develop models that address the perspectives and aspects that constitute the rows and columns, respectively, of the framework.

To this end, we mapped the criteria of Table 3.17 to the rows and columns of a Zachman framework. Criticality to Mission Success is in row 1, the scope of the organization. Business value and benefit-to-cost ratio are in row 2, the *business model*. Architecture is in row 3, the *system model*. Complexity, implementation difficulty, and dependencies are in row 4, the *technical model*. Finally, reuse potential is in row 5, the *detailed representation*. Now for the columns, in a risk analysis, you look at the high-risk entities and ask, "What could possibly go wrong?" Hence risks are in column 1, *What*. Implementation difficulty quantifies how things are done; hence it is in column 2, *How*. Architecture describes the positioning and interconnection of things, so it is in column 3, *Where*. Commitments are made to people; therefore commitment goes into column 4, *Who*, along with customer satisfaction. When it is needed is in column 5, *When*. Business value and benefit-to-cost ratio have to do with the motivation of people and the organization, so they are in column 6, *Why*.

So, although the criteria of Table 3.17 do not fill all 36 cells of a Zachman framework, they do cover each row and each column. Consequently, we can conclude that, the criteria of Table 3.17 should cover the aspirations and needs of most organizations.

3.5.4 Section Summary

Requirements, evaluation criteria, goals, customer needs, capabilities, risks, directives, initiatives, issues, activities, features, functions, and technical performance measures should all be prioritized. Prioritization can be expected to help with budget, schedule, system architecture, customer satisfaction, and risk reduction. Prioritization can be rather simple, using integers from 0 to 10 for the weights and the values, or, it can be robust using normalized weights of importance and scoring functions for the criteria. Sometimes it is useful to create two sets of priorities: one from the customer's perspective, and the other from the contractor's perspective.

Surfacing these conflicting objectives can often help explain and clarify existing misunderstandings. Finally, "In any resource-limited situation, the true value of a service or product is determined by what one is willing to give up to obtain it [35]."

3.5.5 *Homework*

3.5.5.1 Parking Garage Requirements

The requirements from the basic parking garage design are repeated here.

Functional Requirements

FR1. The system shall determine that a car is waiting at the Entrance or Exit Gate.
FR2. The system shall determine the validity of the transponder's ID.
FR3. The system shall store the identity of every car in the garage (from the ConOps).
FR4. The system shall open and close the Entrance and Exit Gates.
FR5. The system shall determine when a car has cleared the Entrance or Exit Gates.

Derived Requirement

FR6. The system shall allow a car to enter the garage only if the transponder's ID is valid AND the database shows that the car is outside of the garage.

Nonfunctional Requirements

NFPR1. When the transponder is valid, the system shall *begin* opening the entrance gate within 2 s of interrogation. This requirement includes the response time of the Sensor, the Reader, and the Controller.
NFPR2. The system shall open and close the entrance and exit gates in 1–3 s.

Prioritize these requirements. Table 3.18 is a short example using a fragment of Table 3.17.

Based on only these three criteria, FR5, NFPR1, and NFPR2 have two pluses and no minuses. Therefore, they should have the highest priority

Table 3.18 Solution using only a fragment of Table 3.17

Criteria	Description and relevant requirements
Customer satisfaction	Features that will increase customer satisfaction should get high priorities
	FR1, FR2, FR4, FR5, NFPR1, and NFPR2
Benefit	Give high priorities to features that increase performance measures
	NFPR1 and NFPR2
Safety	Safety critical features should have high priorities
	FR5

3.6 Section 25. Analyzing the Data

3.6.1 *Frequency Versus Probability*

The measure of risk is the severity of the consequences times the frequency (or probability) of occurrence. Because humans are poor at assessing probabilities [36, 37] and estimated probabilities interact with stated severities [38], we use frequency of occurrence instead of estimated probability. The following example, based on Gigerenzer [39], shows this advantage of using frequencies over probabilities.

3.6.1.1 Bayes' Theorem

We begin with a story. A woman receives news of a positive mammogram. She wonders what is the probability that she has cancer. The following information is available about women in her age group: (a) the probability that a woman tested in this lab has breast cancer is 0.8 %. (b) If a woman has breast cancer, the probability that she will have a positive mammogram is 88 %. (c) If a woman does not have breast cancer, the probability that she will have a positive mammogram is 7 %. This problem seems ideal for application of Bayes' theorem. Let D be the disease. In this case, the disease D is breast cancer. Let TR be the test result. In this case, the test result TR is a positive mammogram. Then

$$P(D|\mathrm{TR}) = \frac{P(\mathrm{TR}|D) \times P(D)}{P(\mathrm{TR}|D) \times P(D) + P(\mathrm{TR}|\overline{D}) \times P(\overline{D})}$$

$$P(D|\mathrm{TR}) = \frac{0.88 \times 0.008}{0.88 \times 0.008 + 0.07 \times 0.992} = 0.09 = 9\%$$

3.6.1.2 How Most People Think

However, humans are not Bayesian. They do not think this way. Therefore, the Bayesian solution is not intuitive to most people. Most people focus on the causally relevant link, "(b) if a woman has breast cancer, then the probability that she will have a positive mammogram is 88 %," which is much more than the 9 % produced by Bayes' theorem. They ignore "(a) the probability that a woman tested in this lab has breast cancer is 0.8 %," because it is "just statistics," not based on causality. And they ignore "(c) If a woman does not have breast cancer, the probability that she will have a positive mammogram is 7 %," because it is not about the category in which we are interested, namely women *with* breast cancer. Most people ignore the evidence statements (a) and (c) and erroneously conclude that most women with positive mammograms have cancer. Figure 3.30 shows how people think.

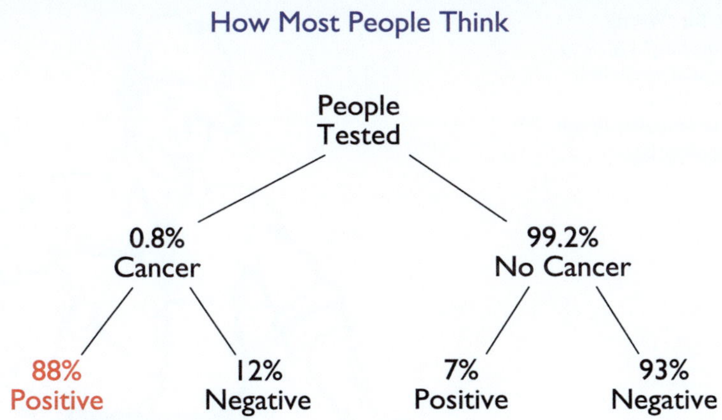

Fig. 3.30 How most people think

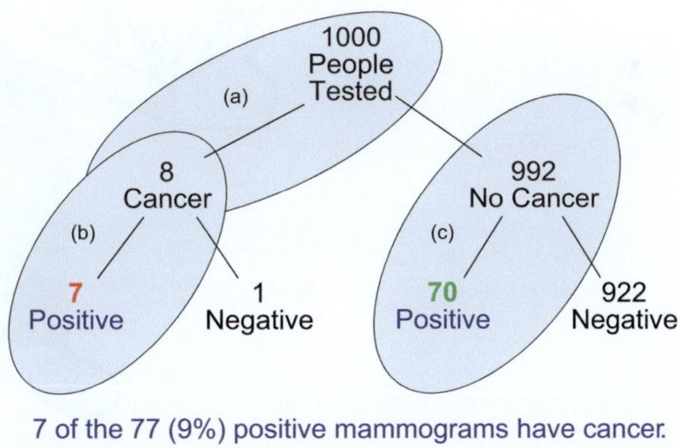

Fig. 3.31 Tree diagram showing frequency of occurrence for the evidence statements (**a**), (**b**), and (**c**) of the breast cancer example

3.6.1.3 Frequency of Occurrence

Now let us look at the frequency of occurrence approach.

(a) Eight of 1000 women tested in this lab have breast cancer.
(b) Of the 8 women with breast cancer, 7 will have a positive mammogram.
(c) Of the remaining 992 women, 70 will have a positive mammogram.

These three pieces of data (a), (b), and (c) can be represented with a tree, as shown in Fig. 3.31.

Imagine a sample of women who have a positive mammogram. What fraction of them has cancer? From this tree, it is easy to see that 7 of the 77 (9 %) positive mammograms have cancer. This tree is easier for most people to understand than Bayes' theorem. The frequency of occurrence approach uses a mental operation that people perform quite well, partitioning a set of cases into exclusive subsets. The successive partitioning of sets is well represented by a tree structure. The Bayesian approach employs probabilities. Consequently, it is difficult for most people, especially those without a formal education in probabilities.

Fig. 3.32 Mr. Wrong Wrench was taught how to use pentagonal wrenches and he uses them everywhere, whether or not they are appropriate

3.6.2 Shape of the Data

"Problem stating is more important than problem solving" Wymore [20]. Accordingly, we present tools not just for solving problems, but more importantly, how to state the problem, study the problem, and analyze the data. In fact, we sometimes spend more time studying the problem rather than designing the solution. We need to find the best tool for the problem, rather than applying our favorite tool to the problem. Before choosing a tool, it is important to determine what the data *look* like. Thereafter, the best tool for the problem can be selected. You need to resist becoming like Mr. Wrong Wrench (Fig. 3.32) and using your favorite tool for every problem. An important part of stating the problem is studying the problem and the shape of the data.

Cost and performance criteria for intercontinental ballistic missiles are shown in Fig. 3.33. These data are distributed nicely on a circle. This is the expected distribution in the multicriterion decision-making literature. CEP stands for circular

Fig. 3.33 Ideal Pareto-optimal distribution of missile capabilities

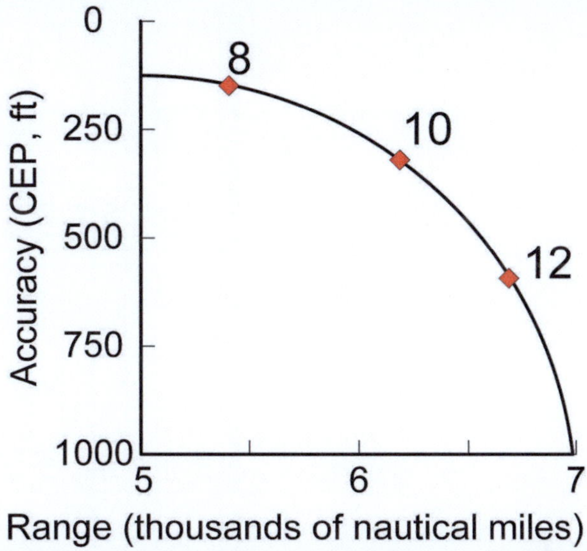

error probable, which is the diameter of a circle that will contain 50 % of the data points. The alternatives shown are for 8, 10, or 12 warheads.

In general, real-world data will not fall neatly onto curves or surfaces such as a circle. But often they may be bounded by such functions. In the operations research literature, such data sets are called convex, although the function bounding them is called concave [40].

Table 3.19 shows another example of using the popular data combining functions. These data were collected for the price of Coca-Cola several years ago. The cost criterion was the reciprocal of price and the performance criterion was the quantity in liters. We made this example very simple by omitting weights and scoring functions.

The important point here is that the different combining functions gave different results. For the product combining function, the largest Trade-off value was for the 3-L bottle, whereas all of the other functions favored the 24 can case. What could be wrong here? For one thing, we did not use scoring functions to normalize the evaluation criteria. A second suspicion is that the data set is oddly shaped. Our input data fit along a hyperbola as shown in Fig. 3.34.

Is this really a strange example, or a common, but relatively unknown fact. To answer, let us look at another example, the force–velocity relationship of human muscle shown in Fig. 3.35.

The equation for this relationship is $(\text{force} + f_0)(\text{velocity} + v_{\text{max}}) = \text{constant}$, where f_0 (the isometric force) and v_{max} (the maximum muscle velocity) are constants. Humans sometimes use one combining function and sometimes they use another. If a bicyclist wants to achieve maximum acceleration, he or she uses the point $(0, f_0)$. However, if there is no resistance and maximum speed is desired, then using the point $(v_{\text{max}}, 0)$ is appropriate. These solutions result from maximizing the

Table 3.19 Choosing amongst alternative soda pop packages

Data		Evaluation criteria		Trade-off values				
Item	Price (dollars)	Cost (dollars^{-1})	Quantity (L)	Sum	Product	Sum minus Product	Compromise with $p=2$	Compromise with $p=10$
1 can	0.50	2.00	0.35	2.35	0.70	1.65	2.03	2.00
20 oz.	0.60	1.67	0.59	2.26	0.98	1.27	1.77	1.67
1 L	0.79	1.27	1.00	2.27	1.27	1.00	1.62	1.27
2 L	1.29	0.78	2.00	2.78	1.56	1.22	2.15	2.00
6 pack	2.29	0.44	2.13	2.57	0.94	1.63	2.17	2.13
3 L	1.69	0.59	3.00	3.59	**1.78**	1.81	3.06	3.00
12 pack	3.59	0.28	4.26	4.54	1.19	3.35	4.27	4.26
24 pack	5.19	0.19	8.52	**8.71**	1.62	**7.09**	**8.52**	**8.52**

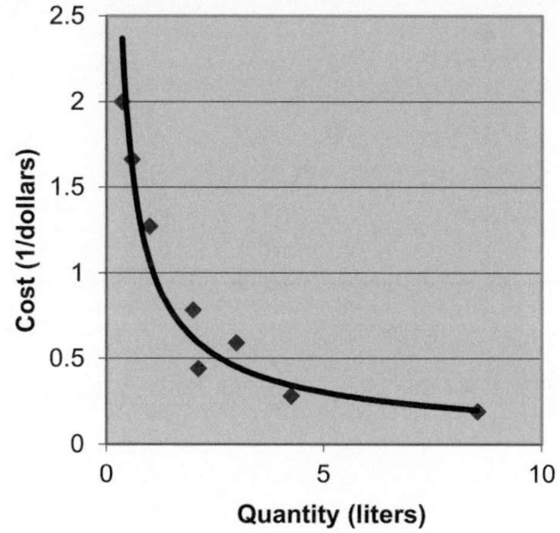

Fig. 3.34 Cost of eight different soda pop packages

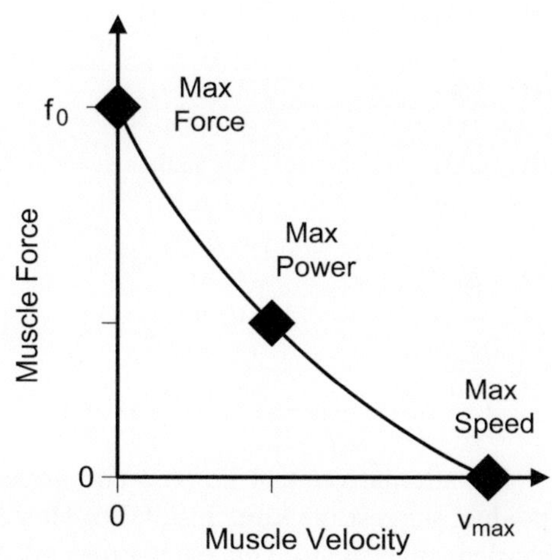

Fig. 3.35 Muscle force–velocity relationship

sum of force and velocity. However, if there is energy dissipation (for example, friction or air resistance) and maximum speed is desired, then bicyclists choose the maximum power point, the maximum product of force and velocity. This shows that the appropriate trade-off function may well depend on the context [8].

The muscle force–velocity relationship fits neatly onto curves such as the hyperbola. This is not always going to be the case. However, when it is not, the data may be bounded by such functions. In the operations research literature, such data sets are called concave, although the function bounding them is called convex [40].

From these examples, you might suspect that the sum and the product combining functions might recommend different alternatives. Let us sum this up it with Fig. 3.36.

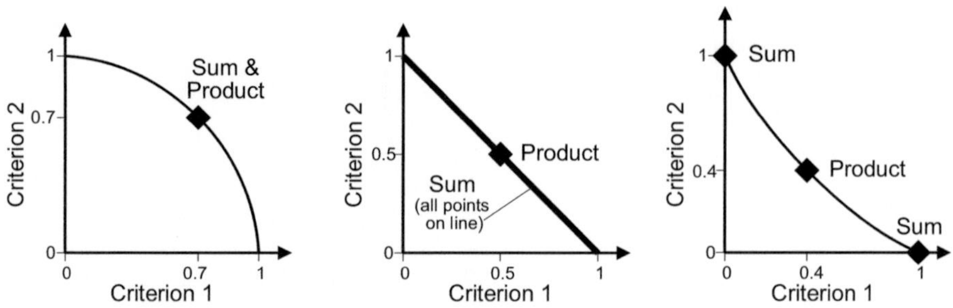

Fig. 3.36 The best designs for three shapes of data distributions

Table 3.20 Four data sets from Anscombe [41]

Data set-1 (x_1, y_1)	Data set-2 (x_2, y_2)	Data set-3 (x_3, y_3)	Data set-4 (x_4, y_4)
(10, 8.04)	(10, 9.14)	(10, 9.14)	(8, 6.58)
(8, 6.95)	(8, 8.14)	(8, 8.14)	(8, 5.76)
(13, 7.58)	(13, 8.74)	(13, 8.74)	(8, 7.71)
(9, 8.81)	(9, 8.77)	(9, 8.77)	(8, 8.84)
(11, 8.33)	(11, 9.26)	(11, 9.26)	(8, 8.47)
(14, 9.96)	(14, 8.1)	(14, 8.1)	(8, 7.04)
(6, 7.24)	(6, 6.13)	(6, 6.13)	(8, 5.25)
(4, 4.26)	(4, 3.1)	(4, 3.1)	(19, 12.5)
(12, 0.84)	(12, 9.13)	(12, 9.13)	(8, 5.56)
(7, 4.82)	(7, 7.26)	(7, 7.26)	(8, 7.91)
(5, 5.68)	(5, 4.74)	(5, 4.74)	(8, 6.89)

For alternatives that lie on a circular curve (left figure), both the sum and the product suggest the same result. For alternatives that lie on a straight line, such as sharing a pie (pie for you and pie for me), the product combining function suggests an equal split, however, the sum combing function cannot provide a recommendation. Finally, for alternatives that lie on a hyperbola, such as various soda pop packages or human muscle, the sum and the product suggest distinctly different points. The product combining function always favors alternatives with moderate scores for all criteria. It rejects alternatives with a very low score for any criterion. Therefore, the product combining function may seem better than the Sum Combining Function. But the Sum Combining Function is used much more in engineering and business. Therefore, different distributions of alternatives in criteria space may produce different preferred alternatives.

Consider another example from a totally different era and domain that conveys the same message. The following example by Anscombe [41] emphasizes looking at the data before modeling the system. Consider the data sets in Table 3.20.

What model would you use for each data set? If you just came out of a class discussing linear regression analysis, then most likely you would use linear

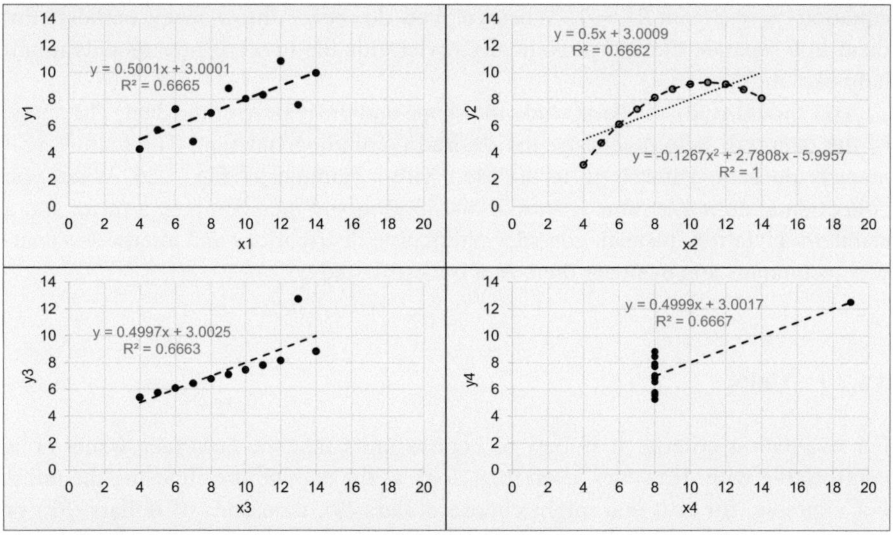

Fig. 3.37 Four data sets all best fit with $y = 0.5x + 3$ with $R^2 = 0.67$, where x is the input and y is the output. The equations are for the straight-line linear regressions

regression analysis. It turns out that linear regression analysis is a good model for all four of these data sets yielding $y = 0.5x + 3$ with $R^2 = 0.67$. The four y-variables have the same mean (7.5) and variance (4.12). Anscombe shows many other statistics that are also the same for all four of these data sets. The correlation coefficient $R^2 = 0.67$ indicates a good fit. Should we stop and submit our results to our customer and claim victory? No! Let's do what we should have done in the first place. Let us *look* at the data. Plotting these data yields Fig. 3.37.

Even though linear regression analysis gives the same good fit ($R^2 = 0.67$) for all four data sets, after seeing the graphs, we would prefer different models. For data set-1, we would stay with $y_1 = 0.5x_1 + 3$ with $R^2 = 0.67$. Data set-2 does not look like a linear function. Therefore, we would choose the parabola $y_2 = 0.13x_2^2 + 2.8x_2 - 6$ with $R^2 = 1$. For data set-3, we would investigate the data point (13, 12.74). We would ask everyone involved in data collection if they could recall anything unusual that occurred when collecting that data point. Whether or not anything was discovered, we would note that the data point (13, 12.74) is an outlier and should not be included in the *statistical* curve fit. Therefore, our model would include *that footnote and* this equation $y_3 = 0.35x_3 + 4$ with $R^2 = 1$. However, it is important to note that we did not throw away the outlying data point; we kept it in the graph. For data set-4, we would investigate the data point (19, 12.5). Specifically, we would ask everyone involved if they could recall anything unusual that occurred when collecting that data point. Whether or not anything was discovered, we would note that the data point (19, 12.5) is an outlier and would not include it in the statistical curve fit. Therefore, our model would be *that footnote and* the equation $y_4 = 8$ with $R^2 = 1$. Whatever you do, *never* throw away outliers: Put them in a separate file and periodically revisit this file to see if new models might help explain them.

You should study the data: study the shape and the texture of the data. The shape of the data will help determine the desired combining function. The texture will identify outliers. Think back to the Man with a Hammer of Fig. 3.25. When you collect data, do *not* assume a normal distribution and then compute a mean and a standard deviation. Instead, consider alternative distributions and alternative analysis techniques and evaluate them in a trade-off study.

3.6.2.1 Units

For evaluation criteria, it is best to choose units that the customer wants or is comfortable with. However, even then, look at the data before choosing the units. For example, for cost you might choose dollars ($), thousands of dollars ($k) or millions of dollars ($M), but only choose after you have studied the data. Choose units so that all alternatives can be compared using the same units. Likewise, frequency of failure might have units of failures per day, failures per month, or failures per year. After studying the data, choose units so that all alternatives can be compared using the same units.

3.6.3 Do Nothing Alternative

In all trade-off studies and risk analyses, it is essential to include a do nothing alternative. There are two interpretations of the do nothing alternative. (1) Evaluate the status quo. (2) Do absolutely nothing. For example, Hot Rod John owns a 1971 Datsun 240Z that has 150,000 miles on it. He might want to select a replacement for it. His alternatives would be a BMW Z4, a Saturn, or do nothing. The two interpretations for do nothing are (1) the status quo, keep the 240Z and (2) do without a car, that is walk or take the bus. The most common of these approaches is to use the status quo as the do nothing alternative.

If the status quo do nothing alternative attains one of the highest scores in the trade-off analysis, then you may have too many (perhaps dependent) cost or risk evaluation criteria, and not enough performance evaluation criteria. Usually the do nothing alternative has low cost and little risk. This hints that perhaps we should have a similar number of performance, cost, and risk evaluation criteria.

If a nihilistic do nothing alternative attains one of the highest scores in the trade-off analysis, then you might have too few performance evaluation criteria. For example, in a trade-off study or a risk analysis for a municipal transportation system, if a pogo stick wins (as in Fig. 4.6), then you have missed some performance evaluation criteria. As a second example, suppose a young couple on a date is selecting a movie to go to and they define their alternatives as comedy, romance, blood, and guts, and skip the movies. If the do nothing alternative wins, then they missed some performance evaluation criteria.

Just as you should not add apples and oranges, you should not combine performance, cost, schedule, and risk evaluation criteria. All the performance evaluation criteria should be combined together; then all the cost evaluation criteria should be combined together; then all the Schedule evaluation criteria should be combined together; and finally all the Risk evaluation criteria should be combined together. Then the performance, cost, schedule, and risk combinations can be combined with clearly stated weights. Weights of ¼, ¼, ¼, ¼ is an adequate default. If the do nothing alternative still wins, then you may have set the weight for performance too low.

3.6.4 Verification and Validation

Validating the system: Building the *right system* means making sure that the system does what it is supposed to do. Validation determines the correctness and completeness of the product, and ensures that the system will satisfy the actual needs of the customer.

Verifying the system: Building the *system right* means ensuring that the system complies with its requirements and conforms to its design.

Validating requirements: Ensuring that the *set* of requirements is correct, complete, and consistent, that a model can be created that satisfies the requirements, that a real-world solution can be built that satisfies the requirements, and that this real-world solution can be tested to prove that it satisfies the requirements. If the client has requested a perpetual motion machine, this is the place to stop the project and stop wasting money.

Verifying requirements: Proving that each requirement has been satisfied. Verification can be done by logical argument, inspection, modeling, simulation, analysis, test, or demonstration. The verification matrix must show a one-to-one mapping between the requirements and the test plan. Its audience is Engineering and the customer.

3.6.5 Testing

It is important to plan system tests early in the design process. For one thing, this will save money. For one particular complex mostly-software system that one of us participated in, each defect caught during design took, on average, three man-hours to correct. Each defect caught in integration took about 20 man-hours to correct. Each defect caught post-delivery took about 40 man-hours to correct.

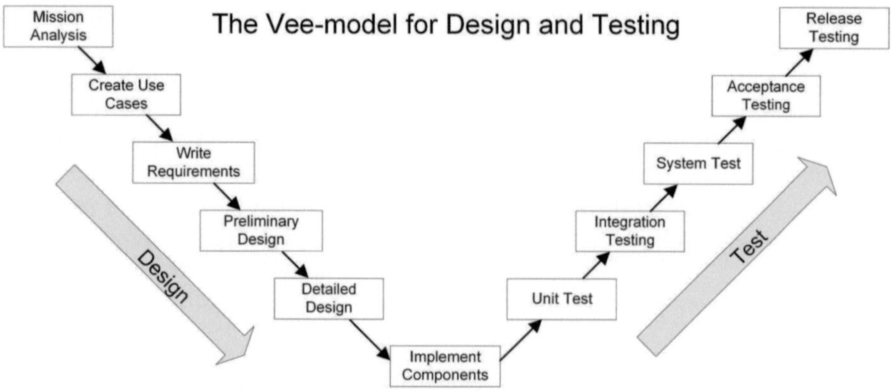

Fig. 3.38 The Vee life cycle model

Figure 3.38 shows the Vee-model for Design and Testing. It shows the design phases on the left down-stroke and the testing on the right up-stroke. Here is a description of the phases in this model.

Mission analysis—read the problem statement, the concept of operations, and the mission statement. Develop a business model and write the operational concept description.

Create use cases—a *use case* is an abstraction of required functions of a system. A use case produces an observable result of value to the user. Each use case describes a sequence of interactions between one or more actors and the system. Use cases have also been called scenarios and stories.

Write requirements—write formal requirements putting them in the specific requirements section of the use cases.

Preliminary design—investigate the architecture by constructing context (object) diagrams or block definition diagrams using the problem domain vocabulary.

Detailed Design—make sequence diagrams for all flows in the use cases. Use these to identify the operations (functions) of the classes.

Implement the components—write the code and build the hardware.

Unit Test—test each component. Use the use cases and the sequence diagrams to write the test procedures. Design system experiments to test each state transition arrow in each state machine. This should test most of the requirements. These tests should be as automated as possible, because they will be rerun in every new iteration.

Integration Testing—test the interfaces between all system components. Write tests for BiST. In the later part of integration testing (sometimes called compatibility testing), you test the interfaces with external systems.

System Test—test system functionality. Verify all requirements. Test operation of the total system.

Acceptance Testing—ensure that the system satisfies the customer's needs. In other words, validate the system. When testing equipment do not use normal subjects. For example, when testing an audiometer, use subjects with abnormal hearing.

They will hear things that normal subjects will not hear. When testing a vision system use subjects with abnormal vision. They will see things that normal subjects will not see. Beta testing by end users who are not a part of development is also a part of acceptance testing.

Release Testing—ensure that the system satisfies the mission. Demonstrate it to the top brass.

Figure 3.38 makes it look like testing does not begin until the Unit Test on the right up-stroke. But this is misleading. The planning for test starts when the case studies are being transformed into requirements. The Cookie system of Section 3 showed how the use cases could produce the functional and nonfunctional requirements. The use cases and the requirements produced the following test plan.

1. Tester turns on the oven and sets the desired temperature to 350 °F.
2. Tester waits until Cookie signals that the oven has preheated to 350 °F.
3. Tester stands 5 ft from the oven, observes the visual display, measures the volume of the auditory signal, measures the actual temperature inside the oven and records the results. After every step, Tester must record the result. We will not keep repeating this statement.

This has tested ReqF3: Cookie shall heat up the oven ~~and cook the food.~~

If a test verifies only a part of a requirement, then the requirement must be decomposed. We indicate partial verification with the ~~strikethrough~~ function.

This has tested ReqF5: Cookie shall emit an audible signal when the oven is preheated ~~and when the baking time is over~~.

This has tested ReqF6: Cookie shall indicate visually when the oven is preheated ~~and when the baking time is over.~~

This has tested ReqNF1: The remaining baking time displayed by Cookie shall be visible to a Student with 20/20 vision standing 5 ft from the oven in a room with an illuminance level between 0 and 1000 lx.

This has tested ReqNF4: The audible signal emitted by Cookie shall have an intensity level of 80 ± 2 dB at a distance of 30 cm.

4. Tester sets the Timer for 10 min.

This has tested ReqF1: Cookie shall provide a mechanism for Student to enter a desired baking time.

5. Tester observes the time on the display.

This has tested ReqF2: Cookie shall be capable of displaying the desired baking time entered by Student.

6. Tester waits until Cookie signals that 10 min is over.
7. Tester stands 5 ft from the oven, observes the visual display, measures the volume of the auditory signal from a distance of 30 cm, notes the desired and actual elapsed time and records the results.

This has tested ReqF5: Cookie shall emit an audible signal when the ~~oven is preheated and when the~~ baking time is over.

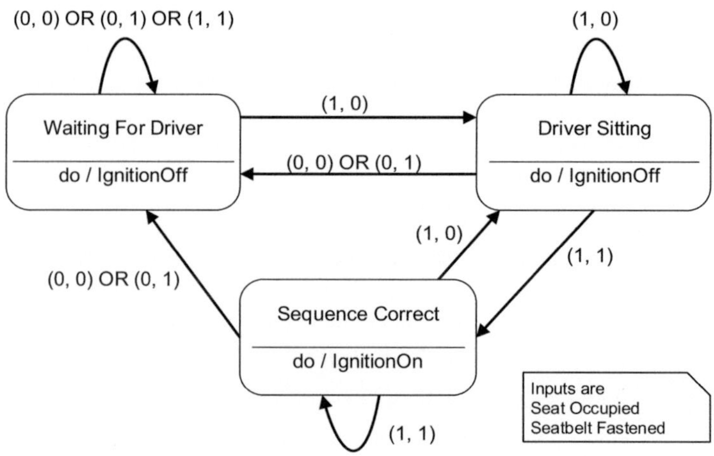

Fig. 3.39 Automobile Seatbelt System

This has tested ReqF6: Cookie shall visually indicate when the ~~oven is preheated and when the~~ baking time is over.

This has tested ReqNF1: The remaining baking time displayed by Cookie shall be visible to a Student with 20/20 vision standing 5 ft from the oven in a room with an illuminance level between 0 and 1000 lx.

This has tested ReqNF4: The audible signal emitted by Cookie shall have an intensity level of 80 ± 2 dB at a distance of 30 cm.

8. Tester turns the oven off.

This test plan has tested ReqF1, ReqF2, ReqF3, ReqF4, ReqF5, ReqF6, ReqNF1, and ReqNF4. It used the precondition and the main success scenario of the Cookie use case. However, you must remember to test all of the flows of all of the use cases.

In the Odd Parity Detector homework problem, we ask you to create test sequences to test your state machine. When you do so, make sure to exercise each state transition arrow. We will now show how to do this using our automobile seatbelt example of Section 9 (Fig. 3.39).

We named our Automobile Seatbelt System Z1. The start state is associated with the driver being outside of the car, the ignition being off, and there being nothing on the drivers' seat. The system is awaiting the driver's action.

Z1 = (SZ1, IZ1, OZ1, NZ1, RZ1), where
the following are the three states of the system
SZ1 = {Waiting For Driver, Driver Sitting, Sequence Correct},
the following are the two inputs to the system and their legal values, which are zero and one
I1Z1 = {0, 1}, /* A 1 at this input port indicates seat occupied */
I2Z1 = {0, 1}, /* A 1 at this input port means seatbelt is fastened */

Table 3.21 Test sequence 1

System experiment for the Automobile Seatbelt System. The initial state is Waiting For Driver					
Time	I1Z1	I2Z1	State	Next State	Output
0	0	0	Waiting For Driver	Waiting For Driver	Ignition Off
1	0	1	Waiting For Driver	Waiting For Driver	Ignition Off
2	1	1	Waiting For Driver	Waiting For Driver	Ignition Off
3	1	0	Waiting For Driver	Driver Sitting	Ignition Off
4	1	0	Driver Sitting	Driver Sitting	Ignition Off
5	1	1	Driver Sitting	Sequence Correct	Ignition Off
6	1	0	Sequence Correct	Driver Sitting	Ignition On
7	1	1	Driver Sitting	Sequence Correct	Ignition Off
8	1	1	Sequence Correct	Sequence Correct	Ignition On
9	0	0	Sequence Correct	Waiting For Driver	Ignition On
10			Waiting For Driver		Ignition Off

Table 3.22 Test sequence 2

System experiment for the Automobile Seatbelt System. The initial state is Waiting For Driver					
Time	I1Z1	I2Z1	State	Next State	Output
0	1	0	Waiting For Driver	Driver Sitting	Ignition Off
1	1	1	Driver Sitting	Sequence Correct	Ignition Off
2	0	1	Sequence Correct	Waiting For Driver	Ignition On
3			Waiting For Driver		Ignition Off

the following is the output of the system and its legal values
OZ1 = {On, Off}, /* An output of On signifies the ignition is enabled */
the next state function comes next. Each element of the next state function shows ((the present state, (the present inputs)), and the next state).
NZ1 = {((Waiting For Driver, (0, 0)), Waiting For Driver),
((Waiting For Driver, (0, 1)), Waiting For Driver),
((Waiting For Driver, (1, 0)), Driver Sitting),
((Waiting For Driver, (1, 1)), Waiting For Driver),
((Driver Sitting, (0, 0)), Waiting For Driver),
((Driver Sitting, (0, 1)), Waiting For Driver),
((Driver Sitting, (1, 0)), Driver Sitting),
((Driver Sitting, (1, 1)), Sequence Correct),
((Sequence Correct, (0, 0)), Waiting For Driver),
((Sequence Correct, (0, 1)), Waiting For Driver),
((Sequence Correct, (1, 0)), Driver Sitting),
((Sequence Correct, (1, 1)), Sequence Correct)},
each element of the readout function shows (the present state, and its outputs)
RZ1 = {(Waiting For Driver, Off), (Driver Sitting, Off), (Sequence Correct, On)}.

To test this system, we create test sequences comprising time and the system inputs. We observe the state and the output (Tables 3.21, 3.22, and 3.23).

Table 3.23 Test sequence 3

System experiment for the Automobile Seatbelt System. The initial state is Waiting For Driver					
Time	I1Z1	I2Z1	State	Next State	Output
0	1	0	Waiting For Driver	Driver Sitting	Ignition Off
1	0	0	Driver Sitting	Waiting For Driver	Ignition Off
2	1	0	Waiting For Driver	Driver Sitting	Ignition Off
3	0	1	Driver Sitting	Waiting For Driver	Ignition Off
4			Waiting For Driver		Ignition Off

Twelve transitions are listed in the next state function NZ1. All of these have been exercised. The output trajectories are correct. However, the equivalence of two state machines cannot be proven by looking at only the outputs, Wymore and Bahill [42]: see Section 17. Therefore, we must make the states observable or add an extra output to the Driver Seated state. This extra output could be small test indicator light or it could enable the radio, etc.

What else should we test? It seems obvious that we should test each function that the system is required to perform. But instead, it will be easier to test each requirement that has test as the verification technique. There should be a one-to-one relationship between functions and requirements.

How do you test a software class? Test each of its functions (operations). However, these functions should all be in sequence diagrams, which are all based on use cases. And the use cases have produced the requirements, as we have just shown. Remember to test all of the flows of all of the use cases. Test the Built-in Self-Test. Test drive the system and apply unusual inputs. Have experts use it. Have typical end users use it. Demonstrate it to the top brass. Use it on unusual people. All of these tests must be made into test cases so that they can be automatically applied with each new iteration.

3.6.6 Anomalous Data

Never throwaway outliers or confusing data, especially if they contradict the status quo. Put them aside and revisit them every few years. Sometime you may find a truly profound explanation, as we have several times. Anomalous data may point out weakness in current models. Future technology improvements will show deficiencies in our present models of humans.

3.6.7 Homework

3.6.7.1 Baseball Statistics

This is an extensive homework problem or team project. It spans Sections 20–25. If you have not already done so, go to the Springer site for this textbook and download the file baseballStatisticsforStudents.xlsx. For different types of baseball statistics see [43].

In the sport of baseball, it seems that batting average and homeruns might be good evaluation criteria. These metrics are phrased so that more is better and they seem to show compensation. There are singles hitters who specialize in placing the ball in the outfield; they should have high batting averages. Then there are power hitters who try to maximize power so that they produce more homeruns. Perhaps batters can trade-off homeruns for batting average. We created an Excel spreadsheet with a couple of dozen common baseball statistics for 60 of the best batters since 1900. Our criteria required that each batter had at least 6000 plate appearances. Then we lowered our threshold for career batting average until it included the batter with the most games, runs, hits, singles, doubles, triples, homeruns, runs batted in, base on balls, sacrifice bunts, stolen bases, least caught stealing, and the least strikeouts. If you make your own spreadsheet with your own criteria, you will get a different set of batters. Our spreadsheet is named BaseballStatisticsForStudents.xlsx and it is available from the publisher. The data came from http://www.baseball-reference.com/play-index/. We normalized their data with Normalized score $= \frac{\text{score}-\text{minimum}}{\text{maximum}-\text{minimum}}$. With this equation, the maximum output value of each normalized criterion was one and the minimum was zero. For the criteria caught stealing (CS) and strikeouts (SO) (where less is best) we inverted the criteria so that more was better with

$$\text{Normalized score} = \frac{\text{maximum} - \text{score}}{\text{maximum} - \text{minimum}}$$

For simplicity, we did not use weights of importance in our solutions. Also we did not use the excellent criteria of annual salary or the criteria of the shady use of anabolic steroids.

Independence

We invented a process to determine if each pair of evaluation criteria was independent and exhibited compensation. First, we plotted the data for each pair. Then we fit a trend line to the data. This was usually a straight line, although we did use some hyperbolas. Then we applied the Excel R^2, which is the square of the Pearson product moment correlation coefficient, to the entire 60-batter database. We decreed that

$$\begin{aligned}
&\text{If } 0 < R^2 < 0.2, \text{ then the pair is independent}\\
&\text{If } 0.2 < R^2 < 0.4, \text{ then the pair is mildly independent}\\
&\text{If } 0.4 < R^2 < 0.6, \text{ then the pair is loosely independent}\\
&\text{If } 0.6 < R^2 < 0.8, \text{ then the pair is mildly dependent}\\
&\text{If } 0.8 < R^2 < 1, \text{ then the pair is dependent}
\end{aligned}$$

Compensation

If the evaluation criteria pair had $R^2 < 0.6$ (this number is arbitrary), we then investigated compensation. We looked at the equation that best fit the data. In this analysis, we are studying optimal players who performed optimally. Therefore, we assumed that each batter had optimized his trade-off values for each pair of evaluation criteria. For example, Babe Ruth hit 714 homeruns and struck out

1330 times in his career. This is a ratio of 1.9, let us round that to two. This means that Babe Ruth valued homeruns about twice as much as he cared about avoiding strikeouts. Ted Williams' ratio was 1.4. The ratio of strikeouts to homeruns for the whole database is about three. Our point is that batters value homeruns about three times as much as they worry about strikeouts and baseball management seems to be willing to pay the same amount for an extra homerun as for three less strikeouts. Therefore, this pair of evaluation criteria shows compensation. On the other hand, between 1905 and 1927 baseball fans were not enamored of homeruns. Therefore, Ty Cobb did not try to hit homeruns. With a ratio of 5.8, he worried about striking out about six times as much as he valued getting a home run.

Homework Questions

Your assignment, should you chose to accept it, is to determine what combining function would be the best for selecting the best major league baseball batters in history? Hint: Study the shape of the data before thinking about combining functions.

Consider the following pairs of evaluation criteria.

Normalized homeruns and normalized strikeouts (HR and SO)
Normalized homeruns and normalized runs batted in (HR and RBI)
Normalized homeruns and ***actual*** batting average (HR and BA)
Normalized runs scored and normalized runs batted in (R and RBI)
Normalized homeruns and normalized successful bunts (HR and SH)
Normalized homeruns and ***normalized*** batting average (HR and BA)
Normalized runs scored and normalized batting average (R and BA)
Normalized caught stealing and normalized stolen bases (CS and SB)
Normalized caught stealing and normalized triples (CS and 3B)

Evaluation criteria should be quantitative, independent, transitive, objective, worded correctly, unaffected by temporal order of evaluation, and exhibiting compensation. Do the above pairs of evaluation criteria exhibit all of these properties?

We have asked you to examine nine pairs of evaluation criteria. How many pairs of evaluation criteria could be made with this spreadsheet data set? Aren't you glad that we did not ask you to evaluate all of them? How many triplets could be made?

Here is an example of what we expect you to do for these nine pairs of evaluation criteria.

Who were the greatest batters of the twentieth century based on the evaluation criteria of Normalized walks (Base on balls, BB) and normalized strikeouts. With *normalized* strikeouts more is better (Fig. 3.40).

This looks like the ideal Pareto-optimal distribution, with the data bounded by the red Pareto-optimal contour. So either the sum or the product combining functions should be fine.

In Table 3.24, the top batters are listed in order by score, highest to lowest. For the Sum combining function four batters were far ahead of all of the others, so we

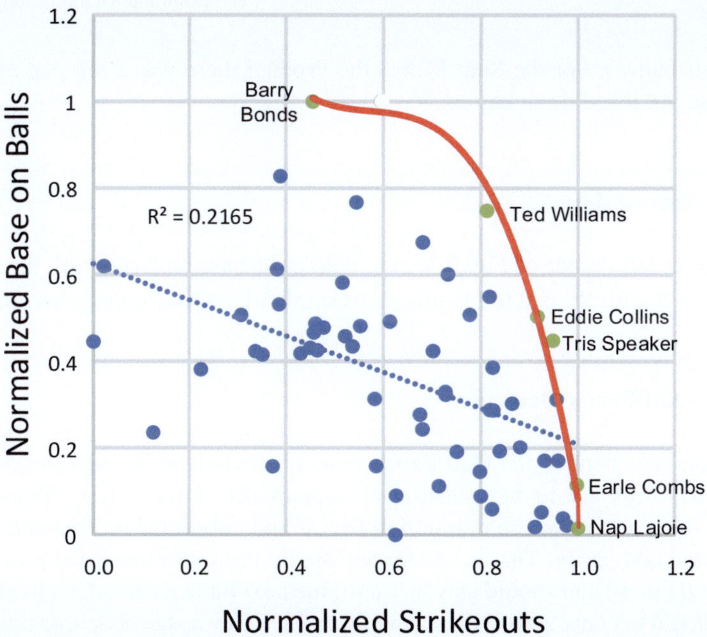

Fig. 3.40 Walks versus strikeouts from our spreadsheet

Table 3.24 The best batters evaluated with base on balls and strikeouts computed with five combining functions

	Evaluation criteria pair (Bases on Balls and Strikeouts)
Combining function	Recommendation
Sum	Ted Williams, Barry Bonds, Eddie Collins, and Tris Speaker
Product	Ted Williams, Barry Bonds, Eddie Collins, and Tris Speaker
Compromise with $p = 2$	Ted Williams, Barry Bonds, Tris Speaker, and Eddie Collins
Compromise with $p = 10$	The batter with the most bases on balls Barry Bonds and The three batters with the fewest strike outs Nap Lajoie, Earle Combs, and George Sisler
Sum minus the product	Barry Bonds, Nap Lajoie, and Earle Combs
Comment	The Sum and the Product combining functions recommend the same batters in the same order. The Compromise with $p = 2$ recommends the same batters but in a different order
Properties	These evaluation criteria are quantitative, transitive, objective, positive numbers, worded correctly (the normalized version has "more is better"), and unaffected by temporal order of evaluation
Independence and compensation.	With an $R^2 = 0.2$, these criteria are independent. We think this pair of criteria shows compensation. Mentally most batters would value a walk and avoiding a strikeout about the same. With a full count, most batters would "take" a close pitch
Recommendations	The Sum, Product, and Compromise with $p = 2$ combining functions seem to make the most sense. They would be the preferred combining functions. The Compromise with $p = 10$ behaves as expected: it gives the batters with the highest scores in each category. The Sum Minus the Product is a little strange

listed four batters. For the Sum Minus the Product there was a big gap after three batters, so we listed three batters.

3.6.7.2 Anscombe

In the upper left corner of Fig. 3.37, the data points are replaced with the equation $y = 0.5x + 3$ with $R^2 = 0.67$. Is this an example of abstraction or generalization?

3.6.7.3 Odd Parity Detector

This is the statement of the Odd Parity Detector homework problem of Section 9. Computer words are to be transmitted sequentially down a wire. Draw a state diagram for a system whose output will be 1 if and only if the present and previous bits showed odd parity. That is, the output should become a one if the previous two bits were 01 or 10 and should stay at 1 until the next bit is received. Conversely, the output should become zero if they were 00 or 11. Draw a state machine diagram for this system or specify Z3 = (SZ3, IZ3, OZ3, NZ3, RZ3). Now, we want you to create test sequences that will verify your design, as we did for the Automobile Seatbelt System.

3.6.7.4 House Heating System

Design a controller for a heating, ventilation, and air conditioning (HVAC) system for a typical Tucson family house. Assume that the outside air temperature varies annually between 25 and 115 degrees Fahrenheit (°F) (−4 to 46 °C) and that the room temperature is supposed to stay between 68 and 73°F (21 and 23 °C). The primary components of this HVAC system are the heater, air conditioner, fan, thermostat, and controller. We will start with the use case.

Name: Heat the House.
Iteration: 6.2
Derived from: concept of operations
Brief description: When it is cold outside, heat the house with a Heater and maintain the roomTemp between heaterLowerLimit (see Rule2) and heaterUpperLimit (see Rule3) degrees Fahrenheit.
Level: user goal
Priority: 9
Scope: An HVAC system for a typical Tucson family house.
Added value: The house is comfortable and the occupants will be more productive doing their homework.
Primary actor: Home Owner.
Supporting actor: Heater.
Frequency: The system operates continuously.

Precondition: The HVAC system is on. The roomTemp is between the heaterLowerLimit and the heaterUpperLimit. The heater is off.

Main Success Scenario

1. The roomTemp falls below heaterLowerLimit.
2a. The controller turns on the Heater {in accordance with Rule1}.
3. The roomTemp rises to the heaterUpperLimit.
4. The controller turns off the Heater {Rule1} {exit use case}.

Unanchored Alternate Flow

2b. Home Owner smells "gas."
2b1. Home Owner turns off the Heater {exit use case}.

Rules

Rule1:. When the Heater is on, turn the Fan on.
When the Heater is off, turn the Fan off.
Rule2:. Heaterlowerlimit default value is 68 °F.
Rule3:. Heaterupperlimit default value is 72 °F.

Specific Requirements

Functional Requirements

FR1:. The controller shall be capable of reading the temperature on the thermostat.
FR2:. The controller shall turn the heater on when the temperature drops below the heaterLowerLimit.
FR3:. The controller shall turn the heater off when the temperature rises above the heaterUpperLimit.

Nonfunctional Requirements

NFR1:. On-off cycles should last at least 15 min.

Author/owner: Home Owner
Date of last change: January 8, 2016

Testing using test vectors

Test Matrix

Using test vectors is a common technique for testing systems. With this technique, the system is set in an initial state, an input event is applied and the next state and the resulting output are observed and recorded. If there are many test vectors, they are arranged in a matrix as shown in Table 3.25. Each row is an individual test. The rows are called vectors. This is not a system experiment where time goes from top to bottom.

Table 3.25 A matrix of test vectors for the heater system

Initial state	Event (inputs)	Next state	Resulting output
Any	roomTemp > heaterUpperLimit	Heater Off	Heater Off
Heater Off	roomTemp > heaterLowerLimit	Heater Off	Heater Off
Heater Off	roomTemp = heaterLowerLimit	Heater On	Heater On
Heater On	roomTemp < heaterUpperLimit	Heater On	Heater On
Heater On	roomTemp = heaterUpperLimit	Heater Off	Heater Off
Any	roomTemp < heaterLowerLimit	Heater On	Heater On
Any	heaterUpperLimit ≤ heaterLowerLimit	Heater Off	SignalError

Test Procedure

The following test procedure specifies how the test vectors will be used.

1. Choose a cold day, e.g., outside temperature < 60 °F.
2. Set the heaterUpperLimit and heaterLowerLimit according to the Test Inputs given below.
3. Turn the heater on.
4. Choose a particular test vector from the test matrix of Table 3.25.
5. Put the system in the initial state.
6. Apply the input event.
7. Observe the next state and the resulting output
8. Observe the room temperature for TBD minutes. This TBD must be resolved before Total System Test.
9. If it goes outside the heaterLimits, record an error.

Test Inputs

The following inputs would be suitable for testing the heater system.

Normal
heaterUpperLimit = 71 °F and heaterLowerLimit = 70 °F

Extremes
heaterUpperLimit = 80 °F and heaterLowerLimit = 78 °F
heaterUpperLimit = 36 °F and heaterLowerLimit = 34 °F

Mistakes
heaterUpperLimit = 70 °F and heaterLowerLimit = 72 °F

Testing using use case scenarios
Now, we want you to use the use case to create a test plan for this system.

3.7 Section 26. Sensitivity Analysis

Students, who are not unfamiliar with the mathematics employed in this section, may skip this section without loss of continuity.

3.7.1 *Simple Definitions*

3.7.1.1 In Mathematics

A *variable* is an object that may assume any given value or set of values. It is also the symbol that represents this object. The inputs and outputs for functions and systems are variables.

A *constant* is an object whose numerical value does not change. Some examples are 3, π, and $\sqrt{2}$

A *parameter* is a variable that is held constant when a function is used. In high school, we were told that coefficients in equations were parameters. Varying a parameter produces a family of functions. For example, in the straight-line function $f(x) = 2x + b$ it is x that varies when we evaluate the function; b is treated as a constant. But if we change b (for instance, let $b = 0$, 1, 2, and 3 in turn), and graph the function (y versus x) we get a family of parallel lines, each of slope 2 but with a different y-intercept, b. We could write this family of functions as $f(x; b) = 2x + b$. In this notation, variables go before the semicolon and parameters go after. This makes it explicit that the function depends on the parameter b just as it does on the variable x. When the function is used as a function, only x is varied; but for example, in a sensitivity analysis it is just as valid to vary b as it is to vary x.

3.7.1.2 In Statistics

Information recorded about a sample of individual people comprises measurements such as blood pressure, temperature, and weight and attributes such as ethnicity, blood type, and evidence of diabetes. Values of these will vary among the subjects; in this context, blood pressure, temperature, blood type, etc. are *variables*. Variables are quantities that vary from person to person. By contrast, *parameters* do not relate to actual measurements or attributes but to quantities defining a theoretical model, such as the mean and standard deviation. For example, if the data set is best fit with a normal distribution, then we have a family of distributions. Each member of the family is defined by its mean and standard deviation, the parameters that specify the particular normal distribution. We will also use parameters as values that can be altered to see what happens to the performance of some system. For example, the performance of our chocolate chip cookie acquisition system will depend upon the parameters baking time and oven temperature. In statistical experiments, the independent variables are called *factors*.

3.7.2 Introduction

In many system designs the specification can be inspected to reveal that the major indices of the system (such as performance, cost, schedule, or risk) are driven by relatively few input variables and system parameters. A common interpretation of Pareto's rule would state that 80 % of the influence could be traced to 20 % of the inputs and parameters. That is, variations of the few prominent inputs or parameters can have a substantial impact on the characteristics of the system. Information of this type can be extremely valuable to the stakeholder, because it not only alludes to inputs and parameters that are central in characterizing the system, but it also conveys to the stakeholder which inputs and parameters are robust to change. This can shed new light on trade-off possibilities. The process of uncovering inputs and parameters that drive the system's properties is called *sensitivity analysis*. This section is based on Smith, Szidarovszky, Karnavas, and Bahill [44].

We recommend that all trade-off studies and risk analyses incorporate sensitivity analysis. The results of a sensitivity analysis can be used to (1) validate a model, (2) flag strange or unrealistic model behavior, (3) suggest new experiments or guide future data collection efforts, (4) highlight important assumptions of the model, (5) adjust numerical values of the parameters, (6) suggest the required accuracy for calculating parameters, (7) guide the formulation of model structure, (8) allocate resources, and (9) pin-point true cost drivers [45]. If you show your customer the requirements that are driving system cost, then he or she may relax a requirement and save money.

In analyzing the sensitivities of a trade-off study or a risk analysis, we are typically interested in (1) those inputs, parameters, and evaluation criteria that are most important and deserve further attention and (2) those inputs and parameters that, when varied, could change the recommended alternative. The first issue can be investigated using a relative-sensitivity measure for each parameter for each alternative. The second can be ascertained by employing a search algorithm using the sensitivity of each parameter [45]. Typical trade-off study or risk analysis parameters include weights at all levels of the trade-off hierarchy, and the scoring function inputs and parameters. Sensitivity analyses can be performed using simple spreadsheet tools.

3.7.3 Motivation

Sensitivity analysis should be performed anytime you create a model, write a set of requirements, design a system, make a decision, do a trade-off study, undertake risk analysis, or want to discover cost drivers. In a sensitivity analysis, you change the values of inputs, parameters, or architectural features and then measure changes in outputs or performance indices. The results of a sensitivity analysis can be used to validate a model, flag unrealistic model behavior, point out important assumptions, help formulate model structure, simplify a model, suggest new experiments, guide future data collection efforts, suggest accuracy for calculating parameters, adjust numerical values of parameters, choose an operating point, allocate resources, detect critical criteria, suggest tolerance for manufacturing parts, and identify cost drivers.

A sensitivity analysis reveals which inputs and parameters are the most important and most likely to affect system behavior and/or model predictions. Following a sensitivity analysis, values of critical parameters can be refined while parameters that have little effect can be either simplified, or ignored. In the manufacturing environment, they can be used to allocate resources to critical parts allowing casual treatment of less sensitive parts. If the sensitivity coefficients are calculated as functions of time, it can be seen *when* each parameter has the greatest effect on the output function of interest. This can be used to adjust numerical values for the parameters. The values of the parameters should be chosen to match the physical data at the times when they have the most effect on the output.

The earliest form of sensitivity analyses that we have discovered are in genetics studies on the pea reported by Gregor Mendel in 1865 [25] and the statistics studies on the Irish hops crops by Gosset writing under the pseudonym Student around 1890 [46]. Since then, sensitivity analyses have been used to: analyze networks and feedback amplifiers; validate social models, engineering models, and physiological models; and target disease treatment. They are also used in numerical computations, expert systems, discrete event simulations, and in techniques that are called response surface methodology. They are also employed in frequency domain experiments, and perturbation analysis. When changes in parameters cause discontinuous changes in system properties, the sensitivity analysis is called singular perturbations. Sensitivity functions are also used in adaptive control systems. In linear programming problems, sensitivities are referred to as marginal values or the shadow price of each constraint and are interpreted as the increase in the optimal solution obtainable by each additional unit added to the constraint.

Statistical techniques can be used when the model is so complex that computing performance indices becomes extremely difficult. For such complex models, the inputs are described with probability distributions and then the induced uncertainly in the performance indices is analyzed. This approach is known as probabilistic sensitivity analysis. We will not deal with such complex simulations in this book. Instead, we present simple techniques that will increase *intuition* and *understanding* of systems.

There are several common ways to perform sensitivity analyses. A partial derivative can be a sensitivity function for a system described by equations. Otherwise, spreadsheets, like Excel, are convenient for doing sensitivity analyses of systems that are not described by equations. In this subsection, we explore several techniques for performing sensitivity analyses.

There are two classes of sensitivity functions: analytic and empirical. Analytic sensitivity functions are used when the system under study is relatively simple, well defined, and mathematically well behaved (for example, continuous derivatives). These functions generally take the form of partial derivatives. They are convenient, because once derived they can often be used in a broad range of similar systems, and are easily adjusted for changes in all model variables and parameters. They also have an advantage in that the sensitivity of a system to a given input or parameter is given as a function of all other inputs and parameters including time or frequency (depending on the model of the system), and can be plotted as functions of these variables.

Empirical sensitivity functions are often just point evaluations of a system's sensitivity to a given parameter(s) when other inputs and parameters are at known, fixed values. Empirical sensitivity functions can be estimated from the point sensitivity evaluations over a range of the values of the inputs and parameters. They are generally determined by observing the changes in output of a computer simulation as model inputs and parameters are systematically varied from run to run. Their advantage is that they are often simpler (or more feasible) than their analytic counterparts, and they can even be determined for an unmodeled physical system. If the physical system is all that is available, the system output is monitored as the inputs and parameters are changed from their normal values.

3.7.4 *Analytic Sensitivity Functions*

This subsection explains three different analytic sensitivity functions. They are all based on finding the partial derivative of a mathematical system model with respect to some parameter. Short examples are then given for each particular form of a sensitivity function.

3.7.4.1 The Absolute-Sensitivity Function

The *absolute sensitivity* of the function F to variations in the parameter α is given by

$$S_{\alpha}^{F} = \left.\frac{\partial F}{\partial \alpha}\right|_{\mathrm{NOP}}$$

where the NOP subscript means that the partial derivative is evaluated at the Nominal Operating Point (NOP) where all the parameters have their normal or nominal values. Of course, the function F must be differentiable with respect to α. The function F may also be a function of other parameters such as time, frequency, or temperature. Absolute-sensitivity functions are useful for calculating output errors due to parameter variations and for assessing the times at which a parameter has its greatest or least effect. Absolute sensitivities are also used in adaptive control systems.

The following two examples show the use of absolute-sensitivity functions. The first shows how to tailor the output of a process, while the second shows how to assess when a parameter has its greatest effect.

Example 1: A Process Model

Assume that the function

$$f(x, y) = z = Ax^2 + By^2 + Cxy + Dx + Ey + F$$

models a process, where x and y are the inputs (raw material) and A to F are models of the system parameters. Let the nominal operating point be

$$(x_0, y_0) = (1, 1),\ \ A_0 = 1, B_0 = 2, C_0 = 3, D_0 = -5, E_0 = -7,$$
$$F_0 = 8,\ \text{producing}\ z_0 = 2.$$

Suppose that the output z is a valuable commodity (perhaps a magic potion). What is the easiest way (the smallest change in one of these eight operating parameters) to increase the quantity of z that is being produced? This sounds like a natural problem for absolute-sensitivity functions.

$$\begin{aligned}
S_A^z &= \tfrac{\partial z}{\partial A}\Big|_{\text{NOP}} = x_0^2 = 1\\
S_B^z &= \tfrac{\partial z}{\partial B}\Big|_{\text{NOP}} = y_0^2 = 1\\
S_C^z &= \tfrac{\partial z}{\partial C}\Big|_{\text{NOP}} = x_0 y_0 = 1\\
S_D^z &= \tfrac{\partial z}{\partial D}\Big|_{\text{NOP}} = x_0 = 1\\
S_E^z &= \tfrac{\partial z}{\partial E}\Big|_{\text{NOP}} = y_0 = 1\\
S_F^z &= \tfrac{\partial z}{\partial F}\Big|_{\text{NOP}} = 1\\
S_x^z &= \tfrac{\partial z}{\partial x}\Big|_{\text{NOP}} = 2A_0 x_0 + C_0 y_0 + D_0 = 0\\
S_y^z &= \tfrac{\partial z}{\partial y}\Big|_{\text{NOP}} = 2B_0 y_0 + C_0 x_0 + E_0 = 0
\end{aligned}$$

Evaluated at $(x_0, y_0) = (1,\ 1)$, the six coefficients are equally sensitive: increase any coefficient by one unit and the output increases by one unit. Since the derivatives with respect to x and y are equal to zero, to analyze sensitivity with respect to these inputs we need to use higher order partial derivatives.

What about interactions? Could we do better by changing two inputs and parameters at the same time? Interaction terms are characterized by, for example, the partial derivative of z with respect to x containing y. The interactions can be bigger than the first-order partial derivatives. You can see that the above partial derivatives with respect to the coefficients contain the operating point coordinates and the partial derivatives with respect to the operating point coordinates depend on the coefficients. Therefore, we should expect the interactions to be significant. Of the 64 possible second-partial derivatives, only the following are nonzero.

$$S^{z}_{x-A} = \left.\frac{\partial^2 z}{\partial x \partial A}\right|_{\text{NOP}} = 2x_0 = 2$$

$$S^{z}_{x-C} = \left.\frac{\partial^2 z}{\partial x \partial C}\right|_{\text{NOP}} = y_0 = 1$$

$$S^{z}_{x-D} = \left.\frac{\partial^2 z}{\partial x \partial D}\right|_{\text{NOP}} = 1$$

$$S^{z}_{y-B} = \left.\frac{\partial^2 z}{\partial y \partial B}\right|_{\text{NOP}} = 2y_0 = 2$$

$$S^{z}_{y-C} = \left.\frac{\partial^2 z}{\partial y \partial C}\right|_{\text{NOP}} = x_0 = 1$$

$$S^{z}_{y-E} = \left.\frac{\partial^2 z}{\partial y \partial E}\right|_{\text{NOP}} = 1$$

$$S^{z}_{x-y} = \left.\frac{\partial^2 z}{\partial x \partial y}\right|_{\text{NOP}} = C_0 = 3$$

$$S^{z}_{x^2} = \left.\frac{\partial^2 z}{\partial x^2}\right|_{\text{NOP}} = 2A_0 = 2$$

$$S^{z}_{y^2} = \left.\frac{\partial^2 z}{\partial y^2}\right|_{\text{NOP}} = 2B_0 = 4$$

From the theory of Young, we know that if the partial derivatives are continuous, then $\frac{\partial z^2}{\partial x \partial A} = \frac{\partial z^2}{\partial A \partial x}$, etc., so we do not have to calculate any more partial derivatives. Five of these mixed second-partial derivatives are bigger than the first-partial derivatives.

At this operating point, $z_0 = 2$. If we increase A by one unit **(this is too large of a step for a sensitivity analysis, but the math is convenient)**, then $A = 2$, this increases $z = 3$, and $\Delta z = 1$. If we increase x by one unit, then $x = 2$, $z = 3$, and again $\Delta z = 1$. However, if we simultaneously increase both x and A by one unit each, $A = 2$, $x = 2$, $z = 7$, and $\Delta z = 5$. The interaction effect is big. This and two other interaction effects are shown in Table 3.26.

Here we have treated the second partial derivative $\frac{\partial^2 z}{\partial y^2}$ as an interaction term. Karnavas, Sanchez, and Bahill [45] explain how this could be the result of a quadratic nonlinearity.

How about the third-order partial derivatives? For example, $\frac{\partial^3 z}{\partial x^2 \partial A} = 2$. Let $x = 3$ and $A = 2$, $z = 15$, and $\Delta z = 13$. This is a bigger output boost than that produced by the sum of changes in any three inputs or parameters. Three of the third-order partial derivatives are greater than zero. One is equal to 1, and two are equal to 2. All of the fourth-order partial derivatives are zero (Table 3.27).

Table 3.26 Effects of Individual and Combined Input and Parameter Changes

Partial derivatives	Normal values	Values increased by one unit	New z values	Δz	Total change in z
$\frac{\partial z}{\partial A}$	$A=1$	$A=2$	3	1	$\sum \Delta z = 2$
$\frac{\partial z}{\partial x}$	$x=1$	$x=2$	3	1	
$\frac{\partial^2 z}{\partial x \partial A}$	$A=1$ $x=1$	$A=2$ $x=2$	7	5	5
$\frac{\partial z}{\partial x}$	$x=1$	$x=2$	3	1	$\sum \Delta z = 3$
$\frac{\partial z}{\partial y}$	$y=1$	$y=2$	4	2	
$\frac{\partial^2 z}{\partial x \partial y}$	$x=1$ $y=1$	$x=2$ $y=2$	8	6	6
$\frac{\partial z}{\partial y}$	$y=1$	$y=2$	4	2	$\sum \Delta z = 4$
$\frac{\partial z}{\partial y}$	$y=1$	$y=2$	4	2	
$\frac{\partial^2 z}{\partial y^2}$	$y=1$	$y=3$	10	8	8

Table 3.27 Effects of individual and combined input and parameter changes

Partial derivatives	Normal values	Values increased by one unit	New z values	Δz	Total change in z
$\frac{\partial z}{\partial A}$	$A=1$	$A=2$	3	1	$\sum \Delta z = 3$
$\frac{\partial z}{\partial x}$	$x=1$	$x=2$	3	1	
$\frac{\partial z}{\partial x}$	$x=1$	$x=2$	3	1	
$\frac{\partial^3 z}{\partial x^2 \partial A} = 2$	$A=1$ $x=1$	$A=2$ $x=3$	15	13	13

3.7.4.2 The Relative-Sensitivity Function

If we want to compare the effects of different inputs and parameters, we should use relative-sensitivity functions. The *relative-sensitivity* of the function F to the parameter α evaluated at the nominal operating point is given by

$$\overline{S}_{\alpha}^{F} = \left.\frac{\partial F}{\partial \alpha}\right|_{\mathrm{NOP}} \frac{\alpha_0}{F_0} \approx \frac{\%\text{ change in } F}{\%\text{ change in } \alpha} = \frac{\frac{\Delta F}{F}}{\frac{\Delta \alpha}{\alpha}}$$

where NOP and the subscripts 0 mean that all functions, inputs, and parameters assume their nominal operating point values. Relative-sensitivity functions are formed by multiplying the partial derivative (the absolute-sensitivity function) by the normal value of the parameter and dividing by the normal value of the function. For statistical models, a relative-sensitivity function can be formed by multiplying the partial derivative by the ratio of the variances. Relative-sensitivity functions are ideal for comparing inputs and parameters, because they are dimensionless, normalized functions. In the field of economics, the lack of dimensions of the relative-sensitivity function is exploited to allow easy comparison of parameters' changes on model outputs even though the parameters may describe widely varying aspects of the model and have different units. Economists refer to the relative-sensitivity function $\overline{S}_B^A$ as the elasticity of B with respect to A, and denote it as $E_{B,A}$.

Example 1: The Process Model (Continued)

Assume that the function

$$f(x, y) = z = Ax^2 + By^2 + Cxy + Dx + Ey + F$$

models the output of a process, where the nominal operating point is

$$(x_0, y_0) = (1, 1), \quad A_0 = 1, B_0 = 2, C_0 = 3, D_0 = -5, E_0 = -7,$$
$$F_0 = 8, \quad \text{producing } z_0 = 2.$$

What is the easiest way (smallest **percent** change in an operating point parameter) to increase the quantity of z that is being produced? Now this problem is appropriate for relative-sensitivity functions.

$$\overline{S}_A^z = \frac{\partial z}{\partial A}\Big|_{\text{NOP}} \frac{A_0}{z_0} = x_0^2 \frac{A_0}{z_0} = 0.5$$
$$\overline{S}_B^z = \frac{\partial z}{\partial B}\Big|_{\text{NOP}} \frac{B_0}{z_0} = y_0^2 \frac{B_0}{z_0} = 1$$
$$\overline{S}_C^z = \frac{\partial z}{\partial C}\Big|_{\text{NOP}} \frac{C_0}{z_0} = x_0 y_0 \frac{C_0}{z_0} = 1.5$$
$$\overline{S}_D^z = \frac{\partial z}{\partial D}\Big|_{\text{NOP}} \frac{D_0}{z_0} = x_0 \frac{D_0}{z_0} = -2.5$$
$$\overline{S}_E^z = \frac{\partial z}{\partial E}\Big|_{\text{NOP}} \frac{E_0}{z_0} = y_0 \frac{E_0}{z_0} = -3.5$$
$$\overline{S}_F^z = \frac{\partial z}{\partial F}\Big|_{\text{NOP}} \frac{F_0}{z_0} = \frac{F_0}{z_0} = 4$$
$$\overline{S}_x^z = \frac{\partial z}{\partial x}\Big|_{\text{NOP}} \frac{x_0}{z_0} = (2A_0x_0 + C_0y_0 + D_0)\frac{x_0}{z_0} = 0$$
$$\overline{S}_y^z = \frac{\partial z}{\partial y}\Big|_{\text{NOP}} \frac{y_0}{z_0} = (2B_0y_0 + C_0x_0 + E_0)\frac{y_0}{z_0} = 0$$

Evaluated at $(x_0, y_0) = (1,\ 1), \overline{S}_F^z = \frac{\partial z}{\partial F}\big|_{\text{NOP}} \frac{F_0}{z_0} = \frac{F_0}{z_0} = 4$, is the largest. Therefore, we should increase F if we wish to increase z.

What about interactions? Could we do better by changing two inputs and/or parameters at the same time? Only the following second-order relative-sensitivity functions are nonzero.

$$\overline{S}^{z}_{x-A} = \left.\frac{\partial^2 z}{\partial x \partial A}\right|_{\text{NOP}} \frac{x_0 A_0}{z_0^2} = \frac{2x_0^2 A_0}{z_0^2} = 0.5$$

$$\overline{S}^{z}_{x-C} = \left.\frac{\partial^2 z}{\partial x \partial C}\right|_{\text{NOP}} \frac{x_0 C_0}{z_0^2} = \frac{x_0 y_0 C_0}{z_0^2} = 0.75$$

$$\overline{S}^{z}_{x-D} = \left.\frac{\partial^2 z}{\partial x \partial D}\right|_{\text{NOP}} \frac{x_0 D_0}{z_0^2} = \frac{x_0 D_0}{z_0^2} = -1.25$$

$$\overline{S}^{z}_{y-B} = \left.\frac{\partial^2 z}{\partial y \partial B}\right|_{\text{NOP}} \frac{y_0 B_0}{z_0^2} = \frac{2y_0^2 B_0}{z_0^2} = 1$$

$$\overline{S}^{z}_{y-C} = \left.\frac{\partial^2 z}{\partial y \partial C}\right|_{\text{NOP}} \frac{y_0 C_0}{z_0^2} = \frac{x_0 y_0 C_0}{z_0^2} = 0.75$$

$$\overline{S}^{z}_{y-E} = \left.\frac{\partial^2 z}{\partial y \partial E}\right|_{\text{NOP}} \frac{y_0 E_0}{z_0^2} = \frac{y_0 E_0}{z_0^2} = -1.75$$

$$\overline{S}^{z}_{x-y} = \left.\frac{\partial^2 z}{\partial x \partial y}\right|_{\text{NOP}} \frac{x_0 y_0}{z_0^2} = \frac{x_0 y_0 C_0}{z_0^2} = 0.75$$

$$\overline{S}^{z}_{x^2} = \left.\frac{\partial^2 z}{\partial x^2}\right|_{\text{NOP}} \frac{x_0^2}{z_0^2} = \frac{2x_0^2 A_0}{z_0^2} = 0.5$$

$$\overline{S}^{z}_{y^2} = \left.\frac{\partial^2 z}{\partial y^2}\right|_{\text{NOP}} \frac{y_0^2}{z_0^2} = \frac{2y_0^2 B_0}{z_0^2} = 1$$

None of these is bigger than $\overline{S}^{z}_{F} = \left.\frac{\partial z}{\partial F}\right|_{\text{NOP}} \frac{F_0}{z_0} = \frac{F_0}{z_0} = 4$.

In summary, using absolute-sensitivity functions, the second-order and third-order terms were the most important, but using relative-sensitivity functions, the coefficient F was the most important. Computing the absolute and relative-sensitivity functions yield different results. The absolute-sensitivity functions show which inputs and parameters have the greatest effect on the output for a *fixed size* change in the inputs and parameters, whereas the relative-sensitivity functions show which inputs and parameters have the greatest effect on the output for a certain *percent* change in the inputs and parameters.

Limitations of the Relative-Sensitivity Function

The relative-sensitivity function is limited in usefulness in analytic studies for three reasons. First, the relative-sensitivity functions are different in the time and in the frequency domains. Second the relative-sensitivity function is the product of two functions (the partial derivative and the original function), and the Laplace transform of a product is not the product of the Laplace transforms, for example

$$L(x(t) \times y(t)) \neq L(x(t)) \times L(y(t))$$

Therefore, you cannot use Laplace transforms to get time-domain solutions. And finally if the function being analyzed is a time function, then it may take on very small values, which would give erroneous results, or it might even take on values of zero, and division by zero is a no-no!.

3.7.4.3 The Semirelative-Sensitivity Function

We have used the absolute-sensitivity function to see when a parameter had its greatest effect on the step response of a system, and we have used the relative-sensitivity function to see which parameter had the greatest effect on the transfer function. Now suppose we wish to compare inputs and parameters, but we want to look at the step response and not the transfer function. What happens if we try to use the relative-sensitivity function to compare parameter effects on the step response? Well, we get into trouble. For a step response, the normal output value y_0 typically varies from 0 to 1 and division by zero is undefined. Furthermore, the relative-sensitivity function gives undue weight to the beginning of the response when y_0 is small. Therefore, let us investigate the use of the *semirelative-sensitivity function*, which is defined as

$$\widetilde{S}_{\alpha}^{F} = \left.\frac{\partial F}{\partial \alpha}\right|_{\text{NOP}} \alpha_0$$

where NOP and the subscript 0 mean that all functions, inputs, and parameters assume their nominal operating point values.

As can be seen by the definition, semirelative-sensitivity functions will have the same shape as absolute-sensitivity functions. They are just multiplied by the constant parameter values. But this scaling allows comparisons to be made of the effects of the various parameters. Of course, when a parameter has a value of zero, its semirelative-sensitivity function is also zero.

Example 1: The Process Model (Continued)

Let us now return to the process model, $f(x, y) = z = Ax^2 + By^2 + Cxy + Dx + Ey + F$. Would semirelative-sensitivity functions be better than relative-sensitivity functions? There would be little difference, because z is not a function of time and we can prevent it from taking on a zero value. We could get the semirelative-sensitivity functions by multiplying the relative-sensitivity functions by z_0, which is 2. Thus, all of the first-order relative-sensitivity functions would be multiplied by 2, which would not change the relative rankings. The second-order relative-sensitivity functions would all be multiplied by 4, once again, no rankings change. Therefore, for the process model we can use either relative or semirelative-sensitivity functions. It makes no difference.

Table 3.28 A generic trade-off study

Criteria	Weight of importance	Alternative-1	Alternative-2
Criterion-1	w_1	sc_{11}	sc_{12}
Criterion-2	w_2	sc_{21}	sc_{22}
Alternative rating		$\text{Sum}_1 = w_1 \times sc_{11} + w_2 \times sc_{21}$	$\text{Sum}_2 = w_1 \times sc_{12} + w_2 \times sc_{22}$

Table 3.29 Trade-off study numerical example

		Alternatives	
Criteria	Weight of importance	Umpire's assistant	Seeing eye dog
Accuracy of the call	0.75	0.67	0.33
Silence of signaling	0.25	0.83	0.17
Sum of weight multiplied by score		0.71 The winner	0.29

Example 3: A Trade-off Study

In modeling dynamic systems, most sensitivity functions are functions of time or frequency, but this is not a prerequisite for sensitivity analyses. For example, let us compute the semirelative-sensitivity functions for a trade-off study. Trade-off studies provide a rational method for choosing among alternatives when simultaneously considering multiple alternatives with many criteria. In the trade-off study of Tables 3.28 and 3.29, the alternative ratings are computed with linear addition of the weighted scores. That is, $\text{Sum}_1 = w_1 \times sc_{11} + w_2 \times sc_{21}$ and $\text{Sum}_2 = w_1 \times sc_{12} + w_2 \times sc_{22}$.

Table 3.29 gives numerical values for one particular trade-off study, The Umpire's Assistant.

Now we want to perform a sensitivity analysis of this trade-off study. We want to discover which inputs and parameters could change the recommendations of the trade-off study. To this end, an appropriate performance index is the rating for the highest rated alternative (alternative-1) minus the rating for second highest rated alternative (alternative-2).

$$\text{PI}_1 = \text{Sum}_1 - \text{Sum}_2 = w_1 \times sc_{11} + w_2 \times sc_{21} - w_1 \times sc_{12} - w_2 \times sc_{22} = 0.420.$$

We want to compute the semirelative-sensitivity functions for this performance index, with respect to the six inputs and parameters. We want to discover which parameter is the most important and plan our future actions.

Table 3.30 Analytic semirelative-sensitivity function values for PI_1, the difference of the alternative ratings

		Alternatives	
Criteria	Weight of importance	Umpire's assistant	Seeing eye dog
Accuracy of the call	$\tilde{S}_{w_1}^{PI_1} = 0.26$	$\tilde{S}_{sc_{11}}^{PI_1} = 0.50$	$\tilde{S}_{sc_{12}}^{PI_1} = -0.25$
Silence of signaling	$\tilde{S}_{w_2}^{PI_1} = 0.16$	$\tilde{S}_{sc_{21}}^{PI_1} = 0.21$	$\tilde{S}_{sc_{22}}^{PI_1} = -0.04$

$$\tilde{S}_{w_1}^{PI_1} = (sc_{11} - sc_{12})w_1 = 0.26$$
$$\tilde{S}_{w_2}^{PI_1} = (sc_{21} - sc_{22})w_2 = 0.16$$
$$\tilde{S}_{sc_{11}}^{PI_1} = w_1 sc_{11} = 0.50$$
$$\tilde{S}_{sc_{21}}^{PI_1} = w_2 sc_{21} = 0.21$$
$$\tilde{S}_{sc_{12}}^{PI_1} = -w_1 sc_{12} = -0.25$$
$$\tilde{S}_{sc_{22}}^{PI_1} = -w_2 sc_{22} = -0.04$$

These sensitivities can be nicely displayed in a table, as in Table 3.30.

The most important input is sc_{11}. Therefore, we need to gather more experimental data and interview more domain experts for this input; we should spend extra resources on this input. The minus signs for sc_{12} and sc_{22} merely mean that an increase in either of these inputs will cause a *decrease* in the performance index.

A problem with this performance index is that if $sc_{11} = sc_{12}$, then the sensitivity with respect to w_1 becomes equal to zero. Numerically this is correct, but intuitively it is wrong. So let us consider a performance index PI_2 that is the rating for just the winning alternative (in this case alternative-1).

$$PI_2 = Sum_1 = w_1 \times sc_{11} + w_2 \times sc_{21}$$

This performance index is more convenient when there are multiple alternatives and their relative rankings change frequently. However, a problem with both PI_1 and PI_2 is that the sensitivity with respect to w_1 does not depend on scores for the non-winning alternatives and we *do* want the sensitivities to depend on the other inputs and parameters. The following performance index solves both these problems.

$$PI_3 = \frac{1}{m}\sum_{i=1}^{n}\sum_{j=1}^{m} w_i sc_{ij}$$

where n is the number of criteria and m is the number of alternatives. This performance index is the sum of all the weight times scores in the whole matrix. With this performance index the semirelative-sensitivity functions are

$$\widetilde{S}_{w_i}^{\mathrm{PI}_3} = \left.\frac{\partial \mathrm{PI}_3}{\partial w_i}\right|_{\mathrm{NOP}} w_i = \frac{1}{m}\sum_{j=1}^{m} w_i \mathrm{sc}_{ij} \text{ and } \widetilde{S}_{S_{ij}}^{\mathrm{PI}_3} = \frac{w_i \mathrm{sc}_{ij}}{m}$$

They can be calculated and displayed directly with a spreadsheet as shown in Table 3.31.

Each cell states the sensitivity of the whole matrix with respect to that element. Here are the general formulae for the first-order sensitivity functions.

$$\mathrm{PI}_3 = \frac{1}{m}\sum_{i=1}^{n}\sum_{j=1}^{m} w_i \mathrm{sc}_{ij}$$

$$\left.\frac{\partial \mathrm{PI}_3}{\partial w_i}\right| = \frac{1}{m}\sum_{j=1}^{m} \mathrm{sc}_{ij}$$

$$\widetilde{S}_{w_i}^{\mathrm{PI}_3} = \left.\frac{\partial \mathrm{PI}_3}{\partial w_i}\right|_{\mathrm{NOP}} w_{i_0} = \frac{1}{m}\sum_{j=1}^{m} w_i \mathrm{sc}_{ij}$$

$$\frac{\partial \mathrm{PI}_3}{\partial \mathrm{sc}_{ij}} = \frac{w_i}{m}$$

$$\widetilde{S}_{\mathrm{sc}_{ij}}^{\mathrm{PI}_3} = \frac{w_i \mathrm{sc}_{ij}}{m}$$

Values for the following second-order sensitivity function are given in Table 3.32.

Table 3.31 Analytic semirelative-sensitivity function values for PI_3, the sum of all weight times scores

		Alternatives	
Criteria	Weight of importance	Umpire's assistant	Seeing eye dog
Accuracy of the call	$\widetilde{S}_{w_1}^{\mathrm{PI}_3} = 0.38$	$\widetilde{S}_{\mathrm{sc}_{11}}^{\mathrm{PI}_3} = 0.25$	$\widetilde{S}_{\mathrm{sc}_{12}}^{\mathrm{PI}_3} = 0.12$
Silence of Signaling	$\widetilde{S}_{w_2}^{\mathrm{PI}_3} = 0.13$	$\widetilde{S}_{\mathrm{sc}_{21}}^{\mathrm{PI}_3} = 0.10$	$\widetilde{S}_{\mathrm{sc}_{22}}^{\mathrm{PI}_3} = 0.02$

Table 3.32 Analytic semirelative-sensitivity function values for the interactions? of PI_3

		Alternatives	
Criteria	Weight of importance	Umpire's assistant	Seeing eye dog
Accuracy of the call		$\widetilde{S}_{w_1-\mathrm{sc}_{11}}^{\mathrm{PI}_3} = 0.25$	$\widetilde{S}_{w_1-\mathrm{sc}_{12}}^{\mathrm{PI}_3} = 0.12$
Silence of signaling		$\widetilde{S}_{w_2-\mathrm{sc}_{21}}^{\mathrm{PI}_3} = 0.10$	$\widetilde{S}_{w_2-\mathrm{sc}_{22}}^{\mathrm{PI}_3} = 0.02$

These cells contain the same numerical values as Table 3.31

$$\frac{\partial^2 \mathrm{PI}_3}{\partial w_i \partial \mathrm{sc}_{ij}} = \frac{1}{m}$$

$$\widetilde{S}^{\mathrm{PI}_3}_{w-\mathrm{sc}_{ij}} = \frac{w_i \mathrm{sc}_{ij}}{m}$$

All of the other sensitivity functions are zero, as shown below.

$$\frac{\partial^2 \mathrm{PI}_3}{\partial w_i^2} = 0$$

$$\widetilde{S}^{\mathrm{PI}_3}_{w_i^2} = 0$$

$$\frac{\partial^2 \mathrm{PI}_3}{\partial w_i \partial w_k} = 0$$

$$\widetilde{S}^{PI_3}_{w_i-w_k} = 0$$

$$\frac{\partial^2 \mathrm{PI}_3}{\partial \mathrm{sc}_{ij}^2} = 0$$

This means any combination of sc_{ij}'s.

$$\widetilde{S}^{\mathrm{PI}_3}_{\mathrm{sc}_{ij}^2} = 0$$

$$\widetilde{S}^{\mathrm{PI}_3}_{\mathrm{sc}_{ij}} = \widetilde{S}^{\mathrm{PI}_3}_{w_i-\mathrm{sc}_{ij}}$$

While these equations may look formidable, they are easy to compute with a spreadsheet. For example, $\sum_{k=1}^{n} w_k \mathrm{sc}_{kj}$ is merely the sum of the weight times scores in column k (Sum_k) and this information already exists in the spreadsheet. Furthermore, because $\widetilde{S}^{\mathrm{PI}_3}_{\mathrm{sc}_{ij}} = \widetilde{S}^{\mathrm{PI}_3}_{w_i-\mathrm{sc}_{ij}}$ and the rest of the second-order sensitivities are zero, Tables 3.31 and 3.32 are complete; they have all of the sensitivities in them.

A trade-off study can be hierarchical. There can be any number of criteria and each of them can have any number of subcriteria. Section 26 discusses the sensitivity analysis of a hierarchical trade-off study.

What about interactions? The nonzero semirelative-sensitivity functions for interactions are

$$\widetilde{S}^{\mathrm{PI}_3}_{w_i-\mathrm{sc}_{ij}} = \frac{w_{i_0} \mathrm{sc}_{ij_0}}{m}$$

Specifically, the semirelative-sensitivity function for the interaction of Wt_1 and S_{11} is

$$\widetilde{S}^{\mathrm{PI}_3}_{w_1-\mathrm{sc}_{11}} = \left.\frac{\partial^2 \mathrm{PI}_3}{\partial w_1 \partial \mathrm{sc}_{11}}\right|_{\mathrm{NOP}} w_{1_0} \mathrm{sc}_{11_0} = \frac{w_1 \mathrm{sc}_{11}}{m} = 0.25$$

Table 3.33 Numerical Example of Interactions for PI_3, the sum of the alternative ratings

Sensitivity function	Normal values	Values increased by 10 %	New PI_3 values	ΔPI_3	Total change in PI_3
$\widetilde{S}^{PI_3}_{w_1}$	$w_1 = 0.75$	$w_1 = 0.82$	0.538	0.038	$\sum \Delta PI_3 = 0.063$
$\widetilde{S}^{PI_3}_{sc_{11}}$	$sc_{11} = 0.67$	$sc_{11} = 0.74$	0.525	0.025	
$\widetilde{S}^{PI_3}_{w_1-sc_{11}}$	$w_1 = 0.75$ $sc_{11} = 0.67$	$w_1 = 0.82$ $sc_{11} = 0.74$	0.565	0.065	0.065

This is large enough that we should investigate its effect. If we increase w_1 by 10 %, $PI_3 = 0.538$, giving $\Delta PI_3 = 0.038$. If we increase sc_{11} by 10 %, $PI_3 = 0.525$, giving $\Delta PI_3 = 0.025$. Now, however, if we increase both w_{11} and sc_{11} by 10 % *at the same time*, $PI_3 = 0.565$, giving $\Delta PI_3 = 0.065$, which clearly is larger than the sum of 0.038 and 0.025 (Table 3.33).

Example 4: Sensitivity Analysis of a Risk Analysis

Let P_j, S_j and R_j be, respectively, the probability of occurrence, the severity, and the risk, for the jth failure mode. The risk of failure is typically measured as the probability (or frequency) of occurrence times the severity (or consequences), $R_j = P_j \times S_j$. For the performance index, we can use the sum of the risks, $PI = \sum_{j=1}^{n} R_j$. (We would get the same basic result if we let the performance index be the difference between the top two risks.) The semirelative-sensitivity of the performance index with respect to the probability of the jth failure mode occurring is $\widetilde{S}^{PI}_{P_j} = S_j \times P_{j_0} = R_j$ and the semirelative-sensitivity function of the performance index with respect to the severity of the jth failure mode is $\widetilde{S}^{PI}_{S_j} = R_j \times S_{j_0} = R_j$. The largest sensitivities are always those for the largest risk. This means that we should spend extra time and effort estimating the probability and severity of the highest ranked risk, which seems eminently sensible.

3.7.5 *Empirical Sensitivity Functions*

When analytic equations are not available for systems, sensitivity analyses can be performed on real-world systems. However, this may require way too much time or too many resources. In such instances, sensitivity analyses can be performed on models of real-world systems. For several models, deriving sensitivity functions is difficult or impossible, so the sensitivity functions are estimated from computer simulations of the models. Specifically, Numerical Estimation is a technique that can be used to assess system sensitivities. Numerical Estimation can be used on physical systems, models of these systems, or computer simulations of these models. Numerical Estimation can be used to estimate any of the previously defined sensitivity functions by properly choosing the inputs and parameters to vary. The estimations are performed by running the system, model, or simulation at the nominal operating point and then at other operating points. Sometimes the changes in parameter values are small and sometimes they are large, which creates unnatural operating points. One problem with using Numerical Estimation is that the number of runs increases geometrically with the number of factors being studied.

3.7.5.1 Numerical Estimation

The Numerical Estimation method is easy to use and evaluate. For many systems, it is the only alternative, if analytic sensitivity functions cannot be derived. In nondeterministic applications where the model is small, it is probably the best empirical method for evaluating sensitivity of a system to parameter changes.

The primary objective of Numerical Estimation is to estimate the sensitivities numerically. In the technique, a function is evaluated at its nominal operating point. Then a parameter is increased or decreased and the function is evaluated at this new point. If we take $\Delta f \approx f'(x_0)\Delta x$, with $\Delta x = x - x_0$, and if f' changes slowly, then $f'(\zeta) \approx f'(x_0)$, with ζ being a point between x and x_0, and f' being the derivative at that point. The difference between these two function values is given by

$$f(x) - f(x_0) = f'(\zeta)(x - x_0)$$

If we want to be more precise, then we should say that

$$f(x) - f(x_0) = f'(x_0)(x - x_0) + \frac{f''(\zeta)}{2!}(x - x_0)^2$$

If $(x - x_0)$ and f'' are small, then the second term on the right is much smaller than the first one. Therefore, usually the difference in function values divided by the difference in the variables estimates the first derivative, $f'(\zeta) = \frac{f(x)-f(x_0)}{(x-x_0)}$. Otherwise f'' has to be taken into account. In practical applications, the third derivative is seldom a concern.

Assume next that f has two variables, x and y, and is twice differentiable. Then

$$\begin{aligned} f(x,y) - f(x_0,y_0) &= f'_x(x_0,y_0)(x - x_0) + f'_y(x_0,y_0)(y - y_0) \\ &+ \frac{1}{2!}\left\{f''_{xx}(\zeta,\eta)(x - x_0)^2 + 2f''_{xy}(\zeta,\eta)(x - x_0)(y - y_0) + f''_{yy}(\zeta,\eta)(y - y_0)^2\right\} \end{aligned}$$

where ζ is between x and x_0 and η is between y and y_0. If either (1) both $(x - x_0)$ and $(y - y_0)$ are small and the second-order derivatives are not large or (2) the second-order derivatives are small, then the first two terms dominate. Otherwise, the second-order terms should be taken into consideration when estimating derivatives empirically and then computing sensitivities.

Example 3: A Trade-off Study (Continued)

Trade-off studies are often implemented with spreadsheet tools, such as Excel. It is important to do sensitivity analyses of such systems. Sometimes partial derivatives can be computed for such systems, but more often than not, the sensitivity analyses are estimated empirically. Typically, the analyst changes a parameter (input value, weight, scoring function parameter, etc.) by a fixed percentage, such as 10 %, and then looks at the change in the metric of interest. Often this is the rating for the top alternative or the difference between the top two recommended alternatives. Then the sensitivity values are computed. Typically, the relative or the semirelative-sensitivity functions are used. The semirelative functions are more robust and they are easier to compute. For example, for a +10 % parameter change the semirelative-sensitivity function approximation is

$$\widetilde{S}_{\beta}^{F} \approx \frac{\Delta F}{\Delta \beta}\beta_0 = \frac{\Delta F}{0.1\beta_0}\beta_0 = 10\Delta F$$

Table 3.34 shows the trade-off study example with S_{11} increased from 0.67 to 0.74. After such changes are made in each of the other five inputs and parameters, the empirically derived semirelative sensitivities can be computed and displayed as in Table 3.35.

It is not surprising that these are the same results that were obtained in the analytic semirelative sensitivity section of this chapter.

But what about the second-order term in the above "definition" of the single-variable derivative? Namely $\frac{f''(\zeta)}{2!}(x - x_0)^2$. If the system in question uses the sum of weighted scores combining function (for example, $\text{Sum}_1 = w_1 \times \text{sc}_{11} + w_2 \times \text{sc}_{21}$ and $\text{Sum}_2 = w_1 \times \text{sc}_{12} + w_2 \times \text{sc}_{22}$), then the second derivatives of a single variable are all zero. So an estimation of the first derivatives that ignores second derivatives will be all right. But this is not true for the product combining function

Table 3.34 Trade-off study with S_{11} increased by 10%

Criteria	Weight of importance	Umpire's assistant	Seeing eye dog
Accuracy of the call	0.75	**0.74**	0.33
Silence of signaling	0.25	0.83	0.17
Sum of weight times score		**0.76**	0.29

Table 3.35 Numerical estimates for semirelative-sensitivity function for PI_3, the sum of the alternative ratings squared, for a plus 10 % parameter perturbation

		Alternatives	
Criteria	Weight of importance	Umpire's assistant	Seeing eye dog
Accuracy of the call	$\widetilde{S}_{w_1}^{\text{PI}_3} = 0.38$	$\widetilde{S}_{\text{sc}_{11}}^{\text{PI}_3} = 0.25$	$\widetilde{S}_{\text{sc}_{12}}^{\text{PI}_3} = 0.12$
Silence of signaling	$\widetilde{S}_{w_2}^{\text{PI}_3} = 0.13$	$\widetilde{S}_{\text{sc}_{21}}^{\text{PI}_3} = 0.10$	$\widetilde{S}_{\text{sc}_{22}}^{\text{PI}_3} = 0.02$

(for example, $F_1 = \mathrm{sc}_{11}^{w_1} \times \mathrm{sc}_{21}^{w_2}$ and $F_2 = \mathrm{sc}_{12}^{w_1} \times \mathrm{sc}_{22}^{w_2}$) or most other common combining functions.

Next, let us take another look at the derivatives of a function of two variables.

$$
\begin{aligned}
f(x,y) - f(x_0,y_0) = \; & f'_x(x_0,y_0)(x-x_0) + f'_y(x_0,y_0)(y-y_0) \\
& + \frac{1}{2!}\left\{ f''_{xx}(\zeta,\eta)(x-x_0)^2 + 2f''_{xy}(\zeta,\eta)(x-x_0)(y-y_0) + f''_{yy}(\zeta,\eta)(y-y_0)^2 \right\}
\end{aligned}
$$

As an example, for the case of this trade-off study, let us examine the second-order terms, those inside the { }, for two reasons. First, we want to know if they are large and should be included in determining empirical estimates for the derivatives and sensitivities. Second, for this particular case, we want to estimate the effects of sensitivity interactions.

The analytic second-order semirelative-sensitivity function for the interaction of Wt_1 and S_{11} is

$$
\widetilde{S}^{\mathrm{PI}_3}_{w_1-\mathrm{sc}_{11}} = \left.\frac{\partial^2 \mathrm{PI}_3}{\partial w_1 \partial \mathrm{sc}_{11}}\right|_{\mathrm{NOP}} w_{1_0}\mathrm{sc}_{11_0} = \frac{w_{1_0}\mathrm{sc}_{11_0}}{m} = 0.25
$$

this is about as large as the largest first-order semirelative-sensitivity function.

To estimate values for the second-order sensitivities, we might be tempted to use the following simple-minded approximation, here shown for 10 % changes in parameter values.

$$
\widetilde{S}^{F}_{\alpha-\beta} \approx \frac{\Delta F \Delta F}{\Delta\alpha\Delta\beta}\alpha_0\beta_0 = \frac{(\Delta F)^2}{0.1\alpha_0 0.1\beta_0}\alpha_0\beta_0 = (10\Delta F)^2
$$

Letting $\alpha = w_1$ and $\beta = \mathrm{sc}_{11}$ and using the trade-off study values in Tables 3.28 and 3.29, incremented by 10 % where it is appropriate, we get

$$
\widetilde{S}^{\mathrm{PI}_3}_{w_1-\mathrm{sc}_{11}} \approx (10\Delta F)^2 = 0.424
$$

However, this does not match the analytic value of 0.25. The lesson here is that perturbation step size $(x - x_0)$ should be small. 5 and 10 % step sizes are probably too big for some systems. We might be able resolve this conundrum by using a smaller perturbation size. What if we reduce the perturbation step size to 0.1 %?

$$
\widetilde{S}^{F}_{\alpha-\beta} \approx \frac{\Delta F \Delta F}{\Delta\alpha\Delta\beta}\alpha_0\beta_0 = \frac{(\Delta F)^2}{0.001\alpha_0 0.001\beta_0}\alpha_0\beta_0 = (1000\Delta F)^2
$$

$$
\widetilde{S}^{\mathrm{PI}_3}_{w_1-\mathrm{sc}_{11}} \approx (1000\Delta F)^2 = 0.393
$$

This is closer but still not equal to 0.25, the number we computed as the analytic semirelative sensitivity. Therefore, we need to improve the mathematical formulation of the estimate of the second-order sensitivity interactions.

In the above simple-minded approximation, we changed both the input and the system parameter at the same time and then compared the value of the function to the value of the function at its nominal operating point. However, this is *not* the correct estimation for the mixed second-partial derivative. To estimate the second-partial derivatives we should start with

$$\frac{\partial^2 f(\alpha_0,\beta_0)}{\partial\alpha^2} \approx \frac{f'_\alpha(\alpha,\beta_0) - f'_\alpha(\alpha_0,\beta_0)}{\Delta\alpha} \text{ with } \alpha = \alpha_0 + \Delta\alpha$$

And by way of approximation for the first-order derivative we have

$$\frac{\partial^2 f(\alpha_0,\beta_0)}{\partial\alpha^2} \approx \frac{\frac{f(\alpha+\Delta\alpha,\beta_0)-f(\alpha,\beta_0)}{\Delta\alpha} - \frac{f(\alpha,\beta_0)-f(\alpha_0,\beta_0)}{\Delta\alpha}}{\Delta\alpha}$$

Simplifying this a little bit we get

$$\frac{\partial^2 f(\alpha_0,\beta_0)}{\partial\alpha^2} \approx \frac{f(\alpha+\Delta\alpha,\beta_0) - 2f(\alpha,\beta_0) + f(\alpha_0,\beta_0)}{(\Delta\alpha)^2}$$

The mixed second-partial derivatives can be obtained similarly.

$$\frac{\partial^2 f(\alpha_0,\beta_0)}{\partial\alpha\partial\beta} \approx \frac{f'_\alpha(\alpha_0,\beta) - f'_\alpha(\alpha_0,\beta_0)}{\Delta\beta} \text{ with } \beta = \beta_0 + \Delta\beta$$

Now we can estimate the two first-order derivatives and obtain

$$\frac{\partial^2 f(\alpha_0,\beta_0)}{\partial\alpha\partial\beta} \approx \frac{\frac{f(\alpha,\beta)-f(\alpha_0,\beta)}{\Delta\alpha} - \frac{f(\alpha,\beta_0)-f(\alpha_0,\beta_0)}{\Delta\alpha}}{\Delta\beta}$$

Simplifying this we get

$$\frac{\partial^2 f(\alpha_0,\beta_0)}{\partial\alpha\partial\beta} \approx \frac{f(\alpha,\beta) - f(\alpha_0,\beta) - f(\alpha,\beta_0) + f(\alpha_0,\beta_0)}{\Delta\alpha\Delta\beta}$$

Now we apply this to the trade-off study of Tables 3.28 and 3.29 with PI_3 as shown in Table 3.36.

Using the values in Table 3.36, we get

$$\frac{\partial^2 PI_3}{\partial w_1 \partial sc_{11}} \approx \frac{0.50062650 - 0.50025125 - 0.50037500 + 0.50000000}{0.00075 \times 0.00067} = 0.5$$

Table 3.36 Values to be used in estimating the second derivative for PI_3

Terms	Parameter values with a 0.1 % step size, that is $\Delta w_1 = 0.00075$ and $\Delta sc_{11} = 0.00067$	Function values
$f(\alpha, \beta)$	$w_1 = 0.75075$ $sc_{11} = 0.67067$	0.50062650
$f(\alpha_0, \beta)$	$sc_{11} = 0.67067$	0.50025125
$f(\alpha, \beta_0)$	$w_1 = 0.75075$	0.50037500
$f(\alpha_0, \beta_0)$	$w_1 = 0.75000$ $sc_{11} = 0.67000$	0.50000000

Now to get the semirelative-sensitivity function we multiply this mixed-second-partial derivative by the normal values w_{1_0} and sc_{11_0}

$$\widetilde{S}^{PI_3}_{w_1 - sc_{11}} = \left. \frac{\partial^2 PI_3}{\partial w_1 \partial sc_{11}} \right|_{NOP} w_{1_0} sc_{11_0} = 0.5\, w_{1_0} sc_{11_0} = 0.25$$

Now, *this* is the same result that we derived in the analytic semirelative sensitivity section of this chapter.

Although we do not use it in this section, for completeness, we now give the formula for estimating the second-order derivative for a function of only one variable: $f''(\alpha) \approx \frac{f(\alpha - \Delta\alpha) - 2f(\alpha) + f(\alpha + \Delta\alpha)}{\Delta\alpha^2}$

There are two important lessons to be learned from this study. First, for a trade-off study using the sum combining function and looking only at first derivatives, anything works. However, for most other sensitivity analyses the perturbation step size needs to be small. Thus, 5 and 10 % perturbations are not acceptable. Second, it is incorrect to estimate the mixed second partial derivative by changing two variables or parameters at the same time and then comparing the resulting value of the function to the value of the function at its nominal operating point; estimating mixed second partial derivatives requires evaluation of four not two numerator terms.

3.7.6 *The Taguchi Sensitivity Technique*

The basic philosophy of the previous sensitivity analysis techniques was to estimate the derivative of a function with respect to many inputs and system parameters at a point of interest. Often this meant changing a parameter by a certain amount and looking at the change in the output.

When doing a Taguchi 3-level Design of Experiments (DoE) [47], for each factor (independent variable) you pick a normal value, a high value, and a low value. If the high and low values are some percentage change, then you are doing a relative (or semirelative) sensitivity analysis. If the high and low values are plus and

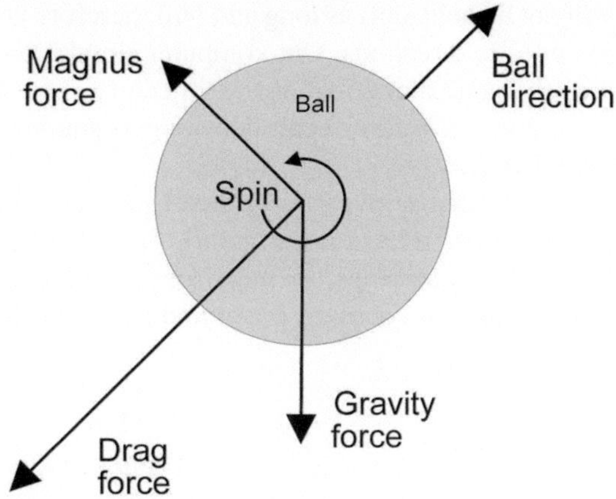

Fig. 3.41 The forces acting on a spinning ball flying through the air. Copyright ©, 2009, Bahill, Baldwin and Ramberg, used with permission

minus a unit, then you are doing an absolute sensitivity analysis. Alternatively, the high and low values could be realistic design options. We will now explain this technique.

3.7.6.1 Forces Acting on a Spinning Ball in Flight (Fig. 3.41)

This section is based on [48]. Three forces affect the ball in flight, as shown in Fig. 3.41: gravity pulls the ball downward, air resistance or drag operates in the opposite direction of the ball's motion, and, if it is spinning, there is also a force perpendicular to the direction of motion. Here are the equations for these forces.

$$F_{\text{gravity}} = m_{\text{ball}} g$$
$$F_{\text{drag}} = 0.5 \rho \pi r_{\text{ball}}^2 C_{\text{d}} v_{\text{ball}}^2$$
$$F_{\text{Magnus}} = 0.5 \rho \pi r_{\text{ball}}^3 \omega_{\text{ball}} v_{\text{ball}}$$

where ρ is air density and C_{d} is the coefficient of drag, which is often approximated as 0.5.

3.7.6.2 Sensitivity Analysis a Spinning Ball in Flight

This section is also based on Bahill, Baldwin, and Ramberg [48]. The distance a ball travels is inversely related the air density. But the explanation for this is not straightforward. The above equations show that both the drag and Magnus forces are directly proportional to the air density. So if air density gets smaller, the drag force gets smaller, this allows the ball to go farther. But at the same time, as air density gets smaller, the Magnus force also gets smaller, which means that the ball will not be held aloft as long and will therefore not go as far. So these two effects are in opposite directions. Our computer simulation shows that the change in the drag force has a greater effect on the trajectory of the ball than the change in the Magnus force does; therefore, as air density gets smaller, the range of a potential home run ball increases.

Air density is inversely related to altitude, temperature, and relative humidity, and is directly related to barometric air pressure. We created the following equation for these relationships. It came from the WeatherLink Software [2009, verified by Bruce Johnson, February 2016] and the CRC Handbook of Chemistry and Physics [1980–1981].

$$\text{Air Density} = \rho = 1.2929 \times \frac{273}{\text{Temp} + 273} \times \frac{\text{Air Pres} - (\mathbf{SVP} \times \mathbf{RH/100})}{760}$$

where Air Density, ρ, is in kg/m^3.

Temp is temperature in degrees Celsius.

Air Pres is the pressure of the air in mm of Hg and is given in Eq. (3.2).

SVP is saturation vapor pressure in mmHg and is given in Eq. (3.3).

RH is relative humidity as a percentage.

However, Al Nathan, http://baseball.physics.illinois.edu [personal communication 2016] pointed out that the term in **boldface** is the partial pressure of the water vapor. But what really matters for the baseball is the density of whatever is there (air plus water vapor) pushing against it. Therefore, the term in boldface should be multiplied by $\left(1 - \frac{\text{The molecular weight of water}}{\text{The molecular weight of air}} = 0.379.\right)$ Therefore, our equation for air density becomes

$$\begin{aligned} \text{Air Density} &= \rho \\ &= 1.2929 \\ &\times \frac{273}{\text{Temp} + 273} \times \frac{\text{Air Pres} - 0.379(\text{SVP} \times \text{RH}/100)}{760} \end{aligned} \tag{3.1}$$

This equation agrees with the exquisite results of Shelquist [49].

This equation uses the absolute (or actual) atmospheric air pressure, which is also called *station pressure* because it is the air pressure at a particular weather station. It can be computed from the U.S. Weather Service sea-level corrected barometric pressure (which is given in newspapers, on television, and on personal computers) with the following formula.

$$\text{Air Press} = \text{Barometric Pressure}\left[e^{\frac{-gM\text{Altitude}}{R(\text{Temp}+273.15)}}\right] \tag{3.2}$$

where g is the Earth's gravitational acceleration (9.80665 m/s^2 at sea level),

M is the molecular mass of air (0.0289644 kg/mol),

R is the Universal Gas Constant (8.31447 J/K mol),

Table 3.37 Air density in some typical baseball stadiums

	Altitude (feet above sea level)	Temperature (°F) average daily maximum in July	Relative humidity, on an average July afternoon (%)	Average barometric pressure in July (inch of Hg)	Air density (kg/m^3)
Denver	5190	88	34	29.98	0.967
Houston	45	94	63	29.97	1.133
Minneapolis	815	83	59	29.96	1.130
Phoenix	1086	104	20	29.81	1.077
San Francisco	0	68	65	29.99	1.201
Seattle	10	75	49	30.04	1.187

Altitude is the altitude of the ballpark in meters,
Temp is the temperature in the baseball stadium in °C
and Barometric Pressure is sea-level corrected barometric pressure

Dozens of equations have been fit to the experimental saturation vapor pressure (SVP) data. Here is one by Buck [50].

$$\mathrm{SVP} = 4.5841\mathrm{e}^{\frac{\left(18.687-\frac{\mathrm{Temp}}{234.5}\right)\times \mathrm{Temp}}{257.14+\mathrm{Temp}}} \tag{3.3}$$

As before, Temp is in degrees Celsius and SVP is in mmHg.

Air density is inversely related to altitude, temperature, and humidity, and is directly related to barometric pressure. For the range of values in major league ballparks, the altitude is the most important of these four input variables. Table 3.37 gives values for a typical late-afternoon summer game, assuming that the stadium roofs are open and there are no storms. For these examples, baseball units are used instead of SI units. Our tactic is to use baseball units (for example, feet, mph, and pounds) for inputs, SI units (for example, meters, kilograms, and seconds) for computations, and baseball units for outputs.

For a potential home run ball, both the drag and the lift (Magnus) forces are the greatest in San Francisco, where the park is just at sea level, and smallest in the "mile high" city of Denver. However, as previously stated, the drag force is more important than the Magnus force. Therefore, if all collision variables (e.g., pitch speed, bat speed, and collision point) are equal, a potential home run ball will travel the farthest in Denver and the shortest in San Francisco.

These values were chosen to show realistic numbers with natural variation. On any given afternoon in July, it is likely that baseball games will be played at the high and low ends of all these ranges.

To understand how the four fundamental input variables, altitude, temperature, humidity, and barometric pressure, determine the air density, these equations were evaluated at 81 experimental points. These points were selected at the low, middle, and high values of the fundamental inputs, or at 3^4 or 81 points. We ran a least squares regression analysis on these 81 equations in four unknowns.

Surprisingly, a simple linear equation explained most of the changes, or variability, in the air density values. The linear algebraic equation for air density obtained by least squares analysis is

$$\begin{aligned}\Delta\text{Air density}(\text{percent change from midlevel}) = \\ -\ 0.0035\ (\text{Altitude} - 2600) \\ -\ 0.2422\ (\text{Temperature} - 85) \\ -\ 0.0480\ (\text{Relative Humidity} - 50) \\ +\ 3.4223\ (\text{Barometric Pressure} - 29.92)\end{aligned} \tag{3.4}$$

where ΔAir density is stated as a percent change from midlevel value of 1.058, Altitude is in feet, Temperature is in degrees Fahrenheit, Relative Humidity is in percent, and Barometric Pressure is in inches of Hg. Equation (3.4) explains 99.3 % of the variation of air density across the 81 setting. This equation can be reexpressed to give the air density in kg/m^3

$$\begin{aligned}\text{Air density} = \rho = 1.058 + 0.01045\{ \\ -\ 0.0035\ (\text{Altitude} - 2600) \\ -\ 0.2422\ (\text{Temperature} - 85) \\ -\ 0.0480\ (\text{Relative Humidity} - 50) \\ +\ 3.4223\ (\text{Barometric Pressure} - 29.92)\}\end{aligned} \tag{3.5}$$

Note that the four factors (independent variables) are in different dimensions with different ranges. Hence, the magnitudes of the coefficients should be interpreted in this light. That is, a coefficient with a larger magnitude does not necessarily mean it has a greater impact on the response. Also, keep in mind that the equations that yield the air density values are deterministic. That is, there is no random variation. Hence, the sum of squares residual is the variation remaining after predicting the response from the linear approximation. There is no pure error, but rather simply lack of fit to the true model. The least squares analysis differentiates between the input variables for the range of the 81 observations as follows. The R-squared values showed that altitude explained 80 % of the variation between the model and the data; temperature explained 13 %, barometric pressure accounted for 4 %, and relative humidity accounted for 3 %.

Since Eq. (3.4) is linear, the impact of each factor can be shown graphically. Figure 3.42 shows the changes in air density that should be expected over the range of values that would be typical for a baseball stadium on an afternoon in July in the United States of America. It shows that altitude is the most important factor, followed by temperature, barometric pressure, and relative humidity. Since the factor ranges given are indicative of their natural variation, larger absolute slopes means stronger effects. These results are for baseball and should not be used for other purposes, such as calculating safe take-off parameters for a small airplane.

In Fig. 3.42, the point for the low value for Altitude (9.44 %) is the percent change from midlevel for the average of the air densities for 27 of the 81 points where altitude takes on its low value. Similarly, the point for the high value for

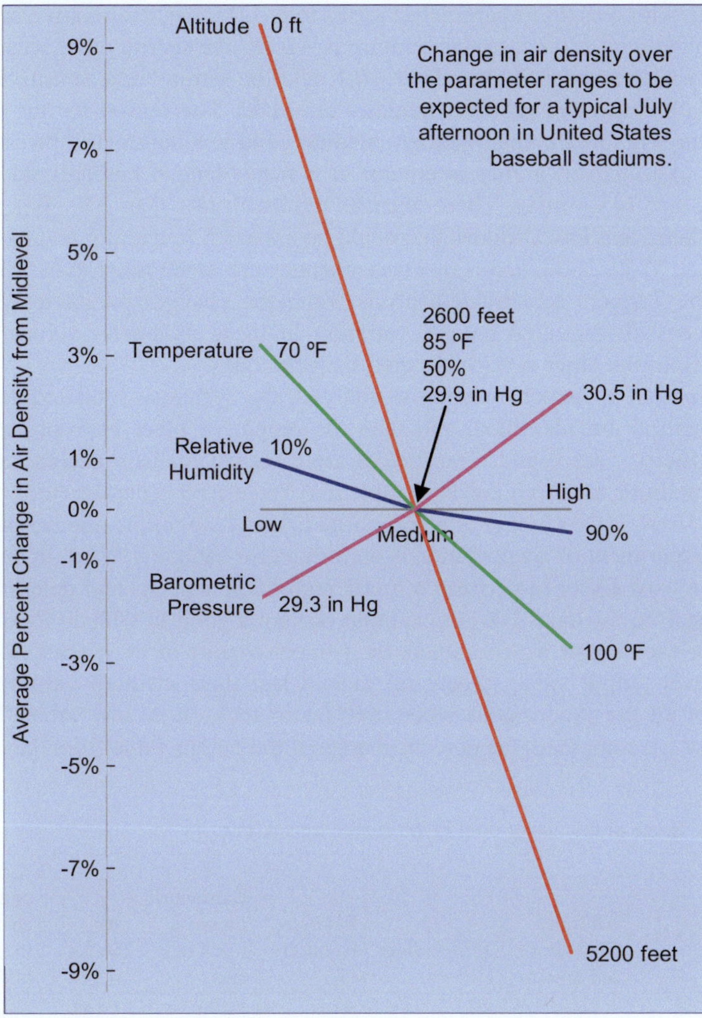

Fig. 3.42 Air density depends on altitude, temperature, barometric pressure, and relative humidity. Copyright ©, 2009, Bahill, Baldwin and Ramberg, used with permission

Barometric Pressure (2.26 %) is the percent change from midlevel for the average of the air densities for the 27 points where barometric pressure takes its high value, etc. From this sensitivity analysis, we can see that altitude is the most important parameter in determining the air density for baseball games. It is followed in importance by temperature, barometric pressure, and humidity, in that order.

Please note that this is a new type of sensitivity analysis. In a traditional sensitivity analysis, each input would be changed by a certain percent and then the resulting changes in the output or other metrics would be calculated [44]. For baseball, if we change each input by 5 % we find that the semirelative sensitivity of air density with respect to barometric pressure is 1.07; the semirelative sensitivity of air density with respect to temperature is −0.18; the semirelative sensitivity of air density with respect to altitude is −0.1; and the semirelative sensitivity of air density with respect to relative humidity is −0.09. The reason for the different results from Fig. 3.42 is that the high, medium, and low barometric pressures that could be expected on a July afternoon in a major league baseball stadium are 775, 760, and 745 mmHg. These changes are much less than 5 %. Whereas, the high, medium, and low altitudes that could be expected in a major league baseball stadium are 5200, 2600, and 0 ft. These changes are much more than 5 %. Stated simply, the Taguchi sensitivity analysis technique changes parameter values in accordance with expected natural variation in these parameter values. This is different from the other sensitivity analysis techniques.

Summary. In a Taguchi sensitivity analysis, the system is modeled, important inputs (factors) are identified, and then the output or other appropriate system behavior metrics are found. The factors are assigned realistic design values for the low, medium, and high levels. A table like Table 3.38 is constructed where the number of rows is the number of levels raised to the power of the number of factors. Then an experiment or a calculation is performed for each row. Next, in our simple model, we used Excel to perform a linear regression analysis and determined the best linear fit to the data. For our example, we got a good fit with an R-squared of 0.993. For each factor we computed, the percent change of the output value from the midlevel output value (where all factors had their medium values) for the average of all the experiments where that factor took on its low value. Then for each factor we computed, the percent change of the output value from the midlevel

Table 3.38 Some of the values used in the sensitivity analysis

	Altitude (feet above sea level)	Temperature (°F)	Relative Humidity (%)	Barometric pressure (inch of Hg)	Air density (kg/m^3)	Air density, percent change from midlevel
Low altitude	0	85	50	29.92	1.158	9.4
Low temperature	2600	70	50	29.92	1.089	2.9
Low humidity	2600	85	10	29.92	1.065	0.7
Low barometric pressure	2600	85	50	29.33	1.037	−2.0
Mid-level	2600	85	50	29.92	1.058	0.0
High barometric pressure	2600	85	50	30.51	1.079	2.0
High humidity	2600	85	90	29.92	1.051	−0.7
High temperature	2600	100	50	29.92	1.027	−2.9
High altitude	5200	85	50	29.92	0.967	−8.6
Highest density	0	70	10	30.51	1.223	15.5
Lowest density	5200	100	90	29.33	0.910	−14.0

output value (where all factors had their medium values) for the average of all the experiments where that factor took its high value. We used these values to make a chart like that shown in Fig. 3.42. Finally, we declared that the factors with the largest slopes were the most important factors in the model; the factors with the smallest slopes were the least important.

3.7.6.3 Application to Baseball

The range of a batted ball is defined as the distance from home plate to the spot where it first hits the ground. Table 3.39 shows the range for a perfectly hit simulated baseball. The pitch is an 85 mph (38 m/s) fastball with 1200 rpm backspin, the ball hits the sweet spot of the bat, which is going 55 mph (24.6 m/s), CoR is 0.55, the ball is launched upward at 34° at 102 mph with 2000 rpm of backspin. The ball is in the air for five and a half seconds. This is a potential home run ball. Reducing the air density by 10 %, for example, from 1.0 to 0.9 increased the range of this potential home run ball by 4 %, 16 ft. Combining these data with the data in Table 3.37 shows that this perfectly hit ball would travel 363 ft in San Francisco and 397 ft in Denver, 34 ft farther.

The pop-up was launched at 82 mph (36.7 m/s) at an upward angle of 58° with a backspin of 4924 rpm. The line drive was launched at 92 mph (41.1 m/s) at an upward angle of 15° with a backspin of 263 rpm.

In this section, average values were used. Of course, ball games are not played at the average values and the actual values are not constant throughout the game. In particular, wind speed and direction could change on a minute-by-minute basis. The effects of prevailing winds or height and distance of the outfield walls were not modeled.

For most games played at Coors Field in Denver there was a light breeze (for example, 5 mph, 2.2 m/s) blowing from center field toward home plate [51]. The outfield walls at Coors Field were farther back than in most stadiums. These two factors reduced the number of home runs by 3–4 %, which nearly compensated for Denver's high altitude. Greg Rybarczyk's data (personal communication, 2009) show that the greatest wind effects in major league stadiums are in San Francisco's AT&T Park where the average is a gentle breeze blowing from home plate into the right-center field stands at 10 mph (4.5 m/s). So the wind helps to equalize Denver and San Francisco.

Table 3.39 Range as a function of air density

Air density (kg/m^3)	Range (ft)			Range (m)		
	Home run	Pop up	Line drive	Home run	Pop up	Line drive
1.3	351	205	114	107	62	35
1.2	363	209	129	111	64	39
1.1	377	214	142	115	65	42
1.0	392	219	158	119	67	48
0.9	408	224	176	124	68	54
0.8	425	230	196	130	70	60

3.7.7 *The* What-If *Sensitivity Technique*

Some systems are not composed of equations or quantitative models. Therefore, the analytic sensitivity functions and numerical estimation techniques are not appropriate. For such systems, we can use the *what-if* sensitivity analysis technique. We simply ask what would happen if a certain event occurred or a certain parameter were changed. For the Cookie acquisition system we could ask, How would the cost of the system change if we changed the required frequency or power of the audible alarm? Or, How would the baking time and oven temperature change if we wanted peanut butter cookies instead of chocolate chip cookies? For the Parking Garage we could ask, What changes would be needed if we decided to let people park for free at night? What if motorcycles are too small to be detected by the gate area sensors? For the generic tactical missile, for the Launch and Fly to Target use case iteration: 3 step 11, what would happen if the missile did not detonate?

There are many other methods for carrying out sensitivity analyses including Response Surface Methodology and Sinusoidal Variation of Parameters.

3.7.8 *Discussion*

Validation. A sensitivity analysis is powerful technique for validating systems. Validation means ensuring that the system suits the customer's actual needs. If a system (and its model) is very sensitive to parameters over which the customer has no control, then it may be the wrong system for that customer. If the sensitivity analysis reveals the most important input or parameter and that result is a surprise, then it may be the wrong system. If a system is more sensitive to its parameters than to its inputs, then it may be the wrong system or the wrong operating point. If the sensitivities of the model are different from the sensitivities of the physical system, then it may be the wrong model.

In a set of requirements, if you delete a requirement, then your completeness measure (perhaps a traceability matrix) should reflect that change. When you make a decision, you should do a sensitivity analysis and then see if changing the most important decision parameters would change your decision. In a trade-off study, domain experts should agree that the most important criteria identified by the sensitivity analysis are indeed the most important criteria. In a risk analysis, experts should agree that the risks identified by the sensitivity analysis as being the greatest are indeed the greatest risks. After you prioritize a set of items, you should do a sensitivity analysis and discover the most important criteria. Then the values for those criteria should be changed to see if they change the prioritization. These are all validation concerns.

Verification. Sensitivity analyses can also be used to help verify systems. Verification means ensuring that the system complies with its requirements and conforms to its design. In a man-made system or a simulation, unexpected excessive sensitivity to any parameter is a verification mistake. Sensitivity to interactions should definitely be flagged and studied; such interactions may be unexpected and undesirable.

The **types and methods** of sensitivity analysis presented in this subsection have ranged from precise mathematical definitions yielding exact sensitivities to empirical methods for estimating sensitivities. Each has its purpose, advantages, and shortcomings.

The precise mathematically defined sensitivity functions yield the maximum information about the sensitivity of a system to its inputs and parameters at the expense of requiring a differentiable set of system equations. These differential equations are written in terms of time or frequency. The resulting sensitivities are functions of all other inputs and parameters as well as time or frequency. Relative-sensitivity functions, which involve a transform, are difficult to compute correctly, because multiplication in the time domain requires convolution in the frequency domain. Consequently, of the analytical sensitivity functions, the semirelative-sensitivity functions are the most useful.

In this subsection, we have shown the importance of using small parameter perturbations to get accurate estimates of the derivatives. However, large step sizes are often used to estimate the robustness using broad range sensitivity analyses. For example, for small step sizes, in our process control model, $f(x, y) = z = Ax^2 + By^2 + Cxy + Dx + Ey + F$, the output z is most sensitive (in a relative sense) to changes in F. But for a step size of 10 (or more), the output z is much more sensitive (relatively) to changes in y.

The sensitivities also depend on the **operating point**. If the nominal operating point is changed from $(x_0, y_0) = (1,\ 1)$ to $(x_0, y_0) = (10,\ 10)$, then the output z becomes most sensitive (relatively) to the input y.

Usually it is desirable for a model to be more sensitive to its inputs than to its system parameters. However, this is not the case if system parameters can be changed easily, for example, by just turning a knob; in contrast, it is very true if the system parameters are hard to change, for example, if they model the temperature or pressure in a large boiler. At the operating point $(x_0, y_0) = (1,\ 1)$, our process control example is not sensitive to the inputs, which means we could twiddle with the inputs forever and not be able to control the output. Therefore, this is not a desirable operating point, if the inputs were clean and easily controlled; however, if the inputs were noisy and hard to control, then this would be an excellent operating point. You would not want your car to be insensitive to changes in the position of your steering wheel. However, you do want your car to be insensitive to the gasoline flow rate when you are filling up your gas tank.

Our process control example $f(x, y) = z = Ax^2 + By^2 + Cxy + Dx + Ey + F$ is insensitive to its inputs near the operating point (1, 1). That is the $\partial z / \partial x$ and $\partial z / \partial y$ are equal to zero at the (1, 1) operating point and higher order derivatives must be used to evaluate the sensitivities. Moreover, these partial derivatives are near zero only near this operating point. Therefore, we should use this operating point only if the inputs are hard to control. Otherwise, we should use a different operating point.

Once the sensitivities of a system to its inputs and parameters are determined, they can be used to guide future research and design. If the sensitivities of a model are quantitatively similar to the sensitivities of the physical system, then the validity of the model is enhanced. Discrepancies can be used to direct improvements of the model or further testing of the physical system.

The examples in this subsection were carefully chosen to be simple, but at the same time succinctly illustrate particular points about sensitivity analyses. At one time, they were models of real-world systems. Then the details and domain knowledge were abstracted out, to leave a mathematical epitome. This technique was perfected by Wymore [17].

The sensitivity (a linear function) to a parameter of a linear system is not necessarily linear. This can take the form of interactions with other parameters or nonlinear direct effects. The principle of superposition in linear systems applies to inputs but not parameters. In most cases, the effects of changing two parameters independently are different than changing those same two parameters simultaneously.

In general, we do not say that a model is most sensitive to a certain input or system parameter. Rather we must say that a particular facet of a model is most sensitive to a particular input or parameter at a particular frequency, point in time, and operating point. The chosen facet most likely will be related to the question that the model was formulated to answer. Choosing the facet is the key to the sensitivity analysis.

Many commercial trade-off study tools perform sensitivity analyses. We believe they implement relative-sensitivity functions. However, it is sometimes difficult to determine what *type* of sensitivity functions they are computing and, therefore, it is difficult to understand how to use their results.

Sensitivity analyses are especially helpful when modeling systems with uncertainty. The sensitivity analysis shows which inputs and parameters are important and which are not. This allows the selective allocation of resources to experiments that will reveal more accurate values for the most important parameters. Sensitivity analyses help us to understand how uncertainties in the inputs and model parameters affect the model and its predictions. While not addressed in this book, sensitivity analyses can also be used to study uncertainty in model architecture, assumptions, and specifications. Sensitivity analyses are used to increase the confidence in the model and its predictions, by providing an understanding of how the model responds to changes in its inputs, the data used to calibrate it, its architecture, or its independent variables.

To show which parameters have the greatest effect on the metric of interest, either relative or semirelative-sensitivity functions may be used. However, if the metric of interest is a function of time, then only semirelative-sensitivity functions should be used.

3.7.9 *Section Summary*

To help validate a model, compute sensitivities for all parameters and inputs. Most systems should be more sensitive to their inputs than to any of their parameters. To help verify a system, compute sensitivities. Excessive sensitivity to any parameter is a verification mistake. After you build a model, write a set of requirements or design a system, you should study it to see if it makes sense. One of the best ways to study a model is through sensitivity analysis.

When estimating first derivatives, the step size should be small and the second derivatives must be computed to ensure valid results. When estimating mixed second-partial derivatives, four numerator terms, not two, must be used. Mixed second-partial derivatives are used to determine if the interaction of inputs and system parameters is important.

This section has described multiple techniques for performing a sensitivity analysis. Sensitivity analyses typically have different purposes and produce different results. Therefore, when presenting the results of sensitivity analysis, it is very important to state what type of sensitivity analysis was performed. For many sensitivity analyses, this would simply be a statement that the absolute, relative, or semirelative-sensitivity functions were computed either analytically or empirically, and the nominal operating point and the step size of the parameter perturbations would be given.

3.7.10 *Homework Problems*

3.7.10.1 Magic Potion

Assume that the function $f(x) = z = A\sqrt{x} + B$ models a process, where x is the input (raw material) and A and B are models of the system parameters. Let the nominal operating point be $x_0 = 1,\ A_0 = 2, B_0 = 3$, producing $z_0 = 5$.

Suppose that the output z is a valuable commodity (perhaps a magic potion). What is the easiest way (smallest **percent** change in an operating point parameter) to increase the quantity of z that is being produced? To help your reasoning you might want to calculate some semirelative-sensitivity functions. What about interactions? Could we do better by changing two inputs or parameters at the same time? Interaction terms are characterized by, for example, the partial derivative of z with respect to x containing A. The interactions can be bigger than the first-order partial derivatives. For this problem, we should expect the interactions to be significant. Of all the sensitivities that you calculated, which is the biggest and therefore the most important?

Table 3.40 Numerical values for a typical baseball in flight

Rho (kg/m^3)	pi	r (m)	C_d	Velocity (m/s)	Omega (rad/s)	$F_{drag}(n)$	$F_{Magnus}(n)$
1.04	3.14	0.037	0.5	42.5	126	2.02	0.44

3.7.10.2 Baseball in Flight

These are the equations for the drag and lift forces on a baseball or softball, as shown in Fig. 3.41. Assume a 95 mph (42.5 m/s) fastball has 20 revolutions per second of pure backspin. Let $C_d = 0.5$. The drag force will be opposite to the direction of motion.

$$F_{drag} = 0.5C_d\rho\pi r_{ball}^2 v_{ball}^2$$

The Magnus force will be perpendicular to the direction of motion.

$$F_{Magnus} = 0.5\rho\pi r_{ball}^3 \omega v_{ball}$$

Find numerical values for the semirelative sensitivity of the drag force and the Magnus force with respect to the (1) air density, ρ, (2) ball radius, r, (3) coefficient of drag, $C_{d,}$ (4) linear velocity of the ball, v, and (5) angular velocity of the ball, ω. What are the units for these sensitivity functions? What does this teach us? You may use the following numerical values, if you wish (Table 3.40; Fig. 3.43).

3.7.10.3 Cubic Equation

Assume that the equation $z^3 = B^3x^2$ models a certain process, *where* x is the input, z is the output, and B is a constant. Find the absolute sensitivities (for example, $\frac{dz}{dB}$ and $\frac{dz}{dx}$) at the nominal operating point of (0, 0). Comment on your result.

3.7.10.4 Square Root Function

Assume that the function $f(x) = z = A\sqrt{x} + B$ models a process, where x is the input (raw material) and A and B are models of the system parameters. Let the nominal operating point be

$$x_0 = 1, \quad A_0 = 2, B_0 = 3, \quad \text{producing } z_0 = 5$$

Suppose that the output z is a valuable commodity (perhaps a magic potion). What is the easiest way (smallest **percent** change in an operating point parameter) to increase the quantity of z that is being produced? To help your reasoning you might want to calculate some semirelative-sensitivity functions.

Fig. 3.43 Thinking about a ball in flight

$$\bar{S}_{\alpha}^{F} = \left.\frac{\partial F}{\partial \alpha}\right|_{\mathrm{NOP}} \alpha_0$$

What about interactions? Could we do better by changing two inputs or parameters at the same time? Interaction terms are characterized by, for example, the partial derivative of z with respect to x containing A. The interactions can be bigger than the first-order partial derivatives. You can see that the above partial derivatives with respect to the coefficients contain the operating point coordinates and the partial derivatives with respect to the operating point coordinates depend on the coefficients. Therefore, we should expect the interactions to be significant. Compute the interaction sensitivities.

3.7.10.5 Process Model

Absolute-Sensitivity Functions

The *absolute sensitivity* of the function F to variations in the parameter α is given by

$$S_{\alpha}^{F} = \left.\frac{\partial F}{\partial \alpha}\right|_{\mathrm{NOP}}$$

where NOP means the partial derivative is evaluated at the Nominal Operating Point (NOP) where all the inputs and parameters have their normal values.

Assume that the function

$$f(x, y) = z = Ax^2 + By^2 + Cxy + Dx + Ey + F$$

models a process, where x and y are the inputs (raw material) and A to F are models of the system parameters. Let the nominal operating point be

$(x_0, y_0) = (3, 3)$, $A_0 = 1, B_0 = 2, C_0 = 3, D_0 = -5, E_0 = -7, F_0 = 8$, producing $z_0 = 26$. Please note that this operating point is different from that used in the text.

Suppose that the output z is a valuable commodity (perhaps a potion). What is the easiest way (the smallest change in one of these eight inputs and parameters) to increase the quantity of z that is being produced? To help your reasoning, you might want to calculate some absolute-sensitivity functions.

$$S_A^z = \left.\tfrac{\partial z}{\partial A}\right|_{\mathrm{NOP}} = x_0^2 = 9$$
$$S_B^z = \left.\tfrac{\partial z}{\partial B}\right|_{\mathrm{NOP}} =$$
$$S_C^z = \left.\tfrac{\partial z}{\partial C}\right|_{\mathrm{NOP}} =$$
$$S_D^z = \left.\tfrac{\partial z}{\partial D}\right|_{\mathrm{NOP}} =$$
$$S_E^z = \left.\tfrac{\partial z}{\partial E}\right|_{\mathrm{NOP}} =$$
$$S_F^z = \left.\tfrac{\partial z}{\partial F}\right|_{\mathrm{NOP}} =$$
$$S_x^z = \left.\tfrac{\partial z}{\partial x}\right|_{\mathrm{NOP}} =$$
$$S_y^z = \left.\tfrac{\partial z}{\partial y}\right|_{\mathrm{NOP}} = 2B_0y_0 + C_0x_0 + E_0 = 14$$

Evaluated at $(x_0, y_0) = (3,\ 3)$, the system is most sensitive to y and then x.

What about interactions? Could we do better by changing two inputs and/or parameters at the same time? Of the 64 possible second-partial derivatives, only the following are nonzero.

$$S^z_{x-A} = \left.\frac{\partial^2 z}{\partial x \partial A}\right|_{\text{NOP}} = 2x_0 = 6$$

$$S^z_{x-C} = \left.\frac{\partial^2 z}{\partial x \partial C}\right|_{\text{NOP}} =$$

$$S^z_{x-D} = \left.\frac{\partial^2 z}{\partial x \partial D}\right|_{\text{NOP}} =$$

$$S^z_{y-B} = \left.\frac{\partial^2 z}{\partial y \partial B}\right|_{\text{NOP}} =$$

$$S^z_{y-C} = \left.\frac{\partial^2 z}{\partial y \partial C}\right|_{\text{NOP}} =$$

$$S^z_{y-E} = \left.\frac{\partial^2 z}{\partial y \partial E}\right|_{\text{NOP}} =$$

$$S^z_{x-y} = \left.\frac{\partial^2 z}{\partial x \partial y}\right|_{\text{NOP}} =$$

$$S^z_{x^2} = \left.\frac{\partial^2 z}{\partial x^2}\right|_{\text{NOP}} =$$

$$S^z_{y^2} = \left.\frac{\partial^2 z}{\partial y^2}\right|_{\text{NOP}} = 2B_0 = 4$$

At the nominal operating point, $z_0 = 26$. If we increase A by one unit (**this is too large of a step for a sensitivity analysis,** but it is convenient because you can do the arithmetic in your head), then $A=2$, this increases $z=35$, and $\Delta z = 9$. If we increase x by one unit then $x=4$, $z=37$, and $\Delta z = 11$. However, if we simultaneously increase both x and A by one unit each, $A=2$, $x=4$, $z=53$, and $\Delta z = 27$. Clearly $27 > 9+11$. This means that the interaction affect is big. This interaction affect is shown in Table 3.41. Fill in Table 3.41 for the interaction of y and B.

Changing both x and A at the same time produces a bigger affect than the sum of the two individual changes.

Table 3.41 Effects of individual and combined parameter changes for some second-order interaction terms with delta of 1.0, where $x = x_0 + \Delta x$ etc.

Functions	Normal values	Values increased by one unit	New z values	Δz	Total change in z
$f(x,A)$	$A=1$ $x=3$	$A=2$ $x=4$	53	27	27
$f(x_0,A)$	$A=1$	$A=2$	35	9	$\sum \Delta z = 20$
$f(x,A_0)$	$x=3$	$x=4$	37	11	
$f(x_0,A_0)$	$A=1$ $x=3$		26	0	
$f(y,B)$	$B=2$ $y=3$	$B=3$ $y=4$			
$f(y_0,B)$	$B=2$	$B=3$			
$f(y,B_0)$	$y=3$	$y=4$			
$f(y_0,B_0)$	$B=2$ $y=3$				

Now, can we use these values to estimate the mixed-second-partial derivative, according to the following formula?

$$\frac{\partial^2 f(\alpha_0,\beta_0)}{\partial\alpha\partial\beta} \approx \frac{f(\alpha,\beta) - f(\alpha_0,\beta) - f(\alpha,\beta_0) + f(\alpha_0,\beta_0)}{\Delta\alpha\Delta\beta}$$

$$\frac{\partial^2 z}{\partial x \partial A} \approx \frac{53 - 35 - 37 + 26}{1} = 7$$

That is the *wrong* answer. By analytic techniques used previously, we found that the correct value is 6. The purposes of this exercise were to allow you to do the computations in your head, so that you could gain intuition and to point out that a step size of 1.0 is too large.

Let us now try a smaller step size, such as 0.01. We can compute values with a model implemented in Excel or we can use the following Taylor series approximation [44].

$$\begin{aligned} f(x,y) - f(x_0,y_0) = {} & f'_x(x_0,y_0)(x-x_0) + f'_y(x_0,y_0)(y-y_0) \\ & + \frac{1}{2!}\left\{f''_{xx}(\zeta,\eta)(x-x_0)^2 + 2f''_{xy}(\zeta,\eta)(x-x_0)(y-y_0) + f''_{yy}(\zeta,\eta)(y-y_0)^2\right\} \end{aligned}$$

Converting to find the value if we change x and A yields

$$\begin{aligned} f(x,A) - f(x_0,A_0) = {} & f'_x(x_0,A_0)(x-x_0) + f'_A(x_0,A_0)(A-A_0) \\ & + \frac{1}{2!}\left\{f''_{xx}(\zeta,\eta)(x-x_0)^2 + 2f''_{xA}(\zeta,\eta)(x-x_0)(A-A_0) + f''_{AA}(\zeta,\eta)(A-A_0)^2\right\} \end{aligned}$$

Now, using the symbols that we used in our absolute sensitivity functions, we can write

$$\begin{aligned} \Delta z = {} & S^z_x(x_0,A_0)\Delta + S^z_A(x_0,A_0)\Delta \\ & + \frac{1}{2!}\{S^z_{x^2}(\zeta,\eta)\Delta^2 + 2S^z_{x-A}(\zeta,\eta)\Delta^2 + S^z_{A^2}(\zeta,\eta)\Delta^2\} \end{aligned}$$

Inserting numbers, we get

$$\Delta z = 10 \times 0.01 + 9 \times 0.01 + \frac{1}{2}\{2 \times 0.0001 + 2 \times 6 \times 0.0001 + 0\} = 0.1907$$

This is the Δz that was put into row 2 column 5 of Table 3.42. The other values in Table 3.42 were calculated with an Excel spreadsheet. You should now do the same for y and B.

Now, let us use these values to estimate the mixed-second-partial derivative, according to this formula [44].

Table 3.42 Effects of individual and combined parameter changes for some second-order interaction terms with delta of 0.01, where $x = x_0 + \Delta x$ etc.

Functions	Normal values	Values increased by delta	New z values	Δz	Total change in z
$f(x, A)$	$A=1$ $x=3$	$A=1.01$ $x=3.01$	26.1907	0.1907	0.1907
$f(x_0, A)$	$A=1$	$A=1.01$	26.0900	0.0900	$\sum \Delta z = 0.1901$
$f(x, A_0)$	$x=3$	$x=3.01$	26.1001	0.1001	
$f(x_0, A_0)$	$A=1$ $x=3$		26.0000	0	
$f(y, B)$	$B=2$ $y=3$	$B=2.01$ $y=3.01$			
$f(y_0, B)$	$B=2$	$B=2.01$			
$f(y, B_0)$	$y=3$	$y=3.01$			
$f(y_0, B_0)$	$B=2$ $y=3$				

Table 3.43 Values to be used in estimating the second partial derivative

Terms	Parameter values with a 0.01 step size	Function values
$f(\alpha, \beta)$	$A = 1.01$ $x = 3.01$	26.1907
$f(\alpha_0, \beta)$	$A = 1.01$ $x = 3.00$	26.0900
$f(\alpha, \beta_0)$	$A = 1.00$ $x = 3.01$	26.1001
$f(\alpha_0, \beta_0)$	$A = 1.00$ $x = 3.00$	26.0000

$$\frac{\partial^2 f(\alpha_0, \beta_0)}{\partial\alpha\partial\beta} \approx \frac{f(\alpha, \beta) - f(\alpha_0, \beta) - f(\alpha, \beta_0) + f(\alpha_0, \beta_0)}{\Delta\alpha\Delta\beta}$$

Using the values in Table 3.43

$$\frac{\partial^2 z}{\partial x \partial A} \approx \frac{26.1907 - 26.0900 - 26.1001 + 26.0000}{0.01 \times 0.01} = \frac{0.0006}{0.0001} = 6$$

This is the correct value that we have already computed analytically. You should now numerically compute $\frac{\partial^2 z}{\partial y \partial B} \approx$.

How about the third-order partial derivatives? For example, $\frac{\partial^3 z}{\partial x^2 \partial A} = 2$. Fill in the following table for y and B (Table 3.44).

Changing x twice and A at the same time produces a bigger effect than the sum of the three individual changes. So the third-order interaction affects are important.

Calculate the following third-order partial derivatives: $\frac{\partial^3 z}{\partial y^2 \partial B} =$ and $\frac{\partial^3 z}{\partial x \partial y \partial C} =$. Unfortunately, they are as big as the biggest first and second-order partial

Table 3.44 Effects of individual and combined parameter changes for some third-order interaction terms, delta = 0.001

Functions	Normal values	Values increased by delta	New z values	Δz	Total change in z
$f(x, A)$	$A = 1$ $x = 2$	$A = 1.001$ $x = 2.002$	26.2916	0.2916	0.2916
$f(x_0, A)$	$A = 1$	$A = 1.001$	26.0900	0.0400	$\sum \Delta z = 0.1402$
$f(x, A_0)$	$x = 2$	$x = 2.001$	26.1001	0.0501	
$f(x, A_0)$	$x = 2$	$x = 2.001$	26.1001	0.0501	
$f(x_0, A_0)$	$A = 1$ $x = 2$		26.0000	0	
$f(y, B)$	$B = 2$ $y = 2$	$B = 2.001$ $y = 2.002$			
$f(y_0, B)$	$B = 2$	$B = 2.001$			
$f(y, B_0)$	$y = 2$	$y = 2.001$			
$f(y, B_0)$	$y = 2$	$y = 2.001$			
$f(y_0, B_0)$	$B = 2$ $y = 2$				

derivatives. So we must consider them. Should we consider the fourth-order partial derivatives?

Now test your intuition. Without calculating any numbers, if the operating point is changed from (3, 3) to (100, 100) what will be the most important inputs or parameters?

Relative-Sensitivity Functions

If we want to compare the effects of different inputs and parameters, we should use relative-sensitivity functions. The *relative-sensitivity* of the function F to the parameter α evaluated at the nominal operating point is given by

$$\overline{S}_{\alpha}^{F} = \left.\frac{\partial F}{\partial \alpha}\right|_{\text{NOP}} \frac{\alpha_0}{F_0} \approx \frac{\%\text{ change in } F}{\%\text{ change in } \alpha} = \frac{\frac{\Delta F}{F}}{\frac{\Delta \alpha}{\alpha}}$$

where NOP and the subscripts 0 mean that all functions and inputs and parameters assume their nominal operating point values.

Assume that the function

$$f(x, y) = z = Ax^2 + By^2 + Cxy + Dx + Ey + F$$

models the output of a process, where the nominal operating point is

$$(x_0, y_0) = (3, 3),\ \ A_0 = 1, B_0 = 2, C_0 = 3, D_0 = -5, E_0 = -7,$$
$$F_0 = 8,\ \ \text{producing } z_0 = 26$$

What is the easiest way (smallest **percent** change in an operating point parameter) to increase the quantity of z that is being produced? To help your reasoning you might want to calculate some relative-sensitivity functions.

$$\overline{S}_A^z = \frac{\partial z}{\partial A}\Big|_{\text{NOP}} \frac{A_0}{z_0} = x_0^2 \frac{A_0}{z_0} = 0.35$$
$$\overline{S}_B^z = \frac{\partial z}{\partial B}\Big|_{\text{NOP}} \frac{B_0}{z_0} =$$
$$\overline{S}_C^z = \frac{\partial z}{\partial C}\Big|_{\text{NOP}} \frac{C_0}{z_0} =$$
$$\overline{S}_D^z = \frac{\partial z}{\partial D}\Big|_{\text{NOP}} \frac{D_0}{z_0} =$$
$$\overline{S}_E^z = \frac{\partial z}{\partial E}\Big|_{\text{NOP}} \frac{E_0}{z_0} =$$
$$\overline{S}_F^z = \frac{\partial z}{\partial F}\Big|_{\text{NOP}} \frac{F_0}{z_0} =$$
$$\overline{S}_x^z = \frac{\partial z}{\partial x}\Big|_{\text{NOP}} \frac{x_0}{z_0} =$$
$$\overline{S}_y^z = \frac{\partial z}{\partial y}\Big|_{\text{NOP}} \frac{y_0}{z_0} = (2B_0 y_0 + C_0 x_0 + E_0)\frac{y_0}{z_0} = 1.62$$

Evaluated at $(x_0, y_0) = (3, 3)$, $\overline{S}_y^z = \frac{\partial z}{\partial y}\Big|_{\text{NOP}} \frac{y_0}{z_0} = 1.62$, is the largest. Therefore, we should increase y if we wish to increase z.

What about interactions? Could we do better by changing two inputs and/or parameters at the same time? Only the following second-order relative-sensitivity functions are nonzero.

$$\overline{S}_{x-A}^z = \frac{\partial^2 z}{\partial x \partial A}\Big|_{\text{NOP}} \frac{x_0 A_0}{z_0^2} = \frac{2x_0^2 A_0}{z_0^2} = \frac{2 \times 3^2 \times 1}{26^2} = 0.03$$
$$\overline{S}_{x-C}^z = \frac{\partial^2 z}{\partial x \partial C}\Big|_{\text{NOP}} \frac{x_0 C_0}{z_0^2} =$$
$$\overline{S}_{x-D}^z = \frac{\partial^2 z}{\partial x \partial D}\Big|_{\text{NOP}} \frac{x_0 D_0}{z_0^2} =$$
$$\overline{S}_{y-B}^z = \frac{\partial^2 z}{\partial y \partial B}\Big|_{\text{NOP}} \frac{y_0 B_0}{z_0^2} =$$
$$\overline{S}_{y-C}^z = \frac{\partial^2 z}{\partial y \partial C}\Big|_{\text{NOP}} \frac{y_0 C_0}{z_0^2} =$$
$$\overline{S}_{y-E}^z = \frac{\partial^2 z}{\partial y \partial E}\Big|_{\text{NOP}} \frac{y_0 E_0}{z_0^2} =$$
$$\overline{S}_{x-y}^z = \frac{\partial^2 z}{\partial x \partial y}\Big|_{\text{NOP}} \frac{x_0 y_0}{z_0^2} =$$
$$\overline{S}_{x^2}^z = \frac{\partial^2 z}{\partial x^2}\Big|_{\text{NOP}} \frac{x_0^2}{z_0^2} = \frac{2x_0^2 A_0}{z_0^2} =$$
$$\overline{S}_{y^2}^z = \frac{\partial^2 z}{\partial y^2}\Big|_{\text{NOP}} \frac{y_0^2}{z_0^2} = \frac{2y_0^2 B_0}{z_0^2} = 0.08$$

Which is the largest of the 17 relative sensitivity functions that you have just computed?

In summary, at the operating point $(x_0, y_0) = (3, 3)$, using absolute-sensitivity functions, _______________ (fill in the blanks) were the most important terms, but using relative-sensitivity functions, _________________________ were the most important. The absolute and relative-sensitivity functions give different results. The absolute-sensitivity functions show which inputs or parameters have the greatest effect on the output for a *fixed size* change in the parameters Whereas the relative-sensitivity functions show which inputs or parameters have the greatest effect on the output for a certain *percent* change.

Semirelative-Sensitivity Functions

We have used the absolute-sensitivity function to see when a parameter had its greatest effect on the step response of a system, and we have used the relative-sensitivity function to see which parameter had the greatest effect on the transfer function. Now, let us investigate the use of the *semirelative-sensitivity function*, which is defined as

$$\widetilde{S}_{\alpha}^{F} = \left.\frac{\partial F}{\partial \alpha}\right|_{\text{NOP}} \alpha_0$$

where NOP and the subscript 0 mean that all functions, inputs, and parameters assume their nominal operating point values.

Let us now return to the process model, $f(x, y) = z = Ax^2 + By^2 + Cxy + Dx + Ey + F$. Would semirelative-sensitivity functions be better than relative-sensitivity functions?

3.7.10.6 Taguchi Sensitivity Technique

Here are some apocryphal data for the Cookie acquisition system of Chap. 1. There were nine observations or experiments using realistic design parameters (Table 3.45).

Linear regression analysis yielded the following best-fit equation for the data.

$$\text{Doneness} = -1 + 0.001467 \times \text{Temperature} + 0.1 \times \text{BakingTime}$$

Create a figure like Fig. 3.42 for these data. Based on this sensitivity analysis, which parameter is most important for ensuring doneness of the cookies, temperature or baking time? Is that the same result that we obtained in Chap. 1?

Table 3.45 Experimental data for the Cookie acquisition system

Results of nine experiments on the Cookie acquisition system		
Doneness, 0–1 with 0.5 being perfect	Temp (°F)	Baking time (min)
0.79	400	12
0.71	350	12
0.64	300	12
0.59	400	10
0.51	350	10
0.44	300	10
0.39	400	8
0.31	350	8
0.24	300	8

3.7.10.7 Effects of Air Density on the Batted Ball

Describe at least five methods for doing a sensitivity analysis on the air density system of Section 3.7.6.2.

3.7.10.8 Which Is Best?

In this section, we studied the absolute sensitivity analysis technique, the relative sensitivity analysis technique, the semirelative-sensitivity analysis technique, the empirical sensitivity analysis technique, the Taguchi sensitivity analysis technique, and the what-if sensitivity analysis technique. Which of these is the best?

3.8 Summary

This chapter presents methods and processes that are commonly used in requirement discovery, trade-off studies, and risk analyses.

In Section 20, we began by discussing how to state a problem correctly. Then we discussed concept exploration, validation, the design of evaluation criteria (with examples), weights of importance (including their derivation), and followed up with homework problems that reinforce the key concepts associated with correctly stating a problem.

In Section 21, we delved into scoring functions. Specifically, we discussed linear normalization and the Wymorian standard scoring functions. We presented guidelines for how to select an appropriate scoring function along with examples for each. We then discussed the mathematical properties of the Wymorian scoring functions and followed that up with homework problems that reinforced the key concepts associated with scoring functions.

In Section 22, we covered combining functions. We noted that different methods exist for combining scores, or values in a trade-off study (or risk analysis) to calculate numerical measures that can be used to compare alternatives. We presented common combining functions (including sums and products). We then shifted to discussing mathematical techniques including classical optimization, Praeto-optimal design, and sequential optimization. We also provided examples of nontechnical combining functions. We concluded this section by reminding the reader that the key idea is to be conversant in studying the problem before choosing the tool or set of tools for the problem. We followed this discussion with homework problems that reinforced the key concepts associated with combining functions.

In Section 23, we covered quantitative measurements. We noted that evaluation criteria, measures, and technical performance measures are all terms that are used to quantify system performance. While these terms are often used synonymously, there are subtle differences. *Evaluation criteria* are used in requirements discovery, trade-off studies, and risk analyses. *Measures* are used to manage a company's problems. *TPMs* are used to mitigate risk during design and manufacturing. We then gave several examples of TPMs, and followed that up with homework problems that reinforced the key concepts associated with the selection of quantitative measures.

In Section 24, we discussed the concept of *prioritization*. We noted that prioritization is applicable to requirements discovery, deriving weights of importance for evaluation criteria used in trade-off studies, and rank ordering of attributes/factors such as risks, goals, customer needs, capabilities, directives initiatives issues, activities, use cases, technical performance measures, features, functions, and value engineering activities. We then reviewed the criteria used to establish priorities. Exemplar criteria include risk, criticality to mission success, customer satisfaction, business value, cost, benefit, benefit-to-cost ratio, safety, stability, reuse potential, and implementation difficulty. While the criteria are meant to be independent, in the real world it is almost inevitable that there will be overlap. We then discussed different ways to combine information and addressed the issue of completeness of coverage. We concluded the section with homework problems that reinforced the concepts associated with prioritization.

In Section 25, we addressed the problem of *analyzing data*. Specifically, we explained the difference between frequency and probability and noted that humans are poor at estimating probabilities but better at gauging frequency of occurrence. We then reviewed Bayes' Theorem and reminded the reader that humans are not Bayesian and therefore, the Bayesian solution is not intuitive to most people. We also emphasized the importance of problem framing and stressed that it is invariably more important than problem solving. We then recommended choosing units for the evaluation criteria that the customer wants or is comfortable with using. We drew the reader's attention to the do nothing alternative (aka Maintain the Status Quo). We then moved on to verification, validation and testing. We defined *validation* as building the right system (i.e., the system does what is supposed to do) and *verification* as building the system right (i.e., the system complies with requirements and conforms to design specification). Validating requirements means ensuring that the *set* of requirements is correct, complete, consistent, and traceable. Verifying requirements means proving that each requirement has been satisfied. We then discussed *testing* and noted that the earlier a defect is caught the less time and money it will take to correct. We gave various examples of testing and concluded with homework problems to reinforce the key concepts associated verification, validation and testing.

In Section 26, we covered sensitivity analysis. We assured readers unfamiliar with the underlying mathematics that skipping this section would not result in a loss of continuity for them because each section is written in stand-alone fashion. We advised the reader that all trade-off studies and risk analyses need to include sensitivity analyses. We also noted that the results of a sensitivity analysis can be used to validate a model, flag strange or unrealistic model behavior, suggest new experiments or guide future data collection efforts, highlight important assumptions of the model, suggest required accuracy for calculating parameters, allocate resources, and pinpoint true cost drivers. We recommended showing requirements that are driving system cost to customers, allowing them to decide whether or not to relax requirements to drive down costs. We described different sensitivity analysis techniques including analytic sensitivity functions, relative sensitivity functions, the semirelative sensitivity functions, empirical sensitivity functions, the Taguchi technique and what-if techniques. We concluded this section by remarking that sensitivity analysis techniques are especially helpful when modeling systems with uncertainty. Sensitivity analysis shows which parameters are important and which parameters are not. We noted that this knowledge enables selective allocation of resources to experiments that have a high likelihood of increasing the accuracy of the most important parameters. We concluded with a set of homework problems that reinforce the key concepts of sensitivity analysis.

References

1. Bahill AT, Dean F (2009) Discovering system requirement. In: Sage AP, Rouse WB (eds) Handbook of systems engineering and management. Wiley, Hoboken, pp 205–266
2. Madni AM (2013) Generating novel options during systems architecting: psychological principles, systems thinking, and computer-based aiding. Syst Eng 17(1):1–9
3. Madni AM (2014) Expanding stakeholder participation in upfront system engineering through storytelling in virtual worlds. Syst Eng 18(1):16–27
4. Kirkwood C (1998) Strategic decision making multiobjective decision analysis with spreadsheets. J Oper Res Soc 49(1):93–98
5. Madni AM, Nance M, Richey M, Hubbard W, Hanneman L (2014) Toward an experiential design language: augmenting model-based systems engineering with technical storytelling in virtual worlds. Procedia Comput Sci 28:848–856
6. Madni AM, Samet MG, Freedy A (1982) A trainable on-line model of the human operator in information acquisition tasks. IEEE Trans Syst Man Cybern 12(4):504–511
7. Buede DM (2009) The engineering design of systems: models and methods, 2nd edn. Wiley, New York
8. Madni AM, Madni CC, Salasin J (2002) 5.4.1 ProACT™: process-aware zero latency system for distributed, collaborative enterprises. In: INCOSE international symposium, vol 12, no 1, pp 783–790
9. Bahill AT, Karnavas WJ (2000) Risk analysis of a pinewood derby: a case study. Syst Eng 3(3):143–155

10. Madni A, Freedy A, Estrin G, Melkanoff M (eds) (1991) Concurrent engineering workstation for multi-chip module product development process. Invited paper presented at QALS Q QE Washington'91 conference and exposition, Washington, DC
11. Saaty TL (1980) The analytic hierarchy process: planning, priority setting, resource allocation. McGraw-Hill, New York
12. Von Neumann J, Morgenstern O (1953) Theory of games and economic behavior, 3rd edn. Wiley, New York
13. Botta R, Bahill AT (2007) A prioritization process. Eng Manag J 19(4):20–27
14. Edwards W (1977) How to use multiattribute utility measurement for social decisionmaking. IEEE Trans Syst Man Cybern 7(5):326–340
15. Keeney RL, Raiffa H (1976) Decisions with multiple objectives: preferences and value tradeoffs. Wiley, New York
16. Watson SR, Buede DM (1987) Decision synthesis: the principles and practice of decision analysis. Cambridge University Press, Cambridge
17. Wymore AW (1993) Model-based systems engineering. CRC, Boca Raton
18. Wayne LW (1994) Operations research: applications and algorithms. Duxbury Press, Boston
19. Keeney RL (1992) Value-focused thinking: a path to creative decisionmaking. Harvard University Press, Cambridge
20. Chapman WL, Bahill AT, Wymore AW (1992) Engineering modeling and design. CRC, Boca Raton, http://sysengr.engr.arizona.edu/pinewood.pdf
21. Buchanan BG, Shortliffe EH (1984) Rule-based expert systems. Addison-Wesley, Reading
22. Steuer RE (1986) Multiple criteria optimization: theory, computation, and applications. Wiley, New York
23. Smith ED (2006) Tradeoff studies and cognitive biases. University of Arizona, Tucson
24. Madni A, Ross AM (2016) Exploring concept tradeoffs. In: Parnell GS (ed) Using tradeoff analyses to identify value and risk. Wiley, New York
25. Stern C, Sherwoods ER (1966) The origins of genetics: a Mendel source book. Freeman, San Francisco
26. Keefer DL, Pollock SM (1980) Approximations and sensitivity in multiobjective resource allocation. Oper Res 28(1):114–128
27. Bahill AT, Dahlberg SO, Lowe RA (1998) 1.2.3 Difficulties in using multicriterion decision making techniques for tradeoff studies selecting amongst alternative concepts. In: INCOSE international symposium, vol 8, no 1, pp 192–197
28. Oakes J, Botta R, Bahill AT (2006) 11.1.1 Technical performance measures. In: INCOSE international symposium, vol 16, no 1, pp 1466–1474
29. Gilb T, Maier MW (2005) 11.4.2 Managing priorities: a key to systematic decision-making. In: INCOSE international symposium, vol 15, no 1, pp 1687–1705
30. Jacobson I, Booch G, Rumbaugh J, Rumbaugh J, Booch G (1999) The unified software development process. Addison-Wesley, Reading
31. Daniels J, Werner PW, Bahill AT (2001) Quantitative methods for tradeoff analyses. Syst Eng 4(3):190–212
32. Bahill AT, Chapman WL (1993) A tutorial on quality function deployment. Eng Manag J 5(3):24–35
33. Ghiya KK, Bahill AT, Chapman WL (1999) QFD: validating robustness. Qual Eng 11(4):593–611
34. Bahill AT, Botta R, Daniels J (2006) The Zachman framework populated with baseball models. J Enterp Archit 2(4):50–68
35. Rechtin E, Maier M (1997) The art of systems architecting. CRC, Boca Raton
36. Gigerenzer G (2002) Reckoning the risk. Penguin, New York
37. Clausen D, Frey DD (2005) Improving system reliability by failure-mode avoidance including four concept design strategies. Syst Eng 8(3):245–261
38. Harris AJ, Corner A, Hahn U (2009) Estimating the probability of negative events. Cognition 110(1):51–64

39. Gigerenzer G (1991) How to make cognitive illusions disappear: beyond “heuristics and biases”. Eur Rev Soc Psychol 2(1):83–115
40. Kuhn HW, Tucker AW (eds) (1951) Nonlinear programming. In: Berkeley symposium on mathematical statistics and probability. University of California Press, Berkeley, CA
41. Anscombe FJ (1973) Graphs in statistical analysis. Am Stat 27(1):17–21
42. Wymore AW, Bahill AT (2000) When can we safely reuse systems, upgrade systems, or use COTS components? Syst Eng 3(2):82–95
43. Lewis M (2003) Moneyball: the art of winning an unfair game. W. W. Norton & Company, New York
44. Smith ED, Szidarovszky F, Karnavas WJ, Bahill AT (2008) Sensitivity analysis, a powerful system validation technique. Open Cybern Syst J 2:39–56
45. Karnavas WJ, Sanchez PJ, Bahill AT (1993) Sensitivity analyses of continuous and discrete systems in the time and frequency domains. IEEE Trans Syst Man Cybern 23(2):488–501
46. Box JF (1981) Gosset, Fisher, and the t distribution. Am Stat 35(2):61–66
47. Pignatiello JJ, Ramberg JS (1985) Off-line quality-control, parameter design, and the Taguchi method-discussion. J Qual Technol 17(4):198–206
48. Bahill AT, Baldwin DG, Ramberg JS (2009) Effects of altitude and atmospheric conditions on the flight of a baseball. Int J Sports Sci Eng 3(2):109–128
49. Shelquist R [cited 2016 February]. https://wahiduddin.net/calc/calc_da_rh.htm
50. Buck AL (1981) New equations for computing vapor pressure and enhancement factor. J Appl Meteorol 20(12):1527–1532
51. Chambers F, Page B, Zaidins C (2003) Atmosphere, weather, and baseball: how much farther do baseballs really fly at Denver’s Coors Field? Prof Geogr 55(4):491–504

Chapter 4
Discovering System Requirements

4.1 Section 27. What Are Requirements?

A requirement is a statement that identifies a capability or function needed by a system to satisfy a customer need. A functional requirement defines what, how well, and under what conditions one or more inputs must be converted into one or more outputs at the boundary in question in order to satisfy a customer need. A customer's need might be to solve a problem, achieve an objective, or satisfy a contract, standard, or specification.

No two systems have identical requirements. However, there is an identifiable process for logically discovering system requirements throughout the system life cycle regardless of system purpose, size, or complexity, Grady [1]. This process is examined in this section: it is based on Bahill and Dean [2].

For small projects, requirements may be written into a text document. For medium-sized projects, requirements might be stored in a spreadsheet. For large projects, requirements should be kept in a requirements database. In all cases, configuration management of requirements is necessary to ensure consistent usage in evolution of requirements and derived requirements.

This section explains the systems requirements process without getting into the specific details of (1) commercial requirements management tools and databases; (2) tools for modeling systems such as UML, SysML, and functional decomposition; (3) methods for flowing down requirements from the system to subsystems and from the subsystems to components; or (4) scripts for producing specific reports such as a Requirements Allocation Sheet (RAS) or a Requirements Verification and Traceability Matrix (RVTM). The terminology used in this section is a generalization of common usage at many companies. It is not patterned after any particular company, branch of government, or tool.

Requirements should be defined using a common language with a restricted vocabulary. Each project should create an online special dictionary with a defined

A.T. Bahill, A.M. Madni, *Tradeoff Decisions in System Design*,
DOI 10.1007/978-3-319-43712-5_4

set of words for use by a spell-checker. Project-specific terms should be defined in a glossary. Acronyms should be expanded and put in a list.

4.1.1 State the Problem

This section is an elaboration of Section 20.1. Before we can effectively discover and develop requirements, we need a clear statement of the problem. Stating the problem at the right level and with the right scope is one of the engineer's most important responsibilities. Problem framing and stating are in fact more important than problem solving, because an elegant solution to the wrong problem is worthless. The problem must be stated clearly and unambiguously at the right level and scope. The problem statement is intended to articulate the customer's needs, project goals, business needs, required system capabilities, system scope, concept of operations, stakeholders, deliverables, and key decisions that need to be made.

The problem should be framed in terms of needed capabilities, without implying a solution or an implementation. For example, in Section 4 we discussed a flood that washed out a bridge across the Santa Cruz River and made it difficult for the Tohono O'odhams at Mission San Xavier del Bac to get to the Bureau of Indian Affairs Health Center. A common way of stating this problem was "We must rebuild the bridge across the Santa Cruz River." However, a better way would be to say, "The Tohono O'odhams at San Xavier Mission need a way to get to their health center."

For examples of stating the problem without reference to specific implementations, it is worth reading some US patents. A patent does not protect the system; it protects the idea behind the system. (A patent is only a license to sue.) Here is an excerpt from the patent on the Bat Chooser™, a system that computes the Ideal Bat Weight™ for individual baseball and softball batters.

> "The invention relates to an instrument for measuring a player's bat speed. Over a number of swings, the bat speed data can be used to determine the maximum ball speed after contact. The optimal bat weight is determined. Baseball and softball are examples in which the use of the instrument would be appropriate. The invention also relates to a process of using the instrument to obtain data on bat speed, plotting the bat speed data, fitting a best fit curve to the data, using the best fit curve of physiological data to obtain the ball speed after contact, plotting the ball speed, determining the maximum-ball-speed bat weight and the optimal bat weight. It is the coupling of the physiological equations to the conservation of momentum equations of Physics which is unique." Bahill and Karnavas [3]

It is common engineering practice to state the problem in terms of capabilities that the system must have or top-level functions that the system must perform. However, it might be preferable to state the problem in terms of the deficiency that must be ameliorated. This stimulates generation and consideration of additional design alternatives (see examples below).

Example 1

Top-level function:	The system shall hold together 2–20 sheets of 8½ by 11 in., 20 lb paper
Alternatives:	Stapler, paper clip, fold the corner, put the pages in a folder

Example 2

The deficiency:	My reports are typically composed of 2–20 sheets of 8½ by 11 in., 20 lb paper. The pages get out of order and become mixed up with pages of other reports, especially when they lie on a car seat with the windows down
Alternatives:	Stapler, paper clips, fold the corner, put the pages in folders, number the pages, put them in envelopes, put them in three-ring binders, throw away the reports, convert them to electronic form, have them bound as books, put them on audio tapes, distribute them electronically, put them on DVDs, put them on microfiche, transform the written reports into videotapes

It is important to realize that you cannot take the first thing your customer says as a correct statement of the problem. It is important to verify the problem statement with the customer, and iterate multiple times until the problem statement is clear, unambiguous, at the right level, and properly scoped. For an excellent (and enjoyable) reference on stating the problem see [4] an excerpt from which is provided in Point 17 of Section 4.1.3.

4.1.1.1 Define Stakeholders

Stakeholders include people and organizations that are a part of the system's environment because the system can potentially provide some benefit or harm to them and, therefore, they have a vested interest in the system. They include end users, operators, bill payers, owners, regulatory agencies, victims, sponsors, maintainers, architects, managers, customers, surrogate customers, testers, quality assurance, risk management, purchasing, and the environment. Stakeholder interests are often in conflict with each other.

There are many stakeholders for a commercial airliner, such as the Boeing 737. The users are the passengers that fly on the airplane. The operators are the crew that flies and manages the plane, and the mechanics that maintain it. The bill payers are the airline companies, such as Southwest and American Airlines. The owners are the stockholders of these companies. The Federal Aviation Administration (FAA) writes the regulations and certifies the airplane. Among others, people who live in the vicinity of the airport are also stakeholders in that they are potential victims of noise and air pollution. If the plane is highly successful, then Airbus (the manufacturer of a competing airplane) is also a potential victim. The sponsor, in this example, would be the corporate headquarters of Boeing.

However, since engineering delivers both the *product* and a *process* for producing it, we need to also include the stakeholders of the process. The users and operators of the process are employees in the manufacturing plant in Renton. The bill payer is the Boeing company. The owners are Boeing stockholders. The regulators include the Occupational Safety and Health Administration (OSHA). Victims include physically injured workers and, according to Deming, those workers who have little control of the output, but who are reviewed for performance [5, 6].

4.1.1.2 Identify the Audience

Before writing any documents, the audience for those documents must be identified. For a requirements document, the audience comprises the customer and the system developer including the designers, producers, and testers.

The customer and the developer inevitably have different backgrounds and needs. Therefore, Wymore [7] suggested two different documents for these two different groups. The *customer requirements document* is a detailed description of the problem in plain language. It is intended for management, the customer, and engineering. The *derived requirements document* is a succinct mathematical description or model of the requirements as described in the customer requirements document. Its audience is engineering. Sometimes the derived requirements are referred to as technical, or design or product requirements. Each derived requirement must be traceable to a customer requirement or a document, such as a vision/mission statement or a concept of operations.

Stakeholder needs, expectations, constraints, and interfaces are transformed into customer requirements. Customer requirements are nontechnical and use the customer's vocabulary. These are transformed into derived requirements that are expressed in technical terms and are used for design decisions.

4.1.1.3 Avoid the Word "Optimal"

The word *optimal* (which goes with optimize) should not appear in the problem statement, because there is no optimal solution for a complex systems problem. Most system designs have multiple performance and cost criteria. Alternative designs satisfy these criteria to varying degrees. Moving from one alternative to another often improves satisfaction of one criterion and worsens satisfaction of another. In other words, there are trade-offs to be made. None of the feasible alternatives is likely to optimize all the criteria. Therefore, we need to settle for less than optimality.

It might be possible to optimize some subsystems, but when they are interconnected, the overall system may no longer be optimal. It is important to realize that the best possible system will not be made by combining optimal subsystems. An All Star team may have optimal people at all positions, but is it likely that such an All Star team could beat the World Champions? For example, in football a Pro Bowl team is not likely to beat the Super Bowl champions. In other words, a team of experts does not make an expert team.

If the requirements called for an optimal system, then it is important to realize that data invariably cannot be provided to prove that any resulting system was indeed optimal. As a result, precious talent, money, and time could be wasted trying to satisfy this rather impossible requirement. Instead, these resources should be used to design an effective system—a satisficing system as was discussed in Section 5. Engineers design systems that satisfice, not optimize [8]. To exploit reuse, they design general-purpose systems, rather than specialized systems. And, of course, no complex system is likely to be optimal for all the people, all the time.

> "You can fool all the people some of the time, and some of the people all the time, but you cannot fool all the people all the time." Abraham Lincoln

It is important to remember that humans are not optimal performers. There is ample evidence to support this well-known, but often ignored, fact. Cheetahs can run faster than humans. Porpoises can swim faster. Dolphins have bigger brains. Bats have wider bandwidth auditory systems. Deer have more sensitive olfaction systems. Pronghorn antelope have sharper vision. For color vision, raptors have four types of cones, rather than man's three. Interestingly, man has not exploited evolution thus far to advance these capabilities. Man remains a versatile generalist. Consider the fact that the frog's visual system has evolved much farther than man's: frogs have individual cells in the superior colliculus that are specialized to detect moving flies. Leaf-cutting ants have had organized agricultural societies millions of years before humans. And yet, even though humans are not optimal in any sense, they still rule the world. Why? Because we live in an ever-changing world in which optimality is less valued than affordable adaptability [9].

However, if it is required that optimization techniques be used, then these methods should be applied only to subsystems and only late in the design process. Total system performance must be analyzed to decide if the cost of optimizing a subsystem does not come at the cost of compromising system performance and effectiveness. Furthermore, total system performance should be analyzed over the whole range of operating environments, potential futures, and trade-off functions, because what is optimal in one environment with one set of trade-offs is more than likely not going to be optimal with other sets [9].

Because of the rapid rate at which technology is advancing, flexibility and adaptability are more important characteristics than optimality [10]. A company could buy a device, and spend many person-years optimizing its inclusion into a system, only to discover that a newly available device can perform the function better and at a lower cost. In addition to the word *optimal*, there are other words such as *maximize*, *minimize*, and *simultaneous* for which we have a similar sentiment.

4.1.2 Example Requirement

A requirement identifies a capability or function that is needed by a system to satisfy customer needs. A functional requirement defines what, how well, and under what conditions one or more inputs must be converted into one or more outputs at the boundary in question in order to satisfy customer needs. The Capability Maturity Model Integration [11] states that a requirement is (1) a condition or capability needed by a user to solve a problem or achieve an objective or (2) a condition or capability that must be possessed by a product to satisfy a contract, standard, or specification. Requirements should state what the system must do, but they should not specify how the system is to do it. This section presents an example of a requirement. In this book, we set all requirements in the *Italic sans-serif font*. The following is a 30-year-old example of a requirement, from Somerville [12].

The graphic editor facility. To assist in positioning items on a diagram, the user may turn on a grid in either centimeters or inches, via an option on a control panel. Initially the grid is off. The grid may be turned on or off at any time during an editing session and can be toggled between inches and centimeters at any time. The grid option will also be provided on the reduce-to-fit view, but the number of grid lines shown will be reduced to avoid filling the diagram with grid lines.

Good points about this requirement: It provides rationale for the items: it explains why there should be a grid. It explains why the number of grid lines should be reduced for the reduce-to-fit view. It provides initialization information: initially the grid is off.

Bad points: The first sentence has more than one component: (1) it states that the system should provide a grid, (2) it describes the grid units (centimeters and inches), and (3) it tells *how* the user will activate the grid. This requirement provides initialization information for some but not all similar items: it specifies that initially the grid is off, but it does not specify the units when it is turned on. Next, lack of a shall statement makes it difficult to verify the requirement.

The following is an improvement of this requirement [12].

5.3.2.1. The Grid Facility

5.3.2.1.1. The graphic-editor grid facility shall produce a pattern of horizontal and vertical lines forming squares of uniform size as a background to the editor window. The grid shall be passive rather than active. This means that alignment is the responsibility of the user and the system shall not automatically align items with grid lines.

Rationale: A grid helps the user to create a neat diagram with well-spaced entries. Although an active grid might be useful, it is best to let the user decide where the items should be positioned.

5.3.2.1.2. When used in the "reduce-to-fit" mode, the logical grid line spacing shall be increased.

Rationale: If the logical grid line spacing were not increased, the background would become cluttered with grid lines.
Specification: Eclipse/Workstation/Defs:Section 12.6.

This requirement definition refers to the requirement specification, which provides details such as units of centimeters and inches and the initialization preferences.

4.1.2.1 Requirements on Requirements

Each requirement shall describe only one function. Requirements shall be unambiguous. Requirements shall be verifiable. The set of requirements shall be prioritized. The set of requirements shall be complete. The set of requirements shall be consistent.

The requirements document must be under configuration control. This means that changes are handled systematically so that a system maintains its integrity over time. It evaluates proposed changes, tracks the status of changes, and maintains an inventory of system and support documents as the system changes. It is applied throughout the system life cycle. It is a fundamental area of the CMMI.

4.1.3 Characteristics of a Good Requirement

What are the characteristics of a good requirement? The IEEE says, "Requirements must be unambiguous, complete, correct, traceable, modifiable, understandable, verifiable, and ranked for importance and stability." Martin Fowler and Michael Fagan both state that if you design your tests when you write your code, then you will have fewer defects. Therefore, each requirement should have a description of the verification procedure. These qualities and a few dozen others are covered in this section. As stated in the introduction, statements about requirements cannot be dogmatic, because on occasion each statement has been rightfully violated many times for good reasons. Several authors have written about the qualities that make a requirement *good* [13–15]. Sneed [16] has described attempts to automate the collection of structured requirements and use natural language processing to create quality metrics for requirements.

1. *Describes What, Not How.* There are many characteristics of a good requirement. First and foremost, a good requirement defines *what* a system must do, but does not specify *how* to do it. A statement of a requirement should not be a preconceived or inadvertently implied solution to the problem to be solved. To avoid falling into this trap, ask why the requirement is needed, and then derive the real requirements. For example, it would be a mistake to require a relational database for requirements [15]. The following requirements state what is needed, not how to accomplish it: *The system shall store requirements. The*

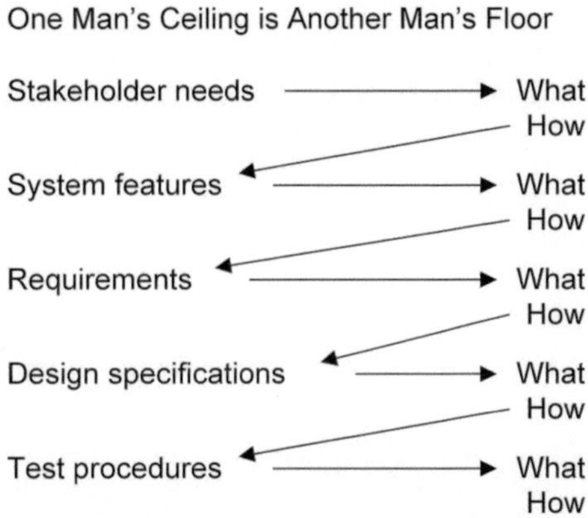

Fig. 4.1 The relationship between *whats* and *hows* at different levels

system shall sort requirements. The system shall add evaluation criteria to requirements. It should be noted that because QFD is often used iteratively to define requirements, the *hows* in one QFD chart become the *whats* in the next, possibly making the above statements confusing. This property could also be confusing, because we say *what* a system is supposed to do, and then the designers decide *how* it should do it. Then this *how* becomes a *what* at the next level down, as shown in Fig. 4.1.

Requirement writers often fall into the implementation trap, when they forget that they are writing requirements and begin to think about *how* to build the system. Requirement statements should not be made confusing by descriptions of design or other implementation-specific information. Sometimes, the implementation information created by requirement writers is of value and should be preserved, but not as requirements. Non-requirement information should be provided as working papers routed to the design team.

Frequently, requirement writers levy requirements on the function rather than on the system. For example, *The Sort function shall. . ..* This practice permits a very convenient sorting of requirements by function. This is all right. However, sometimes requirement writers mistakenly specify exchanges of information between functions as requirements. For example, *The Sort function shall notify the Store function. . ..* Such statements are not requirements, since they do not depend upon external stimuli and generate no results that can be measured at the system boundary. Such language is appropriate for design documents—not requirements.

2. *Atomic (or Unitary or Single Minded).* A requirement should be *atomic*, meaning it is not easily broken into smaller requirements. There should be one concept per requirement. Complex requirements need to be decomposed into individual requirements. A compound statement such as "The System shall perform X and Y and Z" results in an undesirable complex requirement that should be decomposed into three requirements. Requirements should be decomposed to the point where each statement defines stimuli and results associated with a single function (one or more stimuli cause a single result)

for the system or component, considered as a black box. If this test cannot be satisfied, because the requirement addresses more than one function, then the requirement needs to be decomposed.

3. *Allocation*. Each requirement should be allocated to a single entity: except for non-allocated requirements (such as quality of workmanship and electrostatic discharge protection) that are spread across the whole project, like peanut butter. It is acceptable to assign two or more requirements to one physical component. However, it would most likely be a mistake to assign one requirement to two physical components; see Section 6 and [17].
4. *Unique*. A requirement should have a unique label, name, and contents. Avoid repeating requirements. However, some systems engineers claim that designers do not know what is in the requirements specification. Therefore, they deliberately repeat critical safety requirements hoping to increase the likelihood that designers will notice them.
5. *Documented and Accessible*. A requirement must be documented (text, equations, images, databases, etc.) and the documentation must be accessible. In situations where confidentiality is important, each requirement should clearly indicate its classification status. Only individuals with the appropriate clearances and the *need to know* should have access to classified requirements.
6. *Identifies Its Owner*. A good requirement identifies its owner. The requirement's owner must approve of any change in the requirement. This owner could be an individual or a team. The owner has certain responsibilities such as monitoring the associated technical performance measures (TPMs), or risk activities. If the owner is part of the system development team, he or she should know and communicate with the customer's representative that is responsible or most knowledgeable about the requirement or the customer's representative that generated the high-level goal or capability that led to this requirement. The owner is responsible for helping to define and track the tradeoffs of the requirement with other requirements and for monitoring the tradeoff studies that involve the requirement. The owner is also responsible for insuring that requirements management tools that are used properly interpret and track the owner's requirements.
7. *Identifies Its Target*. A good requirement identifies the target of the requirement, that is, whom is it a requirement on? For example the system, the process, the company, the customer. The requirement writer might erroneously construct a requirement levied on the user of the system and not on the system itself: For example, suppose the requirements analyst wrote, *The user shall input a start time for message ingest.* Then someone would have to rewrite this as, *The system shall enable the user to designate a start time for message ingest.* In Section 3.1, we did not write a requirement like "*The Student shall wear an oven mitt when removing the hot cookie sheet from the oven,*" because we are not allowed to write requirements on the user.
8. *Approved*. After a requirement is written, revised, reviewed, and/or rewritten, it must be approved by its owner. Furthermore, each top-level requirement must be approved by the customer.

9. *Traceable*. A good requirement is traceable: it should be possible to trace each requirement back to its source. A requirement should identify related requirements (that is, parents, children, siblings) and requirements that would be impacted by changes to it. A requirements document should have a tree (or graph) structure and this structure should be evident, often in the numbering scheme.
10. *Necessary*. Each requirement should be necessary. Systems engineers should ask, "Is this requirement really necessary?" "Can the system meet the customer's real needs without it?" If yes, then the requirement is not necessary. Avoid over-specifying the system, writing pages, and pages that no one will probably ever read. There are two common types of over-specification: gold plating and specifying unnecessary things. For example, requiring that the outside of a CPU box be gold-plated is not a good requirement, because something far less expensive would probably be just as effective. Also, requiring that the inside of the CPU box be painted pink is probably an unnecessary request. Over-specification (of both types) is how $700 toilet seat covers and $25,000 coffeepots were created [18]. Each requirement should include a statement of its rationale. This might help rule out unnecessary requirements.
    As an example, early in the US space program, NASA contracted for a pen to be used by astronauts in space. Each pen had to write upside down, under water, over grease, and at temperatures of −50 °F to +400 °F. The solution had a sealed cartridge with a sliding float separating the ink from pressurized nitrogen. It cost thousands of dollars. (Today you can buy the Fisher Space Pen for $30.) The Russian solution? Use a pencil.
11. *Complete*. Each requirement must be complete. All conditions under which the requirement applies should be stated. It is all right to have a To Be Determined (TBD) in a requirement. These preliminary values are proposed with the caveat that they be reexamined at a specific review. Therefore, each TBD should have a scheduled resolution date and should be tracked in the risk register.
    Each individual requirement must be complete and the *requirements set* must also be complete. Many times, due to the structure of the requirements set, your computer can identify incompleteness [19]. Putting your requirements into a Zachman Framework will also help identify incompleteness [20].
12. *Is Written Positively*. The requirement, *Do not use wire with Kapton insulation,* would be very difficult to verify for commercial off-the-shelf (COTS) equipment.
13. *Unambiguous*. Can the requirement be interpreted in more than one way? If so, then the requirement should be clarified or removed. For example, time flies like an arrow and fruit flies like a banana.
14. *Is Not Always Written*. It must be noted that all systems will undoubtedly have many *common sense* requirements that will not be written. This is acceptable as long as the requirements really are common sense. An exhaustive list of requirements would take years upon years and use reams of paper, and even then you would probably never finish.

15. *Verifiable*. Quantitative values must be given in requirements. A requirement states a necessary attribute of a system to be designed. The designer cannot design the system if a magnitude is not given for each attribute. Without quantification, system failure could occur because (1) the system exceeded the minimum necessary cost due to overdesign, or (2) it failed to account for a needed capability. Quantitative values for evaluation criteria are also necessary in order to test the product to verify that it satisfies its requirements.

Each requirement must be verifiable by test, demonstration, inspection, logical argument, analysis, modeling, or simulation. Qualitative words like *low* and *high* should be (at least roughly) defined. What is low cost to a big corporation and what is low cost to a small company may be very different. Only requirements that are clear and concise will be easily testable. Requirements with ambiguous qualifiers will probably have to be refined before testing will be possible. Furthermore, the value given should be fully described as, for example, an expected value, a median, a minimum, a maximum, or the like. A requirement such as *Reliability shall be at least 0.999* is a good requirement because it is testable and quantified, and the value is fully described as a minimum. Also the requirement, *The car 's gas mileage should be at least 30 miles per gallon,* is a good requirement as it establishes a performance measure and an expected value. Moody et al. [21] present a few dozen metrics that can be used to evaluate performance requirements.

The verification method and level at which the requirement can be verified should be determined explicitly as part of the development for each of the requirements. The verification level is the location in the system where the requirement is met (for example, the *system level*, the *segment level*, and the *subsystem level*).

Note that often the customer will state a requirement that is not quantified. For example: "The system should be aesthetically pleasing." It is then the engineer's task to define a requirement that is quantified, such as *The test for aesthetics will involve polling two hundred potential users; at least 70 % shall find the system aesthetically pleasing.*

It is also important to make the requirements easily testable. NASA once issued a request for proposals for a radio antenna that could withstand earthquakes and high winds. It was stated that the antenna shall not deflect by more than 0.5° in spite of a 0.5 G force, 100 knot steady winds, or gusts of up to 150 knots. They expected bids around $15 million. But all of their bids were around $30 million. NASA asked the contractors why the bids were so high, and the contractors said that testing the system was going to be very expensive. NASA revised the requirements to *When "hit with a sledge hammer," the antenna shall have a resonant frequency less than 0.75 Hz.* Then they got bids between $12 and $15 million {Eb Rechtin, personal communication, 1996}.

16. *States Its Units of Measurement*. "People sometimes make errors," said Dr. Weiler of NASA. "The problem here was not the error, it was the failure of NASA's systems engineering. That's why we lost the Mars Climate Orbiter

spacecraft." One team used English units (for example feet and pounds) while the other used SI units (meters and kilograms) [22].

17. *Identifies Applicable States*. Some requirements only apply when the system is in certain states or modes. If the requirement is only to be met sometimes, the requirement statement should reflect when. There may be two requirements that are not intended to be satisfied simultaneously, but they could be at great expense.

For example:

1. *The vehicle shall be able to tow a 2000 lb cargo trailer at highway speed (65 mph).*
2. *The vehicle shall accelerate from 0 to 60 mph in less than 9.5 s.*

It would be possible but expensive to build a car that satisfied both requirements simultaneously. Similarly most radios can play AM or FM stations, but not both simultaneously.

For some systems and requirements it is necessary to specify states and for others it is not. Botta, Bahill and Bahill [23], and Section 15 explain when observable states are necessary and when they are not.

Are Your Lights On?

However, as with everything, you can carry the identification of states too far, as illustrated by the following, probably true story [4]. Some time back, the highway department tested a new safety proposal. They asked motorists to turn on their headlights as they drove through a tunnel. However, shortly after exiting the tunnel the motorists encountered a scenic-view overlook. Many of them pulled off the road to look at the reflections of wildflowers in pristine mountain streams and snow-covered mountain peaks 50 miles away. When the motorists returned to their cars, they found that their car batteries were dead, because they had left their headlights on. So the highway department decided to erect signs to warn the drivers to turn off their headlights after exiting the tunnel.

First they tried "Turn your headlights off." But someone said that not everyone would heed the original request to turn their headlights on and for these drivers turning headlights off would not be possible.

So they tried "If your headlights are on, then turn them off." But someone objected saying that that would be inappropriate, if it were nighttime.

So they tried "If it is daytime and your headlights are on, then turn them off." But someone objected saying that that would be inappropriate if it were overcast and visibility was greatly reduced.

So they tried "If your headlights are on and they are not required for visibility, then turn them off." But someone objected saying that many new cars are built so that their headlights are on whenever the motor is running.

So they tried "If your headlights are on, and they are not required for visibility, and you can turn them off, then turn your headlights off." But someone objected . . . .

So they decided to stop trying to identify applicable states. They would just alert the drivers and let them take the appropriate actions. Their final sign said, "Are your lights on?"

18. *States Assumptions*. All assumptions should be stated. Unstated bad assumptions are a major cause of bad requirements.
19. *Usage of Shall, Should, and Will.* A mandatory requirement should be expressed using the word *shall*, for example, *The system shall conform to all state laws.* A trade-off requirement can be expressed using *should* (for example, *The total cost for the car 's accessories should be about 10 % of the total cost of the car).* The term *will* is used to express a declaration of intent on the part of a contracting agency, to express simple future tense, and for statement of fact, for example, "The resistors will be supplied by an outside manufacturer" [1].
20. *Avoids Certain Words*. Words such as *optimize, maximize, and minimize* should not be used in stating requirements, because we could never prove that we had achieved them. Refrain from using the word optimize, because you will not be able to prove that a resulting system is optimal. For the same reason, use words such as maximize and minimize sparingly. They should not be used in mandatory requirements, but they might appear in trade-off requirements. Refrain from using *always* and *never*. Do not use the word *simultaneous* because it means different things to different people. To a physicist, it might mean within a femto second; to a computer engineer, on the same clock cycle; and to a paleontologist studying the extinction of the dinosaurs, within the same millennium. Use of some words is deprecated. To some people, *operational* means high-level description of behavior. To others, it means low-level keystroke-like activities.
21. *Generally Avoid Adverbs, Homographs, Homophones, and Synonyms*. Adverbs are words that modify verbs, usually qualitatively: therefore they are ambiguous. Homographs are words that are spelled the same, but have different pronunciations and meanings, such as the dove dove into the brush. Homophones are words that are pronounced the same, but have different spellings and meanings, such as which doctor is the witch doctor? Synonyms are words that have the same meaning, but different spellings and pronunciations, such as a synonym for synonym is equivalent. The following requirement statement will probably confuse most of you. The technician shall lead the patient wearing the lead vest to the X-ray machine. There X-rays will be taken of the spinal disc: these will be stored on the disc, because they are usually too big for the computer's memory: we only have 1 Mbyte of RAM.
22. *Might Vary in Level of Detail*. The amount of detail in the requirements depends on the intended supplier. For in-house work or work to be done by a supplier with well-established engineering procedures, the requirements can be written at a high level. However, for third-party suppliers and contractors with unknown engineering capabilities, the requirements might be broken down to a fine level of detail.

Requirements also become more specific and detailed with time. In the beginning, requirements describe a generic process that multiple alternative designs could satisfy. As time progresses, and as the problem becomes better understood, the acceptable number of alternatives tends to decrease and the requirements become more detailed.

23. *Contains Date of Approval*. The name of the approver and the date of approval should be included in each requirement.
24. *States Its Rationale*. Although it is seldom done, it is highly desirable to state why each requirement was written and what it was supposed to ensure. This is particularly valuable for making subsequent design changes, for example to deal with obsolete parts or for upgrades. This is important if the original engineers are not available to explain their rationale for how/why they derived requirements as they did.
25. *Respects the Media*. Newspaper journalists quote out of context, and headlines do not reflect the content of their stories. It is important to write each requirement so that it cannot spark undue public criticism of your customer or the project.
26. *Distinguishes Number*. Be careful with modal words such as shall, should, and will, because they remove number distinction. For example, *Moving vehicles shall be prohibited on these premises* could mean *Moving of vehicles is prohibited...* or *Vehicles that move are prohibited.*
27. *Consistent*. The set of requirements should not contain contradictory statements or conditions.
28. *May Use Parameters*. Using parameters rather than fixed numbers will make requirements more reusable. For example, we could write, *The radio shall amplify the signal to produce POWER-OUTPUT between FREQUENCY-RANGE,* where phrases in capital letters are parameters that could be instantiated to numbers such as 115 W or 50–5000 Hz.

4.1.4 Characterization of Requirements

There are several independent, orthogonal characterizations of system requirements. Four of these are types, sources, expressions, and input–output relationships. A brief discussion of each follows.

4.1.4.1 Types of Requirements

There are two types of system requirements: *mandatory* and *trade-off*, as illustrated in Fig. 4.2. The *trade-off requirement* is *The bridge deck should be at the same level as the road surface, with a tolerance of* $\pm 5\ \%$. And the *mandatory requirement* is *The bridge deck shall stretch from bank to bank without a gap between the road and the bridge (95 % is not acceptable).*

Mandatory requirements:

(1) Specify the necessary and sufficient capabilities that a system must have in order to be acceptable
(2) Use the words *shall* and *must*, although the use of *must* is now deprecated
(3) Are passed or failed with no in between (do not use scoring functions)
(4) Should not be included in trade-off studies

The following is a typical mandatory requirement: *The system shall not violate federal, state, local, or tribal laws.* After identifying the mandatory requirements, design engineers propose alternative candidate designs, all of which satisfy the mandatory requirements. Trade-off requirements are then evaluated to determine the *best* designs. Trade-off requirements are sometimes called preference requirements.

Trade-off requirements:

(1) State conditions that would make the customer happier and are expressed with the words *shall* or perhaps *should* (often a significant reward or incentive is attached to how well a performance or trade-off requirement is satisfied)
(2) Should be described with scoring functions (utility functions) as presented in Section 21
(3) Should be evaluated with multicriterion decision-making techniques, because none of the feasible alternatives is likely to optimize all the evaluation criteria
(4) There will be trade-offs among these requirements [24]

The following is a typical trade-off requirement: *Dinner shall contain a mix of items from each of the five major food groups: grains, vegetables, fruits, milk, beer and meat, and beans.*

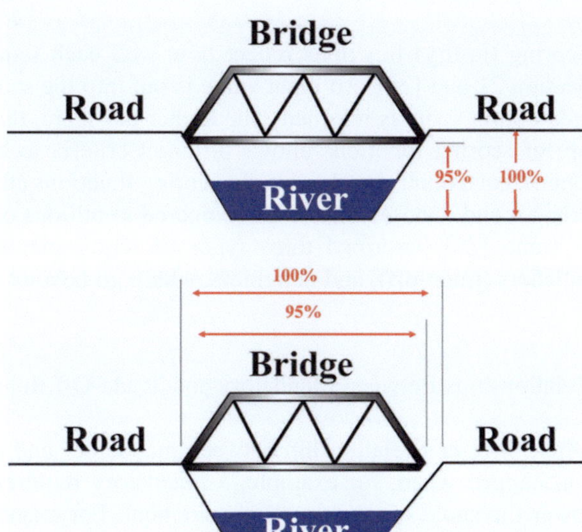

Fig. 4.2 Trade-off (*top*) and mandatory (*bottom*) requirements

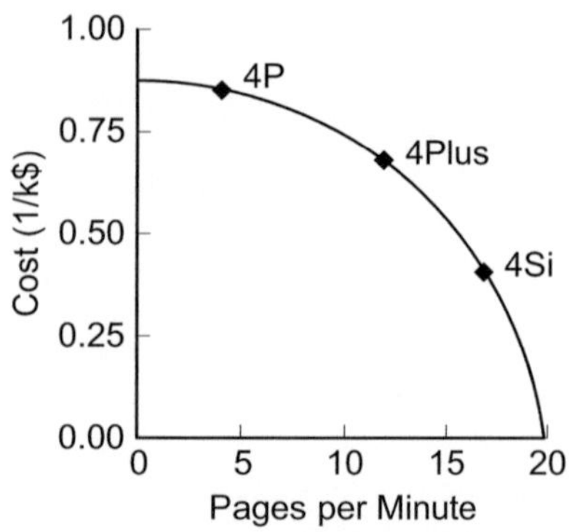

Fig. 4.3 A typical trade-off between trade-off requirements

Figure 4.3 shows an example of such a trade-off in the investigation of alternative laser printers. Many printers were below and to the left of the circular arc. They were clearly inferior and were discarded from further consideration. Three printers lay on the circular arc: they were the best. No printers were in the infeasible region above and to the right of the circular arc. The question now becomes, "Which of these three printers is the best?" With the present data, there is no answer to that question. The customer will have to say which trade-off requirement (or evaluation criterion) is more important before an answer can be obtained. Moving from one alternative to another will improve at least one criterion and worsen at least one criterion. In other words, there will be trade-offs. An arc like this (or a surface when there are more than two criteria) is called a Pareto optimal contour.

Scoring Functions

Scoring (utility) functions reflect how well each requirement has been met; see Section 21 and [25]. An input value is put into the scoring function and a normalized output score is returned. The higher the score, the happier the customer! The use of scoring functions allows different criteria to be compared and traded off against each other. In other words, scoring functions allow apples to be compared to oranges and nanoseconds to be compared to billions of dollars.

Kano [26] described three types of requirements: dissatisfiers (mandatory), satisfiers (trade-off), and delighters, which go beyond what the customer expects.

Relationships Between Mandatory and Trade-Off Requirements

Often there is a relationship between mandatory and trade-off requirements. This can happen when, for example, a mandatory requirement might be an upper or lower threshold of a trade-off requirement. For example, the scoring function of Fig. 4.4 has a trade-off requirement of *The golf ball should weight more than*

Fig. 4.4 A scoring function for the weight of a golf ball

1.6 ounces and a mandatory requirements of *The weight of the golf ball shall not exceed 1.62 ounces.* But sometimes it is better to let the mandatory requirement be the baseline value for a trade-off requirement.

4.1.4.2 There Are Several Sources of Requirements

Thus far, we have discussed requirements by classifying them into two types: trade-off and mandatory. We now classify requirements according to their source. That is, where do requirements come from? We might have called this section *Categories of Requirements* or a *Requirements Taxonomy*.

This section identifies more than two dozen sources of requirements. However, Wymore [7] stated that only the first six sources are necessary: input–output, technology, performance, cost, trade-off, and system test. He says all of the other sources can be put into one of these six. Grady [1] suggested having only five sources: functional, performance, constraints, verification, and programmatic. He asserted that most sources are constraints. The EIA-632 Standard on Systems Engineering calls out only three: functional, performance, and constraints. Project managers contend that there are only three: cost, schedule, and performance, Kerzner [27]. The Unified Modeling Language (UML) employs three categories of requirements: functional requirements, nonfunctional performance requirements, and supplemental requirements that are spread throughout the system.

Functions

Functional requirements describe the functions that the system must provide, for example *Amplify the input signal.*

Input–Output

Most functional requirements describe an input–output relationship. The amplify function could be stated as *The ratio of the output to the input at 10 kHz shall be +20 dB*. Wymore [7] maintains that functional requirements are a subset of input–output requirements. A well-stated input–output requirement describes a function. The above input–output requirement describes the function "Amplify the input signal." The functional requirement *The system shall fasten pieces of paper* is covered by the input–output requirement *The system shall accept 2–20 pieces of 8½ by 11 in., 20 lb paper and secure them so that the papers cannot get out of order.* One function of an automobile is to accelerate. The input is torque (perhaps developed with an engine) and the output is a change in velocity.

Technology

The technology requirement specifies the set of components—hardware, software, and bioware—that is available to build the system. The technology requirement is usually defined in terms of types of components that *cannot be used,* that *must be used,* or both. For example, Admiral Rickover required that submarine nuclear reactor instrumentation be done solely with magnetic amplifiers. Your purchasing department will often be a source of technology constraints.

Performance

Performance requirements include quantity (how many, how much), quality (how well), coverage (how much area, how far), timeliness (how responsive, how frequent), and readiness (reliability, availability). Functional requirements often have associated performance requirements, for example, *The car shall accelerate from 0 to 60 mph in 6.5 s or less*. Performance is an attribute of products and processes. Performance requirements are initially defined through requirements analyses and trade-off studies using customer need, objective, and/or requirements statements.

Cost

There are several types of cost, such as labor, resources, and monetary cost. Examples of cost requirement are, *The maximum purchase price shall be $10,000* and *The maximum total life cycle cost shall be $18,000.*

Trade-Off

Trade-offs between performance and cost are defined as the different relative value assigned to each factor. A simple trade-off might add the performance and cost criteria with associated weights. For example the performance criterion may have a weight of 0.6, and the cost criterion may be given a weight of 0.4 [7, 28]. However, the trade-off should not be limited to performance and cost because there are several other criteria that should be included in the trade-off.

The trade-off requirement is a description of how the data in a trade-off study are going to be combined [25]. The summation of weighted scores [29] is one example:

$$\begin{aligned}\text{Alternative rating} = {} & \text{Weight}_{\text{performance}} \times \text{score}_{\text{performance}} + \text{weight}_{\text{cost}} \times \text{score}_{\text{cost}} \\ & + \text{weight}_{\text{risk}} \times \text{score}_{\text{risk}}\end{aligned}$$

See Section 22. Another example is the benefit-to-cost ratio:

$$\text{Benefit-cost ratio} = w_{\text{Ratio}} \frac{\text{Benefit}^{w_{\text{Benefit}}}}{\text{Cost}^{w_{\text{Cost}}}}$$

System Test

Early in the design process it should be specified how the final system will be tested. The purpose of the system test is to verify that the design and the system satisfy the stated requirements. For example, in an electronic amplifier, a 3 mV, 10 kHz sinusoid will be applied to the input, and the ratio of output to input will be calculated.

Now, we find systems that can be tested and diagnosed over the Internet or over phone lines. A copy machine that does faxes is hooked up to the phone line. Therefore, the supplier could easily query or diagnose the machine on a regular basis over the phone system. They could monitor use and provide just-in-time service for toner and paper. Implementing such testing involves a trade-off of cost with customer convenience and service. This type of testing is especially effective for equipment at remote or unmanned sites.

Built-In Self-Test

Any new design should require that the system test itself. It should perform built-in self-tests (BiST) upon request and whenever it is not servicing one of its actors.

Company Policy

Company policy can often lead to requirements. Learjet Inc. has stated, *We will make the airframe, but we will buy the jet engines and the electronic control systems.* Raytheon should require that all missiles have the capability of being disabled by the Pentagon. (This would prevent Stinger missiles from shooting down our airplanes.) Or, *Our company will not deliver anything to the customer that we are unable to test to some acceptable quality level.*

Business Practices

Corporate business policies might require work breakdown structures, PERT charts, quality manuals, environmental safety and health plans, or a certain return on investment.

Engineering

Engineering might require that every transportable memory device (for example, CD ROM, DVD, memory stick, IPhone) have a Readme file that describes the author, date, contents, software program, and version (for example, Visio 2013 or Excel 2013).

Project Management

Performance, cost, and schedule are requirements that invariably come from project managers. These might include rewards for early completion and penalties for schedule overruns.

Access to source code for all software might be a project management requirement as well. It takes time and money to install new software. This investment would be squandered if the supplier went bankrupt and the customer could no longer update and maintain the system. Therefore, most customers ask for source code especially if they intend to make changes to or extend the software. However, few software houses are willing to provide source code, because it can affect their profits and complicate customer support. When there is a distinct possibility that the supplier might stop supporting a product, the source code should be provided and placed in escrow. This source code remains untouched as long as the supplier supports the product. If and when the supplier ceases to support the product, the customer can acquire the source code and maintain it in-house. Therefore, placing the source code in escrow can be a requirement.

Marketing

The marketing department wants features that delight the customer. Kano [26] calls them exciters. They are features that customers did not know they wanted. In the 1970s, IBM queried customers to discover their personal computer (PC) needs. No one mentioned portability, so IBM did not make it a requirement. Compaq made a portable PC and then a laptop, dominating those segments of the market. In the 1950s, IBM could have bought the patents for Xerox's photocopy machine. But they did a market research study and concluded that no one would pay thousands of dollars for a machine that would replace carbon paper. They did not realize that they could delight their customers with a machine that provided dozens of copies in just minutes. The United States Air Force did not know that they wanted a stealthy airplane until after the engineers explained that they could do it for the F-117.

Manufacturing Processes

Sometimes we might require a certain manufacturing process or environment. We might require our semiconductor manufacturer to have a Class 10 clean room. Someone might specify that quality function deployment (QFD) be used to elicit customer desires (although this requirement would be in bad form, because it states *how*, not *what)*. Recently, minimization of the waste stream has become a common requirement. In a different vein, annealing swords for a certain time and at a certain temperature could be a requirement.

Design Engineers

Design engineers impose requirements on the system. These are the *build to*, *code to*, and *buy to* requirements for products and *how to execute* requirements for processes. These are often called derived requirements, because design engineers derive them from customer requirements.

Reliability

Reliability could be a performance requirement, or it could be broken out separately. It could be given as a mean time between failures or as an operability probability. But, it should be testable. For example do not say, *The system shall be available 99.9 % of the time,* rather say, *The system shall be available 99.9 % of the time in a specified 1-week test period.*

Safety

Some requirements may stem from safety considerations. These may state how the item should behave under both normal and abnormal conditions.

The Environment

Concern for the environment will produce requirements, such as forbidding the use of chlorofluorocarbons (CFCs) or trichloroethylene (TCE).

Ethics

Ethics requires physicians to obtain informed consent before experimenting on human subjects.

Intangibles

Sometimes the desires of the customer will be hard to quantify for intangible items such as aesthetics, national or company prestige (for example, putting a man on the moon in the Apollo project), ulterior motives such as trying to get a foot in the door using a new technology (for example, the stealth airplanes), or starting business in a developing country.

Common Sense

Many requirements are not explicitly stated, because they are implied by common sense. For example, characteristics of the end user are seldom stated. If we were designing a user-computer interface, it would not be stated that the end user would be a human with two hands and ten fingers. Common sense also dictates that *Computers shall pass diagnostic tests after being stored for 24 h at 70 °C (163 °F),* for example during shipping. Furthermore, we do not write that *there shall be no exposed high-voltage conductors on a personal computer*, but it most definitely is a requirement. Many such requirements can be found in de facto standards and certifications.

Laws or Standards

Requirements could specify compliance with certain laws or standards, such as the National Electrical Code, local building codes, EIA-632, ISO-9000, IEEE 1220, ISO-15288, or level 3 CMMI.

The Customer

Some requirements are said to have come from the customer, such as statements that define the expectations of the system in terms of mission, objectives, environment, constraints, and measures of effectiveness. These requirements are defined from a validated needs statement (customer's mission statement), from acquisition and program decision documents, and from mission analyses.

Legacy Requirements

Sometimes the existence of previous systems creates legacy requirements. For example, "Your last system was robust enough to survive a long trip on a dirt road, so we expect your new system to do the same." Many new computer programs must be compatible with COBOL, because so many COBOL business programs are still running. Could you read data archived on a floppy disk? Legacy requirements are often unstated.

Existing Data Collection Activities

If an existing system is similar to the proposed new system, then existing data collection activities can inform discovery of system requirements, because each data item that is collected should be traceable to a specific system requirement. Often it is difficult to make a measurement to verify a requirement. Also, it might be impossible to meet the stated accuracy. Trying to make a measurement to verify a requirement might reveal more system requirements.

Human Abuse

Systems should be required to withstand human abuse, such as standing on a computer case or spilling coffee on the keyboard. Systems should prevent human error, for example, "Are you sure you want to delete that file?" Systems should ameliorate human error if it does occur, for example online real-time spell-checking.

Insightful design can help overcome inadvertent operator errors. A counter example is Microsoft's question, "Would you like to refresh your computer?" If you answer yes, it erases all programs that have been installed since day 1.

Political Correctness

The need to be politically correct mandates many (often expensive) requirements. The political process places constraints that appear unreasonable to formulators of requirements. For example, the process often prohibits use of certain terms and graphics.

Material Acceptance

You must write requirements describing how raw materials, parts, and commercial off-the-shelf items will be inspected when they are received.

Other Sources

There are many other sources of system requirements, such as human factors, the environment (for example, temperature, humidity, shock, vibration), the end user, the operator, potential victims, management, company vision, future expansion, schedule, logistics, politics, the US Congress, public opinion, business partners, past failures, competitive intelligence, liability, religion, culture, government agencies (for example, DoE, DoD, OSHA, FAA, EPA), availability, maintainability, compatibility, service, maintenance, field support, warrantees, need to provide training, competitive strategic advantage, time to market, time to fill orders, inventory turns, accident reports, deliverability, reusability, future expansion, society, standards compliance, standards certification (for example, ISO 9000), effects of aging, user-friendliness, weather (for example, must be installed in the summer), security as in government classification, security as in data transmission, fitting into a certain space, secondary customer, retirement, and disposal.

Considering human factors means designing systems with knowledge of the dimmest lights humans can perceive, the faintest noises they can detect, the heaviest loads they can lift, the colors that colorblind people can see, etc. Furthermore, it also means designing your system so that it can understand what the human does not understand about your system's behavior and adapt to the human's behavior.

4.1.4.3 There Are Several Ways to Express Requirements

For some purposes, the best expression of the requirements will be a narrative in which words are organized into sentences and paragraphs. Such documents are often called operational concept descriptions, or use case models. But all natural descriptions (for example in English) can be expected to have ambiguities, because of both the language itself and the context in which the reader interprets the words. Therefore, for some purposes, the best description of a system will be a list or string of *shall* and *should* statements. Such a list would be useful for acquisition or acceptance testing. However, it is still very difficult to write with perfect clarity so that all readers have the same understanding of what is written. Other modalities that can be used instead of written descriptions include:

- A model
- A prototype
- A user's manual
- Input-out trajectories
- Sequence diagrams
- Use cases
- Computer requirements databases

The main advantage of nonverbal expressions is that they can be rigorous and executable. This helps point out contradictions and omissions. Natural language expressions can be ambiguous. Does "turn up the air conditioner" mean turn up the thermostat (making the room warmer), or turn up the air-conditioning power (making the room colder)? Nonverbal expressions also allow you to perform a sensitivity analysis of the set of requirements to learn which requirements are the real cost drivers [30, 31].

A systems engineering design process that is gaining popularity is model-based systems engineering, Wymore [7] and [32, 33]. In this process, the system engineer first develops a model of the desired system. The model can then be run to demonstrate the desired system behavior. Thus, the model also captures the requirements.

A Prototype Expresses Requirements

A publicly assessable prototype can express system requirements, as they are currently understood. This technique is popular in the software community. Of course, many functions of the final system will not be implemented in the prototype. Instead, there will be a statement of what the functions are intended to do. A publicly assessable prototype is easy to update, and it helps everyone understand what the requirements are.

The purpose of building a prototype is to reduce project risk and/or enhance user acceptance [34]. Therefore, the first functions that are prototyped should be (but usually are not) the most risky functions (Eb Rechtin, personal communication, 1996, also [29]).

Consider Unconventional Alternatives

During concept exploration, encourage consideration of unusual alternatives. Studying unusual alternatives leads to a better and deeper understanding of the requirements by both the engineer and the design engineer. Explore unintended paths through the system workflow. Document unintended functions performed by the system. Investigate unintended uses of the system. Studying models and computer simulations will also help you understand the requirements. Concept exploration is one of the most fruitful phases in requirements discovery.

Creating a User Manual Can Flush Out Requirements

The User Operation Guide should be written by future users early in the system design process [35]. This helps get the system requirements stated correctly and increases user *buy in*.

4.1.4.4 Storytelling and Scenarios

Arguably, the best way to depict the desired behavior of a system is to tell stories about the system [36, 37]. Story-based simulations are an ideal and inclusive way to discover system requirements. In essence, it means creating typical event sequences and decision paths that the proposed system will go through. Typically, a sequence of input values as a function of time (called an input trajectory) is described and an acceptable system behavior that unfolds within a story context is given. In traditional approaches, this system behavior is given as a sequence of output values (called an output trajectory). With traditional approaches, dozens or perhaps hundreds of trajectories are needed for a complete description of the system. Such descriptions of input and output behavior as a function of time have been called behavioral scenarios, trajectories, threads, operational scenarios, logistics, interaction diagrams, and sequence diagrams. The alternate flows help you recognize and deal with unintended as well as intended inputs. Story-based simulations circumvent the need for running a large number of scenarios because it capitalizes on the knowledge of domain experts to create interesting stories that correspond to different types of stress cases [36, 37].

Sequence Diagrams

Sequence diagrams are easy for people to describe and discuss, and it is easy to transform them into a system design. A sequence diagram for an automated teller machine (ATM) is shown in Fig. 4.5. The basis of these diagrams is to list the system's objects along the top of the diagram. Then, with time running from top to bottom, list the messages or commands that are exchanged between the objects [38]. Alternatively, the arrows can be labeled with data that are exchanged between the components or the functions that are performed. These ATM examples were derived using object-oriented modeling. This technique relies on collecting a large number of sequence diagrams. This collection then describes the desired system behavior. Additional scenarios can be incrementally added to the collection. The boxes on top of the lifelines indicate the processes that are active at the same time.

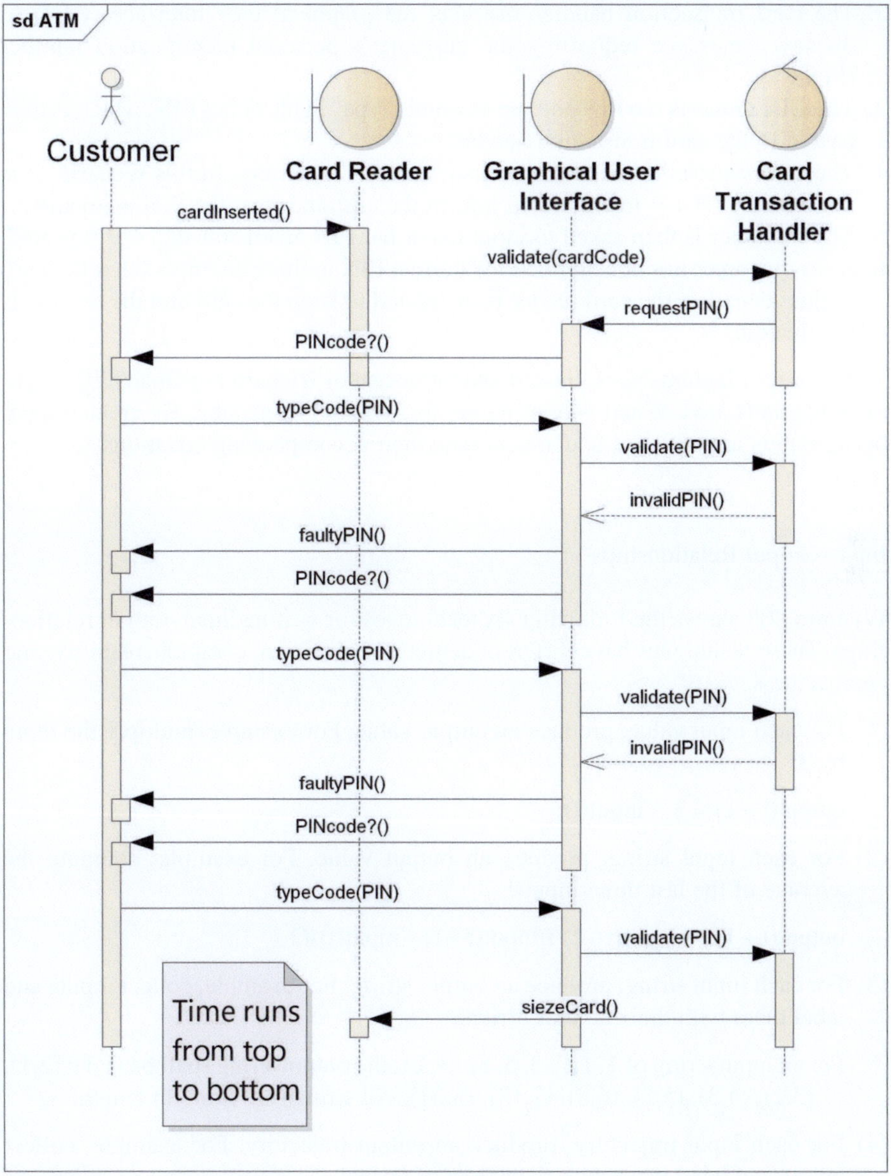

Fig. 4.5 Sequence diagram for an incorrect personal identification number (PIN) sequence of events

Wrong PIN Input Sequence

1. The customer inserts a bankcard; the card reader sends the card's information to the card transaction handler, which detects that the card is valid (if no message is returned, the card is assumed valid). An invalid card will trigger a different sequence.

2. The card transaction handler instructs the graphical user interface (GUI) to display a message requesting the customer's personal identification number (PIN).
3. The GUI requests the PIN and the customer types in his or her PIN, which is then passed to the card transaction handler.
4. The card transaction handler checks if the PIN is correct. In this scenario, it is not, and the GUI is instructed to inform the customer that the PIN is invalid.
5. The customer is then asked to input his or her PIN again and step 4 is repeated.
6. If the customer has not supplied the correct PIN in three attempts (as is the case in this scenario), the card reader is instructed to keep the card and the session is terminated.

Sequence diagrams were created and published by Richard Feynman [39]. Feynman diagrams are a visual way of expressing what is happening. He created them because they are easier to understand than their accompanying equations.

Input–Output Relationships

Wymore [7]} shows the following six techniques for writing input–output relationships. These techniques have different degrees of precision, comprehensibility, and compactness.

(1) For each input value, produce an output value. For example, multiply the input by 3:

output(t + 1) = 3 * input(t)

(2) For each input string, produce an output value. For example, compute the average of the last three inputs:

output(t + 1) = (input(t−2) + input(t−1) + input(t))/3

(3) For each input string, produce an output string. For example, collect inputs and label them with their time of arrival:

For an input string of 1, 1, 2, 3, 5, 8, 13, 21, the output string shall be (1,1), (2,1), (3,2), (4,3), (5,5), (6,8), (7,13), (8,21). All strings are finite in length.

(4) For each input trajectory, produce an output trajectory. For example, collect inputs and label them with their time of arrival.

For an input trajectory of 1, 1, 2, 3, 5, 8, 13, 21, . . . the output trajectory would be (1,1), (2,1), (3,2), (4,3), (5,5), (6,8), (7,13), (8,21). A trajectory may be infinite in length.

(5) For each state and input, produce a next state and next output. For example, design a Boolean system where the output is asserted whenever the input bit stream has an odd number of ones. We designed the following odd parity detector in a homework problem in Section 9.

Z1 = (SZ1, IZ1, OZ1, NZ1, RZ1), where
SZ1 = {Even, Odd}, /* The two states are named Even and Odd. */
IZ1 = {0, 1}, /* A 0 or a 1 can be received on this input port. */
OZ1 = {0, 1}, /* The output will be 0 or 1. */
NZ1 = {((Even, 0), Even), /* If the present state is Even and the input is 0, then the next state will be Even. */ ((Even, 1), Odd), ((Odd, 0), Odd), ((Odd, 1), Even)},
RZ1 = {(Even, 0), (Odd, 1)} /* If the state is Even the output is 0, if the state is Odd the output is 1. */

(6) Most of this section has focused on using qualitative descriptions, which include words, sentences, paragraphs, blueprints, pictures, and schematics.

4.1.5 Attributes of Requirements

In a database or a spreadsheet, each requirement has many fields. Each field has specific contents, such as name or identification number. These fields are called attributes. The requirement template of Table 4.1 has the attributes of text, priority, verification method, computed cost, difficulty to satisfy, link to TPM, derive Reqt. (which is a SysML construct to identify parents and children), owner, date of last change, and comments. Other fields (attributes) that might be in a database include approver, functional area, type of requirement (mandatory or trade-off), source of requirement (engineering, marketing, reliability, safety, security, etc.), assigned to whom, due by date, version, estimated effort to satisfy, allocation to hardware components, and links to risks. In Table 24.1 we listed many more criteria that could be used as attributes in a requirements database.

Satisfaction of a requirement is not a yes or no metric: it grows with the system in time. In Section 23 we presented the technical performance measure (TPM) process, which is a tool used to show how well a system is satisfying its requirements. TPMs are critical technical parameters that a project monitors to ensure that the technical objectives (requirements) of a system are going to be realized. Typically, TPMs have planned values at defined time increments, against which the actual values are plotted. Monitoring TPMs allow trend detection and correction, and help identify possible performance problems prior to incurring significant cost or schedule overruns. To this end TPMs have evaluation criteria and scoring functions.

Most requirements can have evaluation criteria and scoring functions. They show how well each requirement is being satisfied by the system at any given time. They are used to indicate status, stability, and volatility.

In Section 24, we stated that implementation difficulty and estimated cost should be used to prioritize features and requirements. In the above paragraphs, we stated that difficulty in satisfying the requirement and estimated effort to satisfy the requirement were common attributes in a requirements database. We now suggest that those criteria should be combined into one attribute. The cost of satisfying a requirement should be calculated for the high-priority requirements and this would become an attribute. These calculated costs can be compared to determine the cost drivers of the system. These cost drivers can be negotiated with the customer to reduce the overall cost of the system. Furthermore, if the budget permits, the parameters in the scoring functions can be adjusted to perform a sensitivity analysis.

4.1.6 A Requirement Template

Table 4.1 A requirement template

Attribute	Explanation
Identification tag (ID)	The ID tag should reflect the tree structure of the requirements database. It is usually numeric
Name	This should be short, but not be an abbreviation
Text	This must be worded carefully paying strict attention to all the characteristics of good requirements explained in Section 27
Priority	This gives the importance of the requirement with respect to all other requirements in the database. It is often a number between 0 and 10
Verification method	Explain how this requirement will be verified during each phase of the system life cycle. Typical verification methods include inspection, analysis, test, and demonstration
Computed cost	Computing the cost to satisfy a requirement will cost time and money, so it should only be computed for high-priority requirements
Satisfaction difficulty	Estimate how difficult it will be to satisfy this requirement
Link to TPM?	If it is linked to a technical performance measure, specify the linkage
Derive Reqt:	Indicate the parent requirement and possibly children
Owner	A requirement's owner communicates with the customer, defines and tracks trade-offs of the requirement with other requirements, and approves changes in the requirement
Date of last change	This is needed for configuration control
Comment	Comments may explain any of the template fields

4.1.7 Example Requirements

Here are two requirements from the Cookie Acquisition System of Section 1.3.1.

ReqF5: Cookie shall emit an audible signal when the timer has elapsed.
ReqNF4: The audible signal emitted by Cookie shall have an intensity level of 80 ± 2 decibels at 30 cm and a frequency of 440 Hz.

And here are elaborations of three Cookie requirements. ReqNF4 (Table 4.2), ReqNF7 (Table 4.3) and ReqNF8 (Table 4.4).

Table 4.2 An elaboration of ReqNF4

Attribute	Explanation
Identification tag (ID)	ReqNF4
Name	Audible signal for cookies is ready
Text	An audible signal shall inform the student when the cookies are ready. This audible signal shall have an intensity level of 80 ± 2 decibels and a frequency of 440 Hz
Priority	9
Verification method	During design and construction, this requirement will be verified by test. The proposed device will be mounted on a test bench and will be activated in accordance with test instructions. Sound power level in decibels (dB) will be measured at a distance of 30 cm. At final test and during operation, this requirement will be verified by demonstration. An actual oven will activate the audible signal and the sound power level in decibels (dB) will be measured at a distance of 30 cm
Computed cost	The Mallory PK-16N04WA-03Q electromechanical buzzer satisfies these requirements and it costs \$8.39. However, if a higher frequency, such as 3 kHz, is acceptable, then a lower cost piezoelectric buzzer could be used, such as the Mallory PK-21N31PQ at \$2.57
Difficulty	Easy
Link to TPM?	No
DeriveReqt:	This requirement refines *ReqF5: Cookie shall emit an audible signal when the timer has elapsed*
Owner	Engineer
Date of last change	November 26, 2015

Table 4.3 Lost study time requirement from the Cookie Acquisition System

Attribute	Explanation
Identification tag (ID)	ReqNF7
Name	Lost study time
Text	The student shall lose no more than 30 min of study time making the cookies or traveling to the bakery to buy them
Priority	7
Verification method	During design, this requirement will be verified by analysis. At final test and during operation, this requirement will be verified by demonstration
Difficulty	Easy
Link to TPM?	No
DeriveReqt:	None
Owner	Hungry student
Date of last change	November 26, 2015

Table 4.4 Nutrition requirement from the Cookie Acquisition System

Attribute	Explanation
Identification tag (Id)	ReqNF8
Name	Nutrition
Text	If the student eats four cookies, he or she shall consume no more than 520 cal, 24 g of fat, and 72 g of carbohydrates
Priority	5
Verification method	This requirement will be verified by analysis using data from the Internet, for example http://www.pillsbury.com/products/cookies/refrigerated-cookies/chocolate-chip
Difficulty	Controlling weight is difficult
Link to TPM?	no
DeriveReqt:	None
Owner	Hungry student
Date of last change	November 26, 2015

4.1.8 *Homework*

4.1.8.1 Baseball Statistics

What requirements did we have for constructing our baseball statistics spreadsheet? What are the requirements for evaluation criteria of that spreadsheet?

4.1.8.2 Parking Garage Entrance Gate

Decompose the requirements for the motion of the entrance gate arm for the parking garage system of Chap. 1. Then write them in the form of our requirements template of Table 4.1.

4.1.8.3 Parking Garage Interrogation Time Delay

Write requirements for the interrogation time delay of the entrance gate for the parking garage system of Chap. 1. That is, write NFPR1 of Section 11 in the form of our requirements template, Table 4.1.

4.1.8.4 Parking Garage Full Sign

Write requirements for the full sign of the entrance gate for the parking garage system of Chap. 1. That is, write NFPR3 of Section 12 in the form of our requirements template, Table 4.1.

4.1.8.5 Requirements for an 18-Wheel Truck Cab

These are the *actual* requirements for a truck cab (tractor of an 18-wheeler). What is wrong with each one?

3.2.1.2.2.2. Coder decoder assembly (CDA) interface card built in test (BiT)

The CDA interface card consists of software-enabled loopbacks for the CDA interface.

3.3.4.1. Random Vibration

The fully assembled computer shall be subjected to a random vibration test. This testing shall be performed in accordance with the environmental stress screening test plan, CCS-TRPA012.

3.1.2.1.1. Roads

The tractor shall operate on interstate highways, primary highways, secondary paved roads, and unpaved highways.

3.2.1.3. Turn Radius

The turn radius of the vehicle shall be approximately 41 feet.

3.2.2.4. Wheelbase

The wheelbase will be 220 in.

3.2.5.2. Altitude

The altitude shall be from sea level to 10,000 feet with a power derating of 3 % per 1000 feet of elevation above 7500 feet.

3.3.2.1. Electromagnetic Interference (EMI)

The tractor shall continue to operate normally in the presence of EMI disturbances.

4.1.8.6 Famous Failures

The **Lewis Spacecraft** was an Earth-orbiting satellite that was supposed to measure changes in the Earth's land surfaces. It lasted 3 days in orbit. "The loss of the Lewis Spacecraft was the direct result of an implementation of a technically flawed Safe Mode in the Attitude Control System. This error was made fatal to the spacecraft by the reliance on that unproven Safe Mode by the on orbit operations team and by the failure to adequately monitor spacecraft health and safety during the critical initial mission phase" [40].

Did they do requirements development, verification, and validation correctly? Justify your answers with a few sentences.

RMS Titanic had poor quality control in the manufacture of the wrought iron rivets. In the cold water of April 14, 1912, when the Titanic hit the iceberg many rivets failed and whole sheets of the hull became unattached. Would you say that on the Titanic validation, verification, or requirements development were correct or incorrect? Justify your answers.

4.1.8.7 Omega X-33 Watch

Write requirements for the Omega X-33 watch described in the following paragraph.

"If you ever happen to be traveling through space and need to accurately time a 3-min egg, Omega the celebrated Swiss watchmaker has just the timepiece for you. Dubbed the Speedmaster Professional X-33 Mars Watch, it is the first electronic watch to be qualified by NASA for use in spacecraft. All previous space-qualified watches were Omega Speedmaster Professional Moon Watches, which have mechanical movements. (Those older timepieces are still the only ones certified for extravehicular activities.) The X-33's claim to fame is its ability to perform multiple functions. For example, in addition to luminous hands and a conventional analog face, it has a 4-mm liquid-crystal display for displaying Mission Elapsed Time and Coordinated Universal Time, formerly called Greenwich Mean Time. Other functions include day and date display, a countdown timer, a chronograph with split times, and a particularly loud (over 80 dB) alarm, which was developed in response to astronauts' requests. The alarm may be used in conjunction with the Mission Time, Universal Time and countdown timer features. The rugged and reliable X-33 has a titanium case and is rated for use over a temperature range of −20 to +70 ° C. It also withstands air pressure to 300 kPa, immersion to 30 m of water, shocks of up to 3500 g_n, and high levels of humidity and oxygen atmosphere. Under all those conditions, it will be accurate to within −0.3, +0.5 s per day and will maintain integrity of all functions. Easy to use, the Mars Watch has an 8-lx backlight for its digital display, with the option of turning that display off. Another option concerns the date: it may be displayed in conventional (month-day-year) or international (day-month-year) formats. The 30-mm diameter, 13.5 mm-thick watch has a quartz movement powered by a lithium battery that will last 2 years under normal conditions and 3 years if the alarm and backlight are not used. Battery drain is kept under control by stopping the analog hands when the backlight is on, the hands then catch up. Two versions of the X-33 are offered, differing only in their fastening option. The first has a titanium bracelet and sells for $2995, the second comes with a synthetic strap, made mostly out of Kevlar (in red or black), and carries a $2595 price tag. Space gourmets, please note: cooking times for eggs will vary, depending on pressurization of your spacecraft (due to its effect on the boiling point of water)."

4.1.8.8 The Automobile

Assume that your team has been asked to provide advice to someone wanting to buy a new automobile. Define the **boundary** around your automobile system.

List the **interfaces** between your automobile system and its external environment.

Write at least two **requirements** for your automobile. Your requirements will obviously depend on the user and the boundary that you have chosen. Make sure that you include **evaluation criteria.**

Risk analyses are performed during preliminary design to identify problems that could develop because of system failures. This type of analysis is generally oriented to equipment only and does not generally cover the effects of human actions on equipment. The objective is to determine the ways in which equipment can fail and the effects of such failures. System components could fail in ways that are minor, major, critical, or catastrophic failures.

Minor failure: A failure that does not degrade the performance and effectiveness of the system beyond acceptable limits.

Major failure: A failure that degrades the system performance and effectiveness beyond acceptable limits, but can be controlled.

Critical failure: A failure that degrades the system beyond acceptable limits and could create a safety hazard if corrective action is not taken.

Catastrophic failure: A failure that produces system damage that might preclude mission accomplishment or could cause death and personnel injuries.

Pick four system components that could fail in an automobile, describe a particular type of failure, explain the effect of each failure, and specify whether or not they would be minor, major, critical, or catastrophic failures. Fill out a table with headings of Component, Failure Mode, Effect, and Criticality. This is often called a Requirements Traceability Verification Matrix (RTVM).

This **verification** matrix has several of the requirements that you might have derived in previous problems. Each of them must be verified. Select a verification method and provide a short reason why you chose that method. Your answers may be in the form of a word, phrase, sentence, or paragraph, whatever is needed to answer the questions. The column headed "Compliance Measure" names the source of the verification data.

Requirements description	Compliance measure	Verification method				
		Inspection	Analysis	Test	Demonstration	N/A
The automobile shall provide seats for people	Observation					
The automobile shall accommodate people with heights from 5′ to 6′ 8″	Observation					
The automobile shall contain a radio	Window sticker					

(continued)

Requirements description	Compliance measure	Verification method				
		Inspection	Analysis	Test	Demonstration	N/A
The automobile shall pass all federal and state emission standards	Smog emissions label					
The automobile shall attain a highway gasoline efficiency of at least 25 mpg	Gas receipts and odometer readings					
The automobile shall have a 5-year, 50,000-mile warranty	Warranty information booklet					
The automobile shall have a switch for the passenger-side air bag	Owner's manual					
The automobile shall sustain at least 0.9 g of lateral acceleration without substantial sway, drift, or slip	*Road and Track* magazine					
The engine shall produce at least 200 horsepower	*Car and Driver* magazine					
The engine shall produce a peak torque of at least 150 ft lb	*Car and Driver* magazine					
The automobile shall accelerate from 0 to 60 mph in less than 6.0 s	Personal test results					
The automobile shall brake from 60 to 0 mph in less than 130 ft	Personal test results					

4.2 Section 28. Requirements Development and Management

In Fig. 4.6, the size of each icon is proportional to its cost. The length of each object's shadow is proportional to its influence. Discovering requirements costs much less than manufacturing. However, discovering requirements has a big influence on the system: whereas, by the time you get to manufacturing, you can change little. Here is an explanation for each of the icons. Define requirements is a stack of documents. Concept exploration shows a pogo stick. If this satisfies your requirements for a municipal transportation system, then you probably have the requirements wrong. Full-scale design is an engineer. Manufacturing is a big factory. Operations and maintenance is a bus. Retirement and replacement is a futuristic bus.

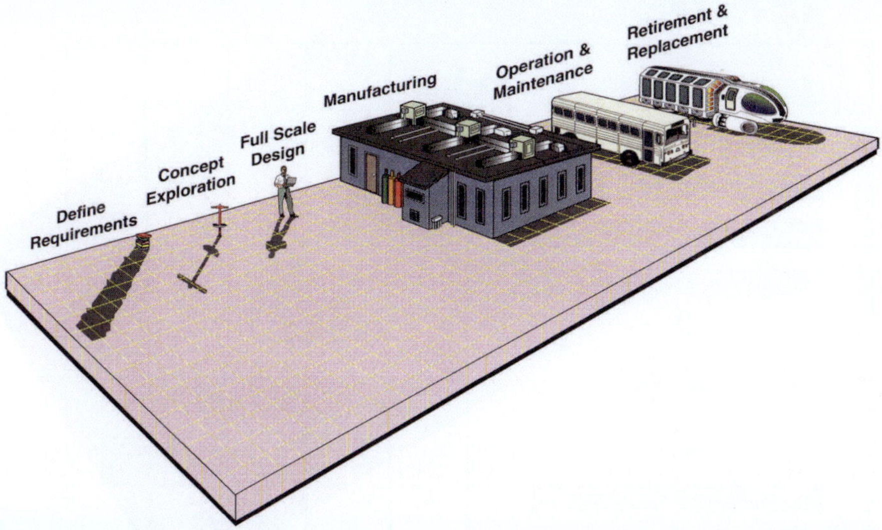

Fig. 4.6 Cost and influence of each phase of the life cycle for a municipal transportation system

4.2.1 Requirements Process

The requirements process shown in Fig. 4.7 should be interpreted figuratively not literally: because it contains only major flows and only two feedback paths. It shows 16 functions, so there should be 16! feedback paths. For example, use cases cannot be written without information about the stakeholders, customer needs, and a problem statement: verification and validation cannot be done without knowledge of the requirements.

The first steps in discovering system requirements are identifying the stakeholders, understanding the customers' needs, stating and scoping the problem, and identifying the needed capabilities of the system.

4.2.1.1 Identify Stakeholders

The first step in developing requirements is to identify the stakeholders. The term *stakeholder* includes anyone who has a right to impose requirements on the system. This includes end users, operators, maintainers, bill payers, owners, regulatory agencies, victims, sponsors, etc. All facets of the stakeholders must be kept in mind during system design. For example, in evaluating the cost of a system, the total life cycle cost and the cost to society should be considered. Frequently, the end user does not fund the cost of development. This often leads to products that are expensive to own, operate, and maintain over the entire life of the product, because the organization funding development saves a few dollars in the development

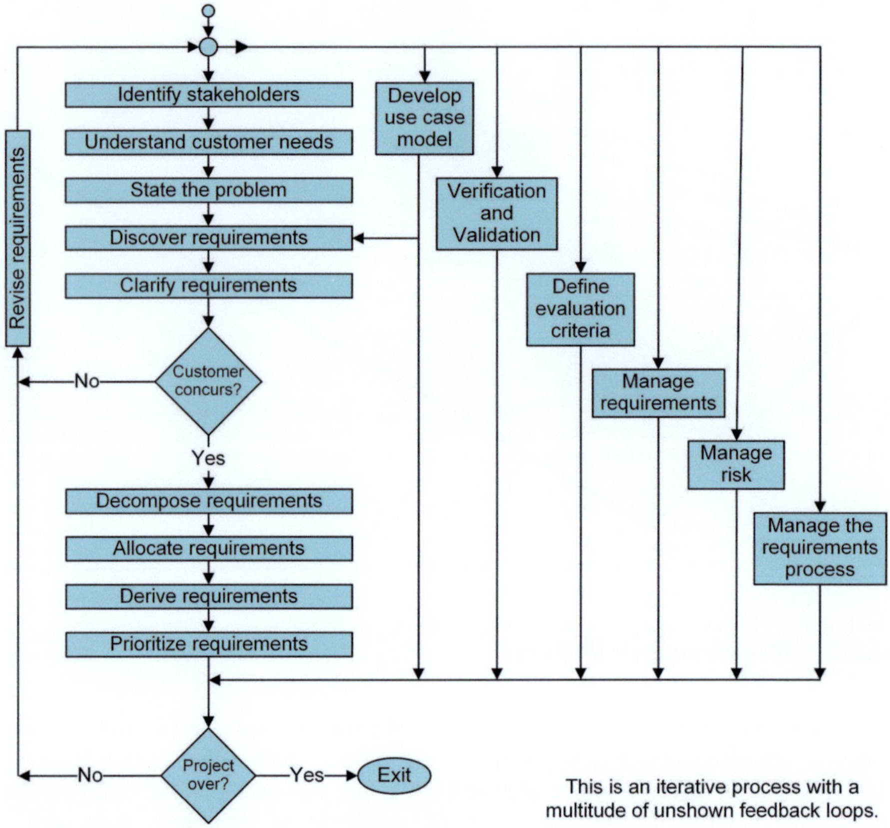

Fig. 4.7 The requirements discovery process

process. Therefore, it is imperative that the design engineer understands and communicates this conflict to the stakeholders. The sponsor and user can then help trade off the development costs against the cost to use and maintain. Total life cycle costs are significantly larger than initial costs. For example, in one of their advertisements, Compaq proclaimed, "80 % of the lifetime cost of your company's desktops comes after you purchase them." In terms of the personal computer, if total life cycle costs were $5000, purchase cost would have been $1000 and maintenance and operation $4000.

For large or complex systems, it is imperative that the design engineer understands the role of each of the major stakeholders and their motivations, interests, constraints, and outside pressures.

4.2.1.2 Understand Customer Needs

System design begins with a complete understanding of customers' needs. If the sponsor, owner, and user are different individuals/organizations, then the design engineer must know and understand the needs of each. For a rental car agency the sponsor could be corporate management, the owner of the cars could be the local airport office, and the user could be the customer renting the car. Each of these three has needs, some in common, some unique. The information necessary to begin a design usually comes from the mission statement, concept of operation, business model, preliminary studies, and specific customer requests. Usually the customer is not aware of what is needed. Design engineers must enter the customer's environment, discover the details, and explain them. Flexible designs and rapid prototyping facilitate identification of details that might otherwise be overlooked. Talking to the customer's customer and the supplier's supplier can also be quite useful. This activity is frequently referred to as mission analysis.

It is the systems engineer's responsibility to ensure that all information concerning the customer's needs is collected. The systems engineer must also ensure that the definitions and terms used have the same meaning for everyone involved. Several direct interviews with the customer are necessary to ensure that all of the customer's needs are stated and that they are clear and understandable. The customer might not understand the needs; he or she may be responding to someone else's requirements. Often, a customer will misstate his or her needs; for example, a person might walk into a hardware store and say he or she needs a half-inch drill bit. But what he actually needs is a half-inch *hole* in a metal plate, and a chassis-punch might be more suitable.

During the first half of the twentieth century, the Indian and Harley Davidson motorcycle manufacturers were fierce competitors. Then during WW II, the US Army asked for a 500 cm^3 motorcycle. Indian made one for them. Harley Davidson thought that the Army would not be satisfied with a 500 cm^3 motorcycle, so they built a 750 cm^3 motorcycle. The Indian motorcycle disappeared for 60 years. Until it was reintroduced as a nostalgic retro item.

4.2.1.3 State the Problem

What is the problem we are trying to solve? Answering this question is one of the design engineer's most important and often overlooked tasks. An elegant solution to the wrong problem has no value.

Early in the process, the customer frequently fails to recognize the scope or magnitude of the problem that is to be solved. The problem should not be described in terms of a preconceived solution. It is imperative that the design engineer help the customer develop a problem statement that is completely independent of solutions and specific technologies. Solutions and technologies are, of course, important; however, there is a proper place for them later in the systems engineering process. It is the design engineer's responsibility to work with the customer, asking the questions necessary to develop a complete picture of the problem and its scope. The US Air Force did not know that they wanted a stealth airplane until after the engineers showed that they could do it. During concept exploration, encourage consideration of unconventional alternatives. This will help you understand the requirements better. Likewise, studying models and computer simulations will help you understand the requirements. Understanding the requirements is one of the most fruitful phases in requirements discovery.

Defining what the system is supposed to do could be done in terms of a hierarchy of functions. But recently it is being stated in terms of a hierarchy of capabilities the system must have.

4.2.1.4 Develop Use Case Models

Modern system design is usually based on use cases. Use cases capture the required functional behavior of the system. A use case is an abstraction of a required behavior of a system. A use case produces an observable result of value to the user. A typical system will have several use cases, each associated with a goal of a particular user. Each use case describes a sequence of interactions between one or more users and the system. A use case narrative is written in a natural language. It is written as a sequence of steps with a main success scenario and alternate flows. This sequence of interactions is contained in a use case report and the relationships between the system and its users is portrayed in use case diagrams. Development of requirements from the use case models is explained in detail in the last section of this section.

4.2.1.5 Discover Requirements

The systems engineer must consult with the customer to discover the system requirements. The systems engineer must involve the customer in the process of defining, clarifying, and prioritizing requirements. It is prudent to involve users, bill payers, regulators, manufacturers, maintainers, and other key stakeholders in the process. Requirements are either discovered or elicited. They are not handed to you. It takes work to discover requirements.

Next, engineering must discover the functions that the system must perform to satisfy its purpose. The system functions form the basis for partitioning the system into subsystems. QFD is useful for identifying system functions [41, 42].

From the foregoing it might appear as if requirements are transformed into functions with a waterfall process. This is not the case. The requirements process is highly iterative with multiple tasks being done in parallel. First, we look at system requirements, and then at system functions. Then we reexamine the requirements and then reexamine the functions. Then we reassess the requirements and again the functions, and so on. Identifying the system's functions provides insights into the system's behavioral requirements.

For each requirement that is discovered, ask yourself why is that requirement needed. This will help you write the rationale for the requirement, prioritize the requirements, and possibly negotiate with the customer to eliminate some requirements.

4.2.1.6 Clarify Requirements

The systems engineer needs to consult with the customer to ensure that the requirements are correct and complete, and then identify trade-off requirements. As with all engineering processes, this process is iterative. The customer should be satisfied that if the requirements are met, then the system will do what it needs to do. This should be done in formal reviews with the results documented and distributed to appropriate parties. These reviews ensure that all the requirements have been met, ensure that the system satisfies customer needs, assess the maturity of the development effort, allow recommending start of the next phase, and facilitate approval to committing additional resources. The systems engineer is responsible for initiating and conducting these reviews.

The system requirements must be reviewed with the customer several times. At a minimum, requirements should be reviewed at the end of the modeling phase, after testing the prototypes, before the start of production, and after testing production units.

It is the job of the systems engineer to push back on the customer to get requirements removed or relaxed. The systems engineer performs a sensitivity analysis on the requirements set to determine the cost drivers. Then the systems engineer shows the customer how cost can be reduced and performance enhanced if certain requirements are changed. The requirements process is a continual interactive dialog with stakeholders.

The main objectives of these reviews are to uncover missing requirements, eliminate unneeded requirements, ensure that requirements have been met, and verify that the system satisfies customer needs. At these reviews, trade-offs are usually made between performance, schedule, and cost. Additional objectives include recommending whether or not to proceed to the next phase of the project and commit additional resources. The results and conclusions of the reviews should be documented. Again, the systems engineer is responsible for initiating and conducting these reviews.

The following definitions of the most common reviews might be useful [43, 44]. No company uses these exact names and reviews. But most companies use similar names and reviews. They are arranged in chronological order. Although these definitions are written with a singular noun, they are often implemented with a collection of reviews. Each system, subsystem, and subsubsystem is reviewed and the totality of these constitutes the indicated review.

Mission Concept Review (MCR) (a.k.a. the Mission Definition Review and the Alternate System Review) is the first formal review. It examines the mission objectives, candidate architecture, top-level functional requirements, measures of effectiveness, interfaces, alternative concepts, and anticipated risks.

System Requirements Review (SRR) shows that the product development team understands the mission and the system requirements. It confirms that the system requirements are sufficient to meet mission objectives. It ensures that the performance and cost evaluation criteria are realistic, and that the verification plan is adequate. It inspects the requirements flow down plan. It checks that the selected requirements tracking and management tool is appropriate. It assesses the reasonableness of the risk analysis. At the end of SRR, requirements are placed under configuration control. Changing requirements after SRR will adversely affect schedule and cost.

In the beginning of a project, requirements can be kept in a word document. However, eventually it will be more convenient to move them to a database. In the proposal phase and up through the system requirements review the requirements might be kept in a spreadsheet. As the number of requirements exceeds 100, most programs will transfer the requirements to a requirements-specific database.

System Functional Review examines the proposed system architecture and design, and the allocation of functions to the major subsystems. It reviews information sharing, interface definitions, and interface limitations such as security blocks. It also ensures that the verification and risk mitigation plans are complete.

Preliminary Design Review (PDR) demonstrates that the preliminary design meets all the system requirements with acceptable risk. The design model is presented. System development and verification tools are identified, and the Work Breakdown Structure is examined. Full-scale engineering design begins after this review.

Critical Design Review (CDR) verifies that the design meets the requirements. The CDR examines the system design in full detail, inspects interface control documents, ensures that technical problems and design anomalies have been resolved, checks the technical performance measures, and ensures that the design maturity justifies the decision to commence manufacturing and coding. Few requirements should be changed after this review.

Test Readiness Review (TRR) is conducted for each hardware and software configuration item prior to formal test. It determines the readiness and completeness of test procedures and their compliance with test plans and descriptions. It inspects test equipment and facilities. It analyzes performance predictions.

Production Readiness Review (PRR)**.** For some systems, there is a long phase when prototypes are built and tested. At the end of this phase, and before production begins, there is a production readiness review. The design engineer is responsible for designing the product and also the process for producing it. This review primarily inspects the requirements for the production system.

Total System Test (TST). At the end of manufacturing and integration, the system is tested to verify that it satisfies its requirements. Technical performance measures are compared to their goals. The results of these tests are presented at the System Acceptance and Operational Readiness Reviews.

Functional Configuration Audit (FCA) verifies that the configuration item meets all the functional requirements. The FCA is a review of the item's performance, to ensure that it not only meets the specification but also that there are no unintended functional characteristics.

Physical Configuration Audit (PCA) compares new item against design documentation to ensure that the physical characteristics and interfaces conform to the product specification.

Figure 4.8 shows the timing of some of these reviews within the total system life cycle [45, 46]. The amount of black ink associated with each activity indicates the amount of time, effort, and money that are consumed.

At various points in the system life cycle and for various reasons, the following reviews are often conducted: startup review, software specification review, test readiness review, peer reviews, expert reviews, and integration readiness review.

At these reviews, it is important to ask why each requirement is needed. This can help eliminate unneeded requirements. It can also help reveal requirements behind the stated requirements. It may be easier to satisfy these requirements, than the stated requirements themselves.

Figure 4.8 is another life cycle model like Figs. 1.13 and 1.14.

4.2.1.7 Decompose Requirements

Requirements decomposition breaks down a requirement into two or more detailed requirements whose joint scope is equivalent to the scope of the original one—just

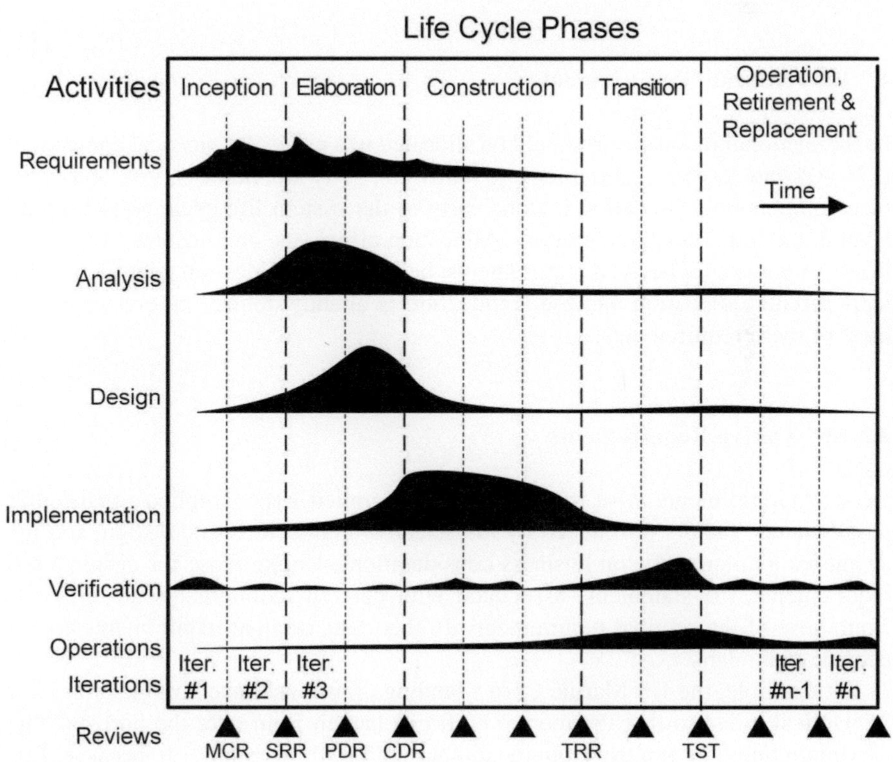

Fig. 4.8 Timing of the major reviews. The amount of black ink on the ordinate axis associated with each activity indicates the amount of time, effort, and money that are consumed. The abscissa is time

expressed with greater specificity or detail. The resulting requirement replaces the original decomposed requirement, which no longer needs to be verified. When a starting written requirement is compound, it needs to be decomposed. Here is an example of requirement decomposition. The customer requirement

When an empty coffee pot is placed in the coffee machine, the camera system shall transmit a digital image to the server.

can be decomposed into the following functional requirements:

1. *The system shall sense when an empty coffee pot is being placed in the coffee machine.*
2. *The system shall trigger the camera when an empty coffee pot is placed in the coffee machine.*
3. *The system shall transmit digital images to the server.*
Stipulation: empty coffee pot means one containing less than 6 fluid ounces.

Then the original requirement can be replaced.

If the requirements are organized in a Zachman Framework [20], then any requirement that spans two or more columns is a candidate for decomposition.

4.2.1.8 Allocate Requirements

Each functional requirement should be allocated to a particular physical component [17]. Allocate system requirements to hardware, software, humans, test, and interface components. Allocation is done early in the system life cycle and at a high level in the requirements hierarchy. Allocation affects the architecture. Allocation is less important for derived requirements, because their allocation goes along with their parents' allocation and the architecture is already defined before we derive most of these requirements.

4.2.1.9 Derive Requirements

Derived requirements arise from constraints, implied issues implied in customer requirements, factors introduced by the selected architecture, and design, and the developer's unique take on business considerations. Unlike those for decomposed requirements, the statements associated with derived requirements are different from those of the original requirements. In this case, original requirements do not become redundant.

For example, the US Marine Corp's landing craft must have a range of at least 25 nautical miles so that the mother craft can launch from over the horizon. The maximum duration of a trip from the mother craft to the beach is 1 h, because trips of over an hour will tend to make most people too seasick to function effectively. Therefore a derived requirement is *The landing craft shall have a minimum speed of 25 knots.*

Design engineers often derive requirements from customer requirements. A more elaborate example for deriving requirements follows. A teenage boy might express the *customer needs statement* this way: *Hey, Dad, We need speakers in the car that will make your insides rumble during drum solos.* The father would translate this into . . . *The performance requirement: For bass frequencies, we need 110 dB of sound output.* Then, the systems engineer would convert this into . . . *The functional requirement: Amplify the radio 's output to produce 115 W in the frequency range 20–500 Hz.* Finally, after a trip to the audio shop, the design engineer would transform this into . . . *The design requirement: Use Zapco Z100S1VX power amplifiers with JL Audio 12 W1-8 speakers.* While this might appear to be a waterfall process at first blush, the requirements process is actually concurrent and iterative.

4.2.1.10 Define Evaluation Criteria

Evaluation criteria are developed and used to assess different designs. Each evaluation criterion must have a fully described unit of measurement. For example, units of power could be horsepower, while units of cost could be dollars. Suppose the evaluation criterion is *acceleration*, then the unit of measurement could be the seconds it takes to accelerate from 0 to 60 mph. There is flexibility in choosing the units of measurement, as long as they are appropriate for the criteria, are fully described, and are used consistently for all designs. The value of an evaluation criterion indicates how effectively a trade-off requirement has been met. For example, the car accelerated from 0 to 60 in 6.2 s. These values are supplied to the scoring functions, which produces the requirements scores, which, in turn, are used to perform trade-off studies [24]. Such measurements are made throughout system development.

4.2.1.11 Manage Requirements

The purpose of requirements management is to (1) track and control requirements, requirement changes, and reallocations; (2) establish system requirements traceability to lower level design requirements and verifications; (3) establish traceability and capture allocations to hardware, software, interfaces, and humans; and (4) detect, identify, quantify, and direct attention to scope or requirements creep.

Requirements have to be under configuration control. Commercial requirements tools help with this function. When a defect report is written and a change in requirements is suggested as a solution, the suggested change must be submitted to a Change Control Board (CCB). If the CCB approves of the change, then requirements are updated at the next baseline change.

4.2.1.12 Verify and Validate Requirements

Verification and validation are not synonyms! Even so, these terms are often confused. Therefore, it is worth examining the definitions of these terms. Dictionary definitions of these terms are:

> verify vt. (1) to prove to be true by demonstration, evidence or testimony; confirm or substantiate (2) to test or check the accuracy or correctness of, as by investigation, comparison with a standard, or reference to the facts
>
> validate vt. to prove to be valid
>
> valid adj. (1) having legal force; properly executed and binding under law (2) well-grounded on principles or evidence; able to withstand criticism or objection, as an argument; sound (3) effective, effectual, cogent, etc. (4) Logic correctly derived or inferred according to the rules of logic. Valid applies to that which cannot be objected to because it conforms to law, logic, the facts, etc.

Verify means to prove that something is true by testing. Validate means to prove that it conforms to law, logic, and the facts, that it is sound, effective, and cogent.

Validating a system: Building the *right system*: making sure that the system does what it is supposed to do. Validation determines the correctness and completeness of the product, and ensures that the system will satisfy the actual needs of the customer.

Verifying a system: Building the *system right*: ensuring that the system complies with its requirements and conforms to its design.

Validating requirements: Ensuring that the *set* of requirements is correct, complete, and consistent, that a model can be created that satisfies the requirements, that a real-world solution can be built that satisfies the requirements, and that it can be verified that such a system satisfies its requirements. If engineering discovers that the customer has requested a perpetual-motion machine, the project should be stopped.

Verifying requirements: Proving that each requirement has been satisfied. Verification can be done by logical argument, inspection, modeling, simulation, analysis, test, or demonstration.

Verification and validation: MIL-STD-1521B (and most systems engineers) and DoD-STD-2167A (and most software engineers) use the words verification and validation in almost the opposite fashion [47]. To add further confusion, ISO-9000 tells you to verify that a design meets the requirements and validate that the product meets requirements. NASA has a different take. According to NASA, verification consists of proving that a system (or a subsystem) complies with its requirements, whereas validation consists of proving that the total system accomplishes its purpose [44]. In the Capability Maturity Model Integration (CMMI), validation (VAL) is system validation, verification (VER) is requirements verification, and requirements development (RD) Specific Goal 3 covers requirements validation. In light of the foregoing, it is necessary that the project team and customer representatives all agree on the definitions of verification and validation from the very beginning of the system development project.

System Validation and Verification

System validation artifacts can be collected from white papers, trade-off studies, phase reviews, life cycle reviews, red team reviews, SRR, PDR, CDR, and field test.

System validation artifacts, which can be collected continuously throughout the life cycle, include results of modeling and simulation and a count of the number of operational scenarios (use cases) modeled. Detectable system validation defects include mismatches between the model/simulation and the real system.

Sensitivity analysis can reveal validation errors. If a system exhibits extreme sensitivity to parameters over which the customer has no control, then it may be the wrong system for that customer. If sensitivity analysis reveals that the most important parameter is a surprise, then it may be the wrong system. If a system is more sensitive to its parameters than to its inputs, then it may be the wrong system or the wrong operating point. If the sensitivities of the model are different from the sensitivities of the physical system, then it may be the wrong model.

Validation defects can be detected at inspection events. To this end, the tester role should be assigned the additional responsibility of collecting validation artifacts. In this regard, the tester should read the vision statement and the concept of operation (ConOps) and specifically look for system validation problems.

Systems are primarily verified at total system test. Sensitivity analyses can also be used to verify systems. In a human-designed system or a simulation, unexpected excessive sensitivity to any parameter is a verification mistake. Sensitivity to interactions should definitely be flagged and studied because such interactions may be unexpected and undesirable.

Requirements Validation and Verification

Validating requirements means ensuring that the *set* of requirements is correct, complete, traceable, and consistent, a model that satisfies the requirements can be created, and a real-world solution can be built and tested to prove that it satisfies the requirements. If the requirements specify a perpetual-motion machine, the project should be stopped.

Here is an example of an invalid requirements set for an electric water heater controller where Temp is the temperature of the water in the tank.

If 70° < Temp < 100°, then output 3000 Watts
If 100° < Temp < 130°, then output 2000 Watts
If 120° < Temp < 150°, then output 1000 Watts
If 150° < Temp, then output 0 Watts

This set of requirements is incomplete; what should happen if Temp < 70°? They are inconsistent; what should happen if Temp = 125°? They are ambiguous, because units are not provided—are those temperatures in degrees Fahrenheit or Celsius?

Detectable requirements validation defects include incomplete or inconsistent sets of requirements or use cases; anything that does not trace to top-level customer requirements (vision); and test cases that do not trace to use case scenarios.

Validation defects can be detected at inspection time. In this regard, the role of tester should be assigned the additional responsibility of conducting for requirements validation. The tester should read the vision statement and concept of operations and specifically look for requirements validation problems, equipped with that knowledge.

Each requirement is verifiable (ordered by increasing cost) by logical argument, inspection, modeling, simulation, analysis, test, or demonstration. Dictionary definitions for these terms are.

Logical argument: a series of logical deductions
Inspection: to examine carefully and critically, especially for flaws
Modeling: a simplified representation of some aspect of a system
Simulation: execution of a model, usually with a computer program
Analysis: a series of logical deductions using mathematics and models
Test: applying inputs and measuring outputs under controlled conditions (laboratory environment)
Demonstration: to show by experiment or practical application (flight or road test). Some sources say that demonstration is less quantitative than test. Demonstrations can be performed on electronic breadboards, plastic models, stereolithographic models, prototypes made in the laboratory by technicians, preproduction hardware made in the plant using developmental tooling and processes, and production hardware using full plant tooling and production processes.

Examples of requirements verification are the following: (1) *The probability of receiving an incorrect bit on the telecommunications channel shall be less than* 10^{-3}. This requirement can be verified by laboratory tests or through demonstration on a real system. (2) *The probability of loss of life on a manned mission to Mars shall be less than* 10^{-3}. This certainly is a reasonable requirement, but it cannot be verified through test. It might be possible to verify this requirement with analysis and simulation. (3) *The probability of the system being canceled by politicians shall be less than* 10^{-3}. Although this may be a good requirement, it cannot be verified with normal engineering test or analysis. It might be possible to verify this requirement with logical arguments.

And then, of course, there was the *Hammer of Verification* that was used to verify that a pope was indeed dead, and not just sleeping! In this case, it was Cardinal Chamberlain who whacked the corpse on the forehead with a hammer to perform verification. Thankfully, this verification technique was last used on Pope Leo XIII in 1903.

There is a trade-off between the cost of verifying a requirement and the value that such verification provides. Good enough is often a forgotten concept in our rush to Six Sigma. And then there are ample Six Sigma. Creative people can always see another feature or a better way. There is a time to send the creativity to the next project and get on with the implementation of the current project. We have seen examples of enormous amounts of money being spent on developing complex computer models for verification that were capable of predicting system performance criteria to several significant digit, when a reasonable prototype and simple measurements could have satisfactorily demonstrated that the performance requirement was exceeded by plus 25 % or minus 15 %. It is usually the case that 10 %. Usually we only need to demonstrate whether a requirement was met or not. We do not have to determine the exact level of accuracy of meeting the requirement.

Traditional requirements verification is conducted by the test organization during the testing phase. However, this approach is costly and ineffective for dealing with the scale and complexity of today's systems. Built-in self-test and automated regression testing are far superior alternatives.

4.2.1.13 Manage Risk

Identifying and mitigating project risk is the responsibility of management at all levels in an organization [48, 49]. Each item that poses a potential risk to project cost, schedule, or performance must be identified and tracked. The following information should be recorded for each identified risk: name, description, type, origin, probability, severity, impact, identification number, identification date, work breakdown structure element number, risk mitigation plan, responsible team, needed resolution date, closure criteria, principal engineer, status, date, and signature of team leader. For the solar oven project, the risks are discussed in Section 23; we identified the following:

(1) Insufficient internal oven temperature was a performance risk. Its origin was design and manufacturing. It had high probability and high severity. We mitigated it by making it a technical performance measure, as shown in Section 23.
(2) High cost of the oven was a cost risk. Its origin was the design process. Its probability was low, and its severity was medium. We mitigated it by computing the cost for every design.
(3) Failure to have the oven ready for testing posed a schedule risk. Its origin was design and manufacturing. Its likelihood was low, but its severity was extremely high. We mitigated this risk by requiring final designs 7 days before the scheduled test date and a preproduction unit 3 days in advance.

Models (or computer simulations) are often employed to mitigate risk. Low-risk aspects of the system should be modeled at a high level of abstraction, while high-risk aspects require higher resolution models.

4.2.1.14 Manage the Requirements Process

The requirements process (Fig. 4.7) should be tailored for each project. Tailoring accommodates special schedule and cost constraints of the project, and incorporates customer-specific needs. The impact of all changes should be studied to determine whether or not they result in improvements to the company's standard requirements process. This task of continually monitoring a process, to make beneficial changes to improve the process, is arguably the most important task of any process. However, this task is often forgotten, once the process is designed.

4.2.2 Fitting the Requirements Process into the Systems Engineering Process

The requirements discovery process (Fig. 4.7) is a subprocess of the systems design process (Fig. 4.9).

Remember that system design is an iterative process. Figure 4.10 shows a spiral life cycle model that suggests four of the iterations. It starts in the center with the customer's need. Then we define requirements, design a model, create the model, simulate the model, and validate the model. Finally, we review all of that work. Then we start a new iteration. We refine the requirements, design a prototype, build the prototype, and test the prototype. Finally, we review all of that work. Then we start the next iteration. In every iteration, the spiral gets thicker as we add more material. The number of cycles and the names in each cycle depend on the size and type of system being designed. However, it all ends with retirement and replacement.

Systems engineering is a fractal process. This metaphor was first proposed by Bill Nickell at Sandia Laboratories in 1994. It means that the same systems engineering process is applied at levels of greater and greater detail. It is applied

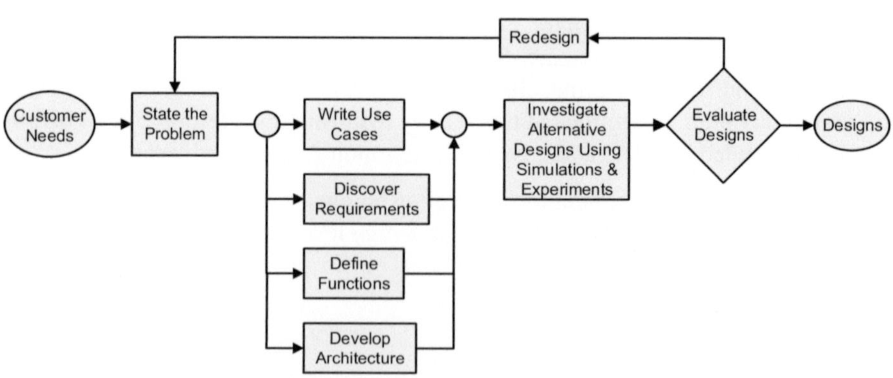

Fig. 4.9 The system design process tailored for the preliminary design phase

Fig. 4.10 The spiral life cycle model of a system

to the system, then to the subsystems, then to the components, etc. Similarly, for a fractal image, the same algorithm is applied at the large structural level, then at the medium-scale level, then at the fine-detail level, etc. as shown in the fractal image of Fig. 4.11 (bottom). This image was created by applying the same computer code over and over again at different scales. In the systems engineering process, there is a vertical hierarchy. It is applied at levels of greater and greater detail [50]: It is applied to the system, then to the subsystems, and then to the components. It is applied to the system being designed, the enterprise in which the system is expected to operate, and the enterprise developing the system. The enterprise developing the system continues to grow and expand as the system development process matures. Then the system development enterprise goes through a metamorphosis as the process moves into the production and verification phases. The system development enterprise is a living organism that grows and undergoes dramatic changes during system development. The systems engineering process is also applied horizontally to alternative-1, alternative-2, alternative-3, and so on, and to component-1, component-2, component-3, and so on. This process is recursive and iterative with much of it done in parallel.

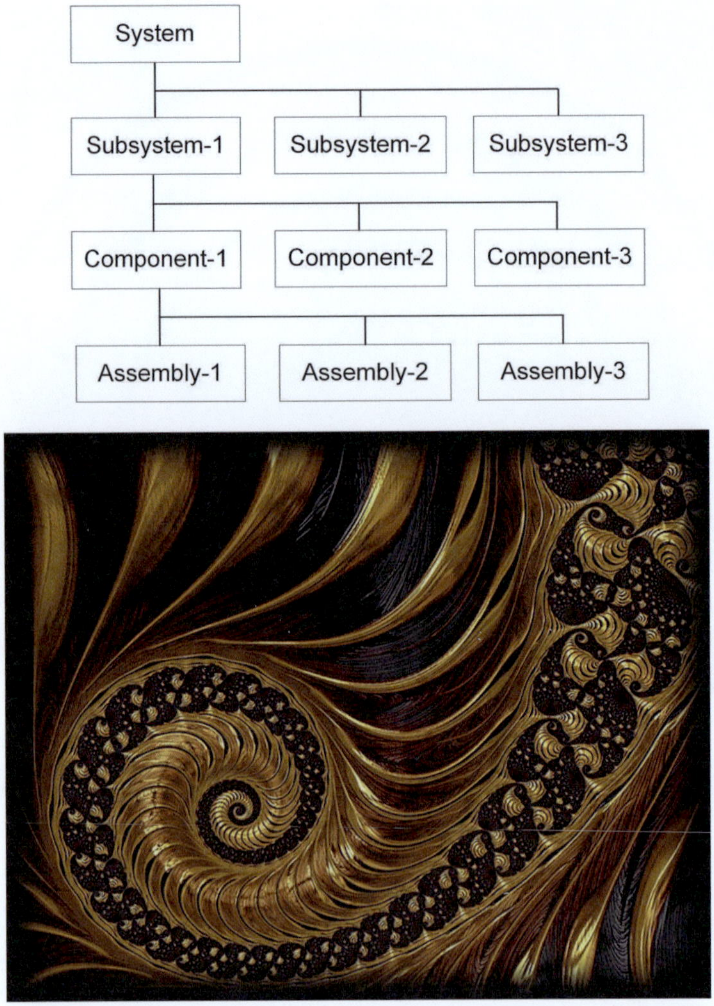

Fig. 4.11 A system hierarchy (*top*) and a fractal image (*bottom*). Creative commons CC0 public domain

The fact that discover requirements (Fig. 4.7) is a subprocess in the system design process (Fig. 4.9) and that the system design process is a subprocess in the systems engineering process of Fig. 4.12 illustrates the hierarchical nature of system engineering. We now explain the repetitive aspect of the system engineering process. In Fig. 4.12, the system design process is applied to preliminary designs, models, prototypes, and the actual system. However, this is not a waterfall process. Each loop in the process is executed multiple times. Execution of the redesign loop (upper left) is quite inexpensive, and should be exercised often. Execution of the redesign loop (lower right) is expensive, but should be exercised nevertheless. However, execution of the redesign loop (from lower right all the way back to

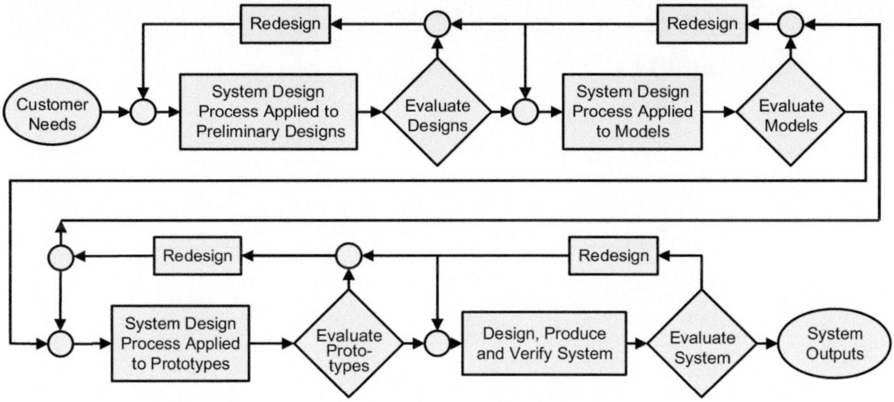

Fig. 4.12 The design portion of the systems engineering process

upper left) is quite expensive and should be exercised sparingly (that is, when it becomes absolutely necessary).

Figure 4.12, of course, shows only a part of the systems engineering process. It omits processes such as logistics, maintenance, risk management, reliability, project management, and documentation [51–53].

Although it may appear that the systems design process is being applied in a *cookie cutter* fashion to the four phases, namely preliminary designs, models, prototypes, and the real system, the emphasis, in fact, is different in each application. In the preliminary design phase, the emphasis is on discovering requirements and defining functions, with some effort devoted to alternative concepts. In the modeling phase, the emphasis is on describing, analyzing, and evaluating alternative concepts, with some effort devoted to rewriting requirements and redefining functions. In the prototype phase, the emphasis is on evaluating the prototype, modeling the highest cost or highest risk components, with some effort devoted to rewriting requirements, redefining functions (and if there are multiple prototypes), describing alternatives, and analyzing alternative concepts. And finally, when applied to the real system, the emphasis is on evaluating the system. When the system design process is applied in each of the four phases, the inputs and outputs are unique and different for each phase. For the preliminary design phase, the input is the customer request, while the output is candidate designs. For the modeling phase, the input is candidate designs, while the output is models and modeling results that are the inputs for the prototype phase, and so on (Fig. 4.13).

4.2.3 Related Items

1. *Requirements have evaluation criteria.* Common evaluation criteria for requirements include priority, rationale, status, risk level, owner, cost, revision,

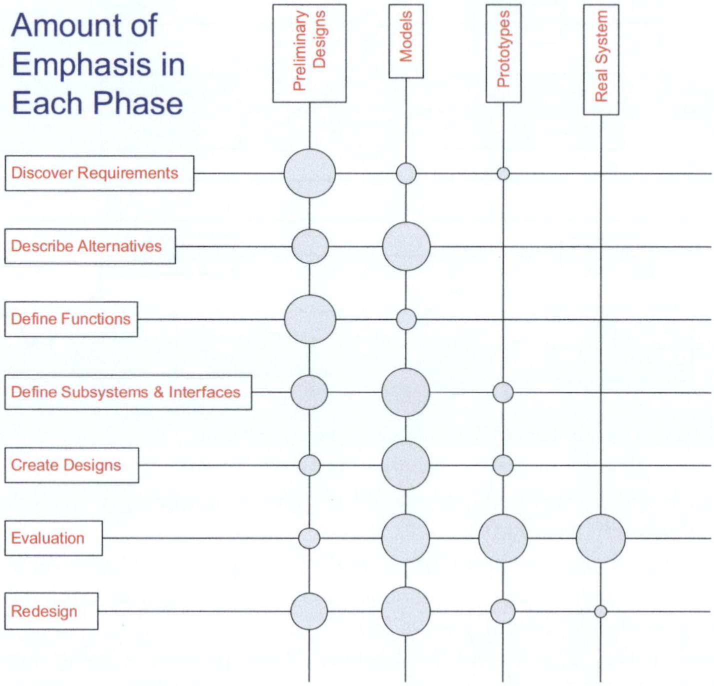

Fig. 4.13 The size of each bubble indicates the amount of time, resources, and effort that are required for each intersection

date requested, associated TPM, effort, date approved, difficulty, trace to (req), stability, trace from (req), origin, and assigned to (person).

2. *Requirements on whom?* Requirements usually describe or constrain the system being designed. For example, when designing an ATM machine we might write, *The ATM machine shall dispense cash within 10 s.* But this might impose a requirement on the central bank: *The central bank shall confirm the account and balance within 5 s of request.* This requires the definition of system boundaries. The importance of boundaries is that requirements on other entities outside the boundaries are not relevant requirements for our system. Statements like "The pilot shall . . ." are operation statements, not requirements [15]. Having identified the target for each requirement, all requirements that are not requirements on our system are gathered and moved to a separate document.
3. *Product or Process*. Requirements can describe the product or the process for producing the product [54]. Product configuration items include product processing, system management, framework, database, catalog processing, hardware, and software. Example process requirements include

A commercial tool shall be used to manage requirements.
The project shall hold the phase reviews described in document xyz.

UML support tools such as Enterprise Architect refer to process documents as the business model.

4. *Requirements versus constraints*. A requirement allows more than one design option (trade-off requirements). A constraint leaves no options (mandatory requirements). A mandated algorithm, database, or technology would be a constraint. Constraints often come from the customer.
5. *High-level requirements*. The system's high-level requirements are contained in the mission statement, the concept of operations, and the business model. They lack the formality of requirements and consequently are better described as goals or capabilities, not requirements. The term *requirements* is reserved exclusively for formal requirements that are typically stored in a requirements database. The high-level documents describe the environment in which the system is expected to operate, and the architecture of the system.
6. *Requirements versus goals*. The mission statement, concept of operations, and business model contain goals, objectives, features, and capabilities. Formal requirements are contained in the specific requirements section of the use cases, supplementary requirements specification, and tradition requirements specification (see Fig. 4.16).
7. *Concept of operations versus operation concept description*. The concept of operations (ConOps) describes the mission of an enterprise and its component systems. It gives a broad outline describing a series of operations that the systems will go through. It states what the systems must do and how they fit into the enterprise. It may include business process models. It is usually supplied by the government or the customer. The operational concept description (OCD) is a lower level description of how an individual system is to be used. It is usually written by the contractor. It is a contract deliverable requirements list (CDRL) validation item. It contains textual narrative, use cases, and diagrams. But when the prime contractor's operational concept description is given to a subcontractor, it becomes the subcontractor's concept of operations. The US DoD has metrics associated with ConOps and OCDs: measures of operational effectiveness (MOE), technical measures of performance (MOP), and design technical performance measurements (TPM).
8. *External versus internal*. Some engineers characterize requirements as external and internal. External requirements are driven by customer need. Internal requirements are driven by company practices and resources. For example, a systems development company might require certain processes or technologies be used in meeting customer requirements. Some customers tend to select a supplier/system developer based on the supplier's internal processes and controls. This suggests that the customer has confidence that the supplier's processes and controls will deliver a quality product.
9. *Requirements versus specifications*. A requirements definition set, usually called requirements, describes the functions that the systems should provide, the constraints under which it must operate, and the rationale for the requirements. It should be written in plain language. It is intended to describe the proposed system to both the customer and the designers. It should be

sufficiently broad so that multiple alternative designs are able to fit into its feasibility space.

The requirements specification, which we usually call the *specification* or *spec*, provides a precise description of the system to be built. It has a formal format and may be written in a technical language. It is intended to serve as the basis of a contract between purchasing and manufacturing. It should narrow the feasibility space to a few points that describe the system to be manufactured.

The set of requirements determines the boundaries of the solution space. The specifications define a few solutions within that space. The requirements say *what*; the specifications say *how*.

10. *Grouping of requirements*. Requirements are organized into categories, subcategories, and so forth. Requirements that are correlated should be grouped together. Suppose a young couple wants to buy a new car. The man says that his most important requirement is horsepower and the woman says that her most important requirement is gas mileage. Although these are conflicting requirements (meaning that they have a negative correlation) they do not pose a problem because the couple's decision about which car to buy will most likely be based on a trade-off between the two conflicting requirements. Now, however, assume that there is another couple where the woman says that her only requirement is safety (as measured by safety claims in advertisements), but the man says that his most important requirements are lots of horsepower, lots of torque, low time to accelerate 0–60 mph, low time to accelerate 0–100 mph, low time for the standing quarter mile, large engine size (in liters), and large number of cylinders. Assume that the man agrees that the woman's requirement is more important than his. In this case, they give safety the maximum weight of importance of 10, and they give his requirements weights of importance of 3 and 4. What kind of a car do you expect they will buy? Ideally, the man's requirements should have been grouped into one subcategory, and this subcategory should have been traded off with the woman's requirement. In summary, similar, but perhaps dependent, requirements ought to be grouped into subcategories.
11. *Requirements hierarchy*. Grouping requirements into categories helps to organize requirements into a hierarchy. The requirements database should have a treelike structure. It should link parent–child relationships and identify siblings. It is also important that requirements at different levels be treated separately [50].

4.2.4 Requirements Volatility

It is not practically possible to get all the requirements right, up front. Requirements tend to be emergent. Customers believe in the "I'll know it when I see it" maxim. Also, customers' priorities change over time. Once initial requirements are satisfied, other requirements become high priority. Requirements volatility per se is not

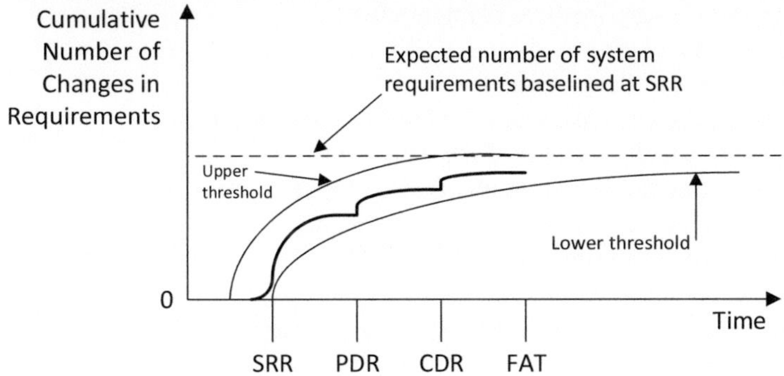

Fig. 4.14 Volatility in systems requirements

bad: in fact, requirements volatility is a good thing: It means systems engineers are doing their job of clarifying, evolving, and deriving requirements.

Some systems engineers have defined requirements volatility as addition, deletion, and modification of the requirements over the system life cycle [55]. With this narrow definition, requirements volatility is always bad. However, important tasks of the systems engineer, especially in the conceptual design phase, are to clarify, simplify, evolve, decompose, and derive requirements, and when possible negotiate to delete some. Therefore, in our metric for the *number of changes in requirements* we have included clarified, simplified, evolved, decomposed, derived, deleted requirements and added requirements. In our case, the former group vastly outnumbered the added requirements and therefore requirements volatility was a good thing. Figure 4.14 shows that most of the volatility occurred around major reviews as engineers scurried to prepare for the review and then fixed problems discovered during the reviews.

Throughout the design process, requirements can be expected to change. That means you need to design the system so that changing requirements will affect as few subsystems as possible. The following principles from [17] and [36] will help with changing requirements. Some of these were discussed in Section 5.

Principles of good design:

- Design for the total life cycle
- Use models to design systems
- Use hierarchical, top-down design
- Work on high-risk items first
- Prioritize where possible
- Control the level of interacting entities
- Design interfaces
- Produce satisficing designs
- Resist premature optimization
- Maintain an up-to-date model of the system

- Develop stable intermediate configurations
- Design resilient systems
- Use evolutionary development
- Understand the enterprise in which the system will operate
- State requirements in terms of *what* not *how*
- Enumerate functional requirements in use cases
- Allocate each function to just one component
- Document all functions
- Provide observable states
- Exploit rapid prototyping in up-front engineering
- Develop iteratively and test immediately
- Employ a modular architecture
- Create libraries of reusable metadata-tagged objects
- Use open standards
- Identify elements that are likely to change
- Write extension points in use cases
- Group data and behavior to simplify design
- Use data hiding
- Write a glossary of relevant terms
- Envelope requirements
- Incorporate design margins
- Design for testability
- Design for evolvability
- Build in preparation for buying
- Create COTS technology-enabled new design process
- Create incentives to achieve behavioral change in people
- Write a lessons learned document

Requirements traceability helps manage change. Business needs drive customer needs that produce user needs, that demand system features, that are described with requirements, that become design specifications, that engineers implement, test, and document (Fig. 4.1). Such traceability helps ensure that when a requirement is changed related requirements reflect the change. For example, if the customer no longer needs a solution for a particular part of a problem, then certain features, requirements, and tests can be eliminated.

Using requirements traceability helps to (1) verify that the system does what it is supposed to do, (2) ensure that the system does only what it is supposed to do, (3) assess the impact of change, (4) find related requirements, and (5) test related requirements. The basic strategy is to trace system requirements into derived requirements, then trace requirements into designs, then trace requirements into test procedures, and finally trace requirements into documented plans.

4.2.5 *Inspections*

Inspections rather than design reviews should be used to detect mistakes. The inspection process is described in [56] . To detect defects, software code, use cases, and requirements are now being regularly inspected. Multiple engineers are each given 250 lines of code or text to study. (The exact number of lines varies with the type of material and the phase in the system life cycle. It should be enough to keep a component engineer busy for 2 h.) Each inspector records the time he or she spent and the number of lines he or she inspected. Each inspection meeting lasts approximately 2 h. At the inspection, there is a chair, a recorder, the author, and several inspectors. Each inspector points out defects/mistakes that he or she discovered. The criticality of each mistake is determined, and it is assigned to a particular individual for resolution. Formal inspections such as these provide a wealth of metrics data for the data analytics group. Inspections save time and money and increase performance. Each defect caught at an inspection takes, on average, 3 man-hours to correct. Each defect caught during the integration phase takes about 20 man-hours to correct. Each defect caught post-delivery takes about 40 man-hours to correct.

4.2.6 *Heuristic Example of Requirements*

Earlier we discussed several ways to express requirements. These included narratives, *shall* and *should* statements, and computer models. Another example is one that uses formal logic notation. A requirement should contain (1) the description of a system function and its output, (2) the name of the system that accepts this output, (3) conditions under which the requirement must be met, (4) external inputs associated with the requirement, and (5) all conditions that determine if the system output is correct [57]. This is organized into a standard template:

The system shall ⟨function⟩,
 for use by ⟨users⟩,
 if ⟨conditions⟩,
 using ⟨inputs⟩,
 where ⟨conditions⟩.

The ⟨function⟩ is usually of the form ⟨verb⟩ ⟨output⟩
Here is an example.

Requirements for an Automated Teller Machine

5.12.0. Transaction Requirements
5.12.1. Requirements Related to the ATM User
 5.12.1.1. Produce Receipt
 5.12.1.2. Dispense Cash

The ATM shall dispense cash

- for use by the ATM user
- if the ATM user requested a withdrawal
- and if the Central Bank verified the account and PIN
- and if the Central Bank validated the withdrawal amount
- and if the ATM cash on hand is greater than or equal to the cash requested
- using the Withdrawal Validation Message from the Central Bank and the Account Verification Message from the Central Bank and the Withdrawal Request from the user
- where the amount of cash produced equals the amount requested
- and where the cash is dispensed within 10 s of the receipt of the Withdrawal Validation Message from the Central Bank.

5.12.1.3. Eject Card

5.12.1.3.1. Eject bankcard at the end of the session

The ATM shall eject the bankcard

- for use by the ATM user
- if the ATM user has inserted a bank card
- and if the ATM user has requested termination of session
- using the bank card and the terminate request
- where the bank card is ejected within 1 s of the receipt of the terminate request.

5.12.1.3.2. Eject unreadable cards

The ATM shall eject the bankcard

- for use by the ATM user
- if the ATM user has inserted a bank card
- and if the bank card does not contain a valid code
- using the bank card
- where the code reading and validation is as specified in bank card specifications, Section w.x.y.

5.12.1.4. Produce Error Messages

5.12.2. Related to the Central Bank

5.12.2.1. Verify Account Message

The ATM shall produce the verify account message

- for use by the Central Bank
- if the ATM user has entered a PIN
- and if the bank card contains a readable code
- using the bank card and user-entered PIN
- where the content and format is as specified in the Central Bank Interface Specification, Section x.y.z.
- and where the message is issued within 1 s of the final digit of the PIN.

The foregoing example illustrates several features of good requirements identified in this section. The numbering scheme manifests the tree structure of this set of requirements: parent, child, and sibling relationships. References are made to the specifications. In each requirement, the customer is identified: for example, the ATM user and the central bank. Multiple sequence diagrams were used to elicit these requirements. Performance evaluation criteria are specified as maximum values (with units provided) and they are testable (for example, cash must be dispensed within 10 s). The requirements state *what*, not *how* (for example, The ATM shall dispense cash). The requirements identify applicable states using the conjunctive *if* clauses. The word choice is correct. What needs to be added is a provision for the underlying rationale.

4.2.7 Sensitivity Analysis

Requirements are seldom described with equations. So the analytic sensitivity techniques of Section 26 are inappropriate. Furthermore, we cannot run experiments on the requirements, so the empirical techniques of Section 26 are also inappropriate. Therefore, we will use the *what-if* technique to perform our sensitivity analysis. Let us now revisit the audible signal requirement for the Cookie Acquisition System (Table 4.5).

The above computed cost analysis was based on low-cost buzzers. We will now do our sensitivity analysis on the higher quality SC series of piezoelectric indicators, because more comparison data are available. We will try to answer these questions. How will the cost be affected, if we want more sound intensity? If we want a different frequency? If we want a different type of input?

Let us start with a baseline unit, the Mallory SC628 series Sonalert, which produces a continuous sound at 2.9 kHz with a minimum sound intensity of 80 dB with a 6 V input (Table 4.6).

The audible signal requirement had a high priority, so it was appropriate to do a sensitivity analysis for it. This analysis showed that cost increases as the required power increases, as input variety increases and as unusual frequencies were required.

4.2.8 Homework

4.2.8.1 The Shadows Figure

In Fig. 4.6, what is the icon for the requirements? What does it mean?
What is the icon for concept exploration? Why is it there? What does it mean?
What is the icon for full-scale design? What does it mean?
What is the icon for manufacturing? What does it mean?

Table 4.5 The audible signal requirement for the Cookie Acquisition System

Attribute	Explanation
Identification tag (Id)	ReqNF4
Name	Audible signal for cookies is ready
Text	An audible signal shall inform the student when the cookies are ready. This audible signal shall have an intensity level of 80 ± 2 decibels and a frequency of 440 Hz
Priority	9
Verification method	During design and construction, this requirement will be verified by test. The proposed device will be mounted on a test bench and will be activated in accordance with test instructions. Sound power level in decibels (dB) will be measured at a distance of 30 cm. At final test and during operation, this requirement will be verified by demonstration. An actual oven will activate the audible signal and the sound power level in decibels (dB) will be measured at a distance of 30 cm
Computed cost	The Mallory PK-16N04WA-03Q buzzer satisfies these requirements and it costs $8.39. However, if a higher frequency, such as 3 kHz, is acceptable, then a lower cost piezoelectric buzzer could be used, such as the Mallory PK-21N31PQ at $2.57
Difficulty	Easy
Link to TPM?	No
DeriveReqt:	This requirement refines *ReqF5: Cookie shall emit an audible signal when the timer has elapsed*
Owner	Engineer
Date of last change	November 26, 2015

Table 4.6 Sensitivity analysis of sound producers

Model	Sound intensity, dB, for a 6 V DC input	Cost ($)	Frequency (kHz)	Voltage type
What if we want a different power output?				
SC628	80	13.13	2.9	DC
SC628N	86	14.57	2.9	DC
SC616	91	16.55	2.9	DC
This shows that as the required power increases, the cost increases				
What if we want a different frequency?				
SC628	80	13.13	2.9	DC
SC628D	67	15.09	1.9	DC
SC628H	78	26.06	4.5	DC
This shows that frequencies different from the 2.9 kHz baseline increase cost				
What if we want to allow AC inputs?				
SC110	68	27.04	2.9	AC/DC
SC110N	80	36.03	2.9	AC/DC
This shows that allowing AC and DC inputs increases cost. It also again shows that increasing required power increases cost				
All of cost quotes are from the same supplier on the same day				

What is the icon for operation and maintenance? Why is it there? What does it mean?

What is the icon for retirement and replacement? Why is it there? What does it mean?

How would this figure change if instead of creating a municipal transportation system, we were designing the replacement for the B-2 bomber?

Where is the light that produced these shadows?

4.2.8.2 Section 28. Figures

In Fig. 4.7, there is a box containing Manage the Requirements Process. What does that mean?

Add the life cycle phases of Fig. 4.8 to Table 4.7.

Figure 4.9 has four processes in parallel: Write Use Cases, Discover Requirements, Define Functions, and Develop Architecture. In which phase (oval) of Fig. 1.13 would each of these go?

Figure 4.10 fits into which phase of Fig. 1.13?

Fill in the following tables to compare the activities and life cycle phases of Figs. 4.8 and 4.11. Point out the differences (Tables 4.8 and 4.9).

Why should Fig. 4.14 have upper and lower thresholds?

Table 4.7 Comparison of some life cycle models

Figure 1.13	Figure 1.14, first spiral	Figure 1.14, second spiral	Figure 4.8
Problem statement	Customer need		
Requirements discovery	Define requirements	Refine requirements	
Alternatives investigation			
System design	Design model	Design prototype	
Implementation	Create model	Build prototype	
Integration and test	Validate model	Test prototype	
Operations and maintenance	Simulate model		
Retirement and replacement	Preliminary design review	Critical design review	

Table 4.8 Activities of Figs. 4.8 and 4.11

Figure 4.8	Figure 4.11
Requirements	
Analysis	
Design	
Implementation	
Verification	
Operations	

Table 4.9 Life cycle phases of Figs. 4.8 and 4.11

The life cycle phases	
Figure 4.8	Figure 4.11
Inception	
Elaboration	
Construction	
Transition	
Operation	
Retirement and replacement	

4.3 Section 29. Hybrid Process for Capturing Requirements

4.3.1 Hybrid Process

So far, this part has adhered to the traditional industry practices for requirements. However, it has not presented a process for actually capturing the requirements; Section 28 was vague. However, today's modern system design is *use case* based. Use cases capture the required functional behavior of the system. So then, why not use the use cases to capture the requirements? This hybrid process for capturing requirements from Daniels and Bahill [58] is presented next.

We begin with a description of a use case and an example. A use case is an abstraction of a required behavior of a system. A use case produces an observable result of value to an actor. An actor is a role that a user (or some other object) plays with respect to the system. A system will typically have multiple use cases, each associated with a goal of a particular actor. Each use case describes a sequence of interactions between one or more actors and the system. A use case narrative is written in a natural language. It is written as a sequence of numbered steps with a main success scenario and alternate flows. This sequence of interactions is contained in a use case report and the relationships between the system and its actors are conveyed through use case diagrams. A simple example of a use case for the status-monitoring portion of a heating, ventilation, and air-conditioning (HVAC) system for a typical Tucson house is presented next [46].

Name: Display System Status
Brief description: Indicate the health of the HVAC system.
Added value: Home owner knows if the system is working properly.
Scope: An HVAC system for a Tucson house
Primary actor: Home owner
Frequency: Typically once a day
Main success scenario:

1. Home owner asks the system for its status.
2. Each component in the system reports its status to the controller.
3. The system accepts this information and updates the system status display.
4. Home owner observes the system status {exit use case}.

Alternate flows go here.
Specific requirement
Functional requirements:

1. *Each component shall report its status to the controller.*
2. *The system shall display the system status.*

Nonfunctional requirement: *The system status shall be displayed within 2 s of request.*
Author: Cactus Jack
Last changed: June 20, 2015

Use cases can be used to derive traditional requirements. A traditional requirements specification, which provides imperatives as textual statements, serves as the primary means for engineers to bound and communicate system capabilities and constraints. However, in recent years the popularity of documenting software requirements with use cases has increased dramatically, with use case-based analysis becoming the foundation of today's software analysis techniques. Due to the success of the use case approach in the software community, systems engineers have begun employing use cases as a tool to document system requirements [58]. Using a common methodology for requirements elicitation helps align systems engineers and software engineers, allowing for an easier transition from system and subsystem requirements into software requirements. We recommend that the combination of shall-statement requirements and use case modeling techniques provides a complementary and synergistic way to document and communicate relevant information to effectively design, develop, and manage complex systems. The resultant quality and clarity of knowledge gained through the combination of two methodologies, use cases, and shall-statement analysis are worth the time spent.

4.3.1.1 Why Write a Requirements Specification?

Requirements are documented so that the characteristics of the to-be system can be discussed and analyzed by engineers and stakeholders to facilitate the creation of a clear, shared vision of the proposed system. Systems engineers and architects use the requirements set to ensure system and enterprise-wide coverage of capabilities and characteristics. Design engineers use the agreed-upon requirements set to derive additional requirements and develop subcomponents that contribute to a system that meets those requirements. Program managers track requirements volatility to measure and monitor risks associated with requirements creep and redefinition. Customers sign off on requirements during acceptance testing to ensure that the capabilities offered by the system meet their needs and intent. It is clear that gathering, discussing, and deriving requirements are an essential step in the engineering process to ensure that the system, once built, actually satisfies stakeholders' needs.

4.3.1.2 Problems with Traditional Requirements

A traditional requirements specification attempts to document the imperative functionality and constraints of a system by enumerating each requirement using *shall* notation through which individual capabilities and constraints are expressed (for example, The system shall ...). The requirements set is typically structured according to functional areas, and each requirement is attributed, possibly traced to and traced from other requirements. Shall statements are used to describe different types of requirements such as functional, performance, security, reliability, availability, and usability.

However, if the requirements specification consists solely of shall-statement requirements, then for even a moderately complex system, the requirements set can become unwieldy, containing hundreds or thousands of requirements. This, by itself, is not a problem, because today's systems are quite complex and often need this level of detail to avoid ambiguity. The problem is that documenting the requirements as a set of discrete and disembodied shall statements without context makes it difficult for the human mind to comprehend the set and fully interpret the intent and dependencies of each requirement. This makes it hard to detect redundancies and inconsistencies. In short, large shall-statement requirement specifications make it difficult for engineers and customers to really understand what the system is supposed to do! Because there is no straightforward way to assimilate the requirements set, requirements are often misinterpreted, redundant, and incomplete [59]. This fact can result in inferior, overly expensive systems and ultimately dissatisfied stakeholders.

4.3.1.3 Use Cases

Use cases have been accepted by the software community as a requirements gathering and documentation instrument that captures system requirements through generalized, structured scenarios that convey how the system operates to provide value to at least one of the system's actors. The primary reason why use cases have become a popular method is the simple and intuitive way in which the system's behavior is described. Use case models serve as a bridge between stakeholders and the technical community. Through a use case model, stakeholders should be able to readily comprehend how the system helps them fulfill their goals. Simultaneously, engineers should be able to use the same information as a basis for designing and implementing a system that executes the use cases.

A set of use cases (collectively referred to as the use case model) is used to describe the fundamental value-added services that users and other stakeholders need from the system. Use cases can be thought of as a structured, scenario-based method to develop and represent the behavioral requirements for a system. The use case approach subscribes to the notion that each system is built to support its environment or actors—be it human users or other systems. Use cases, by definition, describe a series of events that when completed yield an observable result of value to a particular actor [45]. The fundamental concept is that systems designed and developed from use cases will better support their users.

Ultimately, use cases have been shown to be a simple and effective tool for specifying the behavior of complex systems [45]. Therefore, it is no surprise that systems engineers have started using this method for documenting and communicating requirements.

4.3.1.4 Difficulty

Although use cases solve some of the problems with specifying requirements when contrasted with the shall-statement method, there are issues with applying use cases to solve systems engineering problems. There are a few possible reasons for these issues:

1. Use cases don't *look* like a traditional requirements specification.
2. Use cases *feel* somewhat vague.
3. Use cases do not contain *all* of the requirements.
4. The completeness of a set of use cases is sometimes difficult to assess, especially for unprecedented complex systems.

Other contributing factors result because older engineers and customers do not have much experience with use cases. Systems engineers have been accustomed to developing traditional requirement specifications. Use cases are a departure from the past. We fully expect use cases to soon become commonplace. Until then, it is important to educate customers and engineers on the use case concepts and the fact that incorporating use cases into the overall systems engineering process will help reduce risk, and increase the likelihood of delivering the system expected by the customer.

4.3.1.5 Strengths and Weaknesses

Use case modeling has shown strengths in many areas of requirements capture. The dialog between the system and the actor that is expressed in the sequence of events associated with a use case explains clearly how a system reacts to stimuli received by the system's actors. What engineers typically do not realize is that the actual functional requirements are embedded in the sequence of events. In other words, a use case describes the requirements. Whenever a use case's sequence of events indicates that the system performs some function, a functional requirement has been imposed on the system. Each use case is chock full of functional requirements (possibly multiple requirements per sentence!) when viewed in this way. By design, the requirements are an integral part of a use case's natural language story.

A use case's strength is also its weakness. To keep use cases simple, readable, and manageable, they can only tell a fraction of the complete story without becoming unwieldy and difficult to understand [60]. Use cases alone were not meant to capture all the requirements. A use case is good at capturing the functional requirements for a system in an understandable and unthreatening way. However, shall statements, diagrams, tables, equations, graphs, pictures, pseudo code, state machine diagrams, sequence diagrams, use case diagrams, statistics, or other methods must still be used to capture additional requirements and add richness to provide a sufficient level of detail to adequately characterize a system. All of these artifacts may be related to a use case, but they are typically not expressed directly in a use case's natural language story. In some instances, use cases are not the best method to express a system's functional requirements. For example, systems that are algorithmically intense and do not have much interaction with their environment may be better described using some other method—although use cases can be used in these cases too [60, 61].

In summary, use cases are understandable, at the expense of being potentially ambiguous. Shall-statement requirements are typically well defined and often stand alone, where individual capabilities are expressed in a rather abstract way, out of context with the rest of the system's characteristics. The use of shall-statement requirements alone makes it more difficult for engineers and stakeholders to assess one requirement in a virtual sea of disjointed capabilities. Shall statements are very good at precise expression, while use cases are good at communicating requirements in context. This tension promotes the need for a combined use case-shall-statement approach, where the system's behavior is documented precisely and understandably in the same way that it is derived—by analyzing all of the discrete, end-to-end interactions between actor and system that ultimately provide some value to an actor.

4.3.1.6 Hybrid Requirements Capture Process

The specific requirements section of a use case report was originally intended to capture the nonfunctional requirements, which are typically performance requirements on a particular use case step, using shall-statement notation. Daniels and Bahill [58] proposed using the specific requirements section to capture not only nonfunctional but also functional requirements specified in the use case using shall-statement notation. When the shall-statement requirements are captured in this way, they retain their context since they can be readily traced to the use case sequence of events that they were derived from. Using this approach, the comprehensibility of the use case is balanced by precise shall statements.

Engineers will also discover requirements that do not fit nicely within a single use case, but rather they characterize the system in general. These global requirements should be reflected in a separate supplementary requirements specification using shall-statement notation. The supplementary requirements specification is independent of the use case mode. It is intended to contain all requirements that do not apply cleanly to any single use case. Requirements that describe physical constraints, security (physical security, anti-tampering, and information assurance),

Table 4.10 Rating the requirements methods

Quality	Shall-statement method	Use cases	Hybrid process
Necessary	1	3	3
Verifiable	3	1	3
Unambiguous	3	2	3
Complete	2	3	3
Consistent	1	2	3
Traceable	2	2	3

quality, reliability, safety, usability, supportability requirements, or adherence to standards, for example, are good candidates for inclusion in the supplementary requirements specification. These types of requirements are typically not specific to a particular use case. The hybrid process described in this section includes the use of a supplementary requirements specification for capturing these classes of requirements.

The use case model (which contains the individual use case reports) together with the supplementary requirements specification constitutes a way to completely document a system's requirements in an exact, yet understandable manner. In many cases, this alone will be satisfactory for specifying the requirements. However, some customers may still demand that a traditional shall-statement specification be developed. We discuss this in the next section.

Table 4.10 uses some of the qualities of good requirements to compare and contrast the three requirement-gathering methods presented so far: shall statements, use cases, and the hybrid process using both use cases and shall statements. Each technique is given a rating from 1 to 3 for each quality (3 being the best) to assess how well the technique promotes requirements best practices. We believe that the hybrid process encompasses the strengths of each technique, and excels in each quality.

Necessary. The use case method excels here. Since use cases are driven by actor needs, each use case represents a set of capabilities that are of value to the system's stakeholders. It is hard to determine whether a shall-statement requirement is necessary without considering the entire requirements set and additional documentation. The hybrid process, due to its use case-driven foundation, rates highly.

Verifiable. Since shall-statement requirements are concise and discrete, they are, in principle, easier to verify than a use case that comprises multiple requirements in a single sentence. The hybrid process retains the discrete nature of the shall-statement requirements, which makes them easier to verify.

Unambiguous. Shall-statement requirements are more likely to be unambiguous, because they can be written quite tersely and precisely. Use cases, due to their narrative format, are not as well suited to the same accuracy in expression. The context provided by a use case helps remove some of the ambiguity. The hybrid process takes the best of both worlds—context with use cases, and freedom to use more precise wording with shall statements.

Complete. A use case is structured in terms of basic and alternative paths, giving a clear understanding of when a requirement applies, and when it does not. Since use cases are generated by considering the needs of all of the system's actors, it is much less likely that requirements will be missed. Use cases are an excellent way to help ensure that the requirements set is complete. It is difficult to determine whether a shall-statement requirement set is complete without referencing the entire documentation. But many times, due to the structure of the requirements set, you can look for incompleteness [19]. The hybrid process retains the use case method's excellence in complete requirements description.

Consistent. Use cases typically deal with one goal at a time and, therefore, it is easier to separate requirements and reduce the risk of conflicting with other requirements. However, neither method alone excels at ensuring a consistent requirements set. The hybrid process, using the supplementary specifications, provides a mechanism to extract requirements that apply across use cases and specify them in one place. This helps maintain the consistency across the set and reduces duplicity. Admittedly, requirements extracted from use case narratives may need to be refined and aggregated into a comprehensive, nonredundant set of system functional requirements for formal, high-profile projects.

Traceable. Use cases, through their actor-driven derivation, are easily traced to higher level actor goals. However, it is not as straightforward to allocate specific use case requirements to the system components. Shall statements can be traced through a numbering scheme as parent–child requirements. The hybrid process retains the traceability to actor goals through use cases and allows the shall statements to be allocated to system components. The use of object models such as sequence diagrams in the hybrid specification helps allocate responsibilities and requirements to the components of the system.

4.3.1.7 Hybrid Process Defined

There are several methods for gathering requirements. For complex systems, a combined use case and shall-statement approach should be employed to capture a system's requirements. Use cases are easy to understand and provide context and direct traceability to actor needs and interfaces. Shall-statement requirements add the precision necessary to completely and unambiguously specify the system. With any approach, early and frequent interview and review iterations should be conducted with the stakeholder community to ensure that their concerns and desires are addressed.

The hybrid process for capturing requirements proceeds as follows.

1. Using the system vision statement, the concept of operations (ConOps), description of needed system capabilities, and interviews with the customer develop a business model to understand how the system under design fits into the overall enterprise. The business model provides context for the system and can

profoundly increase the quality of later analysis. There is a methodology for defining business models using use case and object-oriented techniques.

2. Discover the system's actors and their goals with respect to the system under design. The business model is a valuable input in accomplishing this task. Define the system architecture.
3. Use the actor goals to sketch out the use case reports, concentrating on use case names and brief descriptions.
4. Iterate on the important or architecturally significant use cases, filling in increasingly more detail in the reports, and capturing alternative sequences of events, preconditions, and postconditions.
5. Extract the functional requirements from each use case's brief description, added value, and flow of events and document them in the specific requirements section of the use case report, or the supplementary requirements specification if they apply to many use cases. Ensure that traceability is maintained.
6. In each use case report, document the nonfunctional requirements, typically performance, in the specific requirements section as they apply to each use case.
7. In parallel with steps 5 and 6, develop a supplementary requirements specification to capture system-wide requirements that do not fit cleanly into individual use case reports.
8. Iterate on the use case set to ensure consistency and completeness as the program progresses.

If a traditional requirements specification (shall-statement requirements only) is to be developed: combine the requirements documented in each use case report's specific requirements section, along with the supplementary requirements specification to generate the traditional requirements specification. This should be primarily a simple copy-and-paste operation. Figure 4.15 shows this hybrid requirements process. Most of this figure is derived from the standard UML use case process [62]. The new task is *extract functional requirements*, which is highlighted in yellow.

The class diagram given in Fig. 4.15 shows how use case, requirement, and specification concepts are related. For a given system, a use case requirements

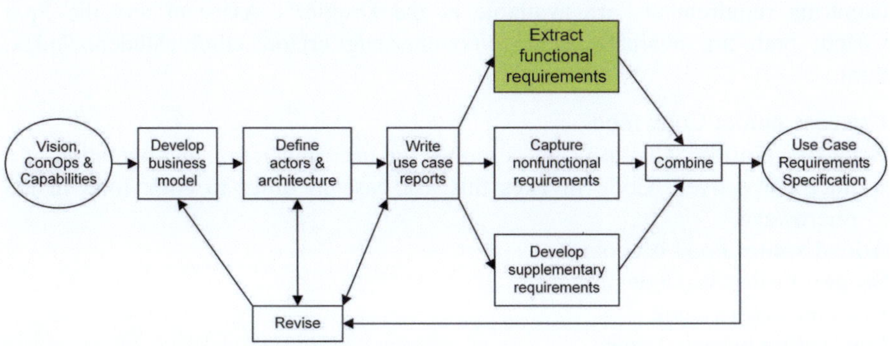

Fig. 4.15 The hybrid process for capturing requirements in use cases

specification contains one use case model and one supplementary requirements specification. A use case model contains one or more use case reports (also referred to by other authors as use case descriptions). Each use case report contains one or more sequences of events (one main sequence of events, and zero or more alternative sequences), as well as one specific requirements section. The sequence of events within the use case report contains the informal functional requirements for that use case, while the specific requirements section contains the nonfunctional requirements and the formal functional requirements extracted from the sequence of events.

The supplementary requirements specification contains system-wide requirements that do not fit nicely within the sequence of events or specific requirements section of one of the use cases. A traditional requirements specification can be generated by combining the functional requirements and nonfunctional requirements from each use case report along with the system-wide requirements from the supplementary requirements specification.

In a traditional systems engineering environment, *test procedures* are detailed and sequential. Sometimes they look like a use case sequence of events. The *requirements specification* may have hundreds of requirements and it paints a picture of the system being designed. It is usually written before the test procedures, but it might not bear a close resemblance. The *design model* captures the architecture and the interfaces. What do these three views of the system have in common? In our hybrid process, they have a common ancestor: the use case requirements specification, as shown in Figs. 4.16 and 4.17.

4.3.1.8 Hybrid Process Example

Here we give a simple example of a use case and a use case diagram (Fig. 4.18) that helps illustrate the concepts presented in this section. This example is based on the Microwave Oven Software Product Line Case Study given in Gomaa [63], which contains relevant some nice use case modeling examples. This one was selected due to its simplicity. Two other examples of implementing this hybrid process for capturing requirements are available at the Umpire's Assistant and the Spin Coach; both are available at http://sysengr.engr.arizona.edu/forStudents/index.html.

Use case name: Cook food.
Brief description: This use case describes the user interaction and operation of a microwave oven. Cook invokes this use case in order to cook food in the microwave.
Added value: Food is cooked.
Scope: A microwave oven.
Primary actor: Cook.
Supporting actors: Timer.
Preconditions: The microwave is waiting to be used.

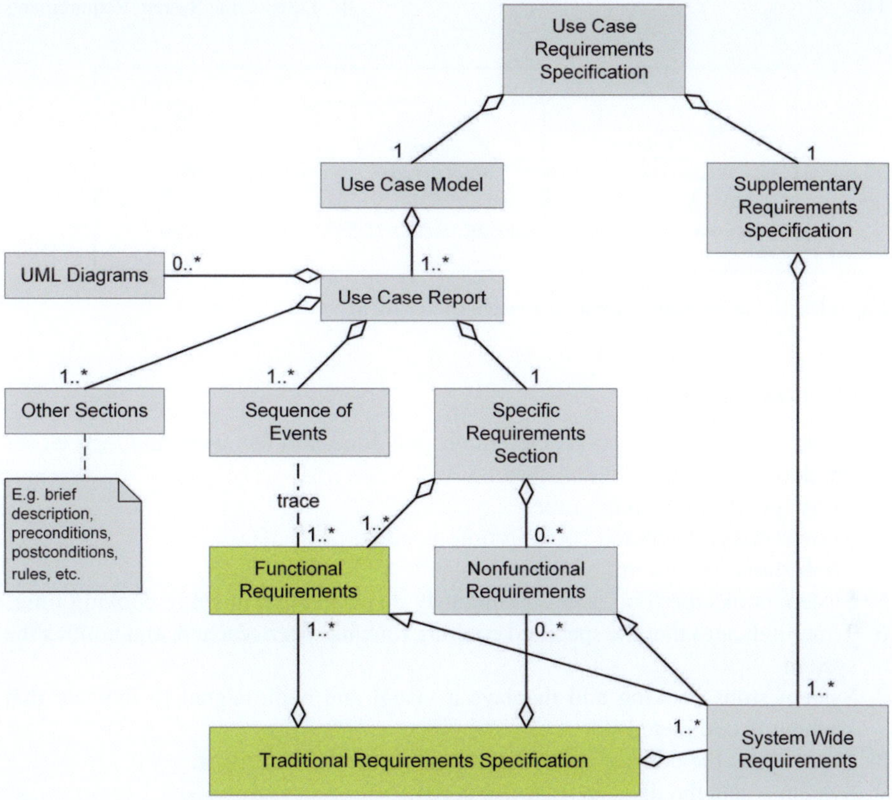

Fig. 4.16 A class diagram depicting the components and interrelationships of a hybrid requirements specification. Lines with *arrowheads* are generalizations. Lines with *diamond heads* are aggregation relationships. The numbers represent multiplicities: * means many

Fig. 4.17 The ancestry model for system documentation

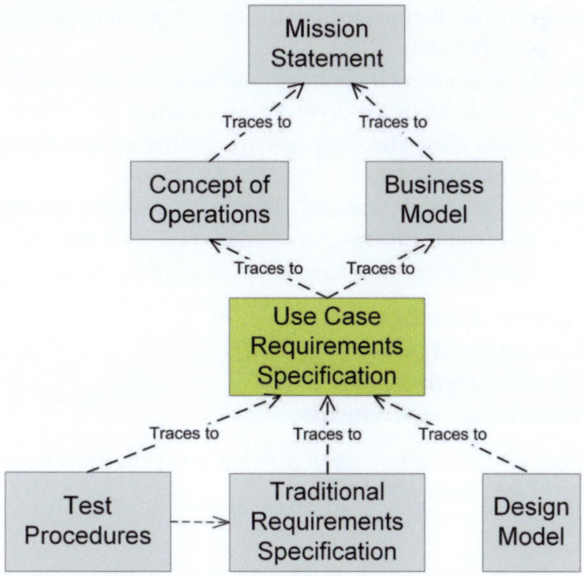

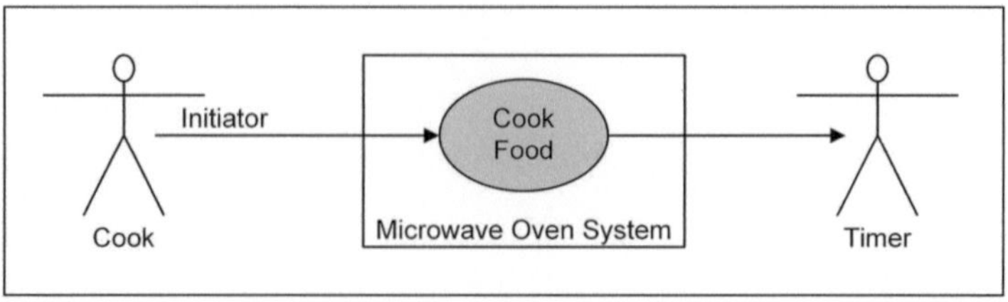

Fig. 4.18 Use case diagram for a microwave oven system

Main success scenario:

1. Cook opens the microwave oven door, puts food into the oven, and then closes the door.
2. Cook specifies a cooking time.
3. System displays entered cooking time to cook.
4. Cook starts the system.
5. System cooks the food, and continuously displays the remaining cooking time.
6. Timer indicates that the specified cooking time has been reached, and notifies the system.
7. System stops cooking and displays a visual and audio signal to indicate that cooking is completed.
8. Cook opens the door, removes food, and then closes the door.
9. System resets the display {exit use case}.

Alternate flows:

4a. If cook attempts to start the system without closing the door, then the system will not start.

4b. If cook attempts to start the system without placing food inside, then the system will not start.

4c. If cook enters a cooking time equal to zero, then the system will not start.

5a. If cook opens the door during cooking, the system will stop cooking. Cook can either close the door and restart the system (continue at step 5) or cook can cancel cooking.

5b. Cook cancels cooking. The system stops cooking. Cook may start the system and resume at step 5. Alternatively, cook may reset the microwave to its initial state (cancel timer and clear displays).

Author: Jesse Daniels
Last changed: October 23, 2015
Specific requirements
Functional requirements:

Req. F1: The system shall provide a mechanism for cook to enter a cooking time {from step 2}.

Req. F2: The system shall display the cooking time entered by the cook {from step 3}.

Req. F3: The system shall cook food using microwave radiation {from step 5}.

Req. F4: The system shall calculate and display the remaining cooking time {from step 5}.

Req. F5: The system shall interface with a timer mechanism such that the system stops when the timer elapses {from step 6}.

Req. F7: The system shall emit an audible signal when the timer has elapsed {from step 7}.

Req. F8: The system shall indicate visually when the timer has elapsed {from step 7}.

Req. F9: The system shall detect when the oven door has been closed {from step 1}.

Req. F10: The system shall not start, if the oven door is open {from step 4a}.

Req. F11: The system shall determine if food has been placed in the oven {from step 4b}.

Req. F12: The system shall not start, if no food has been placed in the oven {from step 4b}.

Req. F13: The system shall stop if the oven door is opened while the system is running {from step 5a}.

Req. F14: The system shall provide a mechanism to cancel a cooking time entered by the cook {from steps 5a and 5b}.

Req. F15: The system shall stop if the cooking time is canceled while the system is running {from steps 5a and 5b}.

Nonfunctional requirements:

Req. N1: The system shall allow cook to enter the cooking time in less than five keystrokes {refinement of Req. F1, obtained from stakeholder interviews}.

Req. N2: The cooking time displayed by the system shall be visible to a person with 20/20 vision standing five feet from the oven in a room with a luminance level between 0 and 100 foot-candles {refinement of requirement F2, obtained from stakeholder interviews}.

Req. N3: The system shall raise the temperature of food in the oven such that temperatures at two distinct locations in the food differ by less than 10 % {refinement of Req. F3, obtained from stakeholder interviews}.

Req. N4: The system shall update the remaining cooking time display every second {refinement of Req. F4}.

Req. N5: The audible signal emitted by the oven shall have an intensity level of 80 $\pm$ 2 decibels {refinement of Req. F7, obtained from stakeholder interviews}.

Req. N6: The system shall detect food items weighing more than 0.5 ounces and with a volume more than 1 cubic inch {refinement of Req. F11, obtained from stakeholder interviews}.

Req. N7: The system shall comply with section 1030 of Title 21-Food and Drugs, Section I—Food and Drug Administration, Department of Health and Human Services, Subsection J: Radiological Health {refinement of Req. F3, specifies required compliance with health standards}.

Req. N8: The system shall provide a minimum of 14 3/4″ width × 8 3/4″ height × 15 3/4″ depth inside cooking area {from step 1, obtained from stakeholder interviews}.

Req. N9: The cooking time shall be adjustable between 1 s and 90 min {refinement of Req. F1, obtained from stakeholder interviews}.

4.3.1.9 Hybrid Process Summary

Use case models and traditional shall-statement requirements are synergistic specification techniques that should be employed in a complementary fashion to effectively communicate and document requirements. Are *use cases* the same as *requirements*? Not exactly—use cases are a vehicle to discover requirements. The requirements are actually embedded within the use case's textual description, making use cases a container for requirements. How do use cases relate to a traditional requirements specification? Use cases provide context for requirements that are documented using shall-statement notation in a traditional requirements specification. These shall-statement requirements can be extracted from a use case's narrative. Where are the requirements actually documented? We suggest that the requirements be documented using the requirements specification structure illustrated in Fig. 4.16. This structure includes (1) a use case model for capturing requirements that are associated with individual use cases and (2) a supplementary requirements specification for capturing system-wide requirements. The specific contribution of the Daniels and Bahill [58] paper was the introduction of a functional requirements part to the specific requirements section. This functional requirements part contains the functional requirements written in formal shall-statement language. The brief description and added value slots of the use cases suggest high-level requirements and goals. This section has presented a comprehensive example that illustrates the hybrid process for combining use case models with traditional shall-statement requirements.

System design often starts with use cases. Then, as shown in this section, requirements can be elicited from the use cases. Sequence diagrams can be drawn to solidify the use case models. Then classes are formed. Functional requirements help the designer to identify the functions or operations of each class. Finally, the relationships between classes are defined. This produces a model upon which the system can be designed.

In summary, the hybrid requirements process captures shall-statement requirements in the use case sequences of events. These functional requirements are inserted in the specific requirements section of the use case report. These functional requirements can also be reflected in the class diagrams and a requirements database to support traditional specs and tracing.

4.3.2 *Homework*

4.3.2.1 Requirements in Use Cases

Describe the purpose of developing use cases.

Compare and contrast the contributions to system development of use cases and specific requirements.

4.3.2.2 Missile System Failure

This was a real event reported by the Vice President of Engineering. The specific names and specifications were invented. A missile has six thrusters, two each to control yaw, pitch, and roll. Up to three of these valves could be opened at the same time. The following table contains some customer requirements for a printed circuit board (PCB) inside the missile. The blah, blah, blahs would be attributes such as ID number, status, priority, TPM links, and performance requirements (Table 4.11).

An engineer then derived the following requirements for the PCB (Table 4.12).

The Type column had two choices, signal or power. Signal traces were 4 mm wide and power traces were 10 mm wide. Signal traces had no special indicator in the Type column. Power traces had a P in the Type column. As you can see, the engineer forgot to mark trace-11 as a power trace (P). As a result, trace-11 was only 4 mm wide, instead of the intended 10 mm wide for power traces. When the missile was flight-tested, two valves opened at the same time (not in the test procedure), trace-11 burned out, and the missile blew up. How could this have been prevented?

Table 4.11 Customer requirements for a missile PCB board

PCB trace number	Requirement
10	Trace-10 shall transmit signals between element 1 and 2. Blah, blah, blah
11	Trace-11 shall supply valves 1–6. Blah, blah, blah. This trace shall have a capacity of X-5 mA. Blah, blah, blah
12	Trace-12 shall deliver power to components 1–4. Blah, blah, blah

X-5 was a specific number

Table 4.12 Derived requirements for a missile PCB board

PCB trace number	Origin	Destination	Type	Length
10	Element-1	Element-2		Stuff
11	Pin-3 IC-2	Pin 4 of valves 1–6		20
12	Stuff	Components 1–4	P	Stuff

The following would be a typical use case for this missile.

Control Missile Attitude

Brief description: Valves 1–6 open and close allowing pneumatic fluid to flow, thereby controlling missile attitude.
Added value: Guidance, navigation, and control (GNC) can control the system state.
Scope: Control of attitude control valves.
Primary actor: GNC.
Supporting actors: Power supply, pneumatic fluid, control valves.
Precondition: The missile is flying and all valves are closed.
Trigger: GNC commands a course change.
Main success scenario:

1a. GNC commands valve-1 to open for X-1 msec, where X-1 will be specified before SRR.
2. Valve-1 opens, drawing X-2 mA from the power supply through PCB trace-11, allowing pneumatic fluid to flow through its orifice.
3. After X-1 msec, valve-1 closes, drawing X-3 mA, thereby stopping the flow of fluid {exit use case}.

Alternate flows:

1b. GNC commands valve-2 to open for X-4 msec.
1b1. Valve-2 opens, drawing X-2 mA from the power supply through PCB trace-11, allowing pneumatic fluid to flow through its orifice.*
1b2. After X-4 msec, valve-2 closes, drawing X-3 mA, thereby stopping the flow of fluid {exit use case}.

{Similar alternative flows exist for valves 3–6.}
Postcondition: Missile has changed course and all valves are closed.
Rules:

1. Valves 1–6 will open and close at the same or at different times.
2. X-1 to X-5 are specific values.

Using this use case, write the requirements for this system.

4.3.2.3 Farmroad-Highway Intersection

There is an intersection between a seldom-used farmroad and a busy highway. A farmer has difficulty getting across the highway, so we are to install a traffic-light system. In Section 11, you designed a controller for this traffic-light system.

Normally, the highway light shall be green and the farmroad light shall be red. When a sensor signals that a tractor is on the farmroad, the highway light shall change to yellow. After a *short-time interval* (normally 10 s), the highway light shall change to red and the farmroad light shall change to green. The farmroad light

Fig. 4.19 Highway-farmroad intersection

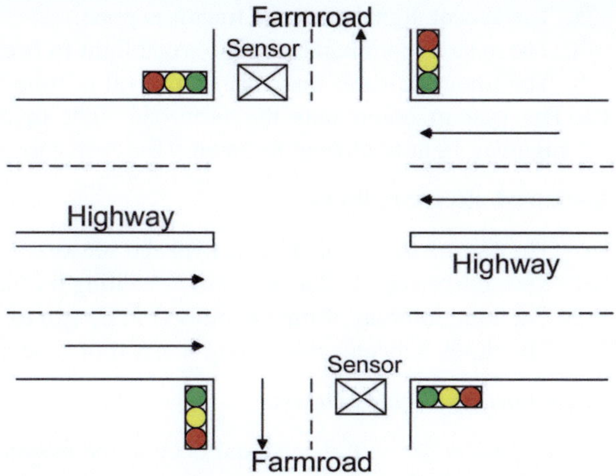

shall stay green until the tractor clears the sensor or after a *long-time interval* (normally 50 s), whichever comes first. Then the farmroad light shall become yellow. After a short-time interval, the farmroad light shall become red and the highway light shall become green. The system shall stay in that state for a least a long-time interval. After a long-time interval, the highway light shall be switched when the sensor detects a tractor. A timer that produces output signals after short-time intervals and long-time intervals will be available. It can be restarted at any time (Fig. 4.19).

This is the use case for traffic-light system of Section 11. Write some requirements for this system.

Name: Cross the highway.
Description: A farmer activates the sensor on the farmroad. The controller changes the lights allowing the tractor to cross and then goes back to the normal highway-green state. The process cycles through green, yellow, and red lights until stopped by malfunction or maintenance.
Level: Low.
Scope: A simple controller for one rural intersection.
Primary actor: Farmer.
Supporting actors: Highway and farmroad lights.
Precondition: The system is in the highway green and farmroad red states.
Main success scenario:

1a. The long interval of time expires (LITE).
2. The farmer drives his tractor over the sensor.
3. The sensor signals that a tractor is waiting on the farmroad.
4. The system commands the highway light to change to yellow.
5. The timer indicates that a short-interval of time has expired (SITE).
6. The system commands the highway light to change to red and then the farmroad light to change to green.

7a. The sensor signals that the tractor is gone.
8. The system commands the farmroad light to become yellow.
9. The timer indicates that a short interval of time has expired.
10. The system commands the farmroad light to change to red and then the highway light to change to green. {Exit use case.}

Anchored alternate flows:

1b. The farmer drives his tractor over the sensor.
1b1. The sensor signals that a tractor is waiting on the farmroad.
1b2. The long interval of time expires (LITE) {go to step 4}.
7b. The timer indicates that a long interval of time has expired.

Unanchored alternate flows:

A1. Whenever the *initialize* signal occurs, the system goes to the highway green and farmroad red states.

Postcondition: The system is in the highway green and farmroad red states.
Rules:

Rule 1: The long-time interval is normally 50 s.
Rule 2: The short-time interval is normally 10 s.
Rule 3: While the farmroad lights are green or yellow the highway lights shall be red.
Rule 4: While the highway lights are green or yellow the farmroad lights shall be red.

Frequency: This cross in the highway use case could be invoked every 2 min, but usually it will be invoked about once every 8 h.
Author/owner: Terry Bahill
Date: April 1, 2016

4.3.2.4 Generic Tactical Missile

Write the requirements that are implied by the main success scenario of the launch and fly to target use case of Section 7.

4.4 Summary

In Chap. 4, we discussed different types of requirements, sources of requirements, and ways to express requirements. We presented a process comprising the various operations/activities that are performed on requirements. These include discover, clarify, revise, develop, decompose, allocate, derive, verify, validate, and manage. We also showed how requirements could be derived directly from use cases.

In Section 27, we defined requirements as statements that identify a capability or function needed by a system to satisfy the needs of the various stakeholders starting with the customer. We noted that a customer's need might be to solve a problem, achieve an objective or satisfy a contract, standard, or specification. We noted that while no two systems have identical requirements, the process of discovering requirements is the same regardless of system purposes, size, or complexity.

We drew attention to the fact that a clear, unambiguous statement of problem at the right level with the right scope is a prerequisite to developing requirements. We also made the point that the problem needs to be stated in an implementation-agnostic fashion, i.e., in a manner that does not imply a particular implementation. We defined stakeholders as individuals or organizations with a vested interest in the system. Examples of stakeholders are end users, operators, bill payers, owner, regulatory bodies, maintainers, architects, managers, as well as individuals/organization that can be favorably or negatively impacted by system-related decisions. We also noted that since engineers deliver both a product and a process for producing it, stakeholders need to include individuals/organizations with a vested interested in both the product and the process. We identified two types of requirements documents: *the customer requirements documents* that describe the problem in plain language, and the *derived requirements document* that offers a succinct mathematical/formal model of the requirements described in customer requirement document. We emphasized that the term "optimal" be avoided when characterizing a solution primarily because of the existence of multiple performance and cost criteria. As importantly, none of the feasible solution alternatives is likely to optimize all the criteria. Consequently, humans tend to design systems that satisfice, not optimize. Furthermore, it is worth reminding ourselves that humans are not optimal performers. Rather, they exhibit creativity and versatility while also exhibiting error-proneness under certain circumstance. Not surprisingly, in today's ever-changing world, optimality is less valued than affordable adaptability [9].

We then gave examples of requirement statements along with a critique of the way the requirement is stated. We presented the characteristics of good requirements, and discussed several mutually independent characterizations of system requirements such as types, sources, expressions, and input–output relationships. We identified two types of system requirements; mandatory and trade-off. We gave examples of each. We also noted that often there is a relationship between mandatory and trade-off requirements. We then discussed the concept of a scoring (i.e., utility) function. We stated that the scoring function reflects how well each requirement is being met. A scoring function transforms an input value into a normalized output score. The higher the score, the more satisfied the stakeholders and customer. A scoring function enables different criteria to be compared and traded off against each other. Simply stated, a scoring function enables the comparison of disparate entities (e.g., apples to oranges, time to dollars).

We then discussed the different sources of requirements, and defined them as categories of requirements. In Section 27, we identified two dozen sources of requirements. We went on to note that only six sources of requirements might be necessary: input–output, technology, performance, cost, trade-off, and system test.

We subsequently described each of these six sources of requirements. We acknowledged that for some purposes the best expression of requirements is narrative text (i.e., a document). Such documents are called operational concept documents, or use case descriptions. However, natural language descriptions can be ambiguous, because of language peculiarities and interpretation context. Therefore, it is advantageous in certain situations to employ "should" and "shall" statements. This structure lends itself well to both acquisition and system testing. To supplement such descriptions that suffer from some level of ambiguity, other modalities can be employed. These include a computer model, a virtual or physical prototype, a user's manual, input–output trajectories, sequence diagrams, use cases, and requirements databases. The main advantages of non-textual representations are that they are rigorous, with some being executable. These properties help illuminate contradictions and omissions in the requirements. Taking advantage of this perspective, model-based systems engineering (MBSE) is steadily growing in popularity. With MBSE, systems engineers first develop an executable model of the desired system, and then run the model to illuminate desired system behavior. As such, the model also captures system requirements.

We then discussed attributes (or characteristics) of requirements. In a spreadsheet or database representation, each requirement has associated fields that represent specific attributes of different types of requirements. For example, a requirements template can have attributes such as text, derived requirements (e.g., using SysML), verify method, link to TPM, priority, and date of last change. Other fields (containing additional attributes) can be included. Examples of additional fields are owner/approver, functional area, requirement type (mandatory/trade-off), requirement source (e.g., engineering, marketing, maintenance), assigned to whom (i.e., responsible individual/organization), satisfaction difficulty, due date, version, estimated effort to satisfy, and allocation to hardware/software components. We then followed up with several examples of requirements using the requirements template format. We gave representative examples of requirements and their attributes for a variety of domains that spanned baseball, parking garages, famous failures, watches, and an automobile.

In Section 28, we described and discussed requirements development and management along with an example. We then reviewed the requirements process in terms of major flows and feedback paths. We then described the requirements discovery process as comprising identify stakeholders, understand customer needs, state the problem, discover requirements, clarify requirements, ensure customer concurrence, decompose requirements, allocate requirements, derive requirements, prioritize requirements, determine project completion status, and revise requirements or exit. Along the way, we defined evaluation criteria, how to manage requirements, how to verify and validate requirements, and how to manage risks.

We then discussed how to fit the requirements process into the system engineering process while acknowledging the iterative nature of the system design process. We addressed the challenges posed by requirements volatility and the emergent nature of requirements in a typical systems engineering effort. We then summarized the principles of good design, and followed that up with a set of homework questions that reviewed the requirements process material.

In Section 29, we described and discussed a hybrid process for capturing requirements. The hybrid process combines traditional industry practices (i.e., shall statements) with use cases. We defined a use case as an abstraction of a required behavior of a system, and a representation that provided an observable result of value to a stakeholder (actor). We gave example of how use cases are represented. We concluded that by combining use cases with shall-statement analysis in a hybrid methodology, we are able to capitalize on the benefits of both use cases and traditional industry practices.

We then discussed the rationale for writing a requirements specification along with typical problems encountered in writing one. We showed how the concepts of use cases from the software engineering community could be harnessed to overcome the challenges in writing a requirements specification. To balance the discussions, we reviewed the strengths and weaknesses of use case modeling also. We summarized the merits and limitations of a use case in the statement, *"use cases are understandable at the expense of being potentially ambiguous."* Similarly, we summarized the merits and limitations of shall statements "*as well-defined but standalone with individual capability requirement expressed in an abstract way, without supporting context.*" Against this backdrop, we went on to make the business and technical case for a hybrid requirement capture process that combines use cases with shall statements. We then defined the hybrid process for capturing requirements, and provided an example. We concluded with a summary of the hybrid process, which provided a convenient segue to homework problems associated with capturing requirements as use cases.

References

1. Grady RB (1992) Practical software metrics for project management and process improvement. Prentice-Hall, Englewood Cliffs
2. Bahill AT, Dean F (2009) Discovering system requirement. In: Sage AP, Rouse WB (eds) Handbook of systems engineering and management. Wiley, New York, pp 205–266
3. Bahill AT, Karnavas WJ (1992) Bat selector. U.S. Patent Number 5,118,102
4. Gause DC, Weinberg GM (1990) Are your lights on?: how to figure out what the problem really is. Dorset House, New York
5. Deming WE (1986) Out of the crisis. Massachusetts Institute of Technology, Center for Advanced Study, Cambridge
6. Latzko WJ, Saunders DM (1996) Four days with Dr. Deming: a strategy for modern methods of management. Long Range Plann 29(4):594–595
7. Wymore AW (1993) Model-based systems engineering. CRC, Boca Raton
8. Simon HA (1957) Models of man: social and rational. Wiley, New York
9. Neches R, Madni AM (2012) Towards affordably adaptable and effective systems. Syst Eng 16 (2):224–234
10. Madni AM, Jackson S (2009) Towards a conceptual framework for resilience engineering. IEEE Syst J 3(2):181–191

11. CMMI for Development (2015) ver 1.3 [cited 2015 December]. http://cmmiinstitute.com/cmmi-models
12. Sommerville I (1989) Software engineering. Addison-Wesley, Reading
13. Young RR (2001) Effective requirements practices. Addison-Wesley, Reading
14. Young R (2004) Criteria of a good requirement. The requirements engineering handbook. Artech House, Norwood
15. Hooks IF, Farry KA (2001) Customer-centered products: creating successful products through smart requirements management. AMACOM Division of American Management Association, New York
16. Sneed H (2014) Automated quality assurance of software requirements. Requirements Engineering Magazine, Germany
17. Bahill AT, Botta R (2008) Fundamental principles of good system design. Eng Manag J 20 (4):9–17
18. Hooks I (1994) Writing good requirements. INCOSE International Symposium. Wiley Online Library, pp 1247–1253
19. Davis R, Buchanan BG (1984) Meta-level knowledge. In: Buchanan BG, Shortliffe E (eds) Rulebased expert systems. The MYCIN experiments of the Stanford Heuristic Programming project. Addison-Wesley, Reading, pp 507–530
20. Bahill AT, Botta R, Daniels J (2006) The Zachman framework populated with baseball models. J Enterprise Archit 2(4):50–68
21. Moody JA, Voorhees F, Bahill AT, Chapman WL (1997) Metrics and case studies for evaluating engineering designs. Prentice Hall PTR, Upper Saddle River
22. NASA (2000) Mars program independent assessment team summary report
23. Botta R, Bahill Z, Bahill AT (2006) When are observable states necessary? Syst Eng 9 (3):228–240
24. Smith ED, Son YJ, Piattelli-Palmarini M, Bahill AT (2007) Ameliorating mental mistakes in tradeoff studies. Syst Eng 10(3):222–240
25. Daniels J, Werner PW, Bahill AT (2001) Quantitative methods for tradeoff analyses. Syst Eng 4(3):190–212
26. Berger C, Blauth R, Boger D, Bolster C, Burchill G, DuMouchel W et al (1993) Kano's methods for understanding customer-defined quality. Center Quality Manag J 2(4):3–35
27. Kerzner H (2015) Project management. Wiley-Blackwell, New York
28. Chapman WL, Bahill AT, Wymore AW (1992) Engineering modeling and design. CRC, Boca Raton
29. Botta R, Bahill AT (2007) A Prioritization Process. Eng Manag J 19(4):20–27
30. Karnavas WJ, Sanchez PJ, Bahill AT (1993) Sensitivity analyses of continuous and discrete systems in the time and frequency domains. IEEE Trans Syst Man Cybern 23(2):488–501
31. Smith ED, Szidarovszky F, Karnavas WJ, Bahill AT (2007) Sensitivity analysis, a powerful system validation technique. Open Cybernet Syst J 2(1):39–56
32. Bahill AT, Szidarovszky F (2009) Comparison of dynamic system modeling methods. Syst Eng 12(3):183–200
33. Madni AM, Sievers M (2016) Model based systems engineering: motivation, current status and needed advances systems engineering
34. Madni AM (1988) HUMANE: a knowledge-based simulation environment for human-machine function allocation. In: Proceedings of the IEEE 1988 national aerospace and electronics conference. Institute of Electrical and Electronics Engineers (IEEE)
35. Shand RM (1994) User manuals as project management tools. II. Practical applications. IEEE Trans Professional Commun 37(3):123–142
36. Madni AM (2014) Expanding stakeholder participation in upfront system engineering through storytelling in virtual worlds. Syst Eng 18(1):16–27
37. Madni AM, Nance M, Richey M, Hubbard W, Hanneman L (2014) Toward an experiential design language: augmenting model-based systems engineering with technical storytelling in virtual worlds. Procedia Comput Sci 28:848–856
38. OMG (2015) Documents associated with the unified modeling language (UML). Version 2.5. http://www.omg.org/spec/UML/2.5/

39. Feynman RP (1949) Space-time approach to quantum electrodynamics. Phys Rev 76(6):769
40. Anderson C, Vanek C, Freeman H, Furlong D, Kirschbaum A, Roy R, et al (1998) Lewis Spacecraft Mission Failure Investigation Board final report
41. Bahill AT, Chapman WL (1993) A tutorial on quality function deployment. Eng Manag J 5 (3):24–35
42. Bickness BA, Bicknell KD (1995) The road map to repeatable success: using QFD to implement change. CRC, Boca Raton
43. Sage AP (1992) Systems engineering. Wiley, New York
44. Shishko R, Aster R (1995) NASA systems engineering handbook. NASA Special Publication
45. Jacobson I, Ericsson M, Jacobson A (1995) The object advantage: business process reengineering with object technology. Addison-Wesley, Wokingham
46. Bahill AT, Daniels J (2003) Using objected-oriented and UML tools for hardware design: a case study. Syst Eng 6(1):28–48
47. Booch G, Henderson-Sellers B, Jacobson I, Mellor S, Rumbaugh J, Wirfs-Brock R (1994) Methodology standards. ACM SIGPLAN Notices 29(10):223–228
48. Bahill AT, Karnavas WJ (2000) Risk analysis of a pinewood derby: a case study. Syst Eng 3 (3):143–155
49. Smith ED, Bahill AT (2007) 7.5.2 Risk analysis. INCOSE Int Symp 17(1):1123–1137
50. Bahill AT, Szidarovszky F, Botta R, Smith ED (2008) Valid models require defined levels. Int J Gen Syst 37(5):553–571
51. Chapman WL, Bahill AT (1996) Design modeling and production. In: Dorf RC (ed) The engineering handbook. CRC, Boca Raton, pp 1732–1737
52. Martin J (1996) Systems engineering guideline. CRC, Boca Raton
53. Rechtin E, Maier M (1996) Systems architecting. CRC, Boca Raton
54. Abadi CD, Bahill AT (2003) The difficulty in distinguishing product from process. Syst Eng 6 (2):106–115
55. Peña M, Valerdi R (2010) Characterizing the impact of requirements volatility on systems engineering effort. Syst Eng 18(1):59–70
56. Bahill AT (2012) Diogenes, a process for identifying unintended consequences. Syst Eng 15 (3):287–306
57. LaPlue L, Garcia R, Rhodes R (1995) A rigorous method for formal requirements definition. In: Systems engineering in the global market place. 22-6
58. Daniels J, Bahill AT (2004) The hybrid process that combines traditional requirements and use cases. Syst Eng 7(4):303–319
59. Bahill AT, Henderson SJ (2005) Requirements development, verification, and validation exhibited in famous failures. Syst Eng 8(1):1–14
60. Alistair C (2001) Writing effective use cases. Addison-Wesley, Reading
61. Jacobson I (2000) Use cases in large-scale systems. Road to the unified process. Cambridge University Press, Cambridge
62. Booch G (2005) The unified modeling language user guide. Pearson Education, India
63. Gomaa H (2004) Designing software product lines with UML: from use cases to pattern-based software architectures. Addison-Wesley Professional, Reading

Chapter 5
Trade-off Studies

5.1 Section 30. Decision Analysis and Potential Mistakes

5.1.1 History

As part of overall responsibilities, a systems engineer or trade-off analyst may have to elicit and capture the values and preferences of a customer (a decision maker) [1]. This is an important step for all stakeholders, to develop confidence in subsequent decisions. However, biases, cognitive illusions, emotions, fallacies, and the use of simplifying heuristics mentioned in Section 18 make the capture of decision maker preferences quite challenging [2]. The timely and judicious use of trade-off studies can help people in making rational decisions. But this approach can only work as long as trade-off studies are not beleaguered with mental mistakes. Fortunately, lessons have been learned in how to minimize mental mistakes during trade-off studies, as we discussed in Section 18.

Trade-off studies, also referred to as trade studies, are an effective method for choosing rationally among alternatives. Trade-off studies involve computation of multiple evaluation criteria in parallel for several alternatives simultaneously. In the absence of a trade-off study, limited attentional capacity invariably leads people to consider criteria sequentially. Tversky and Kahneman showed that humans have a tendency to anchor on certain values before evaluating probabilities [3] or numbers [4]. This tendency leads to mistakes and flawed decisions. These findings appear to be robust and are confirmed by recent research that has shown that different areas of the brain are involved in different types of decisions, and that framing and bias show neurobiological correlates [5]. Framing of hypotheses is a general problem for human decision-making [6]. Fixation on short-term rewards is also a well-documented problem in the absence of a trade-off study [7].

Trade-off studies are probably as old as numbers. The basic procedure of a trade-off study is to account numerically for empirical values and preferences. When this accounting is formalized, it provides a stable base for choosing among alternatives.

A.T. Bahill, A.M. Madni, *Tradeoff Decisions in System Design*,
DOI 10.1007/978-3-319-43712-5_5

Aristotle (384–322 BC) developed logic supported by empirical observation that illuminated the fact that humans are capable of deliberating when making a choice among alternatives. An early description of a trade-off study that "summed weighted scores" is given in a 1772 letter from Benjamin Franklin to Joseph Priestley, Fig. 5.1. The first two sentences state that when people draw out the consideration of multiple criteria over a long period of time, their thoughts about the different criteria become separated. "To get over this" problem, he suggested looking at the pros and cons simultaneously.

The seminal academic research findings on trade-off studies appeared in Keeney and Raiffa [8] and Edwards [9]. Edwards focused on the numerical determination of subjective values, an activity that the cognitive sciences have shown to be difficult. Keeney and Raiffa are best known for their axiomatic derivation of value and utility functions from conditions of preferential or utility independence. In practice, it is expensive and difficult to design and implement experiments that demonstrate these independence conditions, or measure input values among the evaluation criteria of the alternatives. Kindly refer back to Section 21 for a discussion of utility functions. Moreover, a search for alternatives usually produces complex alternatives that are so different from one another that it is difficult to invoke these elegant mathematical theorems. Haimes ([10], p. 208) notes that some discussions of optimality conditions for trade-off approaches are "strictly theoretical, however, since it is difficult or even impossible to implement the utility function approach in practice." This is so because "it is extremely difficult or impossible to actually determine the decision maker's utility function—that is, to assign numerical utilities to the various combinations of the objectives" ([10], p. 209).

The INCOSE Systems Engineering Handbook [11] provides a consensus of the fundamentals of trade-off studies. Trade-off studies are prescribed in industry for choosing and ranking alternative concepts, designs, processes, hardware, and techniques. Today, trade-off studies are broadly recognized and mandated as *the* method for choosing alternatives by simultaneously considering multiple criteria. They are the primary method for choosing among alternatives given in the Capability Maturity Model Integration (CMMI) [12] decision analysis and resolution process.

We note that even if the mathematics and utility curves are done correctly, care needs to be exercised in doing a trade-off study, because it is difficult to overcome mental mistakes. We will discuss mental mistakes in trade-off studies and offer suggestions for ameliorating their occurrence. Systems engineers can exploit this knowledge to sharpen their decision-making, and institutions can incorporate this knowledge in their documented decision-making processes.

5.1.2 Decision Analysis and Resolution

The CMMI [12] is a collection of best practices from over 800 engineering companies. It is based on the recognition that many improvements to our organizations come from process improvements [13]. To this end, the CMMI provides

Dear Sir, London Sept. 19. 1772

In the Affair of so much Importance to you, wherein you ask my Advice, I cannot for want of sufficient Premises, advise you what to determine, but if you please I will tell you how. When these difficult Cases occur, they are difficult chiefly because while we have them under Consideration all the Reasons pro and con are not present to the Mind at the same time; but sometimes one Set present themselves, and at other times another, the first being out of sight. hence the various Purposes of Inclinations that alternately prevail, and the Uncertainty that perplexes us. To get over this, my Way is, to divide half a Sheet of Paper by a Line into two Columns, writing over the one pro, and over the other Con. Then during three or four Days Consideration I put down under the different Heads short Hints of the different Motives that at different Times occur to me for or against the Measure. When I have thus got them all together in one View, I endeavour to estimate their respective Weights; and where I find two, one on each side, that seem equal, I strike them both out. If I find a Reason pro equal to some two Reasons con, I strike out the three. If I judge some two Reasons con equal to some three Reasons pro, I strike out the five; and thus proceeding I find at length where the Balance lies; and if after a Day or two of farther Consideration nothing new that is of Importance occurs on either side, I come to a Determination accordingly. And tho' the Weight of Reasons cannot be taken with the Precision of Algebraic Quantities, yet when each is thus considered separately and comparatively, and the whole lies before me, I think I can judge better, and am less likely to make a rash Step; and in fact I have found great Advantage from this kind of Equation, in what may be called Moral or Prudential Algebra. Wishing sincerely that you may determine for the best, I am ever, my dear Friend, Yours most affectionately.

B Franklin

Dr Priestly

Fig. 5.1 Letter from Benjamin Franklin to Joseph Priestley. Emphasis (in *red*) was added guidance for improving an organization's processes. One of the CMMI process areas is decision analysis and resolution (DAR). This is the big-picture within which trade-off studies fit. In this big-picture, the top-level activity is DAR and trade-off studies are a technique of DAR. Departments select the decision problems that require DAR and incorporate them into their management plans (for example, SEMPs).

DAR is a common process. Common processes are company tools that the user retrieves, tailors, and uses. DAR is invoked throughout the whole program life cycle, whenever a critical decision is to be made. DAR is invoked by IPTs (Integrated Product Development Teams), leads on programs, financial analysts, program core teams, etc. DAR should be invoked in work instructions, in gate reviews, in phase reviews, or with other triggers, which can be used anytime in the system life cycle.

A trade-off study is not an activity that is done once at the beginning of a project. Throughout a project, trade-offs are continually made when creating team communication methods, selecting components, choosing implementation techniques, designing test programs, and maintaining schedule. Many of these trade-offs need to be formally documented. It is important to recognize that in all decisions, something is gained and something is lost. As a systems engineer making decisions, you need to know what they are and make trade-offs deliberately.

The following decision problems can potentially require a formal decision process like DAR: trade-off studies, bid/no-bid, make-reuse-buy, formal inspection versus checklist inspection, tool and vendor selection, cost estimating, incipient architectural design, hiring and promotions, and helping your customer choose a solution.

Figure 5.2 shows a functional flow block diagram for the DAR trade-off study process . The functions (denoted by verb phrases) are in boxes and the inputs and

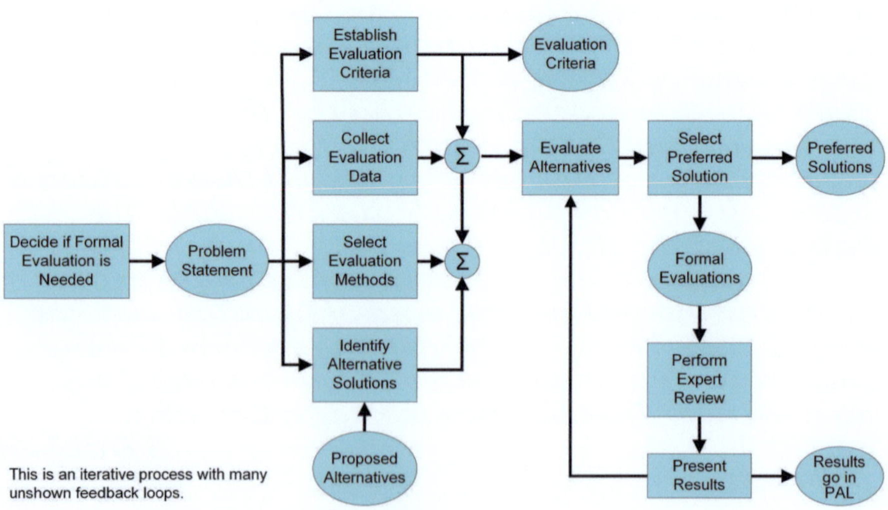

Fig. 5.2 The DAR process

outputs (denoted by noun phrases) are in ovals. We will now walk through this figure and discuss each function. These functions are drawn serially, but they are not performed sequentially. Rather, an iterative process with multiple feedback loops (not shown) is followed. In this designed process, important facets are showing functions performed in parallel and the unshown processes of configuration management and a process to improve the process.

Humans make four distinct types of decisions: (1) allocating resources among competing projects, (2) generating plans, schedules, and novel ideas, (3) negotiating agreements, and (4) choosing amongst alternatives, where the alternatives can be examined in series or in parallel. When examined in series, it is called sequential search. When the alternatives are examined in parallel, it is called a trade-off study. Trade-off studies address a range of problems from selecting high-level system architecture to selecting commercial off-the-shelf hardware and software. Trade-off studies are typical outputs of a formal evaluation processes.

5.1.2.1 Is Formal Evaluation Needed?

The first function in the DAR process (Fig. 5.2) is stating the decision that must be made and then deciding whether or not formal evaluation is needed. In this regard, companies should have polices for when to do formal decision analysis. Criteria might include when the decision is related to a moderate or high-risk issue, when the decision affects work products under configuration control, when the result of the decision could cause significant schedule delays, when the result of the decision could cause significant cost overruns, and when material procurement of 20 % of the parts constitute 80 % of the total material costs.

Guidelines suggesting the need for formal evaluation include: when the decision is selecting one or a few alternatives from a list, when a decision is related to major changes in work products that have been base-lined, when a decision affects the ability to achieve project objectives, when the cost of the formal evaluation is reasonable in comparison to the decision's impact on design-implementation, decisions where technical performance failure may lead to a catastrophic failure, and decisions with the potential to significantly reduce design risk, engineering changes, cycle time, or production costs.

5.1.2.2 Identify Alternatives

The second function in the DAR process is identifying alternative solutions for the problem. You should consider unusual alternatives to test system requirements. *The Memory Jogger* (http://www.goalqpc.com/) explains several tools for creative brainstorming. Refrain from listing alternatives that do not satisfy all mandatory requirements. If a requirement cannot be traded off then it should not be in the trade-off study. Consider use of commercial off-the-shelf and in-house entities.

The make-reuse-buy decision is part of the DAR process. Use killer trades to eliminate infeasible alternatives.

How many alternatives should a trade-off study have? If there are too many, then the trade-off study will be too complicated for decision makers to understand. If there are too few then the trade-off study will not be doing a lot of work. Paraphrasing Miller [14], we can summarize that a trade-off study should have "the number seven plus or minus two" alternatives and evaluation criteria.

At this point, let us pause for a moment to decide what to eat for lunch today. Consider the following decisions that you must make about today's lunch. Is this decision important enough to require a trade-off study? What is the preferred venue? Alternatives include the Student Union, the Big Fat Greek, and McDonalds. Alternative meals include pizza, hamburger, fish and chips, chicken sandwich, beer, tacos, bread, and water. Be sure to consider leftovers in the refrigerator. You should also consider alternative condiments such as ketchup, mustard, salt, and pepper. Another consideration is the time schedule, alternative times may be 11:30 to 12, 12 to 1, or 12 to 1:30. Perhaps the time is irrelevant to you. But for a conference planner, it is very important and worthy of being included in a trade-off study. Did you ever realize how complicated your decision about lunch was?

5.1.2.3 Establish Evaluation Criteria

The third function in the DAR process is establishing and maintaining criteria for evaluating alternatives (see Section 20). Each criterion must have a weight of importance. Each criterion should be linked to a trade-off requirement, that is, a requirement whose acceptable value can be more or less depending on the frame of reference and quantitative values of other requirements (see Section 27). Criteria are arranged hierarchically with the top-level often being performance, cost, schedule, and risk. Program management needs to prioritize (see Section 24) these criteria at the beginning of the project, and ensure that everyone on the project team and the business entity knows the priorities. Companies should maintain a repository of generic evaluation criteria that have been successfully used in the past.

Each criterion would contain the following slots, from Section 20.

Name of criterion
Description
Weight of importance (priority)
Basic measure
Measurement method
Units
Scoring function input
Scoring function type and parameters
Scoring function output
Properties
Trace to

Let us now return to our decision of what to eat for lunch. Let us write some evaluation criteria that will help with the decision. Some suggestions for evaluation criteria are: cost, preparation time, tastiness, novelty, low fat, contains elements from the five food groups, complements Merlot wine, distance to venue, length of line, messiness, and people you would be eating with.

It is worth remembering that evaluating alternatives is an expensive proposition. Therefore, early in the trade-off study, identify the most important requirements that might eliminate many alternatives. Important requirements, that are performance related, are called key performance parameters. These important requirements produce "killer" criteria in the sense that they kill many alternatives. Killer criteria produce killer trades that can often eliminate up to 90 % of the available alternatives. Killer criteria for today's lunch might be vegetarian, nonalcoholic, Kosher, or diabetic.

5.1.2.4 Select Source of Data and Evaluation Method

The fourth function in the DAR process is selecting the source of the evaluation data and the method for evaluating the data. (Yes, this could be treated as two functions, but that is not the way the CMMI does it.) Typical sources for evaluation data include customer statements, expert opinion, historical data, surveys, approximations, product literature, similar existing systems, analysis, models, simulations, experiments, prototypes, and the real system when it is available. These ideas are collectively called the trade space.

Methods for combining data and evaluating alternatives include those in Section 22 as well as the multi-attribute utility technique (MAUT), ideal point, search beam, fuzzy databases, decision trees, expected utility, pair-wise comparisons, analytic hierarchy process (AHP), financial analysis, simulation, Monte Carlo, linear programming, design of experiments, group techniques, quality function deployment (QFD), radar charts, forming a consensus, and trade-off studies.

5.1.2.5 Collect Data

The fifth function in the DAR process is collecting data for evaluating each alternative using the appropriate source (approximations, product literature, analysis, models, simulations, experiments, or prototypes). You must study the data: study the shape and the texture of the data. The shape of the data will determine the desired combining function. The texture will identify outliers.

5.1.2.6 Evaluate Alternatives

The sixth function in the DAR process is evaluating alternative solutions using the evaluation criteria, weights of importance, evaluation data, scoring functions, and combining functions. Evaluating alternative solutions involves analysis, discussion, and review. Iterative cycles of analysis are usually necessary. Supporting analyses, experimentation, prototyping, or simulations may be needed to substantiate both scoring and findings.

5.1.2.7 Select Preferred Alternatives

The seventh function in the DAR process is selecting preferred alternatives. This function involves weighing and combining the results from the evaluation of alternatives. Several combining methods are available. The true value of a formal decision process might not be in enumerating the preferred alternatives, but in stimulating thought processes and documenting their outcomes. A sensitivity analysis helps validate recommendations. The least sensitive criteria should be given weights of zero in future evaluations. This simplifies the trade-off study, while allowing future reviewers to know that you did consider those criteria.

5.1.2.8 Perform Expert Review

The eighth function in the DAR process is performing an expert review. Formal evaluations should be reviewed at regular gate reviews such as SRR, PDR, and CDR and by special expert reviews. Note that we wrote that formal evaluations should be reviewed. We did not write that the *results* of the formal evaluations should be reviewed. Technical reviews started about the same time as Systems Engineering, in 1960. The concept was formalized with MIL-STD-1521 in 1972. Technical reviews are still around, because there is evidence that they help produce better systems at less cost. Technical reviews evaluate the product of an integrated product development team (IPT). They are conducted by a knowledgeable board of specialists that includes supplier and customer representatives. The number of board members should be less than the number of IPT members. However, board expertise should be greater than the IPT's combined experience base. Who should come to the review? Here is a short list: the Program Manager, Chief Systems Engineer, Review Inspector, Lead Systems Engineer, Domain Experts, IPT Lead, Facilitator, Stakeholders for this decision, Builder, Customer, Designer, Tester, PC Server and depending on the decision, the Lead Hardware Engineer and the Lead Software Engineer. But the number of participants should be kept small, because, according to Alpert's law, the value of a review team increases as the square of the number of active participants, but only up to some critical number, after that, it drops precipitously.

5.1.2.9 Present Results to Decision Maker

The ninth function in the DAR process is presenting the results of the formal evaluation to the original decision maker and other stakeholders. These results might be the preferred alternatives, or they could be recommendations to expand the search, re-evaluate the original problem statement, or negotiate goals and capabilities with the stakeholders. A key aspect of these results is a sensitivity analysis.

5.1.2.10 Store Results in a Process Assets Library

The tenth function in the DAR process (Fig. 5.2) is storing the results of the formal evaluations, which were reviewed by the experts, into an organizational Process Assets Library (PAL) or the Project Process Assets Library (PPAL). Evaluation data for trade-off studies come from approximations, analysis, models, simulations, experiments, and prototypes. Each time better data is obtained, the PAL should be updated. Formal evaluations should be designed with reuse in mind.

5.1.2.11 Parking Garage Example

This is a baby example of a trade-off study for the parking garage entry gate of Section 10. The question we will answer is, "What is the best type of entrance gate to use on this type of a garage?"

First, is formal evaluation necessary? This decision (1) is selecting one or a few alternatives from a list, (2) affects the ability to achieve project objectives, and (3) its trade-off study cost is miniscule compared to the revenue the garage will generate. This garage sells 2000 permits per year at an annual cost of $500 each. Thus, it generates one millionThus, it generates one million dollars of revenue per year. Therefore, a formal trade-off study is justified.

We will use these three evaluation criteria.

Name of criterion: Speed of opening.
Description: The duration of the gate movement in seconds will be measured from its first detectable movement until it is stationary at the end of movement.
Weight of importance: 6.
Basic measure: The duration of the gate movement from its first detectable movement until it is stationary at the end of movement.
Measurement method: Tester will use a stopwatch to measure the duration of the gate movement from its first detectable movement until it is stationary at the end of movement.
Units: seconds.
Scoring function input: The domain is 0–10 and the expected values are 1–5.
Scoring function: SSF9 (1, 2, 3, −1, RLS [0–10]).

Scoring function output: 0–1.
Properties: This criterion is related to the constraints of Fig. 1.53 and the equations of Fig. 1.54.
Trace to Section 10 NFPR2.
Owner: Director of Parking and Transportation.
Date of last change: October 26, 2015.

Name of criterion: Footprint.
Description: Square feet of surface area that must be kept free for the gate. This is the obstruction-free region in front and behind the gate. The area in front of the gate must be as big as the largest vehicle that will be allowed into the garage.
Weight of importance: 4.
Basic measure: Maximum square feet of surface area that must be kept free for the gate.
Measurement method: This will be calculated using the first architectural drawings.
Units: square feet.
Scoring function input: The domain is 0–1000 and the expected values are 300–500.
Scoring function: SSF9 (0, 400, 600, −0.005, RLS [0–1000]).
Scoring function output: 0–1.
Trace to concept of operations.
Owner: Director of Parking and Transportation.
Date of last change: October 26, 2015.

Name of criterion: Security.
Description: Difficulty for a transgressor to force unpaid entry.
Weight of importance: 8.
Basic measure: Difficulty for a transgressor to force unpaid entry.
Measurement method: Hire students and loan transportation pool pickup trucks to them. Find garages with similar entry gates. Ask the students to force entry. Tester will assess the difficulty and estimate the damage. Tester will produce a number between 0 and 10.
Units: none.
Scoring function input: The domain is 0–10 and the expected values are 0–10.
Scoring function: SSF3 (0, 5, 10, 0.02, RLS [0–10]).
Scoring function output: 0–1.
Trace to concept of operations.
Owner: Director of Parking and Transportation.
Date of last change: October 26, 2015.

We will consider these alternatives:

- A wooden arm that rotates about a pivot point from a horizontal to a vertical position.
- One or two gates that rotate on hinges like a door in a house.
- A Marine guard with a gun.

Table 5.1 Trade-off study matrix for the parking garage

Evaluation criteria	Weight	Norm criteria weights	Alt 1 Do nothing		Alt 2 rotating arm		Alt 3 swinging gates		Alt 4 Marine guard with gun	
			sc	$w \times$ sc	sc	$w \times$ sc	sc	$w \times$ sc	sc	$w \times$ sc
Speed of opening	6	0.33	1.0	0.33	0.5	0.17	0.3	0.10	0.9	0.30
Footprint	4	0.22	1.0	0.22	0.9	0.20	0.2	0.04	0.9	0.20
Security	8	0.44	0.0	0.00	0.2	0.09	0.5	0.22	1.0	0.44
Alternative rating				**0.56**		**0.46**		**0.37**		**0.94**

sc score, *w* weight, × multiplication sign, *norm* normalized, *Alt* alternative

Please note that the gate does not need to be physical. It could be electronic, electro-magnetic, a force field, etc.

Evaluation data come from expert opinion, similar existing systems, and approximations. The method for combining the data is the sum of weighted scores. The question to be answered is, "What is the best type of entrance gate to use on a university parking garage?" (Table 5.1).

This trade-off study shows that the Marine guard with a gun is the recommended entrance gate for a parking garage. But of course, we did not consider cost. The do nothing alternative is second, which is worrisome. To reduce its score we could add performance criteria such as revenue or the revenue-to-cost ratio. This trade-off study was reviewed by the authors of this book and was stored in their process assets library.

5.1.2.12 Cookie Acquisition System

This is an example of a partial trade-off study for the Cookie acquisition system. The question that this trade-off study will answer is, "What would be the best way for our student to get chocolate chip cookies?" First, is formal evaluation necessary? Our customer, the student, says that this is important. Therefore, we will do a trade-off study. We will use the following three evaluation criteria.

Name of criterion: Audible signal indicating cookies are ready.
Description: An audible signal shall inform the student when the cookies are ready. This signal should have a nominal intensity level of 80 ± 2 dB.
Weight of importance: 9.
Basic measure: Intensity level of an audible signal.
Measurement method: During design and construction, the proposed device will be mounted on a test bench and will be activated in accordance with test instructions. Sound intensity level in decibels (dB) will be measured at a distance of 30 cm. At final test and during operation, an actual oven will activate the audible signal and the sound intensity level in decibels (dB) will be measured at a distance of 30 cm.

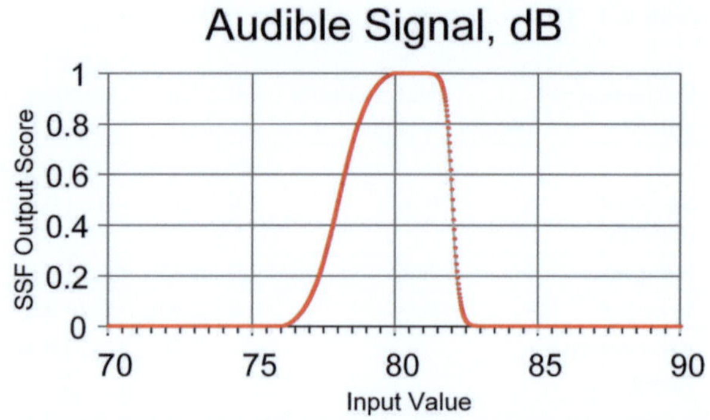

Fig. 5.3 Scoring function for Audible Signal criterion

Units: decibels (dB).

Scoring function input: The measured sound intensity should always be between 70 and 90 dB.

Scoring function: Hill shaped, SSF5, $L = 76$, B1 = 78, $O = 80$, B2 = 82, $U = 84$, S1 = 0.5, S2 = −2, D = RLS[70–90] = plotting range. SSF5(76, 78, 80, 82, 84, 0.5, −2, RLS [70–90]). Here we used the mandatory requirement thresholds of 78 and 82 dB as the baseline values, because we believe the values will improve through the design process (Fig. 5.3).

Scoring function output: 0–1.

Trace to Section 3, ReqF5.

Owner: Engineering.

Date of last change: 1/8/2016.

Name of criterion: Lost study time.

Description: While the student is making cookies, driving to the bakery to buy cookies, or bargaining with his mother to get her to make cookies for him, he will not be studying. This lost study time is the criterion. However, for optimal learning, students do need breaks.

Weight of importance: 7.

Basic measure: Amount of study time lost in getting the cookies.

Measurement method: During design, this lost study time will be calculated by analysis. At final test and during operation, this lost time will be measured during demonstration.

Units: minutes.

Scoring function input: time in minutes.

Scoring function: SSF5(10, 12, 15, 25, 30, 0.4,−0.2, RLS[0–60]), plotting range = IJS[0–30]. Here the mandatory requirement value is the upper threshold. We do not expect the scores for this criterion to increase, so we award no points for alternatives that do not satisfy the mandatory requirement (Fig. 5.4).

Scoring function output: 0–1.

Trace to concept of operations.

Owner: Student.

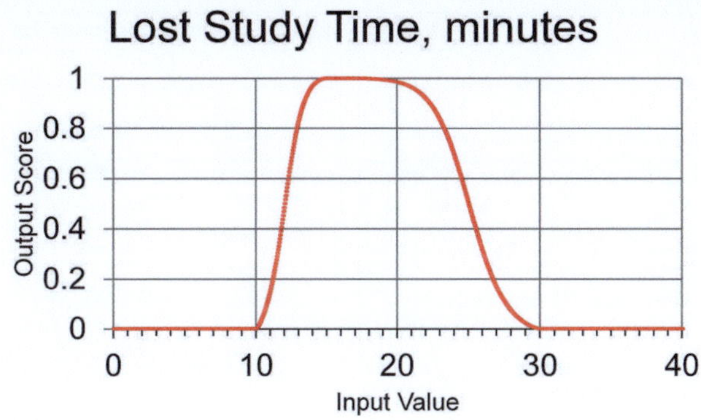

Fig. 5.4 Scoring function for lost study time criterion

Date of last change: 1/8/2016.

Name of criterion: Nutrition.
Description: Four cookies should contain less than 520 cal, 24 g of fat, and 72 g of carbohydrates.
Weight of importance: 5.
Basic measure: calories, grams of fat, and grams of carbohydrates.
Measurement method: Use data from the Internet, for example, http://www.pillsbury.com/products/cookies/refrigerated-cookies/chocolate-chip.
Units: calories, grams of fat and grams of carbohydrates.
Scoring function inputs: RLS[0–1000].
Scoring functions:

Subcriterion calories SSF9(100, 500, 1000, −0.003, RLS[0–1000]).
Subcriterion fat SSF9(10, 25, 50, −0.07, RLS[0–100]).
Subcriterion carbohydrates SSF9(10, 75, 150, −0.02, RLS[0–200]).

Scoring function output: 0–1.
Trace to concept of operations.
Owner: Student.
Date of last change: 1/8/2016.

We will consider these alternatives:

- Ask your mother to bake cookies for you
- Use a tube of Pillsbury refrigerated chocolate chip cookie dough
- Go to the bakery and buy chocolate chip cookies

Evaluation data come from expert opinion and measurements during trips to a grocery store and a bakery. The method for combining the data is the sum of weighted scores (Fig. 5.5). The question to be answered is, "What is the best way for our student to get some chocolate chip cookies?" (Table 5.2).

This trade-off study shows that Mom's cookies are the preferred alternative. The do nothing alternative is ranked high, which is worrisome. This probably happened because we do not have any performance criteria, such as anticipated tastiness.

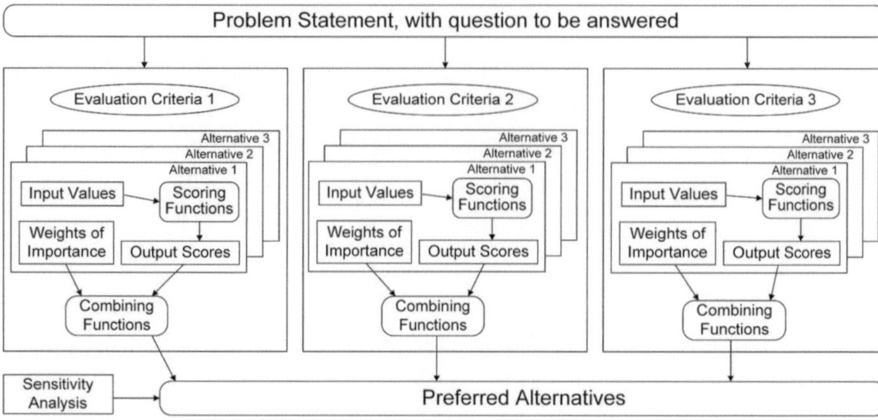

Fig. 5.5 Elements of a trade-off study

This study was criticized because of its lack of performance and cost evaluation criteria. This trade-off study was reviewed by the authors of this book and was stored in their process assets library.

5.1.3 Potential Mental Mistakes in Trade-off Studies

Humans often make mental mistakes when conducting trade-off studies. Smith [15] extracted seven dozen mental mistakes including heuristics, biases, representativeness, anchoring, base-rate neglect, the conjunction fallacy, cognitive illusions, emotions, fallacies, fear of regret, psychological traps, and paradoxes from the field of *behavioral economics* and showed how they could induce mental mistakes in trade-off studies. We studied these mental mistakes and determined whether each one principally affects (1) magnitudes, such as evaluation data, weights, and probabilities or (2) textual descriptions, such as stating the problem, writing criteria, or proposing alternatives.

In Section 18, we described these mental mistakes and offered recommendations for ameliorating their effects on the trade-off study components. We provided a cautionary alert about the likelihood of large-scale occurrences of cognitive biases within trade-off studies. We recommended actions that should help people avoid making specific mental mistakes when conducting trade-off studies. Our intention is to help people in defining trade-off studies to choose among alternatives. We want people to have confidence in their decisions.

We then categorized the mistakes by the trade-off study component affected. The components of a trade-off study are:

- Problem statement
- Evaluation criteria

Table 5.2 Trade-off study matrix for the cookie acquisition system

Hierarchal trade-off study matrix for the cookie system

Criteria	Wt	Norm criteria weights	Norm subcriteria weights	Alt 1 do nothing		Alt 2 ask mom		Alt 3 Pillsbury's chocolate chip cookie dough		Alt 4 buy cookies at bakery	
				sc	$w \times$ sc	sc	$w \times$ sc	sc	$w \times$ sc	sc	$w \times$ sc
Audible signal for cookies are ready	9	0.43		0.0	0.00	0.6	0.26	0.5	0.21	0.4	0.17
Lost study time	7	0.33		1.0	0.33	0.8	0.27	0.5	0.17	0.3	0.10
Nutrition	5	0.24		1.0	0.24	0.3	0.07	0.5	0.12	0.5	0.11
Calories			0.34	1.0		0.3		0.5		0.4	
Fat, grams			0.33	1.0		0.3		0.5		0.5	
Carbohydrates, grams			0.33	1.0		0.3		0.5		0.5	
Alternative rating					0.57		0.60		0.50		0.38
		1.00	1.00								

Abbreviations: *sc* score, *w* weight, × multiplication sign, *norm* normalized, *Alt* alternative

- Weights of importance
- Alternative solutions
- Evaluation data
- Scoring functions
- Normalized output scores
- Combining functions
- Preferred alternatives
- Sensitivity analysis

An important part of the problem statement is stating the question that the trade-off study is trying to answer [16]. For the last two examples, the questions to be answered are, "What is the best type of entrance gate to use on a university parking garage?" and "What is the best way for our student to get some chocolate chip cookies?"

5.1.3.1 Advice

The mental mistakes presented in this section represent errors in judgment to which a trade-off analyst can be expected to fall prey. Recommendations for dealing with these types of mental mistakes exist in the literature [17, 18] and in Section 18.

Humans usually tend to consider alternatives in series, and often hastily choose one alternative after having their attention anchored, or fixated, to only one or a few criteria [16]. Also, humans tend to choose alternatives based on conclusions derived from their favorite theories. Ideally, decision analysts should consider the complete decision trade space, or a complete set of alternatives, and then use a shrinking set of hypotheses brought about by conclusions based on experimentation, empirical data, and data analysis. In other words, we recommend that, for appropriate problems, the decision maker should eschew typical sequential processing of alternatives, in favor of evaluating alternatives in parallel in a trade-off study.

To choose rationally among alternatives, the decision analyst should be aware of mental mistakes, and employ sound debiasing techniques. Advice from subject matter experts should augment individual judgment. Decision analysts should also have a complete understanding of mathematical methods that allow parallelization of human decision processes through trade-off studies, and be able to apply them error-free.

Complex choices among alternatives should not be reduced to "gut-feel decisions." Nonexperts should realize that making important decisions with a holistic, gut-feel approach, without regard to the needed rationality, is risky. To establish rationality, all components of the decision must be carefully considered until clarity results regarding their arrangement and magnitude. This is possible by focusing on the elements individually and then collectively. The higher level decision then becomes a calculation based on a broad base of rationally considered components.

There is considerable interest in complex decisions made in the instant of perception [19]. It should be noted that experts capable of making such judgments have trained for extended periods [20], during which time they have individually, sequentially, repeatedly, comparatively, and rationally examined all components of a decision [21]. Any preconscious parallelization that occurs in an expert's brain is reproduced in the parallel structure of a trade-off study, which is ultimately based on analysis of hard data.

Limited time and resource availability virtually guarantee that no trade-off study will contain all required criteria. Therefore, trade-off studies do not produce *optimal* solutions. Rather, they produce *satisficing* solutions ([22, 23] and Section 5). A trade-off study typically reflects one view of the problem. Different decision analysts might choose different criteria and weights and, therefore, paint different pictures of the problem. In this section, we have not addressed those human decision-making mental mistakes for which we have no suggested corrective action. Examples of these types of mistakes are closed mindedness, lies, conflict of interest, political correctness, and favoritism.

In a trade-off study, it is natural and reasonable to ask, "Have we written down enough criteria? Have we studied a sufficient number of alternatives?" One way to assess the completeness of a modeling study is to complete a Zachman framework for the study [24].

We studied three trade-off studies that had variability or uncertainty statistics associated with each evaluation datum. Two were personal observations and one was published [25]. In these studies, uncertainty statistics were carried throughout the computational process, so that at the end, the preferred alternatives had associated uncertainty statistics. However, these studies employed approaches that were too complicated for most practical purposes. Therefore, in our trade-off studies we purposely do not address propagation of uncertainty throughout the trade-off study or changes in uncertainty or dependencies in the evaluation data during the study in the interest of keeping things understandable.

It is important to recognize that the output of a trade-off study is recommendations which are combined with nonquantitative assessments of organizational vision, goals, culture, and values.

5.1.3.2 Factors Affecting Human Decisions

Some of the factors that affect human decision-making include the decision maker, corporate culture, the decision maker's values, personality types, risk averseness, mental mistakes, information display, wording of questions, context of presentation, the decision, effort required to make the decision, difficulty of making the decision, time allowed to make the decision, needed accuracy of the decision, cost of the decision, and likelihood of regret [2].

Good industry practices for ensuring the success of trade-off studies include having teams evaluate the data, evaluating the data in multiple iterations, and expert review of the results. It is important that expert review teams have reviewers that are external to the project. Additionally, the reviewers need to examine design decisions while also ensuring that required tasks were performed. Unfortunately, reviews are often hampered by failure to allow external reviewers access to proprietary or classified data [26].

5.1.3.3 Purpose of Teaching Trade-off Studies

Emotions, illusions, biases, and use of heuristics make humans far from ideal decision makers. The proper use of trade-off studies can help move decisions from the usual human decision-making lower-right quadrant of Fig. 2.11 to the ideal decision-making upper-left quadrant.

5.1.3.4 Improving the Trade-off Study Process

Here are some suggestions for improving your recommendations to your boss. Inform decision makers about how mental mistakes affect trade-off studies (forewarned is forearmed). Encourage a long-term institutional decision horizon. Use a team approach. Iterate frequently. Institute both expert and public reviews. Finally, try to reduce mental errors by using the aforementioned methods and techniques. For a good decision, avoid asking a human.

Trade-off studies seek to build a mathematical framework for the problem. The goal is a correct, parallel mathematical consideration of all relevant criteria, avoiding misjudgments associated with sequential consideration of criteria in sub-groups. In this regard, cognitive science can help improve the validity and sensitivity of trade-off studies.

5.1.3.5 RMS Titanic, 1912

Another common mental mistake is Invincibility Bias. Many bad decisions can be attributed to the decision maker's sense of invincibility. As noted earlier, teenagers, especially boys, are notorious for thinking: "I won't get caught; I can't get hurt; I will avoid car accidents; I won't cause an unwanted pregnancy; I won't get sexually transmitted disease;" and that, "They can't do that to me." Many other people think, "Nobody is going to steal my identity, I'll be able to quit any time I want to, I don't need sun screen, I won't get skin cancer;" and "I don't have to back up my hard drive, my computer won't crash." The Spanish Armada was thought to be invincible in 1588. In 1941, the Japanese thought they could destroy the US Pacific fleet with a sneak attack on Pearl Harbor without awakening the sleeping giant. The German military thought it was invincible as it stormed across Europe in WW II. And of course, in 1912, the White Star line said that the Titanic was "unsinkable." The invincibility bias will affect the risk analysis, the problem statement, and the selection of criteria for the trade-off study (Fig. 5.6).

The original design for the **RMS Titanic** called for 64 lifeboats, but this number was reduced to 20 before its maiden voyage. This was an avoidable trade-off mistake. The Chief Designer wanted 64 lifeboats, but the Program Manager reduced it to 20 after his advisors told him that only 16 were required by law. The Chief Designer resigned over this decision. The British Board of Trade

Fig. 5.6 Willy Stöwer's 1912 conception of the Titanic's last minutes

regulations of 1894 specified lifeboat capacity: for all ships over 10,000 t, this lifeboat capacity was specified by volume (5500 cubic feet), which could be converted into seats for people (about 1000) or the number of lifeboats (about 16, assuming about 60 people per boat). So, even though the Titanic displaced 46,000 t and was certified to carry 3500 passengers plus crew, its 20 lifeboats complied with the regulations of the time. But let us go back to the design decision to reduce the number of lifeboats from 64 to 20. What if they had performed the following hypothetical trade-off study? In Table 5.3, the weights of importance range from 0 to 10, with 10 being the most important, and the evaluation data (scores) also range from 0 to 10, with 10 being the best. For simplicity, we have not used scoring functions, so the evaluation data are also the scores.

The magnitudes of the Program Manager's and the Chief Designer's alternative ratings are not in themselves important. What is important is that they indicate different preference for alternatives, which result from different sets of weights of importance.

The Program Manager might have had **overconfidence in his subjective choice** of 20 lifeboats. If he had done this trade-off study, might the Program Manager have rethought his decision to use only 20 lifeboats?

Many bad decisions can be attributed to the decision maker's sense of **invincibility**. In 1912, the White Star line said that the Titanic was "unsinkable." If the Program Manager had not also believed this, would he have authorized more lifeboats?

If the Program Manager understood the **Forer effect** (that an analyst might fail to question or rewrite criteria that originated from a perceived authority) might he have reassessed the Board of Trade's regulation for 16 lifeboats?

Table 5.3 An apocryphal trade-off study for the RMS titanic

Evaluation criteria	Weights of importance		Alternatives and their evaluation data			
	Program manager's weights	Chief designer's weights	10 lifeboats	20 lifeboats	30 lifeboats	64 lifeboats
Will it satisfy the Board of Trade regulations? (yes, no)	10	10	0	10	10	10
Amount of deck space required for the lifeboats (ft^2)	2	2	10	8	4	2
Possible perception that the ship is unsafe caused by the presence of a large number of lifeboats	6	2	10	8	4	2
Cost to purchase, install, maintain, launch and operate the lifeboats (£)	9	4	10	8	4	2
Percentage of passengers and crew that could be accommodated in lifeboats, if all lifeboats were launched full of people	1	10	2	4	6	10
Alternative ratings produced by summing the Program Manager's weights times scores			172	**240**	174	144
Alternative ratings produced by summing the Chief Designer's weights times scores			100	204	192	**216**

The Program Manager and the Chief Designer did not do a trade-off study. They merely discussed the 20 and 64-lifeboat alternatives. If they had understood **distinctiveness by adding alternatives** and had done this trade-off study with the addition of the 10 and 30-lifeboat alternatives, is it likely that the Program Manager would have chosen a different alternative?

The mandatory requirement of satisfying the Board of Trade regulations ruled out the 10-lifeboat alternative. A **sensitivity analysis** of the remaining trade-off study shows that the most important parameter from the Program Manager's perspective is the weight of importance for the cost criterion and that the most important parameter from the Chief Designer's perspective is the weight of importance for the "percentage of passengers and crew that could be accommodated in lifeboats" criterion. Therefore, the Program Manager should have spent more time assessing the **magnitude and reliability** of the values of these parameters. In fact, he should have noted the importance of safety regulations, and questioned whether the Board of Trade's regulation for 16 lifeboats was reliable for the new, larger Titanic design.

5.1.3.6 Section Summary

Humans usually consider alternatives sequentially. They often hastily choose one alternative after having their attention anchored to just one or a few criteria. To make good, rational choices among alternatives, a decision maker's awareness of cognitive biases must increase. Limited awareness and possible consequential irrationality can cause erroneous estimation or misuse of trade-off study components. Therefore, trade-off study components should be examined individually. The higher level decision then becomes a calculation resting on a broad base of rationally considered assumptions, components, and calculation methods. Decision makers should understand the mathematical methods that allow the parallelization of human decision processes through trade-off studies.

5.1.4 Homework Problems

5.1.4.1 DAR and Trade-off Studies

What is the difference between DAR and a trade-off study. Hint: DAR is more inclusive.

5.1.4.2 Select Evaluation Methods

DAR's specific process 1.4 is select evaluation methods. Please list at least six methods (or sources of data) to be used to get the evaluation data for a trade-off study.

5.1.4.3 Umpire's Assistant

This is the product position statement for the Umpire's Assistant.

For the baseball umpire who needs to call balls and strikes, the Umpire's Assistant is an intelligent decision system that helps him to call balls and strikes accurately, consistently, and in real-time; unlike unassisted human umpires, our product enforces the same strike-zone standards for all leagues, parks, umpires, batters, and pitchers.

Here are two alternatives for this trade-off study.

Seeing Eye dog. How often have you heard a rowdy fan in the stands yell out, "Ump, You're blind as bat." Well one of the alternatives is making that fan's worst nightmare come true. We will train a Seeing Eye dog to call balls and strikes. At the beginning of the game, the umpire will walk out guided by his Seeing Eye dog and will proceed to call balls and strikes aided by his dog.

The **Umpire's Assistant** could be implemented with multiple cameras.

Here are two evaluation criteria.

Criterion 1: Static accuracy.
Description: Accuracy or precision of calling balls and strikes.
Trace: This criterion traces to functional requirements FR2 and FR3.
Weight of importance: 10.
Basic measure: Percentage of pitches correctly called as a ball or a strike during static testing.
Units: percentage.
Measurement method: The static tests are described in the system test section of Document 3. The result of each test will be recorded by the tester and stored in the system's internal computer. At the end of the static test, it will compute the percentage of correct calls. One incorrect call in the set of 192 tests would yield 99.5 % accuracy.
Input: The domain is 0–100 %. The expected values are 90–100 %.
Output: 0–1.
Scoring function: Monotonic increasing with lower threshold of 96, baseline of 99, baseline slope of 0.1, and upper threshold of 100. SSF3 (96, 99, 100, 1, RLS [0–100]).
Criterion 2: Signaling the umpire. The system shall indicate to the umpire whether the pitch was a strike or a ball. The speed and silence of this signaling is important.

These alternatives and criteria were put into the trade-off study Tables 5.4 and 5.5, where they are combined with linear addition of weighted scores.
Sensitivity Analysis: Now we want to perform a sensitivity analysis of this trade-off study. The criterion function is the final score for the Umpire's Assistant minus the final score for the Seeing Eye dog.

$$F = F_1 - F_2 = w_1 \times \mathrm{sc}_{11} + w_2 \times \mathrm{sc}_{12} - (w_1 \times \mathrm{sc}_{12} + w_2 \times \mathrm{sc}_{22})$$

Table 5.4 Numeric trade-off study table

Evaluation criteria	Weight of importance	Umpire's assistant	Seeing eye dog
Accuracy	0.75	0.67	0.33
Signaling	0.25	0.83	0.17
Sum of weight times score		0.71 The winner	0.29

Table 5.5 Symbolic trade-off study table

Evaluation criteria	Weight of importance	Umpire's assistant	Seeing eye dog
Accuracy	w_1	sc_{11}	sc_{12}
Signaling	w_2	sc_{21}	sc_{22}
Sum of weight times score		F_1	F_2

Compute the semirelative-sensitivity functions for this criterion function, with respect to the six parameters. Which parameter is the most important? What should you do next? What is the significance of the minus sign in some of the semirelative-sensitivity functions?

5.1.4.4 Cookie Acquisition System

Create a trade-off study matrix for the Cookie acquisition system of Section 3. Use these evaluation criteria: anticipated tastiness, price, preparation/travel time, and number of calories. Consider these alternatives: ask mom to bake cookies, buy Pillsbury's chocolate chip cookie acquisition system dough at the grocery store, and buy cookies at the bakery. You do not need to use scoring functions. You should estimate your own values for the scores.

5.1.4.5 Selecting Lunch

Create a trade-off study matrix to help you select today's lunch. Choose evaluation criteria from this list: cost, preparation time, tastiness, novelty, low fat, contains the five food groups, complements Merlot wine, distance to venue, length of line, messiness, and whom you are eating with. Choosing the people you eat lunch with is one of the most important decisions in your career. It establishes communication links, alliances, and helps shape your values. The above criteria are relatively independent and also show compensation; however, they are not objective. We considered these killer criteria for today's lunch: Must be vegetarian, nonalcoholic, Kosher, and diabetic. We considered these candidate meals: pizza, hamburger, fish and chips, chicken sandwich, beer, tacos, bread and water, BBQ ribs, salad, spaghetti and soda and a candy bar out of the vending machine. Be sure that you consider leftovers in the refrigerator. You should use the BIMS spreadsheet. What did you learn from this study? In particular, discuss the results of your sensitivity analysis.

5.1.4.6 Alternative Energy Storage Systems

One technological gap that must be filled to increase the percentage of energy supplied by renewable sources is the need for grid-level energy storage. Utilities will choose from among a broad suite of potential technologies. In a constantly changing technological environment, making the right choice is no simple task. Some of the key grid-level energy storage technologies that are currently in use are described below. [27] is a simple reference and [28] is complex. See also https://en.wikipedia.org/wiki/List_of_energy_storage_projects#Forms_of_energy_storage. Create at least five evaluation criteria for a trade-off study of these alternative energy storage systems. Use the template in Section 20. You will use these criteria in a Section 5.3.7.5 homework problem entitled Alternative Energy Storage Systems.

Pumped-hydro potential energy storage is the most mature and widely utilized energy storage technology in the world, with over 90 GWh of pumped hydro storage in use worldwide. When electricity demand is low, motors pump water from a low reservoir to a higher reservoir. When electricity demand is high, water from the high reservoir flows back to the low reservoir through electric generators.

Hydroelectric dams, such as Hoover Dam, were installed to control floods and generate electricity. They were not intended to be energy storage systems, but they are. They release more water during the daytime when societal demand for electricity is greatest (see Fig. 6.11). They are complementary to renewable electric generating sources, because they can peak when the renewables trough.

Railroad car potential energy storage. When electricity demand is low, railroad cars loaded with dead weight are driven up a steep incline. When demand is low, the railroad cars are drifted down the incline generating electricity. This technology would not work well during periods of excessive snow.

Compressed air energy storage systems pump air into an underground storage formation, such as an abandoned salt cavern or mine. When electricity demand is low, motors compress air and pump it into the cavern. When electricity demand is high, the air is released and as it expands, it drives turbines to make electricity.

Thermal energy storage circulates molten salt. When electricity demand is low, mirrors reflect sunlight onto tanks of molten salt heating them to around 1000 °F. When demand is high, the salt's heat turns water into steam to turn electric-generating turbines. The other alternatives exchange energy with the power grid. This technology accepts solar energy.

Thermal energy coupled to air conditioning. When electricity demand is low, electricity is used to chill water stored in tanks. When demand is high, the water is pumped through countercurrent heat exchangers to air condition buildings. This is inexpensive, but is only useful in the summer.

Hybrid automobiles connected to the grid. When electricity demand is low (for example, at night), the grid charges the batteries in electric cars. This is a big reservoir for energy, but it reduces the life of the batteries. When demand is high, these cars are off the grid, driving home from work. The other technologies time-shift the peak power. This technology does not. It accepts power at night, but does not give power back to the grid.

Electrochemical batteries are receiving so much research that they are certain to be used in electric grids. When demand is low, batteries send negatively charged ions to one side of a physical barrier and positive ions to the other side. When demand is high, batteries send negatively charged ions to the other side of the physical barrier. Examples are lead-acid, molten sodium-sulfur, and lithium ion batteries.

Flow batteries are similar to other rechargeable batteries, except that in flow batteries the electrolytes are liquids that are stored in tanks external to the batteries, they can be much larger and therefore can store much more energy. Examples are vanadium-redox flow batteries.

Flywheels store kinetic energy. When electricity demand is low, the system acts as a motor and accelerates a rapidly spinning cylinder in a vacuum-sealed case. When electricity demand is high, the system acts as a generator and the kinetic energy is converted back into electricity. Energy can only be stored for periods around 15 min.

Superconducting Magnetic Energy Storage uses the magnetic field produced by cryogenically cooled superconductors to store energy. Superconductors lose very little energy in the storage process and the storage and usage is in the same form, electricity. They require a loop 100 miles in diameter.

Do Nothing. Grid-connected household electric generators do not need batteries. The AC electric power grid acts as a perfectly efficient battery that never needs replacing. When the sun shines (or the wind blows), the electric meter spins backwards sending your energy into the grid. When the sun does not shine (or the wind does not blow), your electric meter spins forward. Batteries are about half the cost of a typical renewable-energy electric-generation system.

5.1.4.7 Baseball Statistics

This problem uses results of problem 3.1.6.5 baseball statistics, and also the database located at http://www.springer.com/gp/book/9783319437101#otherversion=9783319437125 and http://sysengr.engr.arizona.edu/forStudents/index.html.

Use several evaluation criteria and create a trade-off study to select the best batters of the twentieth century. Compute the sum and the product combining functions with the following normalized evaluation criteria. The most important factor in a baseball season is winning games. The most important criteria for winning a game is scoring runs. So we want to include runs scored (R) and runs batted in (RBI). These criteria show good compensation. Let us give those weights of eight. Next hits (H) are important because they will put the batter in a position to score runs and it might advance baserunners to positions that are more favorable for scoring runs. Let us give that a weight of five. Batting average (BA) is slightly dependent on hits, but it is very popular with the public, because it does not increase throughout the season. So we will use it and give it a weight of six. We will not use homeruns (HR) because that criterion is dependent on runs batted in. In a close game, the ability to bunt successfully (SH) is important. We will give it a weight of one. Does this set of evaluation criteria satisfy all of the qualities of good evaluation criteria? Remember to normalize your weights. What are your two combining functions? What are the recommendations of these two combining functions?

5.1.4.8 SpinCoach

The following pages are excerpts from and additions to the SpinCoach documents. Point out mental mistakes that you find in them.

Document 1: The Problem Situation

When a spinning object (like a baseball) is put in a moving fluid (like air) it will experience a force that pushes it sideways. Some highly successful baseball players have written that they see this spin of the ball and use this information to predict its future trajectory. But at present, there is no system that can teach high school and college baseball and softball players to predict this spin-induced deflection of the pitch. Our customer needs such a system. The goal of this project is to design a web site that will display images of spinning baseballs, allowing the subject to predict the spin-induced deflection and providing feedback to facilitate learning.

The phrase "a web site that will display images of spinning baseballs," is a solution, not a problem statement.

Document 5: Concept Exploration

Alternative Architectures for the SpinCoach

1. The Status Quo. Some batters can recognize the spin on the ball and predict its spin-induced movement. But they have difficulty verbalizing this capability and teaching it to others. In batting practice, we could have the pitcher announce to the batter "curve" and then throw a curveball: announce "slider" and then throw a slider. This could be done with a human pitcher or a pitching machine.
2. Computer Simulations. Images of spinning balls can be simulated and presented on a computer monitor.
3. Spinning baseballs. Batters can view balls spinning on drills. Videos for such spinning balls are on the web site http://www.sie.arizona.edu/sysengr/baseball/index.html, under the tag line "here is a ten-second video of the four-seam (left) and two-seam (right) fastballs." The user runs the SpinCoach from his or her hard disk. The user must login for each session. The information gathered at login is used to track user performance history.
4. System on CD-ROM. Balls spinning on drills can be photographed and their images stored on CD-ROM disks. Such videos are available at the URL above. These images along with the software program (application) will be transferred from the CD-ROM to the user's hard disk using a license key provided by BICS. There after the user runs the SpinCoach from his or her hard disk. The user must login for each session. The information gathered at login is used to track user performance history. System upgrades will be provided with new CD-ROMs.

4b. Application Sold Online. A slight variant of the CD-ROM alternative is to sell the application through an online store with distribution via Internet downloads.

5. Interactive DVD. Balls spinning on drills can be photographed and their images stored on DVDs. Such videos are available at the URL above. These images along with the software program (application) will be accessed by the user. There will be no user performance tracking. System upgrades will be provided with new DVDs.

6. Web-based Application. Balls spinning on drills can be photographed and their images stored on the BICS web server. Such videos are available at the URL above. These images along with the software program (application) will be on an Internet accessible web site. This system will be based on the Apache web server with web pages written in PHP or HTML and video clips in AVI format. Access to the system will be granted by monthly subscription and login-based authentication control. The user will have a profile in the system and can access this profile from any terminal connected to the Internet. The system will store user information in a database (such as MySQL). This information is used to track user performance history. System upgrades can be made on the web site at any time and will be transparent to the user.
7. Addition to commercial products. Add our module to the commercial off-the-shelf products of the Vizual Edge Performance Trainer™ (VEPT®) (http://www.vizualedge.com/) or the Zepp baseball analysis system http://www.zepp.com/baseball/batting-metrics/.

Evaluation Criteria

Fidelity of Images. How realistic are the images? Are they two or three-dimensional? What is the resolution? What is the color depth? What is the update rate? Will the presentation vary depending on the processor speed or the communications bandwidth? This traces to the Operational Concept Description (OCD) and is also related to validation. Weight of importance is 6.

Varying image size. When the pitcher releases a baseball, the ball is 54 ft from the batter's eye and it subtends 0.26 degrees of arc. When it is two-thirds of the way to the contact point, when the swing must begin, the ball subtends 0.77 degrees of arc. Therefore, our displays should have a changing image size. This traces to the Operational Concept Description (OCD) and is also related to validation. Weight of importance is 4.

Distortion. Are the images distorted? Is the image of the ball circular? Are the images in focus? This traces to the Operational Concept Description (OCD) and is also related to validation. Weight of importance is 7.

Ease of Verification. Verification means proving that our system works correctly. We could help verification by asking, "Can we gather data that shows the player's progress while using the SpinCoach?" Weight of importance is 9.

Ease of Validation. Validity indicates the fidelity of transferring learned knowledge from our product to the real world. We can help validation by asking, "Can we gather data that shows improvements in the player's performance in the real game?" Weight of importance is 9.

Autonomy. Do we control the needed assets, or must we depend on outside sources? Negotiating a contract could take months. Finding and buying equipment might take months. This traces to SR1. Weight of importance is 10.

Killer Trades

The above criteria were used in performing a killer trades study and eliminated many unlisted alternatives. The surviving architectures were

1. The Status Quo
2. Computer Simulations
3. Spinning balls
4. System on CD-ROM
5. Interactive DVD
6. Web-based Application
7. Vizual Edge or Zepp Addition

Trade-off Matrix and Sensitivity Analysis

The trade-off study matrix (Table 5.6) and the sensitivity analysis for these alternatives and criteria suggested that the preferred alternatives are the system on a CD-ROM and the web-based application. The sensitivity analysis of this trade-off matrix shows that the most important parameter is the weight for the schedule criterion.

Throughout the design process, we performed many trade-off studies, either formally or informally. Suggest some that we may have done. You do not need a paragraph of detail for each alternative, a sentence or a phrase will do.

5.1.4.9 U.S. Government Regulations

In 2005, the US government created the Renewable Fuel Standard (RFS). It required that biofuels from corn and other minor sources be blended with transportation fuels. Its goals were to foster rural economic growth, support a robust domestic renewable-fuels sector, encourage national energy independence, promote cleaner air, and lower prices at the pump. How could anyone object to any of these goals? They are better than apple pie and motherhood! However, did the RFS prescribe tests to ensure that these goals were met? Should it have?

5.1.4.10 Ethanol Versus Gasoline

The most important part of a trade-off study is stating the problem. The next most important part is examining alternatives that will be traded off. Failure to identify all trade-off alternatives ensures failure. Let us look at the trade-offs involved in replacing gasoline with ethanol.

Standard political/scientific doctrine states that human activity is increasing greenhouse gases in the atmosphere. This is causing global warming. Some of the

Table 5.6 Trade-off study for the SpinCoach using a sum combining function and semirelative-sensitivity functions with the number of alternatives, $m = 6$

Evaluation criteria	Wt	Norm criteria weights	Norm subcriteria weights	Alt 1	Status Quo	Alt 2	Simulation	Alt 3	CD-ROM	Alt 4	DVD	Alt 5	Web	Alt 6	Vizual edge
				sc	$w \times sc$	sc	$w \times sc$	sc	$w \times sc$	sc	$w \times sc$	sc	$w \times sc$	sc	$w \times sc$
Performance	7	0.35													
Convenience	8		0.21	0.20	0.04	0.70	0.14	0.70	0.14	0.70	0.14	0.80	0.16	0.70	0.14
Fidelity of images	6		0.15	1.00	0.15	0.20	0.03	0.60	0.09	0.60	0.09	0.60	0.09	0.60	0.09
Duration control	5		0.13	0.00	0.00	1.00	0.13	1.00	0.13	0.90	0.12	1.00	0.13	1.00	0.13
Verification	10		0.26	0.20	0.05	0.80	0.21	0.80	0.21	0.10	0.03	0.80	0.21	0.80	0.21
Feedback	10		0.26	1.00	0.26	1.00	0.26	1.00	0.26	1.00	0.26	1.00	0.26	1.00	0.26
Cost	4	0.20													
Product production cost	6		0.30	1.00	0.30	0.50	0.15	0.50	0.15	0.50	0.15	0.50	0.15	0.50	0.15
Shipping cost	5		0.25	1.00	0.25	0.60	0.15	0.60	0.15	0.60	0.15	0.90	0.23	0.90	0.23
Process cost	5		0.25	1.00	0.25	0.50	0.13	0.50	0.13	0.50	0.13	0.50	0.13	0.50	0.13
Controllability	4		0.20	0.00	0.00	0.50	0.10	0.90	0.18	0.90	0.18	0.90	0.18	0.40	0.08
Schedule	9	0.45													
Autonomy	10		0.67	0.90	0.60	0.50	0.33	0.90	0.60	0.90	0.60	0.90	0.60	0.10	0.07
Updateability	5		0.33	0.10	0.03	0.50	0.17	0.90	0.30	0.90	0.30	0.90	0.30	0.20	0.07
Alternative rating					**0.62**		**0.60**		**0.81**		**0.75**		**0.84**		**0.46**

proposed solutions are taxing carbon emissions and switching from gasoline to ethanol fuels for automobiles.

This is a trade-off study, so we must identify the things that will be traded off.

1. Two common metrics to judge the effectiveness of replacing gasoline with ethanol are money and greenhouse emissions. The monetary studies show that with current government subsidies ethanol is about the same cost as gasoline per unit of energy. In this problem, we will look at greenhouse gasses, meaning carbon emissions into the atmosphere. Carbon emissions can be in the form of CO_2 or more complex organic compounds. In this problem, we use units of grams of CO_2 equivalent (Eq) per MJ of energy in the ethanol (g CO_2 Eq/MJ).
2. The two most popular biofuel feedstocks are wheat and corn (called maize in Europe). However, their numerical values are about the same, so we will use the values for corn. Sugarcane is not competitive in the United States.
3. There are three major global models for these greenhouse gasses. They are owned and operated by the California Air Resources Board, the U.S. Environmental Protection Agency (EPA), and the European Union. We have chosen to use the EPA model. Its outputs are given in Table 5.7 from [29].
4. Three processes produce greenhouse gasses. They are given in columns A, B, and C of Table 5.7. Column A gives the amount of atmospheric carbon created by producing the corn. This carbon production includes cultivation, harvesting, transportation, marketing, and the basic plant processes of photosynthesis and respiration shown in the following equation.

$$[\text{sunlight}] + 6H_2O + 6CO_2 \rightleftarrows C_6H_{12}O_6 + 6O_2 + [\text{energy}]$$

The top arrow and sunlight is for photosynthesis and the bottom arrow and energy is for respiration. If a plant is growing, then photosynthesis is larger.

Column B is the CO_2 released by fermentation of corn into ethanol.

$$C_6H_{12}O_6 \rightarrow 2C_2H_5OH + 2CO_2 + 2ATP$$

They accounted for by-products of the fermentation process such as xx.

Column C shows the CO_2 released as vehicle exhaust in the process of combustion.

$$C_2H_5OH + 3O_2 \rightarrow 2CO_2 + 3H_2O$$

Column D shows additional corn production due to both increased yields and new cropland. When there is more demand for ethanol, farmers will plant more corn and they will increase yield by increasing use of water and fertilizer. Plants that are growing absorb CO_2 and expire O_2. So increased production of corn will decrease the amount of carbon released to the atmosphere globally. Hence, the negative sign in Column D.

Table 5.7 Outputs of the EPA model showing global CO_2 carbon emission changes for different phases of corn to ethanol production

	A	B	C	D	E	F	G
Model	Crop production	Fermentation of corn	Vehicle exhaust emissions	Increased global crop production	Reduced release of carbon from crops after consumption	Land use change	Total change in carbon emissions
U.S. EPA	49	36	72	−86	−25	34	79

Diverting corn from animal feed and human food into ethanol production increases the demand, and therefore the price of corn. This, in turn, results in farm acreage being diverted from other food crops to corn production. This reduces the supply of the other food crops and increases their prices as well. So diverting corn into ethanol production means that there is less corn for human consumption and it costs more. This means that poor people around the world will have less food to eat and the food will be of a lower quality (olive, soybean, and vegetable land was shifted to corn). If people and animals worldwide eat less, they will release less carbon into the atmosphere due to respiration and digestion. Column E shows this effect with a negative number.

Finally, farmers might convert forests into cornfields. This would increase carbon in the atmosphere, because forests absorb more carbon than corn does. This land use change is shown in column F.

Searchinger et al. [29] did consider by-products of the fermentation process such as corn oil, corn syrup, and cattle feed. They did not consider carbon emissions resulting from construction of new fermentation plants.

The total change in carbon emissions is 79 g CO_2 Eq/MJ. The EPA then compares this with gasoline, which has 93 g CO_2 Eq/MJ. This does indeed show that replacing gasoline with ethanol reduces net greenhouse gasses in the atmosphere. However, it comes with a price that globally poor people eat less and less well. Notice that without column E, ethanol is worse than gasoline.

The point of this problem is that government models that show the advantages of replacing gasoline with ethanol seldom clearly state the trade-off that poor people are deprived of food.

Do an economic analysis to see if switching automobile fuel from gasoline to ethanol is economically feasible.

5.1.4.11 Some Open-Ended Decisions

Here are some discussion topics that might be amenable for trade-off studies. Should the United Kingdom stay in the European Union? Discuss this from both points of view.

The U.S. Fish and Wildlife Service lists over 1000 endangered species of animals. For simplicity, let us only consider the 250 mammals that live in the United States, including the Key Largo woodrat, Sperm whale, Gila topminnow, Florida panther, Jaguar, Columbia Basin Pygmy Rabbit, Mexican wolf, Chinchilla, Lesser slow loris, Sierra Nevada bighorn sheep, Giant armadillo, Mount Graham red squirrel, and Santa Catalina Island Fox. It is impossible to save them all. Discuss some criteria that might be used to help prioritize this list.

5.2 Section 31. Mental Mistakes Actually Made in Trade-off Studies

Humans often make poor decisions [30]. To help them improve as decision makers, engineering and business professors teach their students to perform trade-off studies. Trade-off studies are broadly recognized as *the* preferred method for simultaneously considering multiple alternatives with several criteria, and as such are recommended and mandated in the CMMI® [12, 31] decision analysis and resolution (DAR) process. The decision-making field of behavioral economics has built up a large body of research on human biases and errors in considering numerical and criteria-based choices. Relationships between experiments in these fields and the elements of trade-off studies show that trade-off studies are susceptible to human mental mistakes. In Section 18, we postulated 28 specific mental mistakes that could affect the ten specific components of a trade-off study. However, we did not offer examples from the system engineering literature. We will now validate our model by documenting specific instances of these mental mistakes in trade-off studies. In this section from Bohlman and Bahill [18], we present real-world examples of a dozen types of mental mistakes that people actually made. Similar psychological traps were found in decision-making with models [32] and additional cognitive biases in decision-making have been described [33–35].

The purpose of this section is to explain through examples certain types of mental mistakes that humans often make when conducting trade-off studies. Many of these mistakes are subtle and hard to avoid. However, it is hoped that systems analysts who read this section will be able to recognize these mistakes in trade-off studies conducted by others, and to avoid them in their own trade-off studies.

5.2.1 Research Methodology

Over the last quarter of a century, we have sought out and collected examples of trade-off studies. We collected 110 university projects, two dozen government contractor trade-off studies (these reports are proprietary and cannot be published), and three publicly accessible trade-off studies such as the San Diego airport site selection study [18].

We then read these reports looking for evidence of the 28 specific mental mistakes postulated in Section 18. We studied 8000 pages of university projects and filled a database with over 800 examples of such mistakes. These examples were put into categories according to the 28 specific mental mistakes that we were looking for. After several iterations, we decided that all of our examples fit into only 12 of these categories. We then rated each example for the perceived heuristic value, meaning the goodness of illustrating each type of mental mistake. The examples with the least perceived value were set aside. The remaining examples were compared with the original documents checking for context and accuracy. They were then rerated for perceived heuristic value. Based on the perceived heuristic value and uniform coverage, we created the 20 development cases that are in this section, and another 50 cases that were used for testing. Finally, we studied the government contractor trade-off studies and the three identified publicly accessible trade-off studies and similarly created case studies.

Often evidence of some of these mental mistakes could not be found in the written documentation. To find evidence for such mistakes, the experimenters would have had to be a part of the data collection and decision-making process. That is why, in this section, we present only 12 of the original 28 mental mistakes. This section presents examples of the following mental mistakes.

- Not stating the problem in terms of stakeholder needs
- Using dependent criteria
- Vague problem statement
- Substituting a related attribute
- Sensitivity analysis mistakes
- Forer effect
- Weight of importance mistakes
- Anchoring and the status quo
- Treating gains and losses equally
- Not using scoring functions
- Implying false precision
- Obviating expert opinion

This section is organized in accord with each of the above headings as follows: (1) the section heading and the first paragraph describe the type of mental mistake, (2) a recommendation for avoiding it is given, (3) the system that the teams were designing is explained (if it is the first time the system has been presented), (4) an excerpt of this type of mistake is presented in *Italic Sans-Serif font*, (5) the mental mistake is explained and relevant paragraphs from [17] are given, and then (6) a suggested rewrite is given in ***bold italic font***. *The mistake excerpts have not been edited*: they are direct quotes; however, some were reformatted. The sections from [17] have been edited.

5.2.2 *Examples of Mental Mistakes in Trade-off Studies*

This section presents examples of a dozen types of common mental mistakes that people have made when conducting trade-off studies. In the following example, the first numbered step given above is written in a numbered sentence set in the Italic font. Then the appropriate part of the example is given. Next, the second numbered step given above is written in numbered sentence two, set in the Italic font. Then another part of the example is given. Next, the third numbered step given above . . . etc.

5.2.2.1 Not Stating Problem in Terms of Stakeholder Needs

(1) *The section heading and the first paragraph describe the type of mental mistake.*

Committing to a class of preconceived solutions (instead of *stating the true stakeholders' needs*) causes inflexibility. Identifying true stakeholder needs can be difficult because stakeholders often refer to both problem and solution domains—whichever comes most naturally. In systems engineering, the initial problem statement must be written before looking for solutions [17, 36, 37].

(2) *A recommendation for avoiding the mistake is given.*

Recommendation: Question stakeholders to determine their values and needs. State the problem in terms of customer requirements or the deficiency to be resolved [38–42]. Later, after a better understanding of evaluation criteria and weights of importance have been gained, find alternative solutions that provide a good match to the requirements.

(3) *The system that the teams were designing is explained (if it is the first time the system has been presented).*

The project was the SpinCoach. When a spinning object (like a baseball) is put in a moving fluid (like air), it will experience a force that pushes it sideways, see the baseball in flight homework problem in Sections 2 and 26 [43–46]. Some highly successful baseball players have said and written that they see this spin of the ball and use it to track the ball with saccadic and smooth pursuit eye movements [47–49]. But at present, there is no system that can teach high school and college baseball and softball players to predict this spin-induced deflection of the pitch. The project was to design and document the design of a system that would help train baseball and softball players to pick up the spin on the ball and predict the spin's effect on the ball's deflection. The system would display spinning balls or images of spinning balls, allowing the subject to predict the spin-induced deflection and provide feedback to facilitate learning. The key architectural decision in this project is whether to design the first version for baseball or softball. Later, we will give an extensive explanation of the alternative architectures (Section 5.2.2.8.1) and the evaluation criteria and (Section 5.2.2.8.2) for the SpinCoach.

(4) *An excerpt of this mistake is presented in* *Italic Sans-Serif font*,

Team Excerpt

The batter believes he can predict the trajectory until the ball and bat connect and, therefore, swing accordingly. Because of the deflection in the last milliseconds of the balls flight however, the batter is not able to adjust his swing in time to compensate and therefore the likelihood for there to be a strike called is increased because of the batter swing. In essence, the pitcher is trying to fake out the batter and get him to swing at the ball or not swing at the ball based on a seemingly predictable trajectory. It appears to the batters, however, that the ball seems to change direction in the last moments of its trajectory and is cause for batters to swing without connecting with the ball.

(5) *The mental mistake is explained and relevant paragraphs from* [17] *might be given.*

Explanation: This problem statement does not mention the batter's needs. It is certainly stating correct things about what the batter and the pitcher are doing. But it fails to describe the batter's needs in terms of the system being designed. If they were designing a different system, then the excerpt might be useful. But they were supposed to be designing a system to help the batter learn the spin-induced deflection of the ball.

(6) *A suggested rewrite is given in* ***bold italic***.

Suggested rewrite: Baseball and softball players need to learn how to recognize the spin of a pitched ball and use that information to predict the spin-induced deflection of the ball.

Second Team Excerpt

A video game trainer could also be used, in which a batter tries to gauge the spin on a video ball and predict where it would end up. This option could be implemented with a CD-ROM. The CD would go through the process of how to see the spin based on video taken from a laboratory. Dr. Terry Bahill, a professor in the Systems and Industrial Engineering (SIE) department at the University of Arizona, has set up a laboratory with equipment to simulate pitches with different types of spins. This laboratory would be a valuable resource in designing experiments to train a player to pick up the spin on a ball.

Suggested rewrite: The top-level system function is to teach batters to recognize the spin of a pitched ball and then use this information to predict the spin-induced movement of the ball.

5.2.2.2 Using Dependent Criteria

Evaluation criteria should be independent. For evaluating humans, height and weight are not independent, gender (male versus female) and intelligence quotient are independent. In selecting a car, the following criteria are not independent: maximum horse power, peak torque, top speed, time for standing quarter mile, engine size (in liters), number of cylinders, and time to accelerate 0–60 mph [17].

Recommendation: Dependent criteria should be grouped together as subcriteria. The seven subcriteria for the car example can be conveniently grouped into the criterion, Power.

Team Excerpt

The maximum current drawn by the system shall not exceed 15 amperes.
The system shall operate on 120 V, 60 Hz electricity.
The system shall not consume more than 1.8 kW.

Explanation: If you know voltage and current, you know the power. Power = Voltage × Current. In a trade-off study, alternatives should be evaluated based on independent evaluation criteria. However, analysts often choose dependent criteria. When scoring these criteria for the different alternatives, having multiple dependent criteria can magnify or diminish the final scores of the alternatives, thus leading to a preference for the wrong alternative.

Suggested rewrite: The system shall operate on 120 volt, 60-hertz electricity. The system shall draw a maximum of 15 amperes.

5.2.2.3 Vague Problem Statement

If a problem statement is vague, proposed solutions could vary greatly, and derive support for very different reasons and in different ways. If a problem statement is poorly written or ambiguous, dissimilar alternative solutions could remain in the solution pool, obfuscating their rational consideration, especially if the rationale for the different psychologically attractive values of the alternative solutions are not well understood [50]. Failing to mention customer needs makes the problem statement seem vague [17].

Recommendation: Stating the problem is the most important and possibly the most difficult aspect of a trade-off study. You should state the problem in terms of the stakeholder's needs. But be sure to state the problem so that it is independent of preconceived solutions. State the problem so that it can be satisfied by a large number of alternative solutions.

The project was the PopUp Coach. Even professional baseball players occasionally find it difficult to gracefully approach seemingly routine pop-ups. McBeath, Nathan, Bahill, and Baldwin [51] describe a set of towering pop-ups with trajectories that exhibit cusps and loops near the apex. For a normal fly ball, the horizontal velocity is continuously decreasing due to drag caused by air resistance. But for pop-ups, the Magnus force (the force due to the ball spinning in a moving airflow) is larger than the drag force (see the baseball in flight homework problem in Section 26); therefore, the horizontal velocity decreases in the beginning, like a normal fly ball, but after the apex, the Magnus force accelerates the horizontal motion [52]. We refer to this class of pop-ups as paradoxical because they appear to misinform the typically robust optical control strategies used by fielders and lead to systematic vacillation in running paths, especially when a trajectory terminates near the fielder. In short, some of the dancing around when infielders pursue pop-ups can be explained as a combination of bizarre trajectories and misguidance by the normally reliable optical control strategy, rather than apparent fielder error. Former major league infielders confirm that our model agrees with their experiences. But at present, there is no methodological system that can teach high school and college baseball and softball players (more specifically catchers and infielders) to track pop-ups. Batting practice and ball games offer few opportunities for a player to learn this skill. Our customer needs a system that will provide frequent convenient opportunities to learn to field pop-ups. The goal of this project is to design and document the design of such a system.

Team Excerpt for the PopUp Coach
The occurrence of pop-ups in major league games is an average of five times per game. This frequency provides reason for effective practice to be devoted to train players in order to hone their ability to catch these routine pop-ups.

Suggested rewrite: For baseball and softball fielders who need to catch pop-ups, the PopUp Coach is a training system that teaches fielders to track and catch pop-ups: unlike present coaches and books, the PopUp Coach explains the trajectory, offers convenient practice opportunities, and complements players' improvement.

Team Excerpt for the SpinCoach
The differences in (baseball) spin result in visually detectable differences in the appearance of the spinning ball as it approaches the batter. Currently, players practice the assessment of spin and trajectory prediction during actual game play and team practice sessions. This process requires coordination between multiple players, and the use of a pitcher capable of delivering a repeatable baseball pitch to the batter.

Suggested rewrite: For the baseball batter who needs to predict the trajectory of the pitch, the SpinCoach is a training system that helps him to recognize the spin on the pitch and predict the ball's spin-induced movement; unlike present coaches and books, the SpinCoach shows the batter how each pitch spins and helps him to recognize this spin.

5.2.2.4 Substituting a Related Attribute

Attribute substitution occurs when a subject, while assessing an attribute, substitutes a related attribute that comes more readily to mind. In effect, people who are confronted with a difficult question sometimes answer an easier one instead ([53], p. 707).

Recommendation: When developing evaluation criteria, be sure to ask, "What do we really want to know?" If this will be difficult to measure, then use attribute substitution, but be sure to tell your readers that you are doing this.

Team Excerpt for the SpinCoach.
Use Case 2.
Name: Learn Spin-induced Deflections.
Iteration: 2.3.
Derived from: Concept of operations.
Brief description: Player uses the SpinCoach and learns to predict the spin-induced deflection of a ball.
Added value: Player will be better able to predict the trajectory of the ball and consequently should have a higher batting average.

Second Team Excerpt for the SpinCoach
5.2.1.2 Effectiveness
The measurement of effectiveness determines the percent increase on the user's batting average over time.

Explanation: The purpose of the SpinCoach is to teach batters to predict the spin-induced deflection of the baseball, but as a measure of success, these teams proposed to record the player's batting average.

Suggested rewrite: We want to teach baseball batters to predict the spin-induced deflection of the baseball. Therefore, we want to know how well they predict the spin-induced deflection, but that is hard to measure; therefore we substitute the change in the player's batting average as a measure of success. (This is attribute substitution; however, we are warning our readers about it.)

After developing prospect theory, Kahneman [54] spent most of his time trying to unify mental mistakes. He suggested that many or most could be explained by attribute substitution. This is a very difficult mistake to avoid. People do it all the time. The point of this section is that analysts should understand what attribute substitution is and avoid making it as a mistake. However, if the analyst really wants to use a substitute attribute, then he or she should go ahead and use it. But be sure to tell that reader that he or she is using attribute substitution.

In this particular design, what we really want to know is "Does training with the SpinCoach teach batters to predict the spin-induced deflection of the baseball?" But this is too hard to measure; therefore, we substitute the change in the player's batting average as a measure of success. But we *do* tell our readers that we are doing this.

Team Excerpt for the PopUp Coach
The ability of a defending baseball team to catch pop-up balls can provide a key advantage needed to win a baseball game. That pop-ups occurs in major league games an average of five times underscores the need for players to improve their ability to catch them. Pop-ups are difficult to catch because their trajectories are irregular and it is not readily predictable where players should stand and catch them. There is no existing training system to improve players' catching percentage.

Second Team Excerpt for the PopUp Coach.
Thus, the need for a system to teach fielders about and offer practice scenarios involving trajectory and spin is necessary if it can help increase fielding percentages and improve the win percentages for players and teams.

Suggested rewrite: For each pop-up, we will compute the player's optimal running path and then compare it to the player's actual running path. We will also compute the optimal running speed at the time of the catch and compare this to the player's actual running speed. These two metrics will be combined to indicate how well the fielder tracked the pop-up.

Attribute substitution is a really tricky mistake because everyone does it. In this design what we really want to know is "Does training with the PopUp Coach teach fielders to use the optical acceleration cancellation algorithm [51] to catch pop-ups?" But we cannot use as a metric the probability of catching a pop-up, because nine of our 12 alternative designs do not use real pop-ups. For these nine designs we think it would be too expensive (if not impossible) to gather enough data for every player to make valid inferences. Furthermore we have no method of measuring the spin of pop-us. Therefore, we think variability in the speed and spin rate of real pop-ups would obscure any evidence that the fielder used the optical acceleration cancellation algorithm to catch pop-ups. Because we could not measure the desired attribute, we substituted a simpler attribute, the running speed and path. Attribute substitution is not always a mistake, often it is deliberate and stated.

As another example, when confronted with a choice among alternatives that should properly be decided by a full trade-off study, there is a strong tendency to substitute a seemingly equivalent yet much simpler decision question in place of the trade-off study process [17].

Recommendation: Sponsors of trade-off studies should realize that a premature reduction of a trade-off study process to a simpler decision question is a common heuristic that prevents consideration of the original multi-objective decision.

5.2.2.5 Sensitivity Analysis Mistakes

Most people are not well trained in the methods of sensitivity analysis. Therefore, they often fail to compute second- and higher order partial derivatives. When estimating partial derivatives, they often use too large a step size. When estimating partial derivatives of functions of two parameters, they often use the wrong formula; they use a formula with two instead of four numerator terms as shown in Section 26. Interactions among parameters in trade-off studies can be important, step sizes for the approximation of effects should be small and second-order derivatives should be calculated accurately. It is expected that only the best-trained personnel will know of such results, illustrating the gap between truth and training [55].

Recommendation: Investments in sensitivity analysis training are necessary and should be made. Perhaps enabling software can substitute for much sensitivity analysis knowledge. [56–58] describe the development and use of sensitivity analyses.

First Team Excerpt

Since the training methods are independent of any resources and a trade-off analysis was not required, no sensitivity analysis will be done for it.

Second Team Excerpt

The I/O performance weights more than the utilization of resources in the trade-off analysis. TW1P0 and TW2P0 values will be switched to determine the sensitivity of the design concepts to weighting and the results are shown in Table 5.8.

Third Team Excerpt

The current trade-off study has an emphasis placed on the performance requirements of the SlugMaster to determine the sensitivity of results to the weighting of the requirements, a plot was constructed as a function of the two weights: I/O Performance and utilization of resources.

Table 5.8 *Second team excerpt*

	Weights are 0.70/0.30	Weights are 0.30/0.70
Concept	Score	Score
1	0.3811	0.7347
2	0.5172	0.3665
3	0.5191	0.3061
4	0.568	0.3416
5	0.563	0.3249
6	0.6275	0.3362
7	0.5739	0.3132

Table 5.9 ***A generic trade-off study***

Evaluation ***criteria***	***Weights of importance***	***Alternative 1***	***Alternative 2***
Criterion-1	w_1	sc_{11}	sc_{12}
Criterion-2	w_2	sc_{21}	sc_{22}
Alternative rating		$\text{Sum}_1 = w_1 \times sc_{11} + w_2 \times sc_{21}$	$\text{Sum}_2 = w_1 \times sc_{12} + w_2 \times sc_{22}$

Table 5.10 ***Trade-off study numerical example***

		Alternatives	
Evaluation ***criteria***	***Weights of importance***	***Umpire's assistant***	***Seeing eye dog***
Accuracy of the call	***0.75***	***0.67***	***0.33***
Silence of signaling	***0.25***	***0.83***	***0.17***
Sum of weight times score		***0.71*** ***The winner***	***0.29***

Explanation: These teams considered at most one parameter, the relative trade-off weight between performance and cost. Their equation was probably something like this.

$$\text{Overall Tradeoff Score} = \text{Perf}_{\text{wt}} \times \text{Perf} + (1 - \text{Perf}_{\text{wt}}) \times \text{Utlization Resources}$$

Suggested rewrite:

Table** 5.10 **gives numerical values for one particular trade-off study, The Umpire's Assistant.

Definition of the semirelative-sensitivity function:

$$\widetilde{S}_{\alpha}^{F} = \left.\frac{\partial F}{\partial \alpha}\right|_{\text{NOP}} \alpha_0$$

***These are the semirelative-sensitivity functions for Tables** 5.9 **and** 5.10.*

$$\widetilde{S}^{\mathrm{PI}_1}_{w_1} = (\mathrm{sc}_{11} - \mathrm{sc}_{12})w_1 = 0.26$$
$$\widetilde{S}^{\mathrm{PI}_1}_{w_2} = (\mathrm{sc}_{21} - \mathrm{sc}_{22})w_2 = 0.16$$
$$\widetilde{S}^{\mathrm{PI}_1}_{sc_{11}} = w_1\mathrm{sc}_{11} = 0.50$$
$$\widetilde{S}^{\mathrm{PI}_1}_{\mathrm{sc}_{21}} = w_2\mathrm{sc}_{21} = 0.21$$
$$\widetilde{S}^{\mathrm{PI}_1}_{\mathrm{sc}_{12}} = -w_1\mathrm{sc}_{12} = -0.25$$
$$\widetilde{S}^{\mathrm{PI}_1}_{\mathrm{sc}_{22}} = -w_2\mathrm{sc}_{22} = -0.04$$

These functions show that the most important parameter is the score for alternative 1 on criteria 1. Sensitivity analyses need mathematical detail. Failure to do the mathematics right produces erroneous results. Section 3.7.5 presents the correct mathematical equations.

5.2.2.6 Forer Effect

According to the Forer effect, previously existing criteria will be adopted if (1) the analyst believes that the criteria apply to the present problem, (2) the criteria are well presented, and (3) the analyst believes in the *authority* of the previous criteria writer. For example, the analyst might fail to question or rewrite criteria from a legacy trade-off study that originated from a perceived authority and is now seemingly adaptable to the trade-off at hand. This is called the Forer effect. Forer [59] gave a personality test to his students. He then asked them to evaluate his personality analyses of them, supposedly based on their test's results. Students rated their analyses on a scale of 0 (very inaccurate) to 5 (excellent) as to how well it applied to them. The average score was 4.26. Actually, Forer had given the same analysis to all the students. He had assembled this analysis of a generally likeable person from horoscopes. Variables that contribute to this fallacy in judgment are that the subject believes the analysis only applies to them, the subject believes in the authority of the evaluator, and the analysis contains mainly positive traits [17].

Recommendation: Spend time considering and formulating criteria from scratch, before consulting and possibly reusing previously written criteria.

Team Excerpt

Team summary of a sensitivity analysis. It shows how the overall tradeoff scores vary for a dozen alternatives as the cost becomes less important and the performance becomes more important.

Explanation: The only parameter being varied here is the relative weight of performance versus cost

The reason that these engineers made this mistake is that their reference book (Chapman, Bahill, and Wymore, 1992) had a primitive sensitivity analysis that only considered one parameter, the trade-off weight between cost and performance. In Section 5.5.4 it stated, "The system is sensitive to trade-off weightings. For example, changing the weights of the trade-off requirement can easily sway the answer. The current trade-off puts heavy emphasis on the I/O performance of the

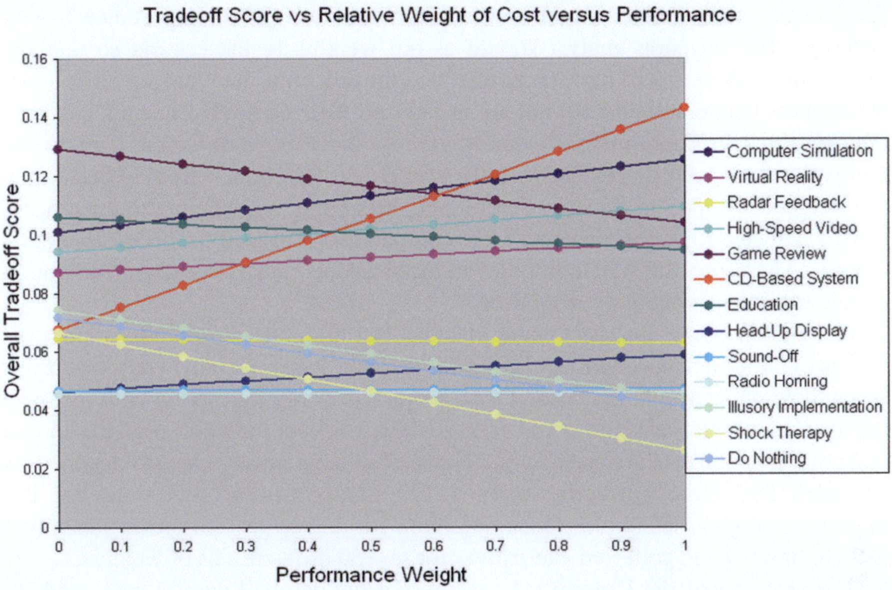

Fig. 5.7 Results of a bad sensitivity analysis

system (0.90) and not on the utilization of resources (0.10). Changing the degree of emphasis can change the results..." We had lectures and homework that described how to take partial derivatives and form the semirelative-sensitivity functions for each parameter in the trade-off study [58]. However, the students repeatedly copied the method of conducting a sensitivity analysis from a legacy trade-off study published in a reference book. Despite warnings about the inadequacy of that sensitivity analysis, these engineers conducted their analyses in that very same way. This was an example of the Forer effect [59]. They failed to question a sensitivity analysis that was presented by a perceived authority and was seemingly adaptable to their own trade-off study (Fig. 5.7).

5.2.2.7 Weight of Importance Mistakes

When a group of people is asked to assign a weight of importance for an evaluation criterion, each person might produce a different value. Different weights arise not only from different preferences, but also from irrational severity amplifiers (Section 17). These include the factors of lack of control, lack of choice, lack of trust, lack of warning, lack of understanding, manmade, newness, dreadfulness, personalization, recallability, and immediacy. Excessive disparities occur when a person assesses a weight of importance after framing the problem differently. An evaluation may depend on how the criterion affects that person, how well that person understands the alternative technologies, the dreadfulness of the results, etc.

As a result, each person might assign a different weight of importance to any criterion. The decision analyst should assign weights to the criteria so that the more important ones will have more effect on the outcome. Weights are often given as numbers between 0 and 10, but are usually normalized so that in each category they sum to 1.0. These methods can be used by individuals or teams. If pair-wise comparisons of preferences between the criteria can be elicited from experts, then the weights of importance can be determined through the analytic hierarchy process (AHP). However, performing pair-wise comparisons can lead to intransitive preferences. Therefore, the AHP computes an inconsistency index to warn if the domain expert is giving intransitive responses.

Recommendation: Interpersonal variability can be reduced with education, peer review of the assigned weights, and group discussions. It is important to note that if you reveal how other people are voting, then people are likely to respond with the most popular answers. It is also important to keep a broad view of the whole organization, so that criteria in one area are considered in light of all other areas. A sensitivity analysis can show how important each weight is. For unimportant weights, move on. For important weights, spend more time and money trying to get consensus; this might include showing the preferred alternatives for several different sets of weights.

The project was the Umpire's Assistant. For the baseball umpire who needs to call balls and strikes, the Umpire's Assistant is an intelligent decision aiding system that helps him or her to call balls and strikes accurately, consistently, and in real-time. Unlike unassisted human umpires, the Umpire's Assistant uses the same strike-zone standards for all leagues, parks, umpires, batters, and pitchers (Tables 5.11 and 5.12).

Team excerpt

Of course, there would be a paragraph explaining each of these short evaluation criteria tags. The abbreviations would be explained in these paragraphs.

5.2.2.8 Anchoring and the Status Quo

The order in which the alternatives are listed has a big effect on the values that humans give for the evaluation data [60–62]. Therefore, you should fill out a trade-off study matrix row by row and the status quo should be the alternative in the first column. This will make the evaluation data for the status quo the anchors for estimating the evaluation data for the other alternatives. This is good because the anchoring alternative is known, is consistent, and you have control over it [17].

Note: The status quo will probably have low evaluation data values (inputs for scoring functions) for performance, cost, schedule, and risk. If the status quo had high performance values, then you probably would not be trying to replace it. The status quo already exists, so (1) it will not be expensive, which gives it a low (good) value for cost, (2) it should not have schedule problems, which gives it a low (good) value for schedule, and (3) it should also be low risk.

Table 5.11 *Umpire's assistant team excerpt*

Utilization of resources evaluation criteria requirements	*Weight value*	*Normalized weight*
1. Available money	*2*	*0.02326*
2. Available time	*2*	*0.02326*
2.1. System design and prototyping by 12/31/05	*2*	*0.02326*
2.2. System verification testing by 2/2/06	*2*	*0.02326*
3. Technological restrictions	*10*	*0.11628*
3.1. To not significantly alter the dynamics of baseball	*9*	*0.10465*
3.2. To comply with local, regional, state, federal laws	*10*	*0.11628*
3.3. To comply with FCC rules	*10*	*0.11628*
4. Adaptability	*8*	*0.09302*
4.1. To comply with standards and specifications of MLB	*8*	*0.09302*
4.2. To comply with standards and specifications of NCAA	*8*	*0.09302*

Explanation: The normalized weights add up to 0.826. They should add up to 1.0 in each category and each subcategory

Table 5.12 *Suggested rewrite for the umpire's assistant*

Utilization of resources evaluation criteria	*Weights of importance*	*Criteria normalized weights*	*Subcriteria normalized weights*
1. Available money	*2*	*0.09*	
2. Available time	*2*	*0.09*	
2.1. System design and prototyping by 12/31/15	*2*		*0.5*
2.2. System verification testing by 2/14/16	*2*		*0.5*
3. Technological restrictions	*10*	*0.45*	
3.1. To not significantly alter baseball dynamics	*9*		*0.31*
3.2. To comply with local, state, and federal laws	*10*		*0.35*
3.3. To comply with FCC rules	*10*		*0.35*
4. Adaptability	*8*	*0.36*	
4.1. To comply with MLB rules	*8*		*0.5*
4.2. To comply with NCAA rules	*8*		*0.5*
Totals		*1.0*	*3.0*

Recommendation: Insert the status quo alternative in the first column. In the first iteration, evaluate the scores left to right, and in the next iteration evaluate them right to left. The more alternatives that exist and the more complicated the decision, the more the status quo will be favored. Do not needlessly increase the number of alternatives in a trade-off study. More alternatives increase the difficulty of the decision. However, in the very beginning of a project it is good to have many alternatives in order to better understand the problem and the requirements. View the problem from different perspectives. Use different starting points. When estimating values for parameters of scoring functions, you should think about the whole range of expected values for the parameters.

Explanation: In a trade-off study matrix the alternatives should be in columns, not rows, because humans find it easier to compare across columns rather than down rows. The alternatives and evaluation criteria should be identified by names, not numbers. The alternatives and evaluation criteria should be explained with sentences and paragraphs. Finally, the do nothing alternative should be in the first column.

Table 5.13 also shows a mistake at the implementation level: the engineers probably intended each weighted score to be its raw score times the metric value, and the solution score to be the summation of the weighted scores in that row. But their Excel worksheet had mistakes: only the do nothing row was correct (Table 5.14).

Throughout this section, we have giving terse tags for the alternatives and criteria. Often we wrote that these tags were not the alternatives and criteria, but

Table 5.13 *Team excerpt for ranking process alternatives*

Criterion	1		2		3		4		5		6		7		Max score
Metric value	1.8		0.7		2.3		1.5		0.7		1		2		100
	Raw	Wtd	Raw	Wtd	Raw	Wtd	Raw	Wtd	Raw	Wtd	Raw	Wtd	Raw	Wtd	Solution
Alternative	Scr	Scr	Scr	Scr	Scr	Scr	Scr	Scr	Scr	Scr	Scr	Scr	Scr	Scr	Score
Solution 1	2	7.2	10	6.3	9	14	3	9	3	2.1	8	2	2	4	41.8
Solution 2	9	3.6	7	7	9	21	9	4.5	2	1.4	2	8	10	20	76.6
Solution 3	3	16	2	4.9	4	9.2	8	14	3	2.1	8	9	1	2	67.3
Solution 4	9	5.4	10	1.4	5	12	9	12	9	6.3	9	8	10	20	76.6
Solution 5	3	16	3	7	5	12	8	14	8	5.6	7	9	2	4	67.3
Solution 6	9	16	3	2.1	5	12	9	11	9	6.3	5	7	10	20	76.6
Solution 7	3	5.4	8	5.6	9	21	7	11	3	2.1	5	5	9	18	67.3
Do nothing	10	18	10	7	10	23	10	15	10	7	10	10	10	20	100

Table 5.14 ***Suggested rewrite of the trade-off matrix for alternative architectures of the SpinCoach***

	Alternatives					
Evaluation *criteria*	*Do nothing*	*Computer simulation*	*CD ROM*	*DVD*	*Web page*	*Video game*
Fidelity of images						
Feedback time						
Product production cost						
Shipping cost						
Updatability						

the alternatives and criteria must be explained in detail in a different part of the documentation. We did not give these detailed explanations because we did not want to interrupt the flow of thinking about the mental mistakes. However, we have now come to a point where it is appropriate to show examples of what the other sections would look like. So here are examples for the SpinCoach.

Alternative Architectures for the SpinCoach

1. The Status Quo. Some batters can recognize the spin on the ball and predict its spin-induced movement. But they have difficulty verbalizing this capability and teaching it to others. In batting practice, we could have the pitcher announce to the batter "curve" and then throw a curveball: announce "slider" and then throw a slider. This could be done with a human pitcher or a pitching machine.
2. Computer Simulations. Images of spinning balls can be simulated and presented on a computer monitor.
3. Spinning Balls. Holes are drilled into baseballs and softballs and the balls are skewered on bolts. These bolts are chucked into drills. The drills are spun at controlled speeds. Videos for such spinning balls are on the web site http://www.sie.arizona.edu/sysengr/baseball/index.html. The user runs the SpinCoach from his or her hard disk. The user must login for each session. The information gathered at login is used to track user performance history.
4. System on CD-ROM. Balls spinning on drills can be photographed and their images stored on CD-ROM disks. Such videos are on the web site http://www.sie.arizona.edu/sysengr/baseball/index.html. These images along with the software program (application) will be transferred from the CD-ROM to the user's hard disk using a license key provided by BICS. There after the user runs the SpinCoach from his or her hard disk. The user must login for each session. The information gathered at login is used to track user performance history. System upgrades will be provided with new CD-ROMs.

4b. Application Sold Online. A slight variant of the CD-ROM alternative is to sell the application through an online store with distribution via Internet downloads.
5. Interactive DVD. Balls spinning on drills can be photographed and their images stored on DVDs. Such videos are on the web site http://www.sie.arizona.edu/sysengr/baseball/index.html. These images along with the software program (application) will be accessed by the user. There will be no user performance tracking. System upgrades will be provided with new DVDs.
6. Web-based Application. Balls spinning on drills can be photographed and their images stored in the cloud or on the BICS web server. Such videos are on the web site http://www.sie.arizona.edu/sysengr/baseball/index.html. These images along with the software program (application) will be on an Internet accessible web site. This system will be based on the Apache web server with web pages written in PHP or HTML and video clips in AVI format. Access to the system will be granted by monthly subscription and login-based authentication control. The user will have a profile in the system and can access this profile from any terminal connected to the Internet. The system will store user information in a database (such as MySQL). This information is used to track user performance history. System upgrades can be made on the web site at any time and will be transparent to the user.
7. Addition to commercial products. Add our module to the commercial off-the-shelf products of the Vizual Edge Performance Trainer™ (VEPT®) (http://www.vizualedge.com/) or the Zepp baseball analysis system http://www.zepp.com/baseball/batting-metrics/.
8. Make it into a video game and sell it to Nintendo, Sony, or Microsoft.

Evaluation Criteria for the SpinCoach

Fidelity of Images. How realistic are the images? Are they two-dimensional or three-dimensional? What is the resolution? What is the color depth? What is the update rate? Will the presentation vary depending on the processor speed or the communications bandwidth? For example, would the system degrade with a low-speed connection to the Internet? This criterion traces to the Operational Concept Description (OCD). Weight of importance is 6.

Feedback Promptness. The system shall provide feedback to the player after each prediction. The system shall provide this feedback to the player within 500 ms of the player's response. This will be a Boolean (yes or no) function. This traces to customer requirement 10. Weight of importance is 10.

Product Production Cost is a measure of how much it will cost in US dollars for BICS to produce one unit of the product. A monotonic decreasing scoring function shall be used ($L=0$, $B=10$, $S=-0.1$, $U=500$). Input range is 0–500 dollars, baseline is 10 dollars, and slope is −0.1. This traces to customer requirement 1. Weight of importance is 6.

The *Shipping Cost* evaluation criterion is composed of shipping weight, shipping expenses, shipping effort, and billing cost per unit. A scoring function is not necessary, if the subcriteria have scoring functions and normalized weights. Our target value is 10 dollars. This traces to customer requirement 1. Weight of importance is 5.

Ease of Update. This criterion evaluates how easy and convenient updates are expected to be. The system shall be continually improved and updated throughout the system life cycle. Corrective maintenance such as bug fixes should be performed within weeks. Adaptive maintenance, which includes revisions necessary to allow the system to run on new or improved hardware and software, should be accomplished in a monthly time frame. Performance and functional updates will be performed yearly. This should trace to the business plan. Weight of importance is 5.

5.2.2.9 Treating Gains and Losses Equally

People do not treat gains and losses equally (Fig. 2.8). People prefer to avoid losses rather than to increase gains. Would you rather get a 5 % discount, or avoid a 5 % penalty? Most people would rather avoid the penalty. Prospect theory [63] suggests that psychologically losses are twice as powerful as gains. In a trade-off study, you will get a different result if the scoring function expresses losses rather than gains [17, 37, 64].

Recommendation: Human unequal treatment of gains and losses suggests that scoring functions in a trade-off study should uniformly express either gains or losses. Principles of linguistic comprehensibility suggest that criteria should be worded in a positive manner, so that more is better. For example, it is preferable to use uptime rather than downtime, mean time between failures rather than failure rate, and probability of success rather than probability of failure. Finally, when employing scoring functions, make sure that more output is better.

Team Excerpt

2.2. Number of complaints
2.3. Number of problems with the system
3.1. Number of accidents per visit
2.5.3. Number of curses per day
5.2.6. Injury—Is it possible for the design to inflict bodily injury on the batter? This is rated by the players on a scale of 1–10 (1 being no bodily harm, 10 being serious injury requiring hospitalization)

Explanation: These criteria are phrased negatively.

Suggested rewrite:
2.2. Customer approval rating (percentage)
2.3. Mean time to failure (MTTF)
3.1. Number of accident-free visits

2.5.3. Time without cursing
5.2.6. Safety—Mean time between injuries

Second Team Excerpt

3.5.2.6. Availability of system
This evaluation criterion measures the availability of the system, in terms of the number of hours per year for which the system is unavailable during working hours due to failure or unscheduled maintenance
3.5.2.7. Safety of system
This evaluation criterion measures the safety of the system in terms of injuries caused by the system

Explanation: Evaluation criteria should be phrased such that "more is better." Availability of System and Safety of System are contradicting terms. The titles are phrased positively, but the descriptions contain the negative words *unavailable* and *injuries*.

Suggested rewrite:
3.5.2.6. System availability
This criterion is the percent of time that the system is available during working hours, average per week
3.5.2.7. System safety

The units for this criterion are the number of days that the system has operated without an injury.

5.2.2.10 Not Using Scoring Functions

Evaluation data are transformed into normalized scores by scoring functions (utility curves) or qualitative scales (fuzzy sets). The shape of scoring functions should ideally be determined objectively, but often, subjective expert opinion is involved in their preparation. Creating scoring function packages takes time and effort [65]. A scoring function package should be created by a team of analysts, and then reevaluated with the stakeholders with each use. Most trade-off studies that we have seen in industry did not use scoring functions. In some cases, scoring functions were explained in the company process, but they were not convenient, hence they were not used [17].

Recommendation: Standard scoring functions (or similar scoring functions, fuzzy sets or utility functions) should be used in trade-off studies. Easy-to-use scoring functions should be referenced in company processes.
Team excerpt:
Many teams just did not use scoring functions.

Evaluation data are transformed into normalized scores by scoring functions as shown in Section 21. The shape of scoring functions should ideally be determined

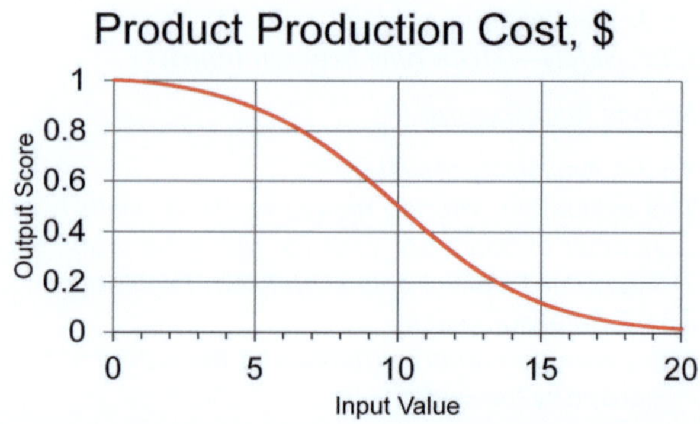

Fig. 5.8 A scoring function for the product production cost evaluation criteria

objectively, but often, subjective expert opinion is involved in their preparation. Scoring functions are also called utility functions, utility curves, value functions, normalization functions, and mappings. Evaluation criteria should always have scoring functions so that the preferred alternatives do not depend on the units used. For example, see what would happen if you were to add values for something that cost about 100 dollars and lasted about a millisecond.

Alt-1 cost a 100 dollars and lasts 1 ms, Sum = 100.001.
Alt-2 only cost 99 dollars but it lasts 2 ms, Sum = 99.002.

The duration does not have an effect on the decision. A simple program that creates scoring functions is available for free at http://www.springer.com/gp/book/9783319437101#otherversion=9783319437125 or http://sysengr.engr.arizona.edu/forStudents/index.html. It is called the Scoring Function toolbox. An example of a scoring function was given in Fig. 5.8 for the product production cost.

Scoring functions must state the units for the input. For example, you can use actual dollar values as input to a cost scoring function. Without scoring functions, the preferred alternative depends on the units used, for example, whether the costs were given in US Dollars or British pounds. With scoring functions, this does not happen.

Implying False Precision

The most common mistake in trade-off studies is implying false precision. For example, a system analyst might ask a subject matter expert to estimate values for two criteria. The expert might say something like, "The first criterion is about 2 and the second is around 3." The analyst puts these numbers into a calculator and computes the ratio as 0.666666667. Clearly, this makes no sense. Even so, nine digits might get used throughout the trade-off study. The Forer effect might explain this. The analyst believes that the calculator is the unquestionable authority in number crunching. Therefore, what the calculator says must be true [17].

Recommendation: Use significant figure methodology. Furthermore, in numerical tables, print only a sufficient number of digits after the decimal place as is necessary to show a difference between the preferred alternatives (Tables 5.15, 5.16, and 5.17).

Table 5.15 *Second team excerpt for bat chooser*

Concept	*Weighted score*	*Ranking*
Concept 1 score	*0.4414*	*7*
Concept 2 score	*0.5420*	*4*
Concept 3 score	*0.2196*	*8*
Concept 4 score	*0.6728*	*2*
Concept 5 score	*0.6558*	*3*
Concept 6 score	*0.4652*	*6*
Concept 7 score	*0.6802*	*1*
Concept 8 score	*0.4860*	*5*

Table 5.16 ***Suggested rewrite for the Bat chooser***

Alternative	***Weighted score***	***Ranking***
Concept 1	***0.44***	***7***
Concept 2	***0.54***	***4***
Concept 3	***0.21***	***8***
Concept 4	***0.67***	***2***
Concept 5	***0.65***	***3***
Concept 6	***0.46***	***6***
Concept 7	***0.68***	***1***
Concept 8	***0.48***	***5***

Table 5.17 *Team excerpt for the SpinCoach*

Performance requirements	*Weight of importance*	*Normalized weight*
1. Accuracy	*8*	*0.235294*
1.1. Spin rate	*10*	*0.384615*
1.2. Launch angle	*8*	*0.307692*
1.3. Launch speed	*8*	*0.307692*
2. Consistency	*7*	*0.205882*
3. Ease of use	*6*	*0.176471*
3.1. Portability	*6*	*0.260870*
3.2. Location	*7*	*0.304348*
3.3. No. of operators	*10*	*0.434783*
4. Opportunity	*8*	*0.235294*
5. Feedback	*5*	*0.147059*

Explanation: The original Value data have one significant digit. Therefore, the normalized weights certainly should not have six digits after the decimal point

Team Excerpt

5.3.1.1. Trade-off scores
Concept 1: Customer chooses the bat
$0.6 \times 0.22500 + 0.4 \times 0.41435 = 0.30074$
Concept 2: Store owner chooses the bat
$0.6 \times 0.25125 + 0.4 \times 0.41435 = 0.31649$
Concept 3: BatChooser chooses the bat
$0.6 \times 0.67500 + 0.4 \times 0.96840 = 0.79236$
Concept 4: BatSelect chooses with the help of the BatChooser
$0.6 \times 0.84280 + 0.4 \times 0.96840 = 0.89304$

Explanation: Presenting five digits after the decimal point obfuscates the equations and does not help to differentiate between the alternatives.

***Suggested rewrite*:**
5.3.1.1. Trade-off scores
Concept 1: Customer chooses the bat
$0.6 \times 0.23 + 0.4 \times 0.41 = 0.30$
Concept 2: StoreOwner chooses the bat
$0.6 \times 0.25 + 0.4 \times 0.41 = 0.32$
Concept 3: BatChooser chooses the bat
$0.6 \times 0.68 + 0.4 \times 0.97 = 0.79$
Concept 4: BatSelect chooses the bat with the help of the BatChooser
$0.6 \times 0.84 + 0.4 \times 0.97 = 0.89$

In Table 5.18, the criteria and subcriteria weights are inserted in separate columns. All of the subcriteria weights for a particular criteria sum to 1.0. All of the criteria weights sum to 1.0.

Table 5.18 ***Suggested rewrite for the SpinCoach***

Evaluation criteria	*Weight of importance*	*Criteria normalized weights*[a]	*Subcriteria normalized weights*[a]
1. Accuracy	*8*	*0.24*	
1.1. Spin rate	*10*		*0.38*
1.2. Launch angle	*8*		*0.31*
1.3. Launch speed	*8*		*0.31*
2. Consistency	*7*	*0.21*	
3. Ease of use	*6*	*0.18*	
3.1. Portability	*6*		*0.26*
3.2. Location	*7*		*0.30*
3.3. Number of operators	*10*		*0.43*
4. Opportunities per hour	*8*	*0.24*	
5. Feedback response time	*5*	*0.15*	

[a]***Significant figures methodology suggests that the normalized weights should only have one significant digit. But here we have used two to make the calculations obvious***

On the other hand, Table 5.13 shows an example not of false precision, but of inconsistent precision. In some columns, Excel is set to display integers while in others it is set to display to one decimal place.

When determining the number of digits that should be printed, you need to consider (1) how many digits are necessary to differentiate between the preferred alternatives and (2) the sensitivity of the final recommendations to the parameters. The most sensitive parameters should be given extra resources and, therefore, perhaps more significant digits [57, 58]. It is important to note that no parameter needs to be more exact than the least accurate parameter (presuming of course that the trade-off matrix is as described in this section, and it does not have a multistep process for estimating any parameters).

As a further complication, when humans state preferences between risky prospects, their judgments are not linear with respect to probability. Humans overweight small probabilities and underweight high probabilities (Fig. 2.9). This has been modeled with several different equations [64, 66].

$$w(p) = \frac{p^{\gamma}}{[p^{\gamma} + (1-p)^{\gamma}]^{\frac{1}{\gamma}}}$$

or

$$w(p) = \frac{\delta p^{\gamma}}{[\delta p^{\gamma} + (1-p)^{\gamma}]}$$

and

$$w(p) = \exp(-\beta(-\ln p)^{\alpha}).$$

In these equations, w is the probability weighting function, p is the probability of a particular prospect, γ, δ, α, and $\beta$are parameters fit to experimental data of individual humans. Now the question becomes, Should a trade-off analyst use equations like these when eliciting information for a trade-off study? The answer depends on the accuracy of the other parameters. In all the design documents that we examined, the weights of importance had only one significant digit. Therefore, correcting for each human's incorrect estimation and use of probabilities is not warranted. In a trade-off study, the number of significant figures should be determined for each parameter. Then resources should not be committed to increasing the number of significant figures for any except the most important and the least precise parameters.

5.2.2.11 Obviating Expert Opinion

An analyst could hold a circular belief that expert opinion or review is not necessary, because no evidence for the need of expert opinion is present. This is especially true if no expert has ever been asked to comment on the trade-off study [17].

Recommendation: Experts should be sought, formally or informally, to evaluate the results of trade-off studies.

The most common mistake in design projects is failing to talk with stakeholders and consult with experts. Universities and local industries are full of experts in the different fields. Yet, very few teams have sought their advice. Why do people fail to seek expert advice? Our subjects offered and rated the following plausible reasons. In each category, the reasons are arranged from the most frequent to the least (Table 5.19).

Table 5.19 Possible reasons for failing to talk with stakeholders, experts, and advisors

Timidity
Perhaps they do not want to inconvenience the wise men or waste their time. However, people are not reluctant to seek the advice of physicians, tax accountants, and lawyers. To overcome timidity, before you talk to an expert, you should formulate your questions and explain your problem in a way that the expert can quickly understand. Tailoring a message and formulating the right questions is hard and must be done iteratively. Before you leave your meeting, you should state what you think the expert said, to make sure that you understood and to prevent having to go back later for clarification
Perhaps they are shy or intimidated by experts
Perhaps they fear that the incompleteness of their project will be interpreted as incompetence
Perhaps they think that a face-to-face meeting would display their naïveté. This is not a problem with e-mails, because most people do not expect e-mails to be thoughtful, coherent and grammatically correct: most students do not edit their e-mails or use a spelling and grammar checker on them
Perhaps they think that seeking advice reveals their ignorance, and that ignorance is shameful
Perhaps they think that consulting experts shows weakness, whereas going it alone shows strength
Perhaps they feel that, because they do not have a charge number, they cannot ask experts in their company for advice
Perceived value
Perhaps they do not realize the usefulness of face-to-face meetings with experts
Perhaps it is a matter of return on investment. Consulting experts takes time and effort. Perhaps these teams thought the improvement in their trade-off studies would not be worth the effort of consulting experts
Perhaps the smart students thought, “We can get an A without wasting our time talking to our advisor”
Perhaps the new technology generation thinks that they can just Google the web and get all the information that experts might provide
Perhaps they noticed that other courses at the university do not provide world-class experts to meet with them, so it must not be important
Perhaps they do not see a direct correlation between their grade and meetings with their advisor
Perhaps they do not perceive added value
Time (obviously time and perceived value will be traded off)
Perhaps they thought that they were too busy; meeting with their advisor would take time and effort; it would be hard to schedule meetings with their advisor. Maybe they were just lazy
Communication
Perhaps they have had no experience initiating a meaningful conversation with a stranger and are therefore reluctant to do so

(continued)

Table 5.19 (continued)

Most of our students communicate with cell phones, twitter, the internet or e-mail. So they are multitaskers, jumping from task to task. Therefore, they are good at multitasking, but their attention spans are perhaps short. So they do not know how to talk face-to-face with an expert
Other
Perhaps they have been taught that engineers work alone: after all, cooperating on exams is frowned upon. However, in the modern industrial environment, engineering is done by teams and when success is important consultants are also hired
Perhaps they are reluctant to change or they don't want to do it someone else's way. If you ask for advice, then you are obligated use the advice you are given
Foreign students said, "It's embarrassing to show weakness in the English language" and "Our culture teaches us to not approach an advisor or mentor"

The medical profession also does not extensively consult with experts. In one study of over 300 breast cancer surgeons [67], about one-fourth typically consulted medical oncologists, radiation oncologists, or plastic surgeons prior to surgery. About two-thirds of the surgeons reported that almost none of their patients participated in patient decision-support activities, such as attending a practice-based presentation, viewing web-based materials, or participating in peer-support programs.

He who trusts in himself is a fool, but he who walks in wisdom is kept safe. (Proverbs 28; 26)

5.2.2.12 Statistical Summary of Mental Mistakes

We examined 110 project reports composed of over 8000 pages of text that had been written over the last two decades and we compiled the following statistics.

Type of mental mistake	Number of mistakes found
Using dependent criteria	75
Not stating the problem in terms of stakeholder needs	62
Vague problem statement	65
Substituting a related attribute	24
Sensitivity analysis mistakes	91
Forer effect	61
Weight of importance mistakes	11
Anchoring and the status quo	69
Treating gains and losses equally	46
Not using scoring functions	32
Implying false precision	59
Obviating expert opinion	85
Other mental mistakes	128

In these 110 trade-off study reports, we found 808 mental mistakes. We put these in a spreadsheet. Then for 99 of these mental mistakes, we made example cases using (1) direct quotes from the original reports, (2) an explanation of the mistake, (3) related paragraphs from [17], and (4) recommended revisions. We used a format similar to that used in this section. Based on the perceived heuristic value and uniform coverage, we selected the two dozen development cases (team excerpts) that are in this section, and another 50 cases that were used for testing. Then 20 Raytheon engineers and 50 university students tried to identify the mental mistakes in the 50 excerpts of the test set. The average agreement was about 80 %.

In general, we found no correlation between the number of mistakes we detected and the students' grades on the reports. There are several reasons for this. First, the reports with the highest grades were usually written better and they were therefore easier to understand. The better we understand something, the easier it is to find and identify mental mistakes in it. Second, the better reports tended to be longer and more complete, and thus there was more opportunity for mental mistakes. On the other hand, the poor reports showed more instances of mistakes of omission, such as failure to use scoring functions, incomplete sensitivity analyses, and failure to get advice from experts.

5.2.2.13 Who Should Care?

Who should care about mistakes made in trade-off studies? Perhaps everyone should. If a trade-off study is skipped or done badly, it could prove quite costly to a company. Consider the San Diego Airport Site Selection trade-off study. This was a large, expensive, publicly accessible trade-off study that contains mental mistakes along the lines we presented in this section. This trade-off study took 6 years and cost 17 million dollars. When its results were presented to the voters in November 2006, the voters turned the proposal down and the $17M was wasted. How could this outcome have been obviated? Let's dig in.

They did a trade-off study, but only four of the ten trade-off study components were utilized: Problem statement, alternate solutions, evaluation criteria, and preferred alternatives.

They used five evaluation criteria: aeronautical, environmental, market, military, and financial. The criteria were arranged hierarchically with subcriteria and subsubcriteria. However, the criteria did not have weights of importance or scoring functions.

They had a dozen alternative sites, including the do nothing alternative. They often added and deleted alternatives. For example, the floating platform in the Pacific Ocean was dismissed early. The Campo and Borrego Springs sites were added late, so these sites had greater visibility in the public mind. However, the Campo and Borrego Springs sites were similar so, because of distinctiveness by addition of alternatives, they faded away.

They did a rudimentary sensitivity analysis looking at changes in their planning parameters at two different demand levels of passengers and cargo. They also did a small sensitivity analysis showing changes in total cost as a function of available funding (without issuing bonds or increasing taxes).

The interim results of the study were continually being leaked or reported in the press. So they certainly received a lot of expert opinions. However, in the end, the voters did not trust the study. The objectivity of the Regional Airport Authority that conducted the study was questioned. It appeared over time that the Authority was more interested in supporting a particular airport site than in explaining the various options for the voters. The Authority was perceived as being pro-business and anti-military. The difficulties of military and civilian joint use were not ameliorated. San Diegans were happy with the status quo: Lindberg Field was good for its community and the Marine Corps Air Station (MCAS) at Miramar was good for its community. The Authority did not show a burning platform or a compelling reason for change. It seemed that they only considered future business growth.

The ballot proposal asked, "Should Airport Authority and government officials work toward obtaining 3000 acres at MCAS Miramar by 2020 for a commercial airport, providing certain conditions are met?" It was turned down 38–62 %. We are not saying that if they had done a more thorough trade-off study, then the ballot proposal would have passed. We are only saying that they could have done a better trade-off study. Or, perhaps the ballot proposal could have been phrased and framed differently.

5.2.2.14 Section Summary

Good industry practices for improving the probability of success of trade-off studies include having teams evaluate the data, evaluating the data in many iterations, and expert review of the results and recommendations. It is important that the review teams have a substantial number of reviewers that are external to the project and that the reviewers consider design decisions as well as simple checking to ensure that tasks were done. Reviews are often hampered by failure to allow external reviewers access to proprietary or classified data. To improve the trade-off study, it is imperative to inform decision makers how mental mistakes are made and how they affect trade-off studies (forewarned is forearmed), encourage a long-term institutional decision horizon, use a team approach with frequent iterations, and institute both expert and public reviews. Finally try to reduce incidence of mental errors by following the recommendations presented in this section.

The literature on decision-making, cognitive science, and behavior economics contains hundreds of experiments showing persistent human mistakes of judgment [17]. Smith [15] postulated 28 specific types of mental mistakes that could affect the ten specific components of a trade-off study. This section has shown examples in a dozen categories of mental mistakes. These are mistakes that systems engineers have actually made while creating trade-off studies. Over 800 examples of such mistakes were found. These examples definitely show that systems engineers do make mental mistakes.

5.2.3 Homework Problems

5.2.3.1 Mental Mistakes

There are many additional homework problems and solutions for Section 5.2 Mental Mistakes Actually Made in Trade-off Studies at http://sysengr.engr.arizona.edu/ in the area for instructors, extrasForPart5. At 10 Mbytes, it is too big for either the textbook or the solutions manual.

5.2.3.2 Accelerated Life-Cycle Testing

On one episode of the TV series MythBusters, they investigated myths concerning pirates. In particular, they wanted to know if pirates could have removed stains from blood, sweat and tears from their clothing with urine. The pirates were said to have one barrel that all the pirates urinated in. After a few months, they would use that stale urine to wash their clothes and take stains out. The MythBusters did not want to wait a few months for the urine to get stale, so they boiled it down to 50 % of its volume. What mental mistake does this illustrate.

5.2.3.3 Medical Advice

These following common statements are examples of which 1 of the 14 mental mistakes studied in this section?

- I have xyz syndrome and I take yyy medicine. I think that you also have xyz syndrome. You should ask your doctor about yyy.
- I looked up your symptoms at a highly respected Internet blog and it said that you should do yyy.
- I saw a television commercial about yyy medicine. It fits your symptoms precisely. You should tell your doctor about yyy.
- I saw a television commercial where a famous person said that yyy is a good medicine. You should try it.

5.2.3.4 What Mistakes Were Made Here?

VISTA Car

	Evaluation criteria	Weight	Normalized weight
1	The average number of successful runs in 5 h was 20 for the whole system test	18	0.161129
2	Happiness value	0.960	0.322581
2.1	Happiness of passengers	19	0.260869
2.1	Happiness of drivers	0.95	0.434782
2.3	Number of accidents	0	0.304347
3	Availability	1	0.258064
4	Reliability	1	0.258064

5.3 Section 32. BIMS, an Example Trade-off Study

Figure 5.2 was a description of the DAR trade-off study process: in it, we discussed the functions (denoted by verbs). Now we switch to a different view—the components of a trade-off study (the nouns).

Components of a trade-off study
- Problem statement
- Concept exploration
- Evaluation criteria
- Weights of importance
- Alternative solutions
- Evaluation data
- Scoring functions
- Normalized scores
- Combining functions
- Preferred alternatives
- Sensitivity analysis

5.3.1 BIMS Problem Statement

The BICS Illuminance Management System (BIMS) was designed to be a nonmilitary example of system design including requirements, trade-off studies, and risk analysis. At the suggestion of John Hayhurst (retired Boeing Vice President), its unique feature was forcing the designers to consider both the sun and the moon in designing an illuminance management system. The complete documentation comprised 65 pages and is available at http://sysengr.engr.arizona.edu/BIMS/index.html.

Design and testing of the BICS Illuminance Management System (BIMS) was described by Bahill [65]. We summarize the findings from that paper in this section. Specifically, we present a portion of the design with a few complete examples.

BIMS is an unprecedented system (which is unusual). This means no similar system has been designed before. In contrast, in Chap. 6 we study an electric utility company. Such companies have hundreds of instances and have evolved for over a century. Obviously describing these two types of systems will be different. BIMS required a lot of low-level preliminary design. Whereas, for the electric utility, it is assumed that readers are familiar with electric utilities and therefore fewer preliminary design details are covered.

System synopsis: For astronomers, who need constant illumination, BIMS is a light and energy management system that is expected to make them more comfortable, more productive, and save money. Unlike conventional lights and blinds, BIMS will automatically control illuminance in the room when rays of the sun or moon change (Fig. 5.9).

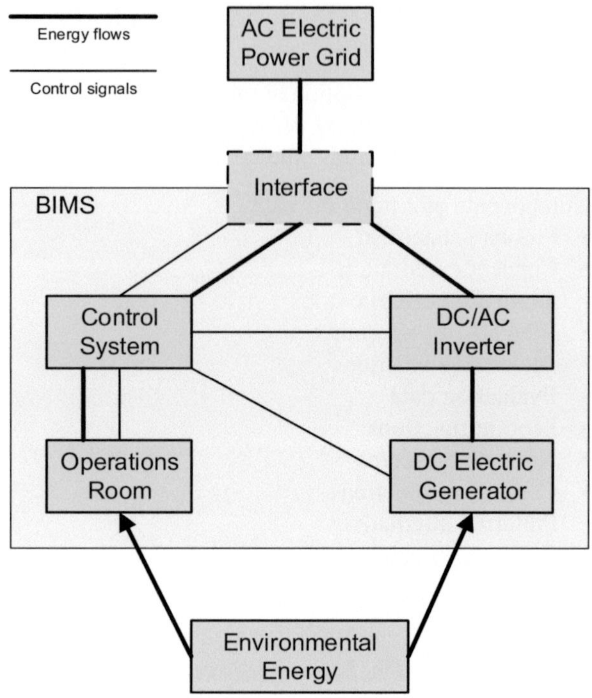

Fig. 5.9 Block diagram showing the scope of the BIMS

The customer (National Optical Astronomy Observatory, NOAO) needs a total light-management system for the operations rooms of telescope enclosures to be built on Mauna Kea in Hawaii and on Cerro Pachón near Vicuña in Chile. However, the system needs to be designed so that it can be adapted to other future structures, such as military C^4 (command, control, communications, and computers) installations.

The system must conserve energy and provide a natural daylight color spectrum. An efficient way of doing this is to have the system use daylight instead of artificial lighting as much as possible. BIMS will be politically correct (environmentally green) because it will use renewable-energy electric-generation. We considered power generation alternatives such as solar panels, wind turbines, and geothermal systems near the Kilauea volcano.

Astronomers use the operations room day and night while observing both the sun and nighttime targets such as stars and galaxies. In the daytime, astronomers need a constant illuminance of 500 lx. The building may use solar panels to generate electricity. So when a cloud temporarily covers the sun, the system cannot increase the power to the lights, in fact, power is reduced. A possible solution for this problem is to open or close window screens or curtains. The operations room will be 7 m deep, 30 m wide, and 4 m high. The east-facing wall will be totally windows. As the sun rises, the lights can be turned on. By mid-morning, lights can be dimmed and the window screens or curtains can be gradually opened. After noon, the window screens or curtains can be opened wide. As the sun sets, the lighting level gradually decreases following a predetermined program.

At night, the astronomers will be continually going in and out of the operations room. There will be no lights outside, because that would interfere with the telescopes. So we want the light inside the operations room to be dim and constant, so that (1) the astronomers do not have to wait minutes while their eyes dark-adapt and (2) the light inside does not leak out and interfere with the telescopes. Therefore, we want the inside illuminance at night to be 0.4 lx. This is roughly the illuminance of a full moon. So the system needs to accommodate phases of the moon, its rising and setting, and clouds that might obscure it.

5.3.1.1 Key Decisions

The system will use renewable-energy electric-generators. This is a political decision, not an economic or scientific decision.

The operations room will use on average 40 kWh/day of electricity.

The average ground wind speed at the telescope site on Mauna Kea is 15-mph. Most wind turbines are rated at higher speeds. Therefore the wind turbines will be derated to 15 mph. Yearly average solar insolation on Mauna Kea is 6 kWh/(m^2 day), which is high. So the solar panels will not have to be derated. If data were available, then both solar panels and wind turbines would be derated for the cold temperatures.

Noncommercial grid-connected electric generators do not need energy storage units such as batteries. Few residential solar panel users have batteries. Therefore, BICS decided that it would not have energy storage units.

The design process starts with use cases. A *use case* is an abstraction of required functions of a system. A use case produces an observable result of value to the user. Each use case describes a sequence of interactions between one or more actors and the system. Here is our first use case.

5.3.2 System Design

5.3.2.1 The First Use Case

Name: Control Illuminance During the Day
Iteration: 4.3
Derived from: ConOps
Brief description: The sun rises and sets, but the BICS Illuminance Management System (BIMS) will keep the illuminance in the operations room constant
Level: high
Priority: This is of the highest priority
Scope: The operations room of a telescope facility on a remote mountaintop and a renewable-energy electric-generator
Added value: Astronomers are more comfortable and more productive

Goal: Maintain specified illuminance in the daytime
Primary actors: Astronomer, Engineer, Tester
Supporting actors: Sun, Clouds and during the night the Moon
Frequency: It will be used every day
Precondition: The system has passed all of its built-in self-tests and Tester or the Engineer has started the system
Trigger: The sun rises

Main Success Scenario

1. The sun rises in the morning.
2. The system turns up the lights and closes the window screens or curtains.
3. The system senses the illuminance in the room with light sensors and adjusts the illuminance with light dimmers and window screens or curtains.
4. The sun "moves" across the sky and the illuminance from the sun starts to increase.
5. The system decreases illuminance from the lights and partially opens the window screens or curtains. The trade-off between these two would be determined by sunlight shining on computer monitors, heating and cooling considerations as well as electricity usage.
6. The sun rises to its zenith.
7. If it will not waste heating or cooling energy, the system opens the window screens or curtains. (It is unusual to have "if statements" in use cases.)
8. The system senses the illuminance in the room with light sensors and adjusts the illuminance with light dimmers and window screens or curtains.
9. The sun starts to set.
10. The system slowly adjusts the illuminance to its nighttime level.
11. The sun sets.
12. Include the Control Illuminance During the Night use case.

Unanchored Alternate Flow, Renewable Energy Drops

1. Electric generation falls due to the wind dropping, waves disappearing, or clouds covering the sun.
2. The system opens the window screens or curtains.
3. The system draws energy from the AC electric power grid.
4. Electric generation resumes due to the wind increasing, waves coming back, or clouds blowing away.
5. The system delivers energy to the AC electric power grid.
6. The system readjusts the light dimmers and window screens or curtains. {return to the main success scenario.}

Postcondition: The system is in the Control Illuminance During the Night use case.
Specific Requirements
Functional Requirements

This is important. From the main success scenario and the alternate flows, we derive the functional requirements for system.

FR1-1:. The system shall predict sunrise, sunset, moonrise, moonset, and the phase of the moon.
FR1-2:. The system shall control the illuminance of the lights.
FR1-3:. The system shall control the opening and closing of Light Blockers.
FR1-4:. The system shall measure the illuminance in the room.
FR1-5:. The system shall buy electricity from and sell electricity to the AC electric power grid.
FR1-6:. The system shall generate electricity.
FR1-7:. The system shall execute Built-in Self-Tests (BiST). (Derived from BICS company policy.)

Nonfunctional Performance Requirements

NFPR1-1:. The system shall maintain the daytime illuminance in the operations room at 500 ± 50 lx ($\approx 50 \pm 5$ fc). Trace to FR1-2, FR1-3, and FR1-4.
NFPR1-2:. The system shall maintain the nighttime illuminance in the operations room at 0.4 ± 0.2 lx ($\approx 0.04 \pm 0.02$ fc). Trace to FR1-2, FR1-3, and FR1-4.

Author/owner: Poli‘ahu, the snow goddess of Mauna Kea.
Last changed: January 8, 2016.

A use case diagram is the table of contents of a use case model. It shows all of the use cases that have been described so far. Figure 5.10 is our first use case diagram.

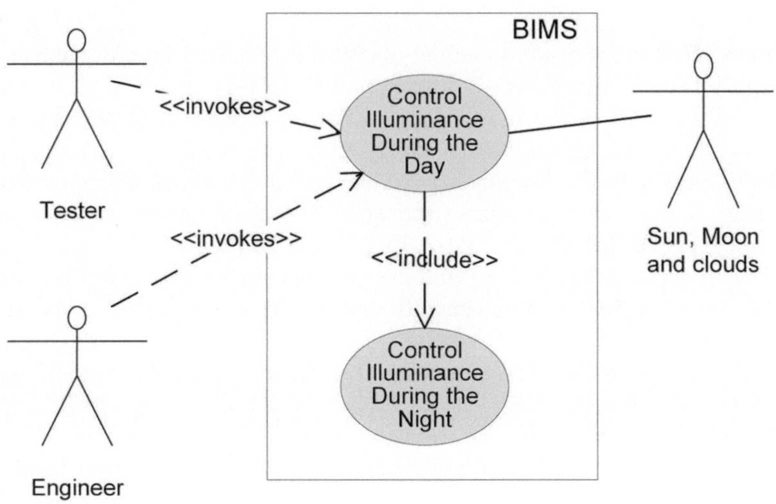

Fig. 5.10 Use case diagram for BIMS

5.3.2.2 The System Diagrams

The system design starts with a description of what the system does. This is described in use cases. These use cases directly provide the functions that the system must perform. These functions become the system requirements. Then the use cases can directly produce the system test plan.

An ancient Chinese proverb, invented by a New York advertisement writer in the 1920s and attributed to Confucius, states "A picture is worth a thousand words." Therefore, we will now describe BIMS with five figures rather than 40,000 words (Figs. 5.11, 5.12, and 5.13).

5.3.2.3 Requirements Verification

The system shall have built-in self-tests. In addition, it will be subjected to the following Logical tests.

5.3.2.4 Test Case 1 for Functional Requirements

{Test using the main success scenario of the Control Illuminance During the Day use case}

1. Before sunrise, Tester starts BIMS and records that it has executed all of its built-in self-test. This verifies FR1-7.
2. The sun rises in the morning. The system turns up the lights and closes the light blockers.
3. Tester records the result including actual and predicted time of sunrise. This verifies FR1-1, FR1-2, FR1-3, FR1-4, and NFPR1-1.
4. The system senses the illuminance in the operations room and adjusts the illuminance with light dimmers and light blockers.
5. Tester measures the illuminance in the operations room, observes that the system is following the sunrise sequence diagram fragment, and records the result. This verifies FR1-2, FR1-3, FR1-4, and NFPR1-1.
6. The sun moves across the sky and the illuminance from the sun increases.
7. The system decreases illuminance from the lights and partially opens the light blockers.
8. At half hour intervals, Tester measures the illuminance in the operations room and records the result. This verifies FR1-2, FR1-3, FR1-4, and NFPR1-1.
9. The sun rises to its zenith.
10. The system reduces the illumination of the lights and opens the light blockers as far as possible, while maintaining the illuminance within its daytime limits.
11. Tester measures the illuminance in the operations room and records the result. This verifies FR1-2, FR1-3, FR1-4, and NFPR1-1.

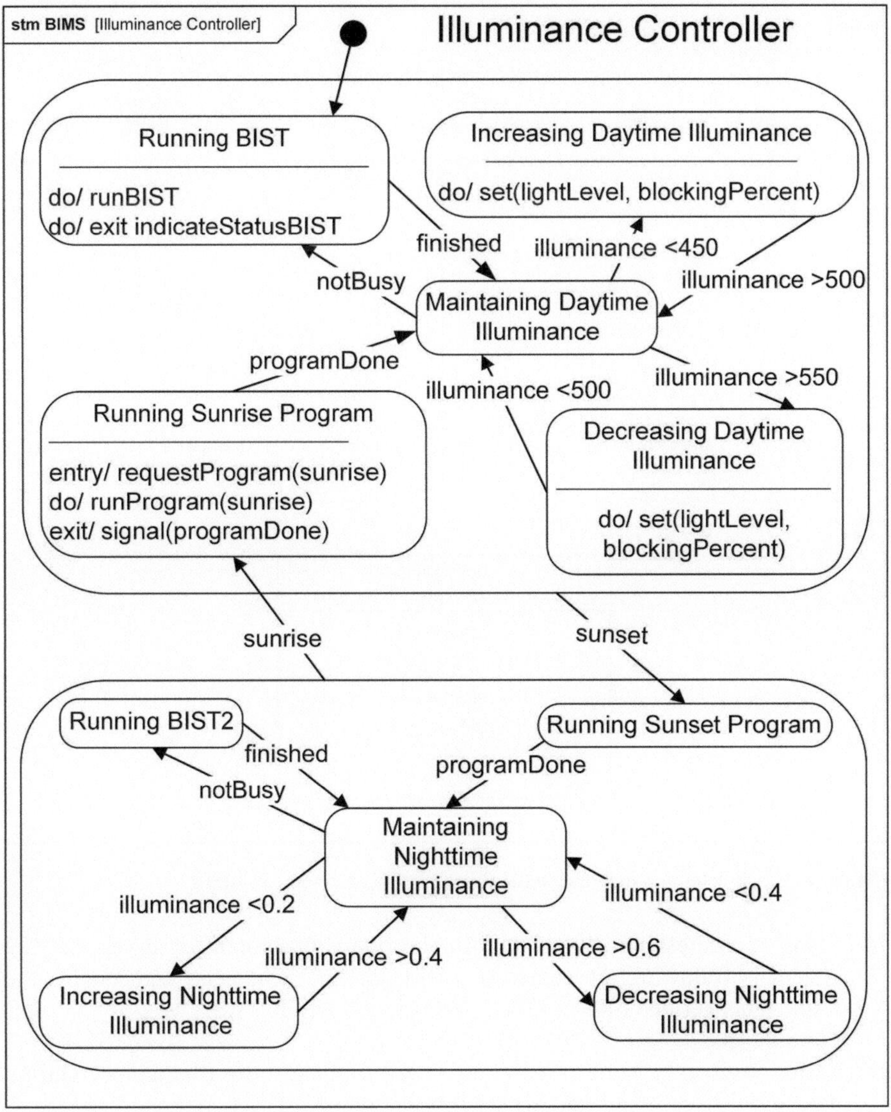

Fig. 5.11 State machine diagram for BIMS

12. The system senses the illuminance in the operations room with light sensors and adjusts the illuminance with light dimmers and light blockers.
13. At half hour intervals, Tester measures the illuminance in the operations room and records the result. This verifies FR1-2, FR1-3, FR1-4, and NFPR1-1.
14. The sun starts to set.
15. The system slowly adjusts the illuminance to its nighttime level.

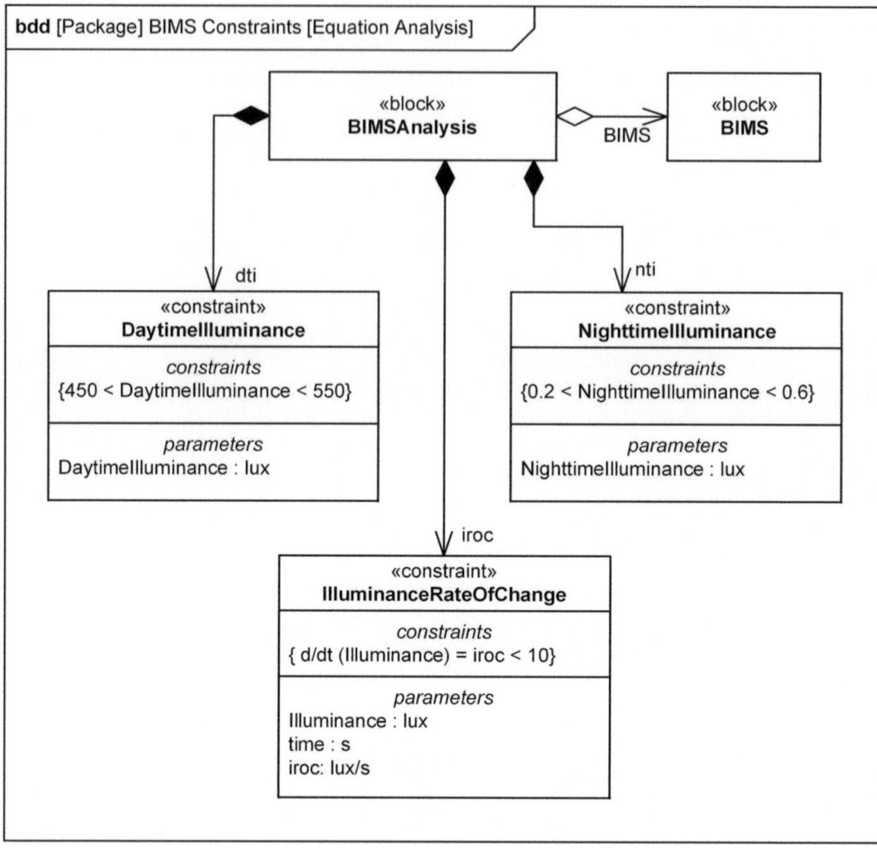

Fig. 5.12 Block definition diagram (**bdd**) showing the constraints on BIMS

16. Tester measures the illuminance in the operations room, observes that the system is following the sunset sequence diagram fragment, and records the result. This verifies FR1-2, FR1-3, FR1-4, and NFPR1-1.
17. The sun sets.
18. Tester observes that the system has shifted to the Control Illuminance During the Night use case and records the result.

{End of test case for main success scenario.}

5.3.2.5 Test for the Renewable Energy Drops Alternate Flow

1. While BIMS is operating properly during the daytime, Tester blocks sun light from the operations room. This could be done by waiting for natural clouds to come by or by unrolling an opaque screen from the roof in front of the operations room window. If wind turbines or other electric generating devices are used, then this statement will have to be changed accordingly.

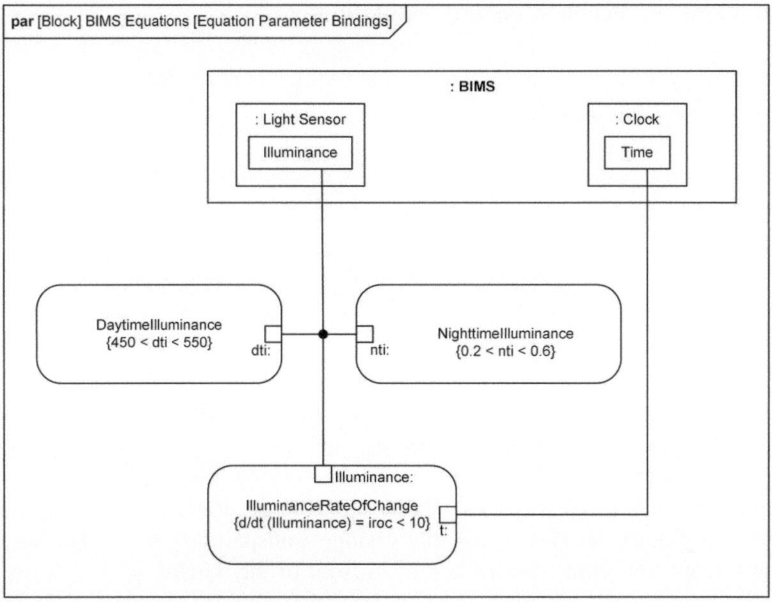

Fig. 5.13 Parametric diagram (**par**) for BIMS

2. The system opens the light blockers.
3. Tester increases the electrical load by turning on ovens, heaters, and stoves.
4. Tester measures the illuminance in the operations room, observes that the system is following the renewable energy drops unanchored alternate flow, and records the result. This verifies FR1-2, FR1-3, FR1-4, and NFPR1-1.
5. The system draws energy from the AC electric power grid.
6. Tester measures the illuminance in the operations room and the flow of energy from the AC electric power grid and records the result. This verifies FR1-2, FR1-3, FR1-4, NFPR1-1, and FR1-5.
7. Tester turns off unnecessary electrical loads and removes the opaque screen from the operations room window.
8. The system delivers energy to the AC electric power grid.
9. Tester measures the illuminance in the operations room and the flow of energy to the AC electric power grid and records the result. This verifies FR1-2, FR1-3, FR1-4, NFPR1-1, FR1-5, and FR1-6.
10. Tester observes that the system returns to the main success scenario and records the result.

{End of test case for Renewable Energy Drops alternate flow.}

This test case procedure has verified FR1-1, FR1-2, FR1-3, FR1-4, FR1-5, FR1-6, FR1-7, and NFPR1-1. Nonfunctional performance requirement NFPR1-2 will be verified in tests based on the Control Illuminance During the Night use case.

5.3.2.6 Test for Verification and Validation

Tester uses the system as in the test for the main success scenario for the Control Illuminance During the Day use case.

Tester observes and records the result.

Tester examines the data and writes a V&V report. This verifies FRz-1 Gather V&V data.

{End of V&V Test}

5.3.2.7 Concept Exploration

In the beginning, we investigate alternatives for the following five subsystems. Later, we perform several trade-off studies.

Light blockers alter the amount of light coming into and out of a room. Examples include blinds, shades, screens, curtains, drapes, shutters, polarized windows, skylights, window films that change transparency with electricity, and opaque screens that unroll from the roof in front of the windows.

Lights provide illumination. Alternatives that should be considered include convention fluorescent tubes, dimmable fluorescent lights, incandescent bulbs, light emitting diodes (LEDs), and compact fluorescent light bulbs (CFLs).

Control systems manage the entire light management system.

Electric generators transform environmental energy into electrical energy. Some common examples include photovoltaic solar panels; wind turbines; ocean currents, waves, and tides; hydroelectric dams; tapping static electricity stored in the atmosphere; and geothermal systems based on the Kilauea volcano.

Energy storage units smooth the electric generation when the environment varies. However, energy need not be stored as electricity. The energy storage system could be electrical (like super capacitors), chemical (like batteries), thermal (like a big rock, a pond, molten salt, or ice), potential (like pumping water between two altitudes or compressed air tanks), or kinetic (like a big flywheel) (Fig. 5.14).

Electric generators transform environmental energy into electrical energy. Some common examples include photovoltaic solar panels; wind turbines; ocean currents, waves, and tides; hydroelectric dams; balloons for tapping static electricity stored in the atmosphere; and geothermal systems based on the Kilauea volcano. For this project, electric generators were connected to the AC electric power grid. Table 5.20 shows the alternative systems that were analyzed.

The question posed in this trade-off study was, "In terms of using renewable-energy to generate electricity for the telescope operations room on top of Hawaii, what is the preferred electric generating system?"

Connection to the AC power grid must comply with the UL1741/IEEE1547 standard. All grid-connected electric generators must have microprocessors, because they must convert the generated DC electricity into sinusoidal AC electricity and they must ensure that their phase is synchronized with the grid before

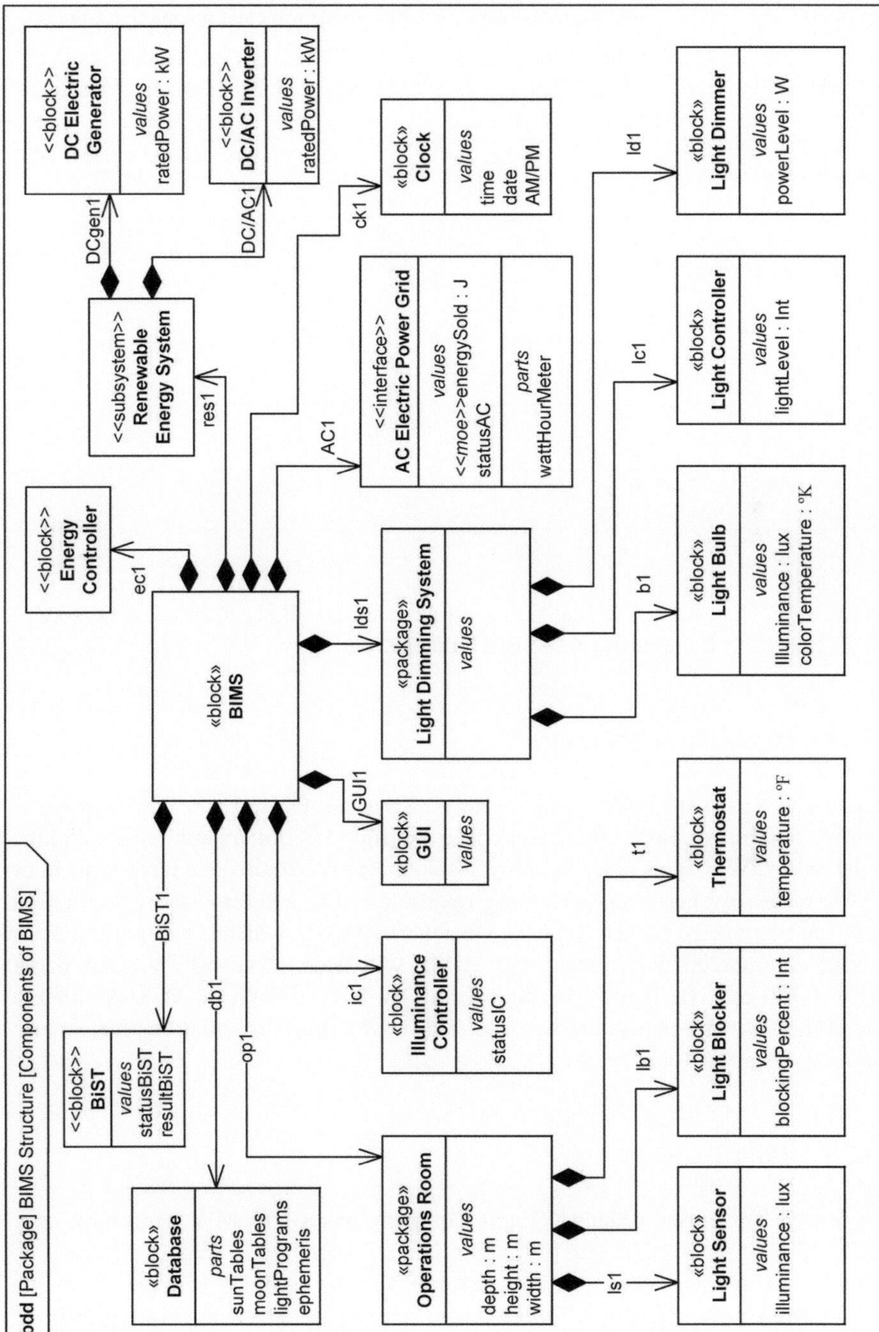

Fig. 5.14 Block definition diagram (**bdd**) showing the structure of BIMS

Table 5.20 Renewable-energy electric-generating systems that were analyzed in the trade-off study

Manufacturer	Rated power (kW)	Purchase price without installation (US dollars)
Grid-connected photovoltaic systems		
Sun Electronics	10.2	$34,000
DMSolar	10.4	$34,000
The Solar Calculator	8.9	$33,000
Grid-connected wind turbines		
Jacobs Wind Systems	7	$55,000
Bergey Windpower	2	$30,000
Abundant Renewable Energy	3.5	$50,000
Southwest Wind Power	2.4	$14,000
Nordtank 65	10.2	$62,000
Other		
LIMPET, Wave Power	100	$260,000
Helium Balloons	6×10^{-12}	$100
The do nothing alternative	0	0

connecting. Therefore, it is likely that all systems are likely to have some sort of BiST. However, the quality of the BiST display can vary.

5.3.2.8 Photovoltaic (PV) Solar

Feasibility: The Topaz Solar Farm east of San Luis Obispo is a 550 MW solar photovoltaic power plant. This farm cost 2.6 billion US dollars and covers 25 km^2 of the desert. With a capacity factor of 23 %, the project delivers a terra Watt-hour (TWh) of energy per year. Assuming operating costs are zero, and construction costs can be amortized over 10 years, we get $0.10/kWh, which is competitive with present commercial electricity rates. BIMS will need about 50 PV solar panel modules (80 m^2). Each will be about 1.6 m by 1 m, 200 W, 24 V, $600, 19 kg. Therefore, PV solar panels are feasible. Table 5.21 shows calculations for generic solar PV systems.

5.3.2.9 Wind Turbines

Feasibility: The Hawaii Electric Light Company has a 10.5 MW wind farm near Upolu Point at the northern tip of the Big Island, where the average wind speed is 20 mph. The average wind speed at the top of the Big Island is 15 mph.

The rated power for most wind turbines is being given at wind speeds of 25–30 mph. The top of Hawaii has an average wind speed of 15 mph. Therefore, we estimated the power of each turbine at 15 mph. The output power of a wind

Table 5.21 Results from the solar calculator for Zip Code 96720 with an electric rate of \$0.30/kWh (residential)

Expected average monthly energy consumption	1200 kWh/month
Generator rated power	8.9 kW
Required solar panel area	900 ft^2
Estimated cost including parts and installation	\$71,000
Federal tax credit	\$21,000
State tax credit	\$17,000
Total purchase price	\$33,000
Expected post installation tax credits	\$32,000
Average monthly savings on the electricity bill	\$360
25 year savings	\$180,000
25 year return on investment	557 %
Break-even point	6.7 years

turbine varies as the speed of the wind cubed, below 10 mph. At 15 mph, it is roughly linear.

For comparison purposes, we compute the price-to-power ratio (\$ per Watt). Each manufacture computes this differently. Therefore, we use the purchase price including all needed elements (PV solar panels, wind turbines, DC/AC inverters, grid connections, tower, and mounting frames). For power, we use the manufactures' rated power adjusted for site location. For wind turbines, this means Estimated Power (kW) at 15 mph average wind speed.

The trade-off parameters for the most feasible alternatives that we investigated are summarized in Table 5.22.

The price-to-power ratio is the purchase price (without installation) divided by the rated power for the PV solar systems or the estimated power at 15 mph for the wind turbines (Fig. 5.15). This is the gray area in Table 5.22.

5.3.2.10 Other Electric Generators

Wave Power

The LIMPET (Land Installed Marine Power Energy Transmitter) 100 kW shoreline wave power plant was installed on the Isle of Islay off the west coast of Scotland in 2000. The plant has been operating remotely since that time and is supplying energy to the AC electric power grid in the United Kingdom. On average, it produces 100 kW. The installed cost is \$2600/kW. So the price-to-power ratio is \$26/W. The estimated sale price of electricity is \$0.057/kWh, which is indeed competitive. It does have BIST. This alternative gets lots of political correctness points.

Table 5.22 Trade-off parameters for BIMS alternatives

	Rated power (kW)	Estimated power (kW) at 15 mph	Output voltage (V)	Purchase price, without installation	Price-to-power ratio ($/W)	Output of SSF9, Fig. 5.15
Do nothing	0	0		0	$0 \times \infty$	0.5
Grid-connected photovoltaic (PV) power systems						
Sun Electronics Model 10250	10.2		208	$34,000	3.3	0.95
DMSolar 10.4 kW system	10.4		208	$34,000	3.1	0.96
Generic solar calculator	8.9		120/240	$33,000	3.7	0.94
Grid-connected wind turbines						
Jacobs Wind Systems model 31–20	20	7	240	$55,000	7.9	0.70
Bergey Windpower 10 kW BWC Excel-S	10	2	240	$30,000	15	0.11
Abundant Renewable Energy (ARE) 442	10	3.5	240	$50,000	14	0.16
Skystream 3.7	2.4	2.4	120	$14,000	5.8	0.85
Nordtank 65 is a big reconditioned windmill that was installed in the ocean	65	10.2	480 3 phase	$62,000	6.1	0.83
Wave power						
Limpet	100				26	0

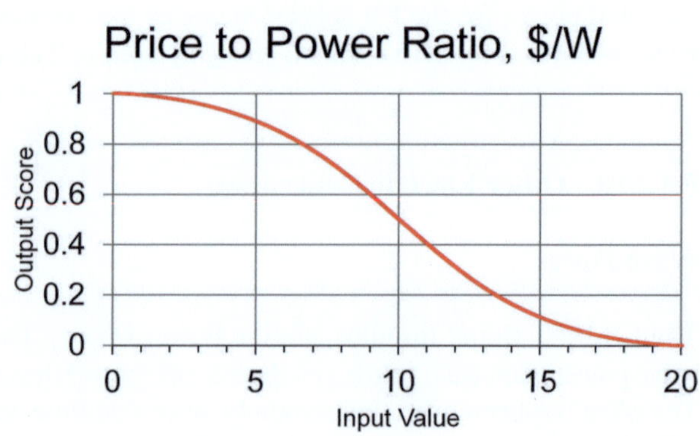

Fig. 5.15 Scoring function for the price-to-power ratio: SSF9, range 0 to 20, L = 0, S = −0.1, B = 10, U = 20

Volcano Power

Feasibility: It is not feasible even for Hawaii. Furthermore, it has zero reusability.

Portable Trailers

TEP has portable trailers with self-sustaining photovoltaic-solar electric-generating systems. They have 24 batteries: each is 12 V, 210 Ah, and cost $350. They put four in series to produce 48 V. The trailer with solar panels cost $80 k and the inverter is $3 k, for a total cost of $91,400. Its rated output power is 1.8 kW.

Balloons Tapping the Electric Potential Gradient in the Atmosphere

There are patents on a system to tap the electric potential between the ionosphere and the Earth's surface. Their plan is to attach a 20-gauge wire to a helium-filled balloon and let it float 30 m above the surface of the Earth. The reason why this might work is that there is almost always lightning somewhere in the world. Lightning (below, in and above the clouds), cosmic rays, and radon in the earth's crust create positive–negative ion pairs. Atmospheric mixing moves these ion pairs. The resulting charge separation creates a potential difference of about 300 kV between the surface of the earth and the bottom of the ionosphere at 80 km. Figure 5.16 is a model for this whole-earth phenomenon.

However, to generate electricity in a region we cannot use the whole earth. We are limited to a column of air in our region. A column of air one meter in diameter as shown in Fig. 5.17 would be a good model.

However, this system might not be feasible, because air is a very good *insulator*. Figure 5.17 shows that the high resistance of air limits the available current to about

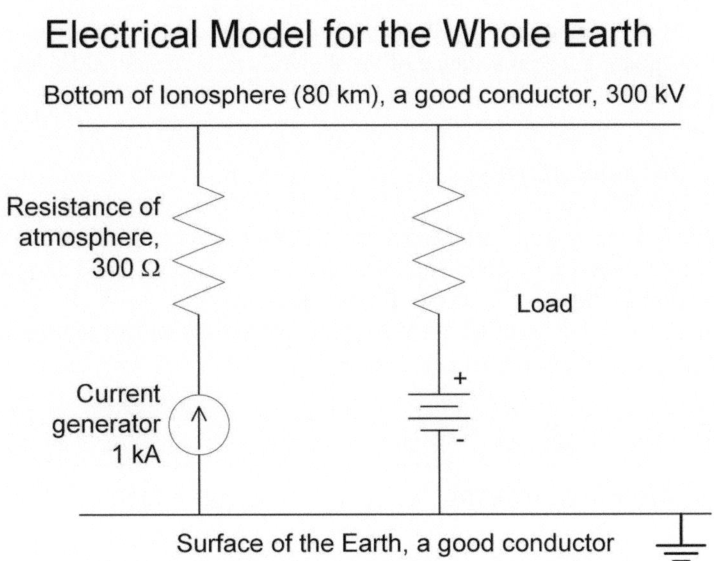

Fig. 5.16 Electrical model for the atmosphere of the Earth

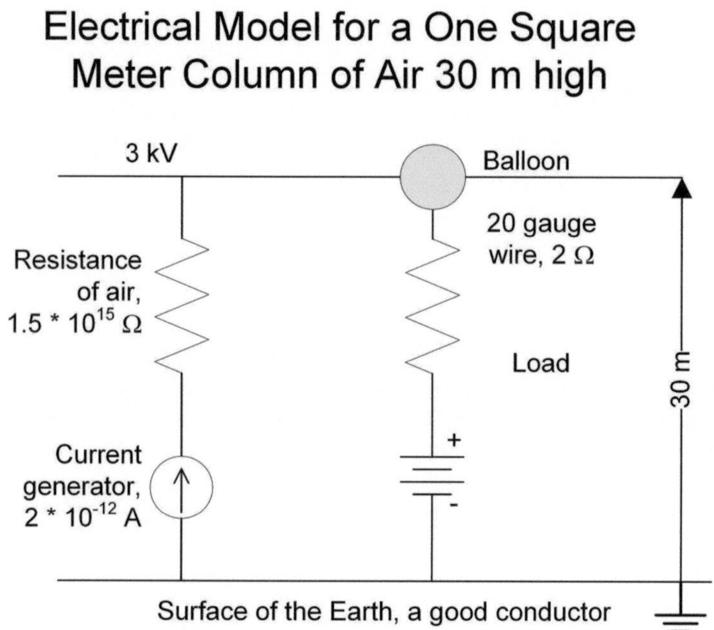

Fig. 5.17 Model for the balloon electric generator

two pico amperes, which produces only 6×10^{-9} W. Our telescope system requires 10 kW.

Figure 5.17 shows a model for a column of air 30 m high and 1 m^2 in area. The 1 m^2 column of air models the balloon as a conductive sphere with cross-sectional area of 1 m squared. The resistance of air shown here is for low altitudes.

5.3.2.11 Do Nothing Option

Feasibility: The do nothing alternative is to buy electricity from the AC electric power grid and not worry about renewable-energy sources. This has political as well as economic impacts. This option was ruled out.

Section 5.3.1.1 and Sections 5.3.2.7 to 5.3.2.11 on alternative architectures are managed by the system architect (Fig. 5.18).

5.3.3 Evaluation Criteria

Evaluation criteria are required for each trade-off study. Some of the criteria may be reusable in multiple trade-off studies.

Fig. 5.18 Zach Bahill's vision of some alternative concepts for BIMS

As stated in Section 20, evaluation criteria must be quantitative, independent, transitive, objective, and worded so that "more is better," independent of the temporal order of evaluation, and show compensation. The evaluation criteria need to be arranged hierarchically. They have scoring functions to normalize output scores. The evaluation criteria for choosing the recommended renewable-energy electric-generating systems is presented next in accord with the evaluation criterion template from Section 20.

The following eight evaluation criteria will be called evaluation criteria in this section. But in Fig. 5.21, they will be called subcriteria, because performance, cost, and company policy are the criteria.

5.3.3.1 Vendor Evaluation

Name of criterion: *Vendor Evaluation.*

Description: Each vendor is evaluated on history in the marketplace, financial stability, support, responsiveness, upgrades, enhancements, custom enhancements, third party support, technology, product maturity, product stability, vendor stability, licensing cost, site licenses, licensing approach, reputation, maintenance approach, complementarity, and whether the vendor is a small disadvantaged business.

Weight of importance (priority): 8.

Basic measure: A number between 0 and 1 that indicates the "goodness" of the vendor. It also warns of specific risks.

Units: none.

Measurement method: The analyst fills out the BICS vendor evaluation spreadsheet.

Input: The analyst assigns cardinal numbers (probably integers) between 1 and 10.

Scoring function: no scoring function is necessary.

Output: 0–1.

Traces to company policy.

5.3.3.2 Interference with Telescopes

Name of criterion: *Possible interference with telescope views.*

Description: Telescopes are built at remote sites to reduce light pollution and visual obstruction. PV solar panels mounted on the ground will not cause additional obstruction. Wind turbines on tall towers might be a problem. There are a dozen telescopes on top of Hawaii, to avoid interfering with any, wind turbines would have to be placed down the hill, where the wind will be slower.

Weight of importance (priority): 10.

5.3.3.3 Rated Power

Name of criterion: *Rated Power.*

Description: Rated power (kW) is the power that the system was designed to generate in steady state at the normal operating point, which is usually near the maximum power point. It is also called generating capacity and nameplate capacity. The normal operating point for the telescope site on top of Hawaii has an average wind speed of 15 mph and an average insolation of 6 kWh/(m^2 day). Therefore, the wind speed operating point power had to be calculated for 15 mph winds. Rated power is the gray area in Table 5.22.

Weight of importance (priority): 4 [68].

Basic measure: the power that the system was designed to generate.

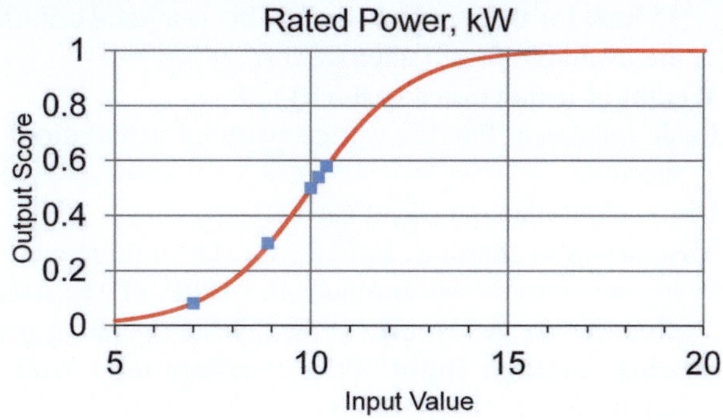

Fig. 5.19 Scoring function for the *rated power* evaluation criterion

Units: kW.
Measurement method: Initially we believe the manufacturers' specifications. During operations, we compute the value of the average power that is actually produced in steady state at the normal operating point.
Scoring function input: Rated power (kW), with a domain of 0 to 1 MW, and expected values between 5 and 25 kW.
Scoring function: SSF1 monotonic increasing, LL = 5, BL = 10, $S = 0.2$, UL = 25, plot limits 5–20.
Scoring function output: 0–1.
Traces to NFPR1-4.

The input to the function of Fig. 5.19 is rated power, which has legal values between zero and one MW. Feasible alternatives will have expected values between 5 and 25 kW. The alternatives listed in Table 5.20 had rated powers between 0 and 100 kW. The baseline value for Fig. 5.19 is 10 kW, which is where the curve has its maximum slope and therefore the greatest differentiating power. Five of the alternatives are shown in Fig. 5.19. The output is normalized to the 0–1 range and more output is better.

5.3.3.4 Total Life Cycle Cost

Name of criterion: *Total Life Cycle Cost* (thousands of US dollars).

The total life cycle cost was computed from the design, purchase, installation, operation and maintenance costs.

5.3.3.5 Price-to-Power Ratio

Name of criterion: *Price-to-Power Ratio.*
Description: The Price-to-Power Ratio is the purchase price (without installation) divided by the rated power for the PV solar systems or the estimated power at

15 mph for the wind turbines. This is a good criterion because comparable data are available for all alternatives.

Weight of importance (priority): 9.

Basic measure: Purchase price (without installation) divided by rated power for the site.

Units: US dollars per Watt ($/W).

Measurement method: Initially we start with manufacturers' specifications. During operations, we compute the value of the average power that is actually produced in steady state at the normal operating point.

Scoring function input: Price-to-power ratio with a domain $0–$100/W and expected values $2–$20/W.

Scoring function: monotonic decreasing, SSF9, L = 0, B = 10, S = −0.1, U = 20. This scoring function was shown in Fig. 5.15.

Scoring function output: 0–1.

Traces to CoR3.

5.3.3.6 Cost of Generated Electricity

Name of criterion: *Cost of Generated Electricity.*

Description: Expected average daily electrical energy cost for one unit in this location. This depends on the amortization period. It will not have good data in early iterations.

Weight of importance (priority): 10.

Basic measure: expected average daily electrical energy cost for one unit in this location.

Units: $/kWh, where $ means US dollars.

Measurement method: computation by BICS analysts.

Scoring function input: Cost of generated electricity with a domain 0–$2/kWh and expected values of $0.05–$0.30/kWh.

Scoring function: monotonic decreasing, LL = 0.02, BL = 0.15, $S = -10$, UL = 0.30 (Fig. 5.20).

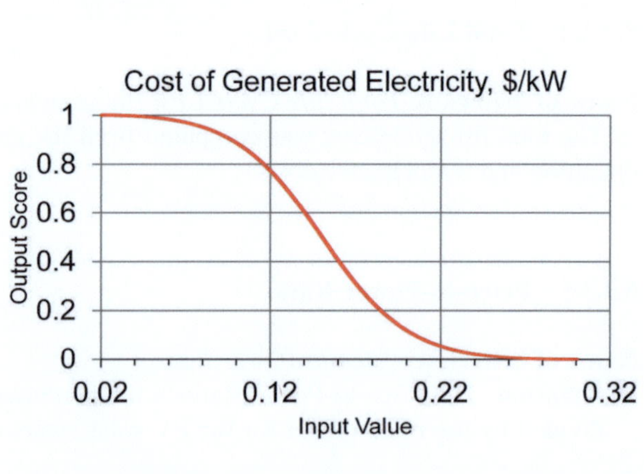

Fig. 5.20 Scoring function for the cost of generated electricity evaluation criterion

Scoring function output: 0–1.
Traces to CoR3.

5.3.3.7 Built-In Self-Test

Name of criterion: *Built-in self-test*:

It was difficult to get data for this criterion because either manufacturers do not have BiST or they do not advertise it. BiST could be passive, where the BiST system monitors outputs and check points and displays status, or BiST could be active, where the BiST system generates signals and applies them to system inputs.

5.3.3.8 Reusability

Name of criterion: *Reusability*:

It is company policy that nondevelopmental products be considered the primary solution to addressing customer needs with customized implementations being secondary solutions. To enhance potential reuse in other present and future BICS systems, a reuse design review with participation of top management shall be scheduled before the systems requirements review. During the design phase, all engineers must flag reuse potential. In the trade-off study, each alternative is given points for use of commercial off-the-shelf (COTS) products, and deducted points for use of custom products. (For example, the components for photovoltaic solar panel and wind turbine systems are COTS, whereas the LIMPET system is custom-made.) Each alternative is given points based on the likelihood that the type of renewable-energy electric-generator could be reused in different geographical locations. For example, solar panels need locations where there is sunshine, whereas wind turbines need locations with wind, and LIMPET systems need locations with ocean waves. Systems based on geothermal energy from the Kilauea volcano are fine for Mauna Kea in Hawaii, but inappropriate for Cerro Pachón in Chile. Each alternative is given points if the analysis and results are likely to be reused in different political environments. Federal, state, and local government subsidies make renewable-energy electric-generation systems economically feasible. So the economic analyses are expected to be different outside of the United States.

5.3.3.9 Unused Criteria

The following criteria were considered, but for the most part, they were not included in the trade-off study. They are included here to assure future analysts that we did consider them.

System size does not trace to the ConOps or a requirement. Height (meters) Tall wind turbines might interfere with telescope visibility. Footprint (cross-sectional area in square meters) Photovoltaic panels could be mounted on the roof. The total roof area of our operations building will be around 500 m^2. BIMS will need less than 100 m^2.

This term estimates the *average daily electricity generation*. It estimates the amount of electricity that a particular system would be expected to generate per day in a particular location (kWh/day), the domain is 0–1000 kWh/day and expected values range from 10 to 100 kWh/day. Average daily electricity generation will not be used in this trade-off study, because there is too much variability in the way different manufacturers define and measure this term.

Capacity Factor: Electric generators do not generate electricity at their maximum rate 24 h a day, 7 days a week. The system must be shut down for maintenance and repairs, and output will be reduced when the loads are small. In addition, renewable-energy systems will have reduced output when, for example, renewable energy drops or there is no wind. The capacity factor is defined for specified period of time: it is the actual electricity generated divided by the energy that theoretically could have been generated if the system ran at maximum capacity for the whole period of time.

$$\text{Capacity factor} = \frac{\text{Actual average daily electricity generated (kWh/day)}}{\text{Rated power (kW)} \times \frac{24\ \text{h}}{\text{day}}}$$

The capacity factor of renewable-energy electric-generators is typically 5–40 %. For photovoltaic solar power, the capacity factor is around 17 %. Capacity factor will not be used in this trade-off study because of the difficulty of obtaining data in the design phase.

Intermittency was a big problem for wind turbines, especially for large wind farms. As the wind waxed and waned huge changes in generated power occurred. However, now with the advent of second by second reporting of wind speed and output power by each individual wind turbine, sophisticated software can angle the blades on individual turbines and balance electric production.

Energy Storage Capacity, Units (kWh), the domain is 0–1000 kWh and expected values range from 10 to 100 kWh. This will not be used in the trade-off study for alternative electric *generation* systems.

System Complexity does not trace to the ConOps or a requirement.

Operating Costs (thousands of US dollars per year) does not trace to the ConOps or a requirement. However, it is included in *total life cycle cost*.

Energy Conversion Efficiency (percent), the domain is 2–50 % and the expected values range from 10 to 20 %. This does not trace to the ConOps or a requirement. It is mentioned because it is a very popular metric.

The *25 year return on investment* and the *break-even point* are valuable evaluation criteria and they are easy to compute.

Even though all alternatives are commercial off-the-shelf (COTS) products *delivery schedules* could be a problem, because Hawaii is so far from the mainland.

5.3.4 *Preferred Alternatives*

The numerical computations for our trade-off study start with prioritizing the evaluation criteria as shown in Table 5.23.

The trade-off study matrix for this project is shown in Fig. 5.21. It shows that the photovoltaic (PV) solar panels have the highest alternative ratings of 0.73, 0.74, and 0.77.

Figure 5.21 is static. The dynamic spreadsheet is available at http://sysengr.engr.arizona.edu/forStudents/index.html.

The values for the cells were derived by a panel of domain experts using the Delphi method. The values came from manufacturers' data, peer reviewed journal papers, and (only for the weather data) Internet web sites. A value was derived for each alternative for each criterion. This value was put into the scoring function for that criterion and the resulting score (sc) was inserted into Fig. 5.21. Each score was then multiplied by its corresponding normalized weight (w) and these products were summed in each column to produce the alternative ratings.

This trade-off study answered the question "In terms of using renewable-energy to generate electricity for the telescope operations room on top of Hawaii, what is the preferred electric generating system?" With that being the question, the preferred alternative was photovoltaic solar panels without battery storage.

Each number in a trade-off study matrix should have an explanation. For example, the score for total life cycle cost for the do nothing alternative was 0.2. Its purchase price was zero, which gave it a high score. But it would sell no energy to the electric company. Therefore, it would not generate revenue. This lowered its

Table 5.23 Prioritization of evaluation criteria

Evaluation criteria	Weight of importance	Trace to
Vendor evaluation	8	Company policy
Interference with telescopes	10	Rob Seaman
Rated power	4	NFPR1-4
Total life cycle cost	6	CoR2
Price-to-power ratio	9	CoR3
Cost of generated electricity	10	CoR3
Built-in self-test (BiST)	10	FR1-7
Reusability	6	Company policy
Physical system size	0	nothing
Expected average daily electricity generation	4	NFPR1-4
Capacity factor	1	NFPR1-4
Constancy of electric generation	0	nothing
Capacity of energy storage units	0	FR1-6
System complexity	2	nothing
Operating costs	1	CoR2
Energy conversion efficiency	0	nothing

Tradeoff Study Matrix for BIMS using a Sum Combining Function and semirelative sensitivity functions with the number of alternatives,																		m=	10				
Evaluation Criteria	Wt	Norm. Criteria Weights	Norm Sub Criteria Weights	Alt 1 Do Nothing		Alt 2 Sun Electronics PV		Alt 3 DMSolar PV		Alt 4 Solar Calculator Generic PV		Alt 5 Jacobs Wind Turbine		Alt 6 Bergey WindPower		Alt 7 ARE 442		Alt 8 Skystream 3.7		Alt 9 LIMPET		Alt 10 Nordtank 65	
				sc	$W \times SC$	sc	$W \times SC$	sc	$W \times SC$	sc	$W \times SC$	sc	$W \times SC$	sc	$W \times SC$	sc	$W \times SC$	sc	$W \times SC$	sc	$W \times SC$	sc	$W \times SC$
Performance	8	0.38																					
Vendor Evaluation	8		0.36	0.9	0.33	0.5	0.18	0.6	0.21	0.8	0.29	0.5	0.18	0.5	0.18	0.5	0.18	0.4	0.15	0.8	0.29	0.8	0.29
Interference with Telescopes	10		0.45	1.0	0.45	1.0	0.45	1.0	0.45	1.0	0.45	0.2	0.09	0.2	0.09	0.2	0.09	0.4	0.18	1.0	0.45	0.1	0.05
Rated Power	4		0.18	0.00	0.00	0.54	0.10	0.58	0.11	0.29	0.05	0.08	0.01	0.00	0.00	0.00	0.00	0.00	0.00	1.00	0.18	0.54	0.10
Cost	4	0.19																					
Total Lifecycle Cost	6		0.24	0.2	0.05	0.5	0.12	0.5	0.12	0.8	0.19	0.5	0.12	0.5	0.12	0.5	0.12	0.5	0.12	0.4	0.10	0.5	0.12
Price to Power Ratio	9		0.36	0.50	0.18	0.95	0.34	0.96	0.35	0.94	0.34	0.70	0.25	0.11	0.04	0.16	0.06	0.85	0.31	0.00	0.00	0.83	0.30
Cost of Generated Electricity	10		0.40	0.2	0.08	0.5	0.20	0.5	0.20	0.8	0.32	0.5	0.20	0.5	0.20	0.5	0.20	0.5	0.20	0.6	0.24	0.6	0.24
Company Policy	9	0.43																					
BiST	10		0.63	0.8	0.50	0.9	0.56	0.9	0.56	0.7	0.44	0.1	0.06	0.1	0.06	0.9	0.56	0.1	0.06	0.5	0.31	0.3	0.19
Reusability	6		0.38	0.3	0.11	0.5	0.19	0.5	0.19	0.8	0.30	0.5	0.19	0.5	0.19	0.5	0.19	0.5	0.19	0.1	0.04	0.9	0.34
Alternative Rating					0.62		**0.73**		**0.74**		**0.78**		0.33		0.28		0.50		0.35		0.57		0.52
PI5=	0.54	1.00	3.00																				

Fig. 5.21 Trade-off study matrix, m = 10 is the number of alternatives. PI5 is the performance index being used. The question being answered by this trade-off study is, "In terms of using renewable-energy to generate electricity for the telescope operations room on top of Hawaii, what is the best electric generating system?"

score to 0.2. Somewhere in each trade-off study, the numerical value in each cell should have an explanation like this.

When filling in a trade-off study matrix, like Fig. 5.21, it is important to pay attention to the common mental mistakes presented in this section. For example, the criteria total life cycle cost, price-to-power ratio, and cost of generated electricity are dependent. But evaluation criteria should be independent. This is the reason for them being subcriteria. Second, there is an uneven level of detail in the description of the photovoltaic solar panels and the LIMPET system. This often occurs for straw man options, but should not exist for serious alternatives.

5.3.4.1 Sensitivity Analysis

In Section 3.7.5, we showed how a sensitivity analysis could be performed with performance indices PI_1, PI_2 and PI_3. However, a problem with these performance indices is that the sensitivity with respect to w_1 does not depend on scores for the nonwinning alternatives and we want the sensitivities to depend on all of the parameters. Therefore, in this section, we show a better performance index that solves this problem. It is especially useful for trade-off studies.

The sensitivity analysis is arguably the most important part of the trade-off study.

5.3.4.2 Hierarchical Trade-off Studies

A trade-off study can be hierarchical. There can be any number of criteria and each of them can have any number of subcriteria. Table 5.24 suggests the structure.

Of course, not all trade-off studies would use the specific criteria of performance, cost schedule, and risk, and many other criteria could also be used. In the following formulae there are k criteria and we index them with l. The first criterion will have n(1) subcriteria. The second criterion will have $n(2)$ subcriteria and the kth criterion will have $n(k)$ subcriteria. Within each criterion ($l = 1, 2, \ldots, k$), we index the subcriteria with i; i will run from 1 to n(l). There are m alternatives and we index them with j. For the criteria weights, we use $\mathrm{CW}^{(l)}$, the subcriteria weights are $w_i^{(l)}$ and the scores are $\mathrm{sc}_{ij}^{(l)}$. In this example, we combine the data with simple linear addition, $\mathrm{Sum}_j = \sum_{l=1}^{k} \mathrm{CW}^{(l)} \sum_{i=1}^{n(l)} w_i^{(l)} \mathrm{sc}_{ij}^{(l)}$.

If the trade-off study is hierarchical, you could create a new sensitivity weights column as follows: remove the hierarchy from the normalized weights column by multiplying each subcriterion by the normalized weight of its super criterion. The resulting sensitivity weights column will then be used in place of the normalized weights column throughout the matrix. But this fails to give the sensitivities for the criteria weights. Therefore, we recommend using this new performance index, PI_5.

Table 5.24 The structure of a hierarchical trade-off study

Evaluation criteria	Normalized criteria weights	Subcriteria	Normalized subcriteria weights	Scores for alternative 1	Scores for alternative 2
Performance	$CW^{(1)}$				
		Subcriteria-1	$w_1^{(1)}$	$sc_{11}^{(1)}$	$sc_{12}^{(1)}$
		Subcriteria-2	$w_2^{(1)}$	$sc_{21}^{(1)}$	$sc_{22}^{(1)}$
		Subcriteria-3	$w_3^{(1)}$	$sc_{31}^{(1)}$	$sc_{32}^{(1)}$
		Subcriteria-4	$w_4^{(1)}$	$sc_{41}^{(1)}$	$sc_{42}^{(1)}$
Cost	$CW^{(2)}$				
		Subcriteria-1	$w_1^{(2)}$	$sc_{11}^{(2)}$	$sc_{12}^{(2)}$
		Subcriteria-2	$w_2^{(1)}$	$sc_{21}^{(2)}$	$sc_{22}^{(2)}$
Schedule	$CW^{(3)}$				
		Subcriteria-1	$w_1^{(1)}$	$sc_{11}^{(3)}$	$sc_{12}^{(3)}$
		Subcriteria-2	$w_2^{(3)}$	$sc_{21}^{(3)}$	$sc_{22}^{(3)}$
Risk	$CW^{(4)}$				
		Subcriteria-1	$w_1^{(4)}$	$sc_{11}^{(4)}$	$sc_{12}^{(4)}$
		Subcriteria-2	$w_2^{(4)}$	$sc_{21}^{(4)}$	$sc_{22}^{(4)}$
		Subcriteria-3	$w_3^{(4)}$	$sc_{31}^{(4)}$	$sc_{32}^{(4)}$
Alternative ratings				Sum_1	Sum_2

$$\text{PI}_5 = \frac{1}{m}\sum_{l=1}^{k} \text{CW}^{(l)} \sum_{i=1}^{n(l)} \sum_{j=1}^{m} w_i^{(l)} \text{sc}_{ij}^{(l)}$$

The first-order partial derivatives and the semirelative-sensitivity functions are

$$\frac{\partial \text{PI}_5}{\partial \text{CW}^{(l)}} = \frac{1}{m} \sum_{i=1}^{n(l)} \sum_{j=1}^{m} w_i^{(l)} \text{sc}_{ij}^{(l)}$$

$$\widetilde{S}_{\text{CW}^{(l)}}^{\text{PI}_5} = \left.\frac{\partial \text{PI}_5}{\partial \text{CW}^{(l)}}\right|_{\text{NOP}} \text{CW}^{(l)} = \frac{1}{m} \text{CW}^{(l)} \sum_{i=1}^{n(l)} \sum_{j=1}^{m} w_i^{(l)} \text{sc}_{ij}^{(l)}$$

$$\frac{\partial \text{PI}_5}{\partial w_i^{(l)}} = \frac{1}{m} \text{CW}^{(l)} \sum_{j=1}^{m} \text{sc}_{ij}^{(l)}$$

$$\widetilde{S}_{w_i^{(l)}}^{\text{PI}_5} = \frac{1}{m} \text{CW}^{(l)} \sum_{j=1}^{m} w_i^{(l)} \text{sc}_{ij}^{(l)}$$

$$\frac{\partial \mathrm{PI}_5}{\partial \mathrm{sc}_{ij}^{(l)}} = \frac{1}{m} \mathrm{CW}^{(l)} w_i^{(l)}$$

$$\widetilde{S}_{\mathrm{sc}_{ij}}^{\mathrm{PI}_5} = \frac{1}{m} \mathrm{CW}^{(l)} w_i^{(l)} \mathrm{sc}_{ij}^{(l)}$$

The second-order partial derivatives are

$$\frac{\partial^2 \mathrm{PI}_5}{\partial \left(\mathrm{CW}^{(l)}\right)^2} = 0$$

$$\frac{\partial^2 \mathrm{PI}_5}{\partial \left(w_i^{(l)}\right)^2} = 0$$

$$\frac{\partial^2 \mathrm{PI}_5}{\partial \left(\mathrm{sc}_{ij}^{(l)}\right)^2} = 0$$

The mixed second-order partial derivatives and the semirelative-sensitivity functions are

$$\frac{\partial^2 \mathrm{PI}_5}{\partial \mathrm{CW}^{(l)} \partial w_i^{(l)}} = \frac{1}{m} \sum_{j=1}^{m} \mathrm{sc}_{ij}^{(l)}$$

$$\widetilde{S}_{\mathrm{CW}^{(l)} - w_i^{(l)}}^{\mathrm{PI}_5} = \frac{1}{m} \mathrm{CW}^{(l)} \sum_{j=1}^{m} w_i^{(l)} \mathrm{sc}_{ij}^{(l)}$$

$$\frac{\partial^2 \mathrm{PI}_5}{\partial \mathrm{CW}^{(l^*)} \partial w_i^{(l)}} = 0 \text{ for } l^* \neq l$$

$$\frac{\partial^2 \mathrm{PI}_5}{\partial \mathrm{CW}^{(l)} \partial \mathrm{sc}_{ij}^{(l)}} = \frac{w_i^{(l)}}{m}$$

$$\widetilde{S}_{\mathrm{CW}^{(l)} - \mathrm{sc}_{ij}^{(l)}}^{\mathrm{PI}_5} = \frac{1}{m} \mathrm{CW}^{(l)} w_i^{(l)} \mathrm{sc}_{ij}^{(l)}$$

$$\frac{\partial^2 \mathrm{PI}_5}{\partial w_i^{(l)} \partial w_{i^*}^{(l^*)}} = 0 \quad \text{for all } i^* \text{and } l^*$$

$$\frac{\partial^2 \mathrm{PI}_5}{\partial w_i^{(l)} \partial \mathrm{sc}_{ij}^{(l)}} = \frac{\mathrm{CW}^{(l)}}{m}$$

$$\tilde{S}^{\mathrm{PI}_5}_{w_i^{(l)}-\mathrm{sc}_{ij}^{(l)}} = \frac{1}{m}\mathrm{CW}^{(l)} w_i^{(l)} \mathrm{sc}_{ij}^{(l)}$$

$$\frac{\partial^2 \mathrm{PI}_5}{\partial w_i^{(l)} \partial \mathrm{sc}_{i^*j}^{(l^*)}} = 0 \text{ for } i^* \neq i \text{ or } l^* \neq l$$

It is important to note that for PI_5

$$\tilde{S}^{\mathrm{PI}_5}_{w_i^{(l)}-\mathrm{sc}_{ij}^{(l)}} = \tilde{S}^{\mathrm{PI}_5}_{\mathrm{CW}^{(l)}-\mathrm{sc}_{ij}^{(l)}} = \tilde{S}^{\mathrm{PI}_5}_{\mathrm{sc}_{ij}}$$

$$\tilde{S}^{\mathrm{PI}_5}_{\mathrm{CW}^{(l)}-w_i^{(l)}} = \tilde{S}^{\mathrm{PI}_5}_{w_i^{(l)}}$$

This means that the interaction terms do not have to be calculated! They are the same as the first-order sensitives. Therefore, without calculating high-order partial derivatives, we can be assured that we will not be blind-sided by interaction terms. Using PI_5, in general, the criteria weights are the most important parameters with the subcriteria weights being the next most important. The interaction terms and the individual input scores are the least important. Of course, the most important subcriterion weight is usually more important than the least important criterion weight, etc. These generalizations are in concert with our collective intuition about trade-off studies.

Please do not fear the apparent mathematical complexity of this performance index, because it is all implemented in a spreadsheet that is available at http://sysengr.engr.arizona.edu/forStudents/index.html. Figure 5.22 is a static version of the sensitivity analysis from that spreadsheet.

There is a one-for-one correspondence between the trade-off study spreadsheet of Fig. 5.21 and the sensitivity analysis spreadsheet of Fig. 5.22. In other words, the largest number in Fig. 5.22 is 0.224 for the sensitivity of PI_5 with respect to the *evaluation criteria weight* for *company policy*. It is in row 12 column C of Fig. 5.22. Now, if you look at row 12, column C of Fig. 5.21, you will find the *normalized value of the criteria weight* for *company policy*. Of the individual scores, the largest number in Fig. 5.22 is 0.024 for the sensitivity of PI_5 with respect to the score for BiST for alternatives 2, 3, and 7. One of them is in row 13 column Q of Fig. 5.22. If you look at row 13, column Q of Fig. 5.21, you will find the score for BiST for Alt 7 ARE 442.

The sensitivity analysis performed on this trade-off study shows that the most important category was the weight for company policy, the most important subcategory was the weight for BiST, and the most important scores were for BiST (for alternatives 2, 3, and 7). Therefore, extra time was spent discussing these among the team and with top management. The purpose of such discussions should be to determine whether performance, cost, or company policy should have the highest weight. Extra resources need to be expended consulting with management and the customer about the relative weights of the subcriteria. Extra research,

Sensitivity Matrix showing semirelative sensitivity values for PI5 = the hierarchical sum of all products/m

Sensitivity values		Sensit Criteria Weights	Sensit Sub Criteria Weights	Alt 1 Do Nothing	Alt 2 Sun Electronics PV	Alt 3 DMSolar PV	Alt 4 Solar Calculator Generic PV	Alt 5 Jacobs Wind Turbine	Alt 6 Bergey WindPower	Alt 7 ARE 442	Alt 8 Skystream 3.7	Alt 9 LIMPET	Alt 10 Nordtank 65
				Sen	Sen	Sen	Sen	Sen	Sen	Sen	Sen	Sen	Sen
Performance		0.214											
Vendor Evaluation			0.087	0.012	0.007	0.008	0.011	0.007	0.007	0.007	0.006	0.011	0.011
Interference with Telescopes			0.106	0.017	0.017	0.017	0.017	0.003	0.003	0.003	0.007	0.017	0.002
Rated Power			0.021	0.000	0.004	0.004	0.002	0.001	0.000	0.000	0.000	0.007	0.004
Cost		0.103											
Total Lifecycle Cost			0.022	0.001	0.002	0.002	0.004	0.002	0.002	0.002	0.002	0.002	0.002
Ratio of Price to Power			0.041	0.003	0.007	0.007	0.006	0.005	0.001	0.001	0.006	0.000	0.006
Cost of Generated Electricity			0.040	0.002	0.004	0.004	0.006	0.004	0.004	0.004	0.004	0.005	0.005
Company Policy		0.224											
BiST			0.142	0.021	0.024	0.024	0.019	0.003	0.003	0.024	0.003	0.013	0.008
Reusability			0.082	0.005	0.008	0.008	0.013	0.008	0.008	0.008	0.008	0.002	0.014

Fig. 5.22 Sensitivity analysis of BIMS

analysis, and modeling can be done to acquire scores that are more reliable for BiST for alternatives 2, 3, and 7. Interference with telescopes was the second most important evaluation criterion.

5.3.4.3 Different Combining Functions

In what follows, we compare the results of the sum combining function, the product combining function, and the compromise combining function with $p = 1$, $p = 2$ and $p = \infty$.

First, we present the results of the sum combining function from Fig. 5.21, which are presented in Table 5.25.

The photovoltaic solar systems are the preferred alternatives.

Results of the product combining function shown in Table 5.26 eliminate alternatives that have a zero on any criteria. Therefore, the do nothing alternative and the small wind turbines are eliminated, because either their rated power is below the 5 kW lower threshold or their price-to-power ratio is below threshold.

Table 5.25 Sum combining function

Rank ordered alternatives	Alternative ratings
Alt 4 Generic PV	0.78
Alt 3 DMSolar PV	0.74
Alt 2 Sun Electronics PV	0.73
Alt 1 Do Nothing	0.62
Alt 9 LIMPET	0.57
Alt 10 Nordtank 65	0.52
Alt 7 ARE 442	0.50
Alt 8 Skystream 3.7	0.35
Alt 5 Jacobs Wind Turbine	0.33
Alt 6 Bergey WindPower	0.28

Table 5.26 Product combining function

Rank ordered alternatives	Alternative ratings
Alt 10 Nordtank 65	0.55
Alt 4 Generic PV	0.32
Alt 3 DMSolar PV	0.21
Alt 2 Sun Electronics PV	0.19
Alt 5 Jacobs Wind Turbine	0.06
Alt 1 Do Nothing	0
Alt 6 Bergey WindPower	0
Alt 7 ARE 442	0
Alt 8 Skystream 3.7	0
Alt 9 LIMPET	0

Table 5.27 Compromise combining function with $p = 2$

Rank ordered alternatives	Alternative ratings
Alt 2 Sun Electronics PV	0.53
Alt 3 DMSolar PV	0.53
Alt 4 Generic PV	0.53
Alt 1 Do Nothing	0.49
Alt 7 ARE 442	0.42
Alt 9 LIMPET	0.42
Alt 10 Nordtank 65	0.36
Alt 8 Skystream 3.7	0.26
Alt 5 Jacobs Wind Turbine	0.23
Alt 6 Bergey WindPower	0.21

Table 5.28 Compromise combining function with $p = 1000$

Rank ordered alternatives	Alternative ratings
Alt 2 Sun Electronics PV	0.56
Alt 3 DMSolar PV	0.56
Alt 7 ARE 442	0.56
Alt 1 Do Nothing	0.50
Alt 4 Generic PV	0
Alt 5 Jacobs Wind Turbine	0
Alt 6 Bergey WindPower	0
Alt 8 Skystream 3.7	0
Alt 9 LIMPET	0
Alt 10 Nordtank 65	0

For the product combining function, the sensitivity to the performance criteria weight was much higher than for the sum combining function. The other sensitivities were less distinct.

The compromise combining function with $p = 1$ is the same as the sum combining function shown in Fig. 5.21. The results of using the compromise combining function with $p = 2$ are shown in Table 5.27.

These results and sensitivities are about the same as for the sum combining function of Fig. 5.18.

The results of using the compromise combining function with $p = 1000$ are different from the other results as shown in Table 5.28.

For the results of Table 5.28, BiST is the most important criteria. So that score determines the rank order of the alternatives. At this point, it is imperative to ask the decision maker if these results are worth keeping. They show how important BiST is, but they are probably not useful for ranking alternatives, because the results depend on only one criteria, BiST.

Tables 5.25, 5.26, 5.27, and 5.28 reveal a surprising consensus. Alternatives 2, 3, and 4 seem to be the preferred alternatives. The variations are due to the emphasis on BiST by the compromise combining function with $p = 1000$. It pushed alternative 4 down a few places. And the product combining function of Table 5.26 pushed

Table 5.29 Consensus preferred alternatives

Rank ordered alternatives	Alternative ratings
Alt 4 Generic PV	0.78
Alt 3 DMSolar PV	0.74
Alt 2 Sun Electronics PV	0.73

Table 5.30 Final preferred alternatives

Alternatives	Ratings
DMSolar PV	0.74
Sun Electronics PV	0.73

the solar panel systems (alternatives 2, 3 and 4) below alternative 10 because their rated powers are small. Using another built in sanity check, we note that the do nothing alternative never broke into the top three (Table 5.29).

However, the Generic PV system is not a commercially available system. It is an ideal that was used to indicate what could be expected with improvements in the technology.

We examined three dozen evaluation criteria and narrowed them down to eight. We examined dozens of alternative architectures and narrowed them down to ten. Then we created the trade-off study matrix shown in Fig. 5.21. We performed a sensitivity analysis on this trade-off study matrix: it is shown in Fig. 5.22. This sensitivity analysis included interactions of criteria. We applied four combining functions to these data. We decided that the sum combining function was the best for this and most other trade-off studies. Therefore, the final recommendation that we will give to our decision maker is given in Table 5.30.

For final recommendations to the top decision makers, two or three alternatives are about right. Decision makers do not want a single alternative, because that means they are excluded from the decision loop. On the other hand, five alternatives are too many, because that makes their decisions quite difficult. Decision makers cannot keep that many alternatives in their minds at the same time. Also, we included the ratings column, because decision makers like to use quantitative data to make decisions.

5.3.5 *Risk Analysis*

Risk analysis is the topic of Chapter 6. However, to provide a foretaste of that section, we will now introduce a simple example of a risk analysis of the BIMS project.

Risk is an expression of the potential harm or loss associated with an activity executed in an uncertain environment. Risk usually implies harm to the system being designed or its primary users, not to unintended consequences on external systems outside the scope of the system being designed. Astronomical observatories face heightened scrutiny from environmentalists due to their prominent siting.

To ameliorate their opposition, BIMS is expected to use renewable-energy electric-generators. This is a political decision, not an economic or scientific decision. In the risk analysis of Table 5.31, the greatest risk is that environmental activists are likely to try to prevent funding and construction of the facility. Table 5.31 present risk analyses from the perspective of the designer and owner of the telescope project.

The greatest risk is that environmental activists are likely to try to prevent funding and construction of the facility. Therefore, public relations personnel need to be apprised about this risk. The second biggest risk is that a similar system has already been patented. Therefore, this risk needs to be attended to as soon as possible. Specifically, BICS's lawyers need to be contacted to begin a patent search.

To help validate our risk analysis, we will now look at several other types of risks. This discussion will not be limited to the top of Hawaii.

5.3.5.1 Environmental Risks

It is important to realize that large raptors can get electrocuted when their wings touch two power lines at the same time. Birds and bats can be killed if they fly into the blades of a wind turbine. Therefore, the wind turbine blades can be designed to make a noise that would deter birds and bats. On the other hand, many people are concerned about wind turbine noise.

Solar panels could destroy habitats and modify bird migration patterns. Panels mounted flat on the earth's surface could look like a lake to migratory birds and throw them off course. As importantly, panels mounted on roofs could reflect light into the eyes of car drivers or airline pilots.

Trips up and down the mountain for installation and maintenance can be expected to change the environment. The electromagnetic spectrum of each telescope on the mountain must be determined. Then BIMS equipment needs to be tested for emissions in all these spectra.

5.3.5.2 Economic Risks

Factors that could affect the economic risks of attaching solar electric generating equipment to the AC power grid include interest rates, continuation of federal, state, and utility company rebates, amortization period, federal stimulus funds, and federal regulations.

5.3.5.3 Risks That Have Already Been Addressed

Electric utilities have conducted risk analyses for several years. Whenever they identified a risk, they took action to minimize its effects. A few examples are offered below.

Table 5.31 Failure modes and effects analysis for BIMS

Failure mode	Potential effects	Relative likelihood[a]	Severity of failure	Estimated risk
Performance				
The volcano might erupt	Acts of God are not normally included in a risk analysis. But in this case, we should ensure that our system is connected to the USGS Volcano Hazards Program, at the Hawaiian Volcano Observatory in case we need to evacuate the mountaintop	0.001	0.9	0.0009
The electric generators might not supply enough electricity in cold temperatures, at high altitudes	We would not be able to sell power to the AC Electric Power Grid	0.01	0.02	0.0002
Telescopes operate below the temperature of the outside air, so telescope enclosures are typically kept cold inside	HVAC demands will be different from for a typical office building (reuse)	0.9	0.001	0.0009
Commercial AC electric power grid may fail for hours at a time	We may have to incorporate backup diesel electric generators	0.01	0.1	0.001
Cooling fans in electronic equipment do not work as well in the thin air at high altitudes	Cooling fans in electronic equipment might have to be replaced with higher capacity fans	0.5	0.001	0.0005
Cost				
It is expensive to build in such a remote location. Skilled labor may be more expensive than expected	We could overrun our budget	0.9	0.001	0.0009
Construction workers working at 13,800 ft (4200 m) will have trouble breathing	This will make the workers less efficient and they will need medical observation	1	0.001	0.001
Something could interrupt the material supply lines across the ocean and up the mountain	We could overrun our budget and schedule	0.5	0.001	0.0005
Electric power is expensive on Hawaii, because they have no natural sources of fossil fuels. Their average electricity rate is \$0.2/kWh to \$0.3/kWh compared to the national average of \$0.1/kWh	This could increase construction and operations costs. It will also affect the balance between electric lighting cost and HVAC costs. This is also an advantage, because it will help us compete with commercial electricity rates	1	0.001	0.001

(continued)

Table 5.31 (continued)

Failure mode	Potential effects	Relative likelihood[a]	Severity of failure	Estimated risk
Cost of generating electricity in this location is more than the cost of buying it	We would abandon this part of the project	0.05	0.5	0.025
Business				
The project might lose political support and Congress or the NSF could stop funding	The project would be terminated	0.01	1	0.01
Observatories face heightened community scrutiny due to their prominent siting	Proactively seeking accommodation with environmental concerns is one ingredient to a successful project	0.5	0.5	**0.25**
A similar system has already been patented	We would have to license patent rights. Although it would cost at least $5000, we should initiate a patent search. If no similar patent is found, then we should initiate a patent application. We should immediately contact BICS's lawyer and ask if putting this document on my web site constitutes disclosure	0.8	0.1	0.08
One of your teammates might refuse to do any work	You will have to fire that teammate and everyone else will have to work harder (process)	0.15	0.2	0.03

[a]Likelihood that the event would happen within the next year
Risk is estimated as the Relative Likelihood times Severity of Failure
These values were estimated by Terry Bahill September 13, 2015

There can potentially be strong opposition from environmentalists, neighbors, or businesses around the areas where solar panels are to be installed. They might claim that solar panels are unsafe, or ugly. Home Owners Associations (HOAs) could potentially prohibit or strongly discourage PV systems. This was considered a long time ago and the some states passed laws making it illegal for HOAs to impose rules against photovoltaic systems.

If homeowners tried to install photoelectric systems themselves, they could get electrocuted. Installation looks like a relatively simple task, but high voltages are involved. If a business or a home is being retrofitted with electricity saving or electricity producing devices, then electric utility will pay about one-third of the purchase price. But to qualify for the program, the system needs to be installed by a *certified* professional. This would dissuade people from installing the system themselves because they would then forego the rebate.

Electric companies might refuse to buy electricity from homeowners. Federal law requires electric companies to buy electricity from their customers.

Photovoltaic panels contain no toxic chemicals or heavy metals. Smashing a panel would not release toxic gases, produce short circuits on the power grid, or create unusual flying debris.

5.3.6 Section Summary

Trade-off studies (also referred to as trade-off studies in systems engineering) are an integral part of systems engineering. Their timely and judicious use can help systems engineers in making rational decisions throughout the system life cycle. As additional information becomes available in various stages of systems engineering, trade-off studies can be repeated. Trade-off studies span a variety of concerns in the system life cycle including performance, cost, schedule, risks, legacy subsystems, a variety of quality evaluation criteria, and technologies.

It is important to recognize that the goodness of trade-off studies are a function of reasonableness of assumptions, availability of required data, and methods used to mitigate the occurrence of human mistakes. The latter is especially a matter of concern because not attending to them early and throughout can riddle trade-off studies with mental mistakes.

This section has covered the trade-off study process in detail with a particular focus on human mistakes and measures that can be taken to obviate them. The trade-off process consists of: (1) determining whether or not a formal evaluation is needed; (2) establishing evaluation criteria; (3) identifying alternatives; (4) selecting sources of data and evaluation method; (5) collecting data; (6) evaluating alternatives; (7) selecting preferred alternatives; (8) performing expert review; (9) presenting results to decision makers; and (10) persistently storing results in a Process Assets Library.

Mental mistakes arising from human biases and limitations are the bane of trade-off studies. These mistakes are attributed in the literature to various forms of erroneous heuristics and biases including representativeness, anchoring, base-rate neglect, and the conjunction fallacy. In addition, other factors such as cognitive illusions, emotional state, a variety of fallacies, fear of regret, psychological traps, and paradoxes from the field of behavioral economics can induce mental mistakes in trade-off studies. A key concern when it comes to trade-off studies is determining the degree to which these biases and other factors affect the various components of trade-off studies. In this regard, this section presents the salient aspects of prospect theory. Prospect theory breaks down subjective decision-making into a preliminary screening stage, and a secondary evaluation stage. In the screening stage, values are considered subjectively from a reference point established by the subject's context prior to making decisions. In this stage, losses are weighted more heavily than gains. In the evaluation stage, a final value for every prospect (alternative) is calculated. Subsequent refinements in this theory include subjective probability assessment. Kahneman advanced a super-heuristic, attribute substitution. This super-heuristic subsumes several heuristics and biases such as representativeness, anchoring, base-rate neglect, and conjunction fallacy. In other words, the latter are instances of evaluation criterion substitution. This section recommends specific remedies that help to avoid mental mistakes during the conduct of trade-off studies. The section also discusses various factors that affect human decisions in some detail.

The section presents several illustrative examples in painstaking detail to give the reader a handle on the level of detail needed for a meaningful trade-off study that exploits the recommendations in this section to circumvent the incidence of human mistakes.

5.3.7 Homework Problems

5.3.7.1 BIMS Diagrams

In the block definition diagram showing the structure of BIMS (Fig. 5.14), the AC electric power grid interface block has units of Joules, however the listed part is watt-Hour Meter. Why the difference? For the BIMS block, what do the arrows with solid black diamond endings tell us? In the parametric diagram for BIMS (Fig. 5.13), we see that $450 < \text{dti} < 550$. What is dti? The real system up on Mauna Kea needs a value for dti so that it can compute measures of effeteness. Where does it get a value for dti? Also in this diagram, we have, $\text{iroc} < 10$. What is iroc? Where does the system up on Mauna Kea get a value for it?

5.3.7.2 Mental Mistakes in BIMS

The following are excerpts from the BIMS documentation. What mental mistake is being exhibited by the chain of thought shown in the following?

The risk analysis told us that the greatest risk is that environmental activists will try to prevent funding and construction of the facility. We must alert our public relations people about this risk, now.

The Concept of Operations (ConOps) states that BIMS will be politically correct (environmentally green) because it will use renewable-energy electric-generators, such as solar panels or wind turbines, but you should consider other electric generation alternatives like geothermal systems using the Kilauea volcano.

Customer requirement 1–5. The system shall buy electricity from and sell electricity to the AC electric power grid.

Table 5.32 Rated power nonfunctional performance requirement

Identification tag (Id)	Nonfunctional performance requirement 1–4
Name	Rated Power
Text	The electric generating system shall have a minimum rated power of 10 kW
Rationale	Expected Average Daily Electricity Generation in this location with this type of plant must be at least 40 kWh/day. To account for environmental variability, the capacity factor of renewable-energy electric-generators is typically 5–40 %. For photovoltaic solar power, we estimate a capacity factor of 17 %. Therefore, the minimum rated power must be 10 kW $\text{RatedPower} = \frac{\text{AverageDailyElectricityGeneration}}{\text{CapacityFactor} \times \frac{24\,\text{h}}{\text{day}}} = \frac{40\text{kWh/day}}{\frac{1}{6} \times \frac{24\,\text{h}}{\text{day}}} = 10\,\text{kW}$
DeriveReqt:	Functional requirement 1–6
Verify method	Inspection
Priority	High
Date of last change	November 26, 2015

Customer requirement 1–10. The system shall generate electricity at a cost that is competitive with commercial electricity costs at that location, after federal and state subsidies (Table 5.32).

5.3.7.3 Wind Turbines and Photovoltaic Solar Panels

The top part of Fig. 5.23 shows two different distributions of wind speeds. One is a uniform distribution with a mean of 5 mph and a range of 4.5–5.5 mph. The other is a uniform distribution with a mean of 5 mph and a range of 0–10 mph. In this range of wind speeds, the output power of a wind turbine is proportional to the wind speed cubed, output power $= k(\text{wind speed})^3$. Which of these distributions would produce more power? You may justify your answer with equations or words.

The bottom of Fig. 5.23 shows a highly simplified model for the variation in incident sunshine due to clouds blocking sunlight from the solar panels. On the same axes, sketch representative power out of the PV solar panels. You may justify your answer with equations or words.

Why are we careful to specify that we are talking about "photovoltaic" solar panels?

5.3.7.4 Compromise Combining Function

In Table 5.28 we used the compromise combining function with $p = 1000$. Do you think that the results would be any different if we chose $p = 1$, 10, 100, 10,000, or ∞? Explain how you arrived at your conclusion.

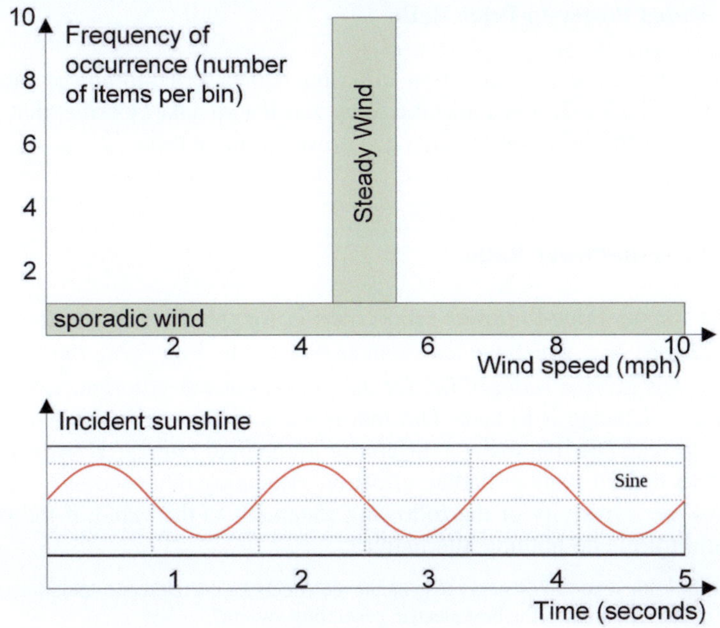

Fig. 5.23 Two wind distributions and a sunshine function

5.3.7.5 Alternative Energy Storage Systems

In the BIMS system design, we considered energy storage, but we did not include that aspect in this textbook. However, we would now like *you* to consider energy storage for the purpose of time shifting the time of generation and consumption. See Fig. 6.11. This is a continuation of Section 5.1.4.6 homework problem entitled alternative energy storage systems, in which you created five evaluation criteria for eleven alternative energy storage systems. Now we want you to create a trade-off study matrix for these alternative energy storage systems. Do not spend time searching for good data. Just guess values off the top of your head. Comment on your preferred alternatives, the position of your do nothing alternative, and the sensitivities. You may use your criteria or ours. If you are using our spreadsheet then you will have to expand it to accommodate 11 alternatives.

5.3.7.6 Springer Site

Go to the Springer site for this textbook http://www.springer.com/gp/book/9783319437101#otherversion=9783319437125 or to http://sysengr.engr.arizona.edu/forStudents/index.html and download the file BIMSforStudents.xlsx.

5.3.7.7 Rated Power-to-Price Ratio

In the text, we gave an evaluation criterion named price-to-power ratio from [65]. Now we think that was a mistake. Why was it a mistake? Correct that mistake now. Write a better criterion for the rated power-to-price ratio.

5.3.7.8 Price-to-Power Ratio

The value for the price-to-power ratio criterion for the do nothing alternative of Table 5.22 was $\frac{0}{0} =$ which we represent as $0 \times \infty$. In Fig. 5.21, the do nothing alternative was given a score of 0.5 for the price-to-power criterion. Do you think that was fair? Change it to zero. Did that make a difference in the final ratings? Change it to one. Did that make a difference in the final ratings? However, do you think that its default number had an effect on evaluating other scores?

What is the sensitivity of the following statement to the value of the price-to-power ratio for the do nothing alternative?

> In terms of using renewable-energy to generate electricity for the telescope operations room on top of Hawaii, what is the best electric generating system?

Solar PV panels are rapidly getting more efficient, while wind turbines are improving more slowly. In future years, which technology will dominate?

5.3.7.9 Rated Power Deration

New data from Jacob's Wind Turbine Industries Corp. shows that their 31-20 wind turbine system only produces 4.4 kW at 15 mph. What would this change in Fig. 5.21 and in your recommendations?

5.3.7.10 Cell Contents

What is in cell T1 of Fig. 5.21? Cell B16 contains the value of the performance index PI5 for this spreadsheet. Of what use is it? Does it change with the combining function? What are the numbers in cells C16 and D16? Are the sensitivities values in cells C21 to D31 of Fig. 5.22 linear mappings of C4 to D11 in Fig. 5.21?

5.3.7.11 Sensitivities

Is it true that the sensitivity values in Fig. 5.22 are zero if and only if the associated scores in Fig. 5.21 are zero?

5.4 Section 33. Trade-off Study of a Ship-to-Air Missile

This section is based on a 120-page design document created by Leigh Brooks, Justin Glier, Celia Jackson, and Darius Wu, ©, 2016, used with permission. Bob Sklar was their Brain Trust Advisor. This section only shows excerpts of their design document. These excerpts keep the team's vernacular. This section may be skipped without loss of continuity.

5.4.1 *Problem Situation*

Our top-level system function is to design a ship-to-air missile to defend our navy ships from incoming airborne missiles.

5.4.2 *History of the Problem*

There is a need to protect United States Navy ships from missile attacks, and thereby protect the security of the American people. As offensive military capability continues to evolve, so must our ability to defend against threats to the United States' naval supremacy. Missiles, unlike unguided bombs and rockets, are accurate and precise. Currently, engineers are finding ways to take the missile technology to the next level by furthering the four fundamental elements assembled in a missile: armament (explosives), guidance and control (antenna and radar), propulsion (thrust), and airframe (structure). The problem that needs to be addressed is to design a short-range missile to be launched from a ship's launch system according to the customer's need.

5.4.3 *The Customer*

5.4.3.1 Owners and Users

The United States Navy is looking for solutions to defend their ships from enemy missiles within a short range. The design of the short-ranged missile must fit the current launch system onboard ships. Safety is the number one priority.

5.4.3.2 Bill Payers

The system will be purchased by the United States Navy.

5.4.3.3 Users

The crew aboard the ship is directly responsible for firing the missile. Their safety shall be addressed.

5.4.3.4 Beneficiaries

Crew members of navy ships, the Navy and the American people will be beneficiaries of this system due to increased survival of the ships' crew members, lasting usefulness of the navy ships and enhanced security of the United States.

5.4.3.5 Victims

The victims of this system will be hostile forces implementing missile attack as a means of destroying America's naval assets.

5.4.4 Use Case Models

The following use cases were written for this system. In this section, we will only look at a few of them.

Business Use Case
Use Case 1: Heightened readiness alert
Use Case 2: Detect incoming target
Use Case 3: Prelaunch sequence
Use Case 4: Launch missile
Use Case 5: Target tracking
Use Case 6: Steering commands
Use Case 7: Fly towards target
Use Case 8: Target destruction
Use Case 9: Anti-tamper

5.4.4.1 Business Use Case

Name: Design a Short-Range Missile to Defend Navy Ships
Iteration: 4.2
Mission: Protect Navy ships from short-range airborne missiles and enemy aircraft
Brief description: The missile will have the ability to destroy an incoming missile targeted at Navy ships
Added value: Navy ships are defended against enemy missiles, thereby defending the security of the United States and its allies

Scope: U.S. Navy ships
Primary actor: JDLC Defense Company
Supporting actors: U.S. Navy
Precondition: U.S. Navy recognizes the need to defend navy ships from enemy missiles that enter the security range and are targeting a navy ship or surrounding ships

Main Success Scenario

1. U.S. Navy realizes the need to procure a weapon system that will defend navy ships against short-range enemy targets.
2. U.S. Navy places a contract with JDLC Defense Company to design, fabricate, test, and certify the defense missile.

3a. JDLC Defense Company designs, fabricates, tests, and certifies the JDLC missile.

4. U.S. Navy deploys the JDLC missile on navy ships {exit use case}.

Alternative Flow

3b. U.S. Navy budget is cut for the JDLC missile.
3b1. JDLC missile program is canceled {exit use case}.

Specific Requirements

Nonfunctional Requirements:
{FBR-1} JDLC Defense Company shall provide the *resources* required to design, fabricate, test, and certify defense missile.
{FBR-2} JDLC Defense Company shall provide the *facilities* necessary to design, fabricate, test, and certify defense missile.
Author/owner: JDLC Defense Company.
Last changed: March 19, 2016.

5.4.4.2 Use Case 3: Prelaunch Sequence

Name: Prelaunch Sequence.
Iteration: 4.2.
Brief description: Crew members perform the necessary steps to arm the missile and perform a built-in self-test (BiST) prior to launch. The target information is transferred from Fire Control to the missile in preparation for launch. Two missiles are involved in every launch a primary missile and a backup missile.
Primary actor: Crew Member.
Supporting actors: Fire Control.
Precondition: Initial threat assessment has been performed and the threat has been confirmed.

Main Success Scenario

1. Crew member commands primary and backup missiles to perform BiST.
2a. Primary and backup missiles send BiST (healthy) signals to the crew member.
3. Crew member sends the ARM command to the primary and backup missiles.
4. Ship system recalculates target parameters and sends information to Fire Control.
5. Fire Control continuously transfers target information to both primary and backup missile processors {exit use case}.

Alternative Flow

2b. Primary missile returns BiST (fail) to the crew member.
2b1. The backup missile becomes the primary missile and the crew member chooses another backup missile {exit use case}.

Specific Requirements

Functional Requirements

{FR 3-1} Missile shall be set in the ARM state from the crew member's shipboard panel.
{FR 3-2} Missile processor shall receive commands from Fire Control prior to missile launch.
{FR 3-3} The missile processor shall transfer initial missile trajectory to the missile autopilot.
Post condition: Primary and backup missiles are in the ARM state and have returned BIST (healthy) signals.
Author/owner: JDLC Defense Company.
Last changed: March 19, 2016.

5.4.4.3 Use Case 4: Launch Missile

Name: Launch Missile.
Iteration: 4.2.
Brief description: Crew member commands Fire Control to launch the missile. Missile is propelled from the launch platform and is guided to the target.
Primary actor: Fire Control, Crew Member.
Supporting actors: Target.
Precondition: BiST has executed on the primary and backup missiles and returned healthy signals. The missile propulsion systems (primary and back up) are armed. Target information has been transferred to the primary and backup missiles.

Main Success Scenario

1. Crew member sends Release Consent to primary missile.
2. Crew member sends launch signal to Fire Control.

3. Fire Control sends launch command to primary missile.
4. Fire Control commands the primary missile to switch to internal power.
5. Primary missile returns missile ready indication to Fire Control.
6. Fire Control sends rocket motor ignition command to primary missile.

7a. Rocket motor ignition occurs and missile leaves the canister.

8. The away-switch tells Fire Control that the missile is unattached to the launch platform.

9a. Crew member monitors the ship's radar to ensure that missile is on course. {exit use case}.

Alternative Flow

7b. If the primary missile launch is a failure, then prepare to launch backup missile.

7b1. If enemy target is still within intercept range, then launch the backup missile {exit use case}.

9b. If the ship's radar determines improper primary missile flight, then abort that missile and launch the backup missile {exit use case}.

{Editor's note: It is unusual to have if-statements in use cases.}

Specific Requirements

Functional Requirements

{FR 4-1} The missile shall receive commands from Fire Control {From step 3}.

{FR 4-2} The missile shall contain an internal power source {From step 4}.

{FR 4-3} The missile shall return a missile ready indication to Fire Control {From step 6}.

Post condition: A missile has been launched from the ship.

Author/owner: JDLC Defense Company.

Last changed: March 19, 2016.

5.4.4.4 Use Case 9: Anti-tamper

Name: Anti-tamper.

Iteration: 4.2.

Brief description: Anti-tamper is a security method employed in hardware designs to prevent individuals without the "need-to-know" from stealing hardware designs by reverse engineering. Should someone attempt to access classified hardware, protective measures will destroy the circuitry, thus preventing the theft of protected information.

Primary actor: Missile Autopilot, Missile Seeker Processor.

Supporting actors: Any other key elements of the circuitry that are proprietary or classified by the U.S. Department of Defense.

Precondition: The missile design team receives the requirement from the customer to include anti-tamper methods in the design.

Main Success Scenario

1. JDLC missile design team encloses classified hardware within a sealed container.
2. The anti-tamper hardware container is opened.
3. Anti-tamper system destroys the protected circuit.

Specific Requirements

Functional Requirements

{FR 9-1} Classified hardware shall be enclosed in a sealed container to prevent unauthorized access.

{FR 9-2} A destructive mechanism shall surround the classified hardware and associated circuitry.

Post condition: Unauthorized access is prevented.

Author/owner: JDLC Defense Company.

Last changed: March 19, 2016.

5.4.5 Concept Exploration

5.4.5.1 System Design Concept

The design concepts are to provide a cost-effective and accurate short-range guided missile launched from a vessel that will defend ships from anti-ship missiles. Trade-off studies were performed on the following alternative concepts.

5.4.5.2 Status Quo

The Navy's Rolling Airframe missile (RAM) Block 1 was used as the status quo. The RAM is the Navy's current defense against short to mid-range targets. The RAM is felt to be incapable of consistently neutralizing the latest generation of missile threats, therefore the Navy would like for JDLC Defense Company to research the potential for a new missile design.

RAM Block 1. General characteristics

- Primary Function: surface-to-air missile
- Length: 2.79 m (9 ft 2 in.)
- Diameter: 127 mm (5.0 in.)
- Fin span: 434 mm (1 ft 5.1 in.)
- Speed: Mach 2.0+
- Warhead: 11.3 kg (24.9 lb) blast fragmentation
- Launch Weight: 73.5 kg (162 lb)
- Range: 9 km (5.6 mi)

- Guidance System: three modes—passive radio frequency/infrared homing, infrared only, or infrared dual mode enabled (radio frequency and infrared homing)
- Unit Cost: one million dollars
- Date Deployed: 1992
- Launch Platform: Turret-mounted
- Target Tracking: Combination IR/RF
- Propellant: Solid fuel
- Warhead: Proximity fuze
- Method of Steering: one pair of steering canards

5.4.5.3 Launch Platform

The purpose of a launching system is to place a weapon into a flight path as rapidly as the situation demands. Launching must occur at the right moment so that the weapon system may function effectively. The launching system must safely withstand the propulsive forces of the weapon, and it must be highly reliable in order for the weapon to achieve the kill probability for which it is designed.

Concept 1: Vertical Launch System

The Vertical Launch System (VLS) is a fixed, vertical, multi-missile storage, and firing system that allows Navy vessels to launch multiple missiles simultaneously. This allows for fast reaction to multiple threats with concentrated, continuous firepower. The VLS is a canister launching system, which provides for rapid-fire launch capability against hostile threats.

Concept 2: Rail System

This system is a fixed position launch platform mounted on the ship above deck. It involves loading the missile via a serpentine belt system that delivers and loads the missiles into the rail system from a missile magazine below deck.

Concept 3: Turret-Mounted System

The turret-mounted launch platform is similar to the rail system model except that it can rotate 360°. This allows the weapon system to point the missile in the direction of the incoming target.

5.4.5.4 Missile Target Tracking

Concept 1: High-Powered RF Source

This system uses a high-powered radio frequency (RF) source that neutralizes electronic signals that cross its path.

Concept 2: Laser Tracking

The launch pad sends out a laser beam and locks onto its target. The guided missile then locks onto the laser reflection and attacks target.

Concept 3: Triangulation Method

The triangulation method uses multiple laser beams to determine target position and velocity. These beams might come from Fire Control, other ships, or airplanes.

Concept 4: Infrared (IR)

An infrared sensor locks onto the heat source of its target. High-intensity heat flux from the enemy target's motor provides the means to reliably track the target.

Concept 5: Radio Frequency (RF)

A radio frequency guided missile uses radar signals emitted by the missile's RF onboard transmitter, which is then reflected back from the radar cross section of the target. The radar locks onto the energy reflected back from the target.

Concept 6: Combination IR/RF

The combination IR/RF missile uses the radar signals emitted by the missile's RF onboard transmitter and also uses the heat sensors from its target to track the movement.

5.4.5.5 Propellant

Concept 1: Solid Fuel

A simple solid rocket motor consists of a casing, nozzle, grain (propellant charge), and igniter. The grain behaves like a solid mass, burning in a predictable fashion and producing exhaust gases. The nozzle dimensions are calculated to maintain a design chamber pressure, while producing thrust from the exhaust gases. Once ignited, a simple solid rocket motor cannot be shut off. More advanced solid rocket motors cannot only be throttled but also can be extinguished and then reignited by controlling the nozzle geometry or through the use of vent ports.

Concept 2: Liquid Fuel

In a liquid propellant rocket, the fuel and oxidizer are stored in separate tanks and are fed through a system of pipes, valves, and turbo pumps to a combustion chamber where they are combined and burned to produce thrust. Liquid propellant engines are more complex than their solid propellant counterparts; however, they offer several advantages. By controlling the flow of propellant to the combustion chamber, the engine can be throttled, stopped, or restarted. Also, most liquid propellants are cheaper than solid propellants.

5.4.5.6 Warhead

Concept 1: Proximity Fuze

The missile detonates in the general proximity of the target. The fuze on the missile is given a timing mechanism to be set based on its mission. If a target has a high closing velocity, a short fuze delay will be set. The warhead may also be triggered by the missile's radar.

Concept 2: Hit-to-Kill

In the hit-to-kill method, the warhead destroys its target by imparting kinetic energy upon impact. This method uses only the force of the direct collision to destroy the target. The warhead does not include any energetic material to destroy the target. Therefore, the warhead must be able to score a direct hit.

5.4.5.7 Method of Steering

Concept 1: Thrust Vector Control

The thrust vector control method controls the flight of the missile by controlling the direction of thrust. In many situations, high maneuverability and a short turn radius are vitally important elements in the performance of missile. A thrust vector control (TVC) system provides lateral forces at the rear end of the missile to accomplish this capability.

Concept 2: Fixed Wing/Fin

This method would include a set of canard fins at the forward end of the missile and a set of fixed fins near the center of gravity of the missile. The primary advantage of canard control is better maneuverability at low angles of attack, although canards tend to become ineffective at high angles of attack. Since canards are ahead of the center of gravity, they cause a destabilizing effect and require large fixed tails to keep the missile stable.

Concept 3: Servo Controlled Fin Movement

This system would include planar control fins that are mounted parallel to the missile body axis. These fins rotate back and forth to generate forces in the horizontal and/or vertical planes that cause the vehicle to yaw right and left, pitch up and down, or roll as the missile maneuvers towards a target. Missile fins are turned using servomechanisms that rotate the fin to a new deflection angle. The size of these servos is dictated by the amount of torque it takes to rotate the fin about its axis.

5.4.5.8 Airframe Material

Concept 1: Composite Material

The missile section is made out of composite alloy materials.

Concept 2: Aluminum Material

The missile section is made out of aluminum giving it a lighter weight.

Concept 3: Steel Material

The missile section is made out of steel giving it a heavier weight.

5.4.6 *Launch Platform Trade-off Study*

The top-level evaluation criteria for the launch platform trade-off study are performance, cost, schedule, and safety.

5.4.6.1 Performance

We gave performance an importance weight of 15 in Table 5.33. This evaluation criterion provides the ability to meet the overall customer requirements. Intercepting enemy targets within the specified engagement range enhances the capabilities of the ship's arsenal.

The following evaluation subcriteria were evaluated for performance. The weights assigned were based on the overall criticality:

Table 5.33 Weights of importance for the launch platform evaluation criteria

	Evaluation criterion for launch platform	Derived from requirement	Raw weight	Top level criteria weights	Subcriteria weights
1	*Overall performance*		15	0.288	
1.1	Launch time	CR2.5.3, CR2.5.12, DR3.4.1.1	6		0.176
1.2	Reliability	CR2.3.4.1, DR3.4.3.1	10		0.294
1.3	Durability	CR2.3.4.1, DR3.4.3.1	9		0.264
1.4	Compatibility with current technology	CR2.4.3.4, DR 3.4.1.1	9		0.264
2	*Cost*	CR2.3.1.1, DR3.5.5.1	8	0.153	
2.1	Design/development cost	Not given	8		0.333
2.2	Production cost		8		0.333
2.3	Maintenance cost		8		0.333
3	*Schedule*	CR2.3.2.1, DR3.5.6.1	9	0.173	
3.1	Ability to meet design schedule	CR2.3.2.1, DR3.5.6.1	7		0.411
3.2	Producability of design	CR2.3.2.1, DR3.5.6.1	10		0.588
4	*Safety*	DR3.5.3.2, CR2.5.9	20	0.384	1.0
	Sum			1.0	4.0

Launch Time: Time in seconds from launch request to missile-away. This is also called the *launch cycle*.

Reliability: Qualitative assessment of mean time between failures or required maintenance. The system must be reliable; therefore, the weight assigned to this evaluation criterion in Table 5.33 was a 10.

Durability: Robustness of system when exposed to various environments on the ship. This includes weather, vibration, and shock encountered due to enemy weapons impacting the launch platform.

Compatibility with Current Technology: Costs and upgrade requirements that would be involved to implement the system on the ship.

5.4.6.2 Cost

Cost received an overall weight of 8. Although it was given the lowest overall weight of the four primary Evaluation Criteria of the project, it is certainly still important. The other characteristics—performance, schedule, and safety—provide the means to deliver the weapon to defeat the particular threat. Cost does not provide the means to meet the customer requirements. Cost is what the customer needs to consider as far as procuring the weapon system. It is not however the primary concern when considering the overall requirements of the system.

Given the project's desire and goal to deliver a cost-efficient weapon, cost will be attended to in the effort of the remaining three evaluation criteria of performance, schedule, and safety.

At the end of the day, cost becomes the burden necessary to deliver the weapon used to defeat the enemies' capability. The program will strive to provide the most economical cost possible for the weapon design. All design, manufacturing, and maintenance aspects will be evaluated to insure that no surprises with respect to cost will present themselves in the future. The customer will be provided with the most accurate cost estimate capable and all aspects of the program will be controlled to insure that expenses meet estimates.

Cost for the program was evaluated based on estimates of three primary areas of concern: material, manpower, and hours. Each of the alternatives was compared with the others and the advantages and disadvantages were discussed, then the cost was given a score between 0 and 1, where 1 was the least costly and 0 was the most costly. {Editor's note: this is where scoring functions *should* have been used, to flip the cost upside down.} Costs were estimated for the design/development, production, and maintenance phases.

5.4.6.3 Schedule

Schedule received an overall weight of 9 and was evaluated by the following evaluation criteria:

Ability to Meet Design Schedule: The ability to satisfy all design requirements and complete the design verification process in time to support weapon delivery. Scores were between 0 and 1.

Producibility of Design: The ease of manufacturing the design. Scores were between 0 and 1.

5.4.6.4 Safety

The overall safety of the system is of most importance to the Navy customer, therefore the safety criterion received a high overall weight of 20. Many strenuous safety tests would have to be performed on the missile to ensure the safety of the men aboard the navy ships and the surrounding areas. The overall weighting of safety assures that ship's personnel and the ship survivability depends on a safe weapon being installed on the ship's platform, so safety is the primary concern for any weapon brought on board a Navy vessel. Scores were between 0 and 1.

5.4.6.5 Weights of Importance

Weights of importance were assigned to evaluation criteria as shown in Table 5.33. Then the weights for the criteria were normalized to sum to one. Then the weights for the subcriteria were normalized so that they summed to one in each category.

Next, we give the parameters necessary to evaluate the evaluation criteria using the scoring functions and graphs given in following sections.

5.4.6.6 Scoring Functions

Overall performance, cost, and schedule do not need scoring functions, because their subcriteria are normalized. Therefore, we will not use them. In this section, we will only use the following four scoring functions.

Launch Time

Output score: LPScore2 = SSF5 (LPL2, LPB2, LPO2, LPB22, LPU2, LPS1, LPS2)
Lower Threshold: LPL2 = 6

{Editor's note: In this acronym, LP stands for launch platform, L stands for lower threshold and 2 stands for the second performance criterion, i.e., Launch Time.}

Baseline 1: LPB12 = 8

{Editor's note: In this acronym, LP stands for launch platform, B1 stands for the first baseline and 2 stands for the second performance criterion, i.e., Launch Time.}

Optimal (middle) Point: LPO2 = 8.5
Baseline 2: LPB22 = 9.5
Upper Threshold: LPU2 = 10
Baseline 1 Slope: LPS12 = 2
Baseline 2 Slope: LPS22 = −2
Domain = RLS [0–10]

Therefore, output score: LPScore2 = SSF5(6, 8, 8.5, 9.5, 10, 2, −2, RLS[0–10]) (Fig. 5.24)

Reliability

Output score: LPScore3 = SSF3(5, 8, 10, 0.5, RLS[0–10]) (Fig. 5.25)

Durability

Output score: LPScore4 = SSF3(5, 8, 10, 0.5, RLS[0–10]) (Fig. 5.26)

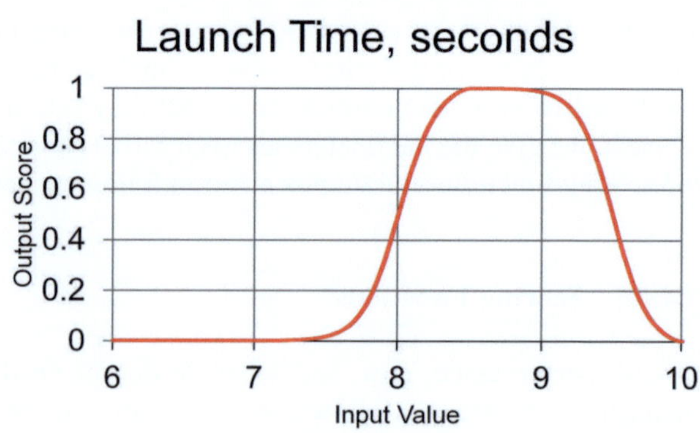

Fig. 5.24 Scoring function for the launch time evaluation criterion, time in seconds from launch request to missile-away

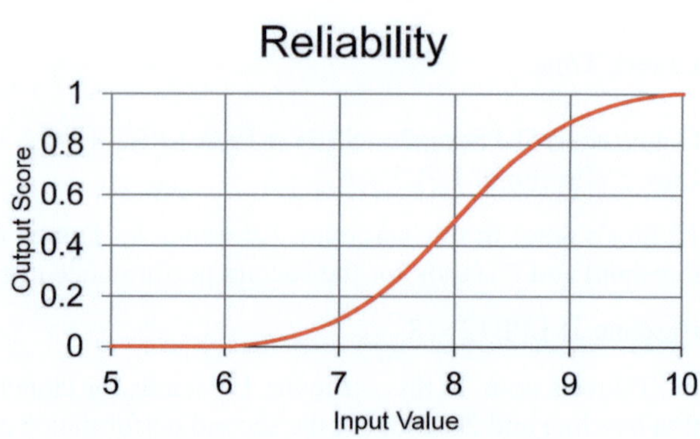

Fig. 5.25 Scoring function for the reliability evaluation criterion

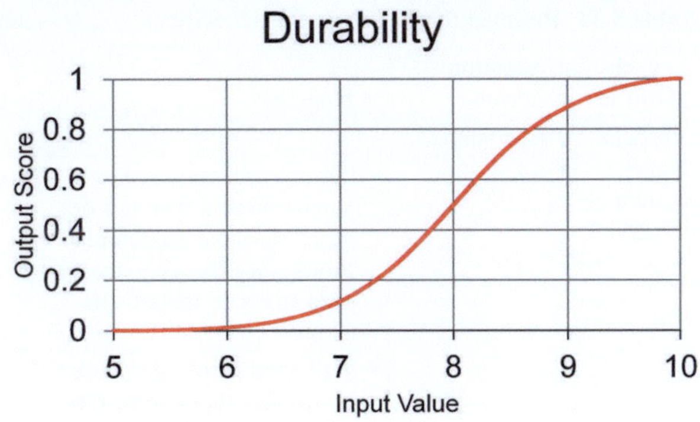

Fig. 5.26 Scoring function for the durability evaluation criterion

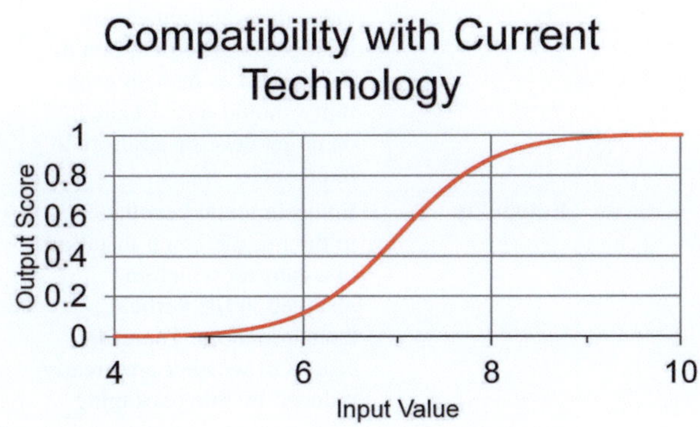

Fig. 5.27 Scoring function for the compatibility with current technology evaluation criterion

Compatibility with Current Technology

Output score: LPScore5 = SSF3(4, 7, 10, 0.5, RLS[0–10]) (Fig. 5.27)

5.4.6.7 Rationale for the Launch Platform Evaluation Criteria Scores

Data given in Tables 5.34, 5.35, and 5.36 for this trade-off study were obtained in interviews with subject matter experts and research. The scores in the right columns were the team consensus. For the performance subcriteria, the domain was 0–10 and scoring functions were used to normalize the data. For cost, schedule, and safety the domain was restricted to 0–1, therefore scoring functions were not needed.

5.4.6.8 Launch Platform Trade-off Study Results

As shown in Fig. 5.28 the vertical launch system (VLS) was the preferred alternative. It provided the overall greatest design capability for this requirement.

Table 5.34 Rational for the scores for the vertical launch system alternative

Launch platform	Evaluation criterion	Advantages	Disadvantages	Score
Vertical launch system (VLS)	*Performance*			
	Launch time	• Shorter time required to launch missile than rail or turret-mounted launch platforms since the VLS does not need to rotate towards its target	• Missiles contained in VLS require time to egress and clear the launcher	9
		• Will contain a large number of missiles that can be readied and launched in a short period of time. This provides the war fighter with multiple engagement potential as well as backup capability should initial launch on target develop a failure to intercept		
	Reliability	• Environmental benefits due to the missile being installed in a canister which in installed in the vertical launch module. Therefore, effects of weather are greatly reduced by this packaging	• Complexity of system degrades overall system reliability	7.5
	Durability	• Improved Gas management system to accommodate more powerful rockets	• Complexity of the system diminishes durability	7
		• Much of launch platform is contained within the hull of the ship, thus preventing external variables to affect system operation		
	Compatibility with current technology	• Open architecture for maximum flexibility and adaptability		8
		• Backwards compatibility for existing missiles		
		• Currently used in ships		
		• Minimal upgrades required to system		
	Cost			
	Design/development cost		• Due to the possibility of a "hang fire," when a missile ignites but is not released from the vertical launch system canister, missile and	0.268
	Production cost			0.268

(continued)

Table 5.34 (continued)

Launch platform	Evaluation criterion	Advantages	Disadvantages	Score
			canisters are generally of a high strength and corresponding high mass design, which results in a more costly design and fabrication effort	
			• Hard launched weapons require a boost motor to initiate the launch of the weapon from the canister, thereby adding additional mass and length to the combined launch system and weapon assembly	
			• Increased cost due to design of associated missile canister that must be mounted in the VLS	
	Maintenance cost	• Separate launcher and weapon functionality allows easier weapon upgrades and maintenance • Reduced maintenance, needing fewer personnel	• A greater cost due to the fact that the missile is encased in its own canister, requiring additional maintenance and upkeep	0.268
	Schedule			
	Ability to meet design schedule	• Rotating launchers with missile handling rails introduce both *complexity* and launch reaction time *delays*—Vertical Launch Systems eliminate these characteristics	• Additional hardware is required to implement the overall system	0.5
	Producibility of design		• Additional items to manufacture	0.044
	Safety	• Weapon is housed in a canister and in the vertical launch platform, which provides isolation from potential accidental excursions due to both on ship personnel and enemy attack	• More safety testing is required to ensure no interference is caused between the canisters due to the presence of the numerous missile effluxes' which could adversely affect the rapid launch of multiple weapons	0.8

Table 5.35 Rational for the scores for the fixed rail system alternative

Launch platform	Evaluation criterion	Advantages	Disadvantages	Score
Fixed rail system	*Performance*			
	Launch time	• Missile launch command is direct and of very short time duration	• Incapable of positioning missile on the intercept bearing	4
	Reliability	• Less complex system requiring less maintenance and functional activity		8
	Durability	• Simple design requires less maintenance	• Exposure to the elements above deck for both missile and rail launch hardware could result in maintenance issues	4
	Compatibility with current technology	• System has been used for many years on board ships		8
	Cost			
	Design/development cost	• Inexpensive to adapt missile to current system	• Design for maneuverability of missile to engage targets on intercept bearings other than in the missile field of view would require higher stress loads on the airframe, thus requiring more complicated designs	0.5
	Production cost	• Simple design requires less production cost		0.732
	Maintenance cost	• Overall reduced maintenance cost due to simplicity	• Portions of the system exposed to sea environments would require additional maintenance	0.116
	Schedule			
	Ability to meet design schedule	• Adaptability to current rail systems should allow for ability to meet design schedule		0.794
	Producibility of design	• Simple design allows for shorter manufacturing time		0.269
	Safety		• Missile is transferred from below deck magazine to rail system thus requiring additional handling, creating the potential for malfunction	0.3
			• Missile is exposed the elements as well as ship's personnel and potential enemy fire	

Table 5.36 Rational for the scores for the turret-mounted rail system alternative

Launch platform	Evaluation criterion	Advantages	Disadvantages	Score
Turret-mounted rail system	*Performance*			
	Launch time	• Similar to rail system with the added advantage of the ability to position missile launch on target bearing		8.5
	Reliability		• Potential issues with turret command and positioning	6.5
	Durability		• Exposure to the elements above deck for both missile and rail launch hardware could result in maintenance issues	3
	Compatibility with current technology		• Ship's interface for turret positioning commands would be required	5
	Cost			
	Design/development cost		• Turret design would be more expensive than rail and VLS due to need to develop turret system	0.3
	Production cost		• Additional components in the overall design increase cost	0.4
	Maintenance cost		• Portions of the system exposed to sea environments would require additional maintenance	0.3
	Schedule			
	Ability to meet design schedule	• Proven technology provides the ability to design a compatible system	• Additional complexity due to turret design will add time to the overall schedule	0.6
	Producibility of design		• Additional components of the turret system will add time to manufacturing schedule • Platform design necessary for the turret increases demands on manufacturing	0.5
	Safety		• A possible safety issue is accidentally shooting back towards the ship • Missile is transferred from below deck magazine to rail system thus requiring additional handling, creating the potential for malfunction • Missile is exposed the elements as well as ship's personnel and potential enemy fire	0.2

Tradeoff Study Matrix using sum combining function and semirelative sensitivity functions with 4 alternatives											
Evaluation Criteria, for the Launch Platform	Wt	Norm. Criteria Weights	Norm Sub Criteria Weights	Alt 1 Do Nothing, Status quo, RAM		Alt 2 Vertical Launch System		Alt 3 Rail System		Alt 4 Turret-Mounted System	
				sc	$W \times SC$	sc	$W \times SC$	sc	$W \times SC$	sc	$W \times SC$
Performance	15	0.288									
Launch Time	6		0.176	1.000	0.176	0.985	0.173	0.000	0.000	1.000	0.176
Reliability	10		0.294	0.100	0.029	0.268	0.079	0.500	0.147	0.042	0.012
Durability	9		0.264	0.100	0.026	0.116	0.031	0.000	0.000	0.000	0.000
Compatibility with Current Technology	9		0.264	1.000	0.264	0.884	0.233	0.884	0.233	0.012	0.003
Cost	8	0.153									
Design/Development Cost	8		0.333	0.600	0.200	0.268	0.089	0.500	0.167	0.042	0.014
Production Cost	8		0.333	0.500	0.167	0.268	0.089	0.732	0.244	0.116	0.039
Maintenance Cost	8		0.333	0.100	0.033	0.268	0.089	0.116	0.039	0.042	0.014
Schedule	9	0.173									
Ability to Meet Design Schedule	7		0.411	0.900	0.370	0.500	0.206	0.794	0.326	0.206	0.085
Producability of Design	10		0.588	0.400	0.235	0.044	0.026	0.269	0.158	0.004	0.000
Safety	20	0.384	1.000	0.200	0.200	0.800	0.800	0.300	0.300	0.200	0.077
Alternative Rating					0.385		**0.537**		0.377		0.109
PI5=	0.35	1.00	4.00								

Fig. 5.28 Trade-off study matrix for the launch platform alternatives

The do nothing alternative is the second most preferred design, because its costs are so low.

5.4.6.9 Sensitivity Analysis

The sensitivity analysis of this trade-off study indicates that the most important evaluation *criterion* is safety. Our customer, the U.S. Navy believes that safety is critical to ensure ship's force and structure is maintained. Any detriment in the weapon's safety could adversely affect the ship's operational capability. Safety is tantamount to the ship, the battle group, the Navy, and to the US military at large. Proper design of the missile propellants, active electrically initiated devices, and overall sound directional stability provide the means to accomplish the reliability necessary to deliver a safe weapon system (Fig. 5.29).

The most important *subcriterion* is compatibility with current technology. The current electrical and mechanical interfaces are functional in the present design. Any modifications to the existing launcher platforms aboard the navy ships would involve greater design modifications resulting in increased cost and schedule delays. Therefore, this evaluation criterion has been weighted accordingly and was confirmed by the sensitivity analysis.

The most important *score* by a big margin is for the safety of the vertical launch system. Seeing this caused us to go back and discuss this element with our customer and our team.

Sensitivity Matrix showing semirelative sensitivity values for PI5 = the hierarchical sum of all products/m											
Sensitivity values		Sensit Criteria Weights	Sensit Sub Criteria Weights	Alt 1 Do Nothing, Status quo, RAM		Alt 2 Vertical Launch System		Alt 3 Rail System		Alt 4 Turret-Mounted System	
				Sen		Sen		Sen		Sen	
Performance		0.114									
Launch Time			0.038	0.013		0.012		0.000		0.013	
Reliability			0.019	0.002		0.006		0.011		0.001	
Durability			0.004	0.002		0.002		0.000		0.000	
Compatibility with Current Technology			**0.053**	0.019		0.017		0.017		0.000	
Cost		0.045									
Design/Development Cost			0.018	0.008		0.003		0.006		0.001	
Production Cost			0.021	0.006		0.003		0.009		0.001	
Maintenance Cost			0.007	0.001		0.003		0.001		0.001	
Schedule		0.061									
Ability to Meet Design Schedule			0.043	0.016		0.009		0.014		0.004	
Producability of Design			0.018	0.010		0.001		0.007		0.000	
Safety		**0.132**		0.019		**0.077**		0.029		0.007	

Fig. 5.29 Sensitivity analysis for the launch platform alternatives

5.4.7 Homework Problems

5.4.7.1 State Diagram

Draw a state machine diagram for the use case Launch and Fly to Target, Iteration 2 of Section 7.

5.4.7.2 Use Case 3: Prelaunch Sequence

There is a step in the main success scenario that states that the backup missile sends a healthy indication signal the crew member. What if the backup missile is sick. Should we proceed? How would you modify the use case to accommodate this?

5.4.7.3 Use Case 4: Launch Missile

It is difficult to write postconditions for use cases, because they must be true no matter which flows were followed. What is wrong with this postcondition?

Post condition: missile has been launched from the ship.
Rewrite it so that it is correct.

5.4.7.4 Thrust Vector Control

Explain a vector thrust control system.

5.4.7.5 Fixed Wing/Fin Method

Explain the use of canard fins.

5.4.7.6 Performance Data

The launch platform trade-off sheet has these scores for the launch time. Alt-1 = 0.985, Alt-2 = 0, and Alt-3 = 0.176. What must have been the inputs to the scoring function, including units? What is responsible for this time delay?

5.4.7.7 Missile Propellant Trade-off Study

We have just presented a trade-off study for the launch platform alternatives. Now, your task is to do a trade-off study of the propellant alternatives. Meaning that the domain experts filled out Tables 5.33, 5.34, 5.35, 5.36, 5.37, 5.38, and 5.39 and Figs. 5.28, 5.29, 5.30, and 5.31. But we erased their results in Tables 5.37, 5.38, and 5.39. We now want you to read their advantages and disadvantages and guess the results they arrived at. Then create a trade-off study matrix using these values. Make sure that you discuss the do nothing alternative and the results of the sensitivity analysis.

The top-level evaluation criteria for the propellant trade-off study are performance, cost, schedule, and safety.

Table 5.37 Weights of importance for the launch platform evaluation criteria

	Evaluation criterion for propellant	Derived from requirement	Raw weight	Top level criteria weights	Subcriteria weights
1	*Overall performance*		15		
1.1	Thrust capability	DR3.5.3.1, CR2.5.1	9		
1.2	Reliability	CR2.5.8, CR2.5.7	12		
2	*Cost*	CR2.3.1.1, DR3.5.5.1	8		
2.1	Design/development cost		8		
2.2	Production cost		8		
2.3	Maintenance cost		8		
3	*Schedule*	CR2.3.2.1, DR3.5.6.1	9		
3.1	Ability to meet design schedule	CR2.3.2.1, DR3.5.6.1	7		
3.2	Producibility of design	CR2.3.2.1, DR3.5.6.1	10		
4	*Safety*	DR3.5.3.2, CR2.5.9	20		
	Sum				

Table 5.38 Rational for the scores for the liquid fuel propellant alternative

Propellant	Evaluation criterion	Advantages	Disadvantages	Score
Liquid fuel	*Overall performance*			
	Thrust capability	• Method of burn is easily controlled		
		• Propellant can be ignited and extinguished multiple times via propulsion control system. This provides the means to establish velocities and accelerations when needed versus one single burn typically the case for solid propellants		
	Reliability	• Once ignited the propellant burns evenly, smoothly, and cleanly	• A chance for leakage and loss of volume during storage	
	Cost			
	Development cost	• The engineering for the liquid containing pressure vessel is considerably simpler than the engineering for solid rocket motors	• Due to the highly toxic concerns of liquid fuels for liquid rocket motors, the personnel safety concerns must be accounted for in the design	
	Production cost		• Handling and storage, and integration costs are increased due to the overall magnitude of safety concerns with respect to fluids	
	Maintenance cost		• Long term maintenance cost due to leakage and potential hazards to the ship	
	Schedule			
	Ability to meet design schedule	• Liquid propellants have been used on multiple missile designs in the past and have a history of design rigor, understanding and overall performance. This understanding of design concepts provides the ability to meet the missile design schedule.		

(continued)

Table 5.38 (continued)

Propellant	Evaluation criterion	Advantages	Disadvantages	Score
	Producibility of design		• Producibility of design can be accomplished with the acknowledgement of the fact that on a navy vessel, safety issues will increase cost dramatically, based on personnel and ship's structure concerns	
	Safety	The idea of using liquid propelled missiles on ships is anathema to the U.S. Navy	• Typical liquid propellant used in missile rocket motor designs are hypergolic in nature and thus are safety concerns for personnel and ship's structure	

Performance

We gave performance an importance weight of 15. This evaluation criterion provides the ability to meet the overall customer requirements. Intercepting enemy targets within the specified engagement range enhances the capabilities of the ship's arsenal. The following evaluation subcriteria were evaluated for performance. The weights assigned for the following were based on the overall criticality:

Thrust Capability: The characteristic whereby the missile is propelled, measured by acceleration.

Reliability: Assessment of mean time between failures or required maintenance. The system must be reliable; therefore, the weight assigned to this evaluation criterion in Fig. 5.37 was a 10.

Cost

Cost received an overall weight of 8. Although it was given the lowest overall weight of the four primary evaluation criteria of the project, it is certainly still important. The other characteristics—performance, schedule, and safety—provide the means to deliver the weapon to the ships to defeat the particular threat. Cost does not provide the means to meet the customer requirements. Cost is what the

Table 5.39 Rational for the scores for the solid fuel propellant alternative

Propellant	Evaluation criterion	Advantages	Disadvantages	Score
Solid fuel	*Overall performance*			
	Thrust capability		• Solid fuels cannot be ignited and extinguished to create multiple thrust cycles, thus limiting to a single burn for intercept capability.	
	Reliability	• The ability of solid rockets to remain in storage for long periods, and then reliably launch at a moment's notice, makes them the design of choice for many military applications	• Any voids in solid propellants could cause decreased acceleration velocity of acceleration performance	
		• Once ignited the propellant burns evenly, smoothly, and cleanly		
	Cost			
	Development cost		• Propellant Bladder is more critical in the overall propellant density to ensure smooth, equivalent burn throughout the rocket motor performance	
			• Any voids or separations in propellant material would cause degradation in overall rocket moor performance, therefore requiring higher design effort	
			• Environmental ground testing would be more extensive to ensure propellant packaging complies with specified performance requirements	
	Production cost	• Solids can also provide high thrust for relatively low cost		
		• Once design is understood, production costs would be much less than those of liquid fuels		

(continued)

Table 5.39 (continued)

Propellant	Evaluation criterion	Advantages	Disadvantages	Score
		since the concerns for safety issues are greatly diminished		
	Maintenance cost	• Once design is understood, maintenance costs would be much less than those of liquid fuels since the concerns for safety issues are greatly diminished		
	Schedule			
	Ability to meet design schedule	• Solid propellants have been used on multiple missile designs in the past and have a history of design rigor, understanding and overall performance. This understanding of design concepts provides the ability to meet the missile design schedule		
	Producibility of design	• Safety and maintenance issues are greatly reduced with respect to liquid propellants		
		• Rocket motors are loaded with propellant and bladders are sealed with minimal regard to safety issues		
	Safety	• Rocket motors are loaded and there are no concerns about leakage or potential toxic issues with respect to personnel or ship's structure	• All personnel must be trained and certified in the handling of explosive material since any hazardous material handling must consider special safety handling procedures	

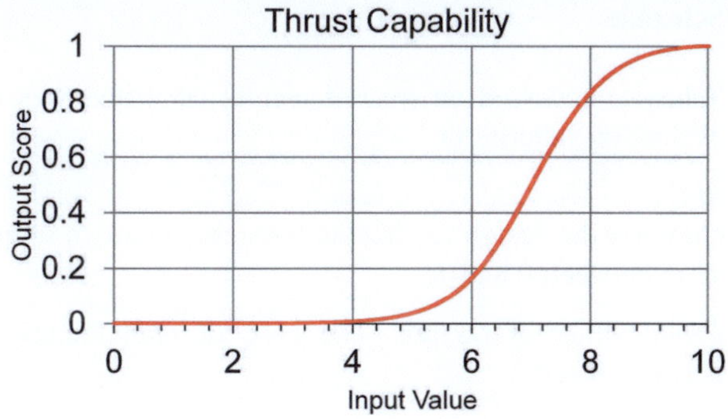

Fig. 5.30 Scoring function for the thrust capability evaluation criterion

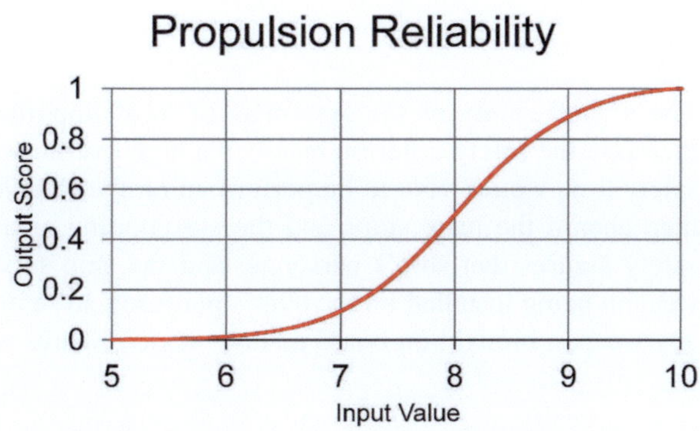

Fig. 5.31 Scoring function for the reliability evaluation criterion

customer needs to consider as far as procuring the weapon system. It is not however the primary concern when considering the overall requirements of the system.

Given the project's desire and goal to deliver a cost-efficient weapon, cost will be attended to in the effort of the remaining three evaluation criteria of performance, schedule, and safety.

At the end of the day, cost becomes the burden necessary to deliver the weapon used to defeat the enemies' capability. The program will strive to provide the most economical cost possible for the weapon design. All design, manufacturing, and maintenance aspects will be evaluated to insure that no surprises with respect to cost will present themselves in the future. The customer will be provided with the most accurate cost estimate capable and all aspects of the program will be controlled to insure that expenses meet estimates.

Cost for the program was evaluated based on estimates of three primary areas of concern: material, manpower, and hours. Each of the alternatives was compared with the others and the advantages and disadvantages were discussed, then the cost was given a score between 0 and 1, where 1 was the least costly and 0 was the most costly. {Editor's note: this is where scoring functions *should* have been used, to flip the cost upside down.} Costs were estimated for the design/development, production, and maintenance phases.

Schedule

Schedule received an overall weight of 9 and was evaluated by the following evaluation subcriteria:

Ability to Meet Design Schedule: The ability to satisfy all design requirements and complete the design verification process in time to support weapon delivery. Scores were between 0 and 1.

Producibility of Design: The ease of manufacturing the design. Scores were between 0 and 1.

Safety

The overall safety of the system is of most importance to the Navy Customer, therefore the safety criterion received a high overall weight of 20. Many strenuous safety tests would have to be performed on the missile to ensure the safety of the men aboard the navy ships and the surrounding areas. The overall weighting of safety assures that ship's personnel and the ship survivability depends on a safe weapon being installed on the ship's platform, so safety is the primary concern for any weapon brought on board a Navy vessel. Scores were between 0 and 1.

Weights of Importance

Weights of importance were assigned to evaluation criteria as shown in the following table. Now you should normalize the weights for the criteria so that they sum to one. Next, you should normalize the weights for the subcriteria so that they sum to one in each category.

Next, we give the parameters necessary to evaluate the evaluation criteria using the scoring functions and graphs given in following sections.

Scoring Functions

Overall Performance

Overall performance, cost, and schedule do not need scoring functions, because their subcriteria are normalized.

Thrust Capability

Output score: PScore2 = SSF3 (PL2, PB2, PU2, PS2, PD)
Output score: PScore2 = SSF3(0, 7, 10, 0.4, RLS[0–10]) (Fig. 5.30)

Reliability

Output score: PScore3 = SSF3(PL3, PB3, PU3, PS3, PD)
Output score: PScore3 = SSF3(5, 8, 10, 0.5, RLS[0–10]) (Fig. 5.31)

Rationale for the Propellant Evaluation Criteria Scores

Data in Tables 5.38 and 5.39 for the propellant trade-off study were obtained in interviews with subject matter experts and research. The scores in the right columns were the final team consensus. For the performance subcriteria, the domain was 0–10 and scoring functions were used to normalize the data. For cost, schedule, and safety, the domain was restricted to 0–1, therefore scoring functions were not needed.

Now we want you to use these values to create a trade-off study matrix. Make sure that you discuss the do nothing alternative and the results of the sensitivity analysis.

Denouement

Trade-off studies are a subset of the decision analysis and resolution process shown in Fig. 5.2. After studying several trade-off studies, we summarize the trade-off study process with Fig. 5.32. It comprises the following activities.

State the problem
Describe alternative solutions
Create evaluation criteria
Develop weights of importance
Select scoring functions
Choose combining function
Collect evaluation data
Produce trade-off matrix
Perform sensitivity analysis
Identify important parameters
Discuss *do nothing* alternative
Analyze results
Identify preferred alternatives
Conduct expert review
Present results to DM and put in PAL

These activities are explained in the following paragraphs.

The problem statement specifies the system's mission, explains the customers' needs and expectations, states the goals of the project, defines the business needs, prescribes the system's capabilities, and delineates the scope of the system. For a high-level system synopsis the following template might be useful.

System synopsis template	
For	(target customer)
who	(statement of need or opportunity)
the (product name)	is a (product category)
that	(statement of key benefits, that is, the compelling reason to buy)
unlike	(primarily competitive alternative)
(our product)	(statement of primary differentiation)
Put product specific information inside the parentheses ()	

The system synopsis can be written as a paragraph, like this: For astronomers who need constant illumination, BIMS is a light and energy management system that will make them more comfortable, more productive, and will save money. Unlike conventional lights and blinds, BIMS will automatically control the illuminance in the room when the sun's or the moon's light rays change.

At the top of the trade-off study, the specific problem being studied should be stated succinctly. For example, the BIMS trade-off study answered the question "In terms of using renewable-energy to generate electricity for the telescope operations room on top of Hawaii, what is the preferred electric generating system?"

The concept exploration document is used to study alternative solutions through brainstorming, modeling, simulation, prototypes, and finally the complete system. In early iterations, this exploration helps produce the problem statement and the incipient system architecture. This document is rewritten continually as more information becomes available. If your system is not unprecedented, then your alternatives should include the competition.

The trade-off study analyst and the customer must identify the customer's preferences and values. Such measures are called evaluation criteria. Evaluation criteria are specific items that need to be quantified to determine how well the system under study satisfies the design requirements from the stakeholder's point of view. Evaluation criteria must be quantitative, independent, transitive, objective, worded so that *more is better*, independent of the temporal order of evaluation and show compensation. Evaluation criteria should be arranged hierarchically. Evaluation criteria are required for each trade-off study. Some criteria may be reusable in multiple trade-off studies.

Weights of importance establish priorities among evaluation criteria and levels in the evaluation criterion hierarchy. Evaluation criteria with higher importance are given more weight in the overall evaluation of the system. As a first approximation, the customer assigns a number between 1 and 10. The weights are then normalized in each category. The key point here is to get the decision maker to think about these issues. The exact numbers are less important.

For complex systems with multiple evaluation criteria, it may be difficult or impossible to directly evaluate each evaluation criterion in raw units and combine them to elicit useful data. This difficulty is due to the lack of a common basis for combining and comparing (for example, comparing hard disk space in Terabytes to processor speed in Gigahertz). Standard scoring functions provide a simple technique through which this issue can be resolved.

Combining functions combine scores or values to calculate a numerical measure that can be used to compare alternatives. They combine data at all levels of the evaluation criterion hierarchy. At the lowest level, when dealing with individual evaluation criteria, the scores are given as outputs of the scoring functions associated with each of the evaluation criteria, and the weights are based on expert opinion or customer preference. At higher levels in the hierarchy, the scores are no longer derived from scoring functions, but are derived by combining scores at lower levels. Again, weights at higher levels in the hierarchy are based on expert opinion or customer preference, perhaps from a different category of stakeholder. Combining values at the top level of the hierarchy yields the overall numerical designation used to compare alternative systems. Overall, the sum combining function was the best.

Data are collected for evaluating each alternative using the appropriate source, for example, approximations, product literature, analysis, journal papers, models, simulations, experiments, and prototypes. Each number in a trade-off study matrix should have an explanation. Here is an example, for the total life cycle cost of the *do nothing* alternative, the purchase price was zero, which gave it a high score (near one). However, it would sell no energy to the electric company. Therefore, it would not generate revenue. This lowered its total life cycle cost score to 0.2.

A trade-off study matrix should be implemented with a spreadsheet. The alternatives go in the columns and the evaluation criteria go in the rows. The weights and the data values go in the central rectangle. The order in which the alternatives are listed has a big effect on the values that humans give for the evaluation data. Therefore, the status quo (the *do nothing* alternative) should be in the first column. This will make the evaluation data for the status quo the anchors for estimating the evaluation data for the other alternatives. This is good because the anchoring alternative is known, is consistent, and you have control over it. Values for each alternative for each criterion are put into their scoring functions and the resulting scores are inserted into the spreadsheet. Each score is then multiplied by its corresponding normalized weight and these products are summed in each column to produce the alternative ratings. Now comes the most important part. The values for the cells are evaluated by a panel of domain experts using the Delphi method. The panel goes through the matrix row by row discussing and comparing the values to other values in that row. This process is repeated many times.

Use the spreadsheet to compute sensitivities. Mathematics is not necessary, because all of the components of the sensitivity functions are in the main trade-off matrix. Look for interactions and nonlinearities. Semirelative-sensitivity functions are the best. A sensitivity analysis identifies the most important parameters in a trade-off study, often these are the cost drivers that are worthy of further investment. A sensitivity analysis completes a trade-off study by gauging the robustness of the preferred alternatives and the rationality of the overall study. A sensitivity analysis helps validate the trade-off study.

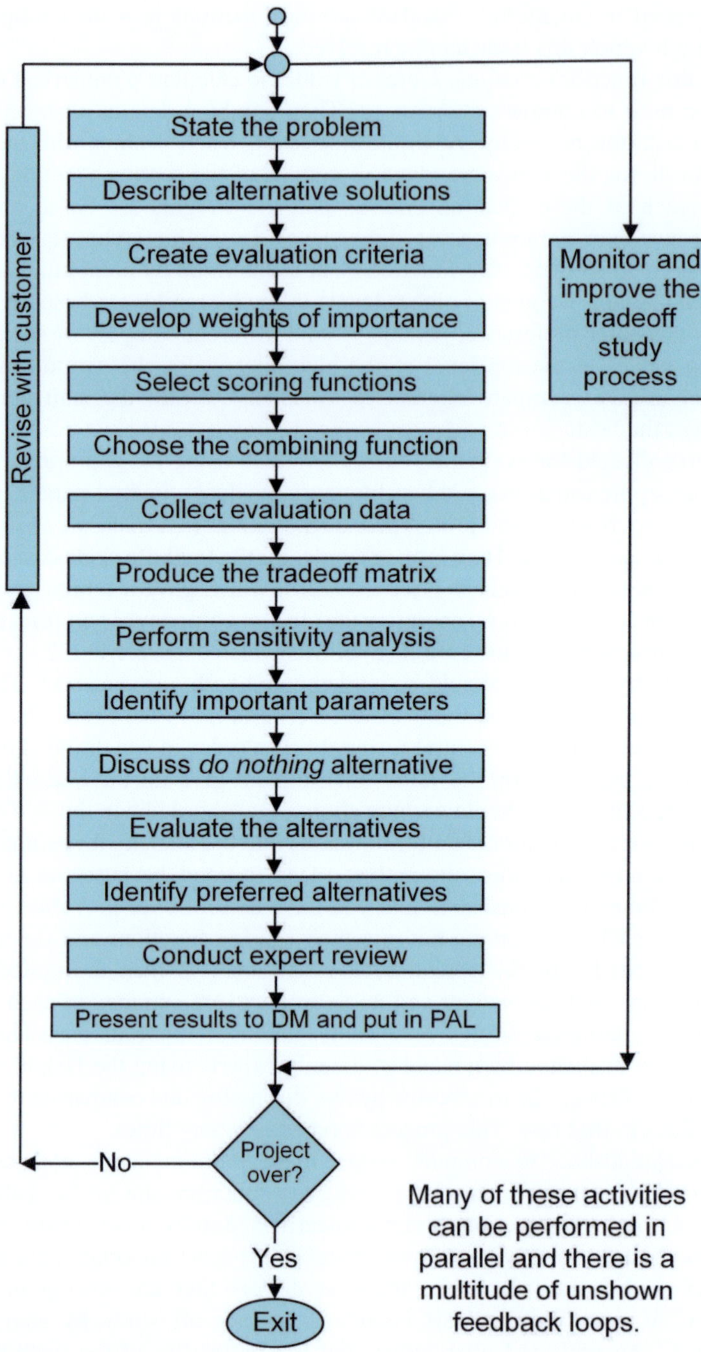

Fig. 5.32 The trade-off study process

Use the sensitivity matrix to point out the most important evaluation criterion, the most important subcriterion, and the most important score. Use this information to reevaluate these parameters. Then discuss them with your customer.

The status quo (*do nothing* alternative) will probably have low evaluation data values (inputs for scoring functions) for performance, cost, schedule, and risk. If the status quo had high performance values, then you probably would not be trying to replace it. The status quo already exists, so (1) it will not be expensive, which gives it a low (good) value for cost, (2) it should not have schedule problems, which gives it a low (good) value for schedule, and (3) it should also be low risk. If the *do nothing* alternative scores high, you should suggest a cause. Often this is caused by having too many dependent cost figures of merit.

Evaluate the alternatives. Review the inputs: evaluation criteria, weights of importance, evaluation data, scoring functions, and combining functions. Review the outputs: the sensitivity analysis, the *do nothing* alternative, and the alternative ratings. Evaluating alternative solutions involves analysis, discussion, and review. Iterative cycles of analysis are necessary. Supporting analyses, experimentation, simulations, or prototypes may be needed to substantiate both the scoring and the results.

Identify preferred alternatives. This action involves weighing and combining the results from the evaluation of alternatives. The true value of a formal decision process might not be in enumerating the preferred alternatives, but in stimulating thought processes and documenting their outcomes. Depending on how many alternatives were included in the trade-off study, two or three alternatives might be recommended to the decision maker.

Trade-off studies should be reviewed at regular gate reviews and at special expert reviews. The whole trade-off study should be reviewed, not just the results. Technical reviews are conducted by a knowledgeable board of specialists that includes supplier and customer representatives. The number of board members should be less than the number of team members. However, board expertise should be greater than the team's combined experience base.

Present the results of the trade-off study to the original decision maker (DM) and other stakeholders. These results might be the preferred alternatives or they could be recommendations to expand the search, re-evaluate the original problem statement or negotiate goals and capabilities with the stakeholders. The results of the trade-off study should be stored in the organizational Process Assets Library (PAL). Evaluation data for trade-off studies come from analysis, models, experiments, and prototypes. Each time better data is obtained, the PAL should be updated. Trade-off studies should be designed with reuse in mind.

5.5 Summary

In Section 5.5, decision analysis and resolution (DAR), we covered trade-off studies, mental mistakes when conducting trade-offs, and followed that up with an exemplar trade-off study for an illumination management system.

We began Section 5.1 by recognizing the role of trade-off studies as an effective, rational method for choosing among alternatives. We cautioned that forgoing a trade-off study inevitably leads to sequential consideration of criteria because of limited human attentional and cognitive capacity. Sequential consideration leads to key criteria being displaced in time, often leading to flawed decisions and outright mistakes.

The basic concept of a trade-off study is to numerically account for empirical values and preferences. When such accounting is formalized within a structured framework, it produces a sound foundation for choosing among alternatives.

DAR is common process within the CMMI rubric that can be tailored and used for a variety of domains. DAR is invoked by Integrated Project Teams (IPTs) on programs to serve a variety of needs. Decision problems such as trade-off studies, bid/no-bid, make-reuse-buy, formal versus checklist inspection, tool and vendor selection, cost estimation, and incipient architectural design all require and can benefit from DAR. A trade-off study is an ongoing, deliberate process that is repeated when new evidence becomes available. A trade-off study is typically undertaken by a collaborative team to balance competing objectives and attributes en route to reaching meaningful consensus. The types of decision made by human decision-making teams include: allocating resources among competing projects; generating novel ideas, plans, and schedule; negotiating; and choosing among alternatives. Alternatives can be examined in series or in parallel. When alternatives are explored in parallel, it is called a trade-off study. Trade-off studies come into play in a variety of situations ranging from selecting high-level system architecture to choosing COTS hardware or software. Trade-off studies are part of a formal evaluation process. Formal evaluation is typically needed when one or more alternatives are selected from a set of alternatives, when a decision is related to a major change to products that have been baselined, when the decision bears on the ability to achieve project objectives, when the cost of the evaluation is affordable and reasonable in comparison to the decisions impact on design implementation, when technical failure can potentially lead to catastrophic outcomes, and when the decision has the potential to dramatically reduce design risk and engineering changes, while also slashing cycle times and production costs.

The likelihood of mental mistakes when conducting a trade-off analysis is well recognized. These mistakes stem from cognitive biases, fallacies, psychological traps, cognitive illusions, inappropriate use of heuristics, and base-rate neglect. We identified specific measures and actions that decision makers should take to avoid making mental mistakes.

In Section 5.2, we reviewed specific mental mistakes that occur in trade-off studies. We discussed each type of mistake along with, its implications, and provided a statistical summary of mental mistakes based on a study of 110 projects. We discovered 808 mental mistakes in the 110 trade-off study reports. We provided the findings from an analysis of sample cases involving trade-off studies. We then reviewed a large, expensive, publicly available San Diego Airport Site selection trade-off study. This study, which took 6 years and cost $17 M, proved to be unconvincing. The voters turned the airport proposal down leading to $17 M down the drain. We then examined the problem to determine whether or not and how a different outcome could have been achieved. We identified the various mistakes in this trade-off study and identified two measures that could have potentially obviated this wasted effort: (1) do a more thorough job of conducting the trade-off study and (2) frame and phrase the ballot proposal differently. Then, in Section 5.3, we undertook a non-military application. We identified the key decisions in this trade-off study and presented specific use cases. We then explored different alternatives including the do nothing option. We then moved on to comprehensive risk analysis and discussed both environmental and economic risks.

We reminded the reader that trade-off studies are an integral part of systems engineering and an ongoing activity. Whenever new evidence negates prior assumptions, a trade-off study needs to be redone.

We concluded with a solemn reminder that mental mistakes, arising from cognitive biases and limitations, are the bane of trade-off studies and that these consideration need to be explicitly addressed during the conduct of a trade-off analysis.

References

1. Madni AM (2014) Expanding stakeholder participation in upfront system engineering through storytelling in virtual worlds. Syst Eng 18(1):16–27
2. Madni AM, Sage AP, Madni CC (2005) Infusion of cognitive engineering into systems engineering processes and practices. In: 2005 I.E. international conference on systems, man and cybernetics. Institute of Electrical & Electronics Engineers (IEEE)
3. Tversky A, Kahneman D (1974) Judgment under uncertainty: heuristics and biases. Science 185:1124–1131
4. Tversky A, Kahneman D (1981) The framing of decisions and the psychology of choice. Science 211:453–458
5. De Martino B, Kumaran D, Seymour B, Dolan RJ (2006) Frames, biases, and rational decision-making in the human brain. Science 313(5787):684–687
6. Petroski H (2003) Framing hypotheses: a cautionary tale. Am Scientist 91:18–22
7. Ainslie G (2001) Breakdown of will. Cambridge University Press, New York
8. Keeney RL, Raiffa H (1976) Decisions with multiple objectives: preferences and value tradeoffs. Wiley, New York
9. Edwards W (1977) How to use multiattribute utility analysis for social decision making. IEEE Trans Syst Man Cybern SMC-7:326–340
10. Haimes YY (2004) Risk of extreme events and the fallacy of the expected value. In: Sage AP (ed) Risk modeling, assessment, and management. Wiley, Hoboken
11. INCOSE Systems Engineering Handbook v2a (2004) [March 2005]

12. CMMI for Development (2015) ver 1.3 [cited 2015 December]. http://cmmiinstitute.com/cmmi-models
13. Madni AM (2000) Thriving on change through process support: the evolution of the process edge enterprise suite and team edge. Inform Knowledge Syst Manage 2(1):7–32
14. Miller GA (1956) The magical number seven, plus or minus two: some limits on our capacity for processing information. Psychol Rev 63(2):81–97
15. Smith ED (2006) Tradeoff studies and cognitive biases. PhD dissertation, University of Arizona, Tucson
16. Madni AM (2013) Generating novel options during systems architecting: psychological principles, systems thinking, and computer-based aiding. Syst Eng 17(1):1–9
17. Smith ED, Son YJ, Piattelli-Palmarini M, Bahill AT (2007) Ameliorating mental mistakes in tradeoff studies. Syst Eng 10(3):222–240
18. Bohlman J, Bahill AT (2013) Examples of mental mistakes made by systems engineers while creating tradeoff studies. Stud Eng Technol 1(1):22–43
19. Gladwell M (2005) Blink: the power of thinking without thinking. Little Brown & Company, New York
20. Klein G (1998) Sources of power: how people make decisions. MIT Press, Cambridge
21. Zimmerman BJ (2006) Development and adaptation of expertise: the role of self-regulatory processes and beliefs. In: Erickson A, Charness N, Feltovich P, Hoffman R (eds) Handbook of expertise and expert performance. Cambridge University Press, New York, pp 707–724
22. Simon HA (1955) A behavioral model of rational choice. Quart J Econ 59:99–118
23. Simon HA (1957) Models of man: social and rational. Wiley, New York
24. Bahill AT, Botta R, Daniels J (2006) The Zachman framework populated with baseball models. J Enterprise Archit 2(4):50–68
25. Ullman DG, Spiegel B (eds) (2006) Trade studies with uncertain information. In: 16th annual international symposium of the international council on systems engineering, Orlando
26. Leveson NG (2002) System safety engineering: back to the future. Massachusetts Institute of Technology, Cambridge
27. Lydersen K (2015) Power stash. Discover 71–73
28. Zakeri B, Syri S (2015) Electrical energy storage systems: a comparative life cycle cost analysis. Renew Sustain Energy Rev 42:569–596
29. Searchinger T, Edwards R, Mulligan D, Heimlich R, Plevin R (2015) Do biofuel policies seek to cut emissions by cutting food? Science 347(6229):1420–1422
30. Madni AM (2011) Towards a generalizable aiding-training continuum for human performance enhancement. Syst Eng 14(2):129–140
31. Chrissis MB, Konrad M, Shrum S (2011) CMMI for Development®: guidelines for process integration and product improvement. Published Mar 10, 2011 by Addison-Wesley Professional. Part of the SEI Series in Software Engineering series
32. Marquard JL, Robinson SM (2008) Reducing perceptual and cognitive challenges in making decisions with models. In: Kugler T, Smith JC, Connolly T, Son YJ (eds) Decision modeling and behavior in complex and uncertain environments. Springer, New York, pp 33–55
33. Sage AP (1981) Behavioral and organizational considerations in the design of information systems and processes for planning and decision support. IEEE Trans Syst Man Cybern 11 (9):640–678
34. Smith ED, Piattelli-Palmarini M, Bahill AT (2008) Cognitive biases affect the acceptance of tradeoff studies. In: Kugler T, Smith JC, Connolly T, Son YJ (eds) Decision modeling and behavior in complex and uncertain environments. Springer, New York, pp 227–249
35. Smith ED, Bahill AT (2010) Attribute substitution in systems engineering. Syst Eng 13 (2):130–148
36. Wymore AW (1993) Model-based systems engineering. CRC, Boca Raton
37. Smith ED, Son YJ, Piattelli-Palmarini M, Bahill AT (2008) Cognitive biases affect the acceptance of tradeoff studies. In: Kugler T, Smith JC, Connolly T, Son YJ (eds) Decision modeling and behavior in uncertain and complex environments. Springer, New York

38. Bahill AT, Dean FF (1999) Discovering system requirements. In: Sage AP, Rouse WB (eds) Handbook of systems engineering and management. Wiley, New York, pp 175–220
39. Bahill AT, Dean F (2009) Discovering system requirement. In: Sage AP, Rouse WB (eds) Handbook of systems engineering and management. Wiley, New York, pp 205–266
40. Hooks IF, Farry KA (2001) Customer-centered products: creating successful products through smart requirements management. AMACOM, New York
41. Daniels J, Bahill AT (2004) The hybrid process that combines traditional requirements and use cases. Syst Eng 7(4):303–319
42. Hull E, Jackson K, Dick J (2005) Requirements engineering. Springer, London
43. Bahill AT, Baldwin D, Venkateswaran J (2005) Predicting a Baseball's path. Am Sci 93 (3):218
44. Baldwin D, Bahill AT, Nathan A (2007) Nickel and dime pitches. Baseball Res J 35:25–29
45. Baldwin D, Bahill AT (2008) Mechanics of Baseball pitching and batting, Applied Biomedical Engineering Mechanics. Informa, London, pp 445–488
46. Bahill AT, Baldwin DG, Ramberg JS (2009) Effects of altitude and atmospheric conditions on the flight of a baseball. Int J Sports Sci Eng 3(2):109–128
47. Bahill AT, Stark L (1975) Neurological control of horizontal and vertical components of oblique saccadic eye movements. Math Biosci 27(3–4):287–298
48. Bahill AT, Stark L (1979) The trajectories of saccadic eye movements. Sci Am 240 (1):108–117
49. Bahill AT, LaRitz T (1984) Why can't batters keep their eyes on the ball. Am Sci 72 (3):249–253
50. Keeney RL (1992) Value-focused thinking: a path to creative decision making. Harvard University Press, Cambridge
51. McBeath MK, Nathan AM, Bahill AT, Baldwin DG (2008) Paradoxical pop-ups: why are they difficult to catch? Am J Phys 76(8):723
52. Bahill AT, Baldwin DG (2007) Describing baseball pitch movement with right-hand rules. Comput Biol Med 37(7):1001–1008
53. Kahneman D (2003) A perspective on judgment and choice: mapping bounded rationality. Am Psychol 58(9):697–720
54. Kahneman D, Frederick S (2002) Representativeness revisited: attribute substitution in intuitive judgment. In: Gilovich T, Griffin D, Kahneman D (eds) Heuristics and biases. Cambridge University Press, New York, pp 49–81
55. Smith ED, Szidarovszky F, Karnavas WJ, Bahill AT (2007) Sensitivity analysis, a powerful system validation technique. Open Cybernet Syst J 2(1):39–56
56. Hsu FK, Bahill AT, Stark L (1976) Parametric sensitivity analysis of a homeomorphic model for saccadic and vergence eye movements. Comput Programs Biomed 6(2):108–116
57. Karnavas WJ, Sanchez PJ, Bahill AT (1993) Sensitivity analyses of continuous and discrete systems in the time and frequency domains. IEEE Trans Syst Man Cybern 23(2):488–501
58. Smith ED, Szidarovszky F, Karnavas WJ, Bahill AT (2008) Sensitivity analysis, a powerful system validation technique. Open Cybernet Syst J 2:39–56
59. Forer BR (1949) The fallacy of personal validation: a classroom demonstration of gullibility. J Abnorm Soc Psychol 44(1):118–123
60. Piattelli-Palmarini M (1994) Inevitable illusions: how mistakes of reason rule our minds. Wiley, New York
61. Shafir E, Tversky A (1992) Thinking through uncertainty: nonconsequential reasoning and choice. Cogn Psychol 24(4):449–474
62. Tversky A, Shafir E (1992) The disjunction effect in choice under uncertainty. Psychol Sci 3 (5):305–309
63. Kahneman D, Tversky A (1979) Prospect theory: an analysis of decision under risk. Econometrica 46(2):171–185
64. Abdellaoui M (2000) Parameter-free elicitation of utility and probability weighting functions. Manage Sci 46(11):1497–1512

65. Bahill AT (2010) Design and testing of an illuminance management system. ITEA J 31 (1):63–89
66. Bleichrodt H, Pinto JL (2000) A parameter-free elicitation of the probability weighting function in medical decision analysis. Manage Sci 46(11):1485–1496
67. Katz SJ, Hawley ST, Morrow M, Griggs JJ, Jagsi R, Hamilton AS et al (2010) Coordinating cancer care. Med Care 48(1):45–51
68. Botta R, Bahill AT (2007) A prioritization process. Eng Manage J 19(4):20–27

Chapter 6
Risk Analysis and Management

6.1 Section 34. Risk Analysis

6.1.1 History of Risk Analysis

Risk is an expression of the potential harm or loss associated with an activity executed in an uncertain environment. Starting in 1662, Arnauld and Nicole, priests living in the convent of Port-Royal in Paris, wrote that risk had at least two components, "Fear of some harm ought to be proportional not only to the magnitude of the harm, but also to the probability of the event [1, p. 273]." This is the first use of the phrases *magnitude of harm* and *probability of the event*. Granted, there are some ancient Greek, Chinese, and biblical sources that express the concept of risk, but they do not have these phrases. Furthermore, it is unlikely that any older source has these phrases, because it was only in 1654 that Arnauld's friend Blaise Pascal invented *probability* [2] and the Oxford English Dictionary cites the earliest use of the word *risk* as a few years after that.

The risk matrix was introduced in the 1970s [3, 4]. Use of the risk matrix was standardized by MIL-STD 882B in 1984 [5]. In the following decades, risk plots (graphs, charts, figures) showing frequency of occurrence versus the severity of consequences were used in risk assessments of nuclear power systems [6, 7] and in Civil Engineering projects [8]. Use of the risk plot was popularized by the Defense Systems Management College [9].

However, in spite of its popularity, the execution of this technique has potential flaws, among them are

1. Failing to differentiate between levels and categories of risk
2. Using estimated probability of the event rather than its frequency of occurrence
3. Using ordinal numbers instead of cardinal numbers for severity
4. Using different ranges for severity and frequency
5. Using an inappropriate combining function
6. Using linear scales instead of logarithmic scales

A.T. Bahill, A.M. Madni, *Tradeoff Decisions in System Design*,
DOI 10.1007/978-3-319-43712-5_6

7. Explaining only intermediate risks while seeming to ignore high and low risks
8. Ignoring interactions between risks
9. Ignoring high-likelihood low-severity and low-likelihood high-severity outliers
10. Ignoring risk severity amplifiers
11. Confusing risk with uncertainty

The purpose of this chapter is to elucidate these methodological mistakes and point out ways to ameliorate them.

Risk analysis is an important part of engineering. There are general guidelines for how it should be done, but there is no *one correct way* to do a risk analysis [10–13]. There is an extensive history of mankind's efforts to understand and manage risk [14]. This section presents the most common industry risk analysis technique summarized by Bahill and Karnavas [15] and shows problems associated with it. There are many levels of risk: (1) system risk including performance, cost, and schedule of the product, (2) project risk, (3) business risk including financial and resource risks to the enterprise, and (4) safety, environment, and risks to the public. The effect of a risk in one category may become the source of risk in another category. Interrelationships between the most common of these risks are shown in Fig. 6.1. The low-level system risks of performance, cost, and schedule comprise the higher level project risks, which might in tern constitute enterprise risks. Most industry risk analyses only present the bottom level explicitly.

Risks can be placed on sequential trees or graphs, in order to make their interrelationships clear. Within these structures, the probability of a risk occurring will be dependent on predecessor risks. Another technique for categorizing risks is

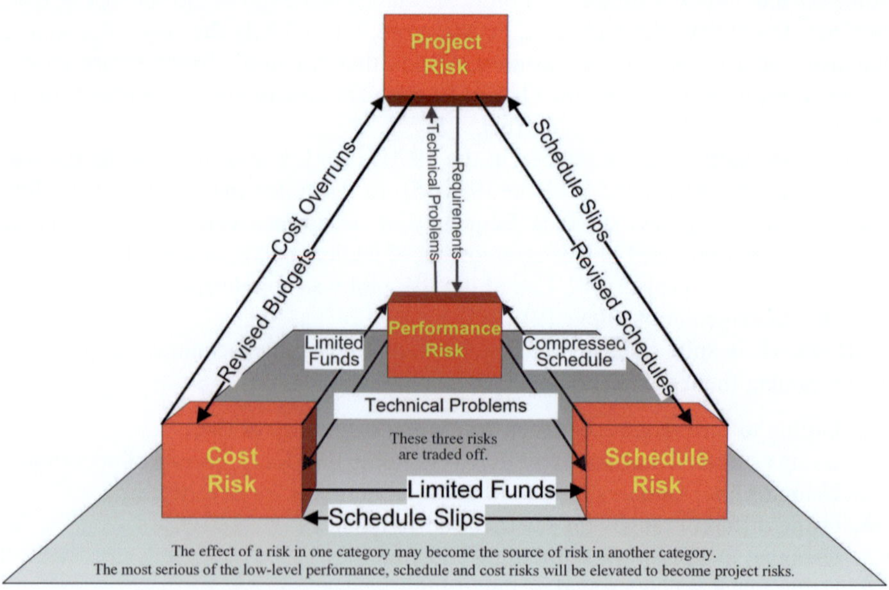

Fig. 6.1 Interrelationship of risks

to use a framework to differentiate risks by different framework areas. Risk analyses can then be conducted solely within framework areas, or only across framework areas that are adjacent to one another.

Good risk management will not prevent bad things from happening. But when bad things happen, good risk management will have anticipated them and will reduce their negative effects.

6.1.1.1 Frequencies Are Better than Probabilities

In Section 25, we showed that frequency of occurrence is a better metric than probability because humans estimate probabilities so poorly, as was shown in Section 17. Therefore, in this section we will define risk as the product of frequency of occurrence multiplied by the severity of the consequences.

6.1.2 Risks in a Pinewood Derby

This section presents the industry standard technique for risk analysis, namely a graph or a plot with frequency of occurrence on one axis and severity of the consequences on the other. It points out the following common errors in implementing this technique: failing to differentiate between types of risk, using estimated probability rather than frequency of occurrence, using ordinal instead of cardinal numbers, using different ranges for severity and frequency, using an inappropriate combining function, using linear instead of logarithmic scales, explaining only intermediate risks, and confusing risk with uncertainty. Then this section points out ways to ameliorate these mistakes. Additionally, it presents an algorithm for deriving severity scores. It also discusses many factors that affect human perception of risk. Finally, it discusses the mitigation benefit versus cost trade-off that is always made in risk management.

Since the 1950s, over 100 million Cubs Scouts have built five-ounce, wooden cars and raced them in Pinewood Derbies. Bahill and Karnavas [15] ran a Pinewood Derby for their Cub Scout pack for 5 years. Each year they did a risk analysis, identified the most severe risks, and ameliorated them. For example, in the first 2 years, the most severe risks were lane biases and placing the cars into the wrong lanes. To ameliorate these risks, they switched to a round-robin race format. (A trade-off study of alternative designs, for example, single elimination, double elimination, and round robin, helped them understand the system.) The round-robin allowed each scout to race often, to race throughout the whole event, and to race more of their friends. Six rounds were used, because that gave each car two races in each lane and still kept the whole event of appropriate duration. The effects of lane biases were eliminated, because each car ran in each lane the same number of times. Placing the

Table 6.1 Pinewood Derby performance risk matrix

Failure mode	Consequences or potential effects	Frequency (failures per 1000 heats)	Severity	Risk
Temporary failure of starting gate or finish line sensors	Heat must be re-run	10	5	50
Lane biases	Some cars have unfair advantage	1000	0	0
Collisions between cars	Heat must be re-run	20	5	100
Mistakes in judging or recording	Wrong car is declared the winner	0.1	1000	100
Human mistakes in				
Weighing cars	Some cars have unfair advantage	30	10	300
Allowing modifications	Some cars gain unfair advantage	15	10	150
Placing cars in wrong lanes	Wrong car is declared the winner	20	1000	**20,000**
Lowering the starting gate	Cars can be broken, race could be unfair	1	100	100
Resetting finish line switches	Heat must be re-run	10	5	50
Wasting time	People get annoyed	10	100	1000

Risk is defined as frequency multiplied by severity

cars into the correct lanes is discussed later. Risk is typically divided into four broad categories: performance, cost, schedule, and project [15]. In Table 6.1, we look at performance, which is often called technical performance. In Table 6.1, risk is defined as frequency multiplied by severity.

6.1.2.1 Rationale for the Performance Failure Frequencies

Typically, in each Pinewood Derby, about 250 heats (races) were run in a little over 4 h, for an average of one heat per minute. Data were collected in 4 years. So failures are given in terms of failures per 1000 heats or failures per 1000 min.

Temporary failure of starting gate or finish line sensors. This happened a little more than twice a year, yielding 10 failures per 1000 heats.

Lane biases. All Pinewood Derby tracks probably have one lane faster than another. So the frequency is 1000 failures per 1000 heats.

Collisions between cars. Originally, about one out of ten heats had collisions: this was unacceptably high. Collisions were probably caused by imperfections where two sections of track were joined. When a wheel hit such an obstruction, the car bounced out of its lane. By carefully aligning and waxing the joints, the failure rate was reduced to 20 failures per 1000 heats.

Mistakes in judging or recording the results. This was never observed after we switched from human to electronic judging. With computer control and electronic judging, this failure mode is very unlikely. Failures were estimated at 0.1 failures per 1000 heats.

Human mistakes in weighing cars. Every year an over-weight car probably snuck through. This car would then have run in, on average, about seven heats, meaning 30 failures per 1000 heats.

Human mistakes in allowing modifications (such as adding graphite or weight) after inspection. Each year one scout probably added graphite in the middle of his heats, so the frequency of an unfair heat would be 15 out of 1000.

Human mistakes in placing cars in the wrong lanes. This was observed five times in the first derby: therefore 20 failures per 1000 heats.

Human mistakes in lowering the starting gate. In the first derby, this happened once.

Human mistakes in resetting finish line switches. This happened about ten times per 1000 heats.

Human mistakes wasting time. This happened about ten times per 1000 heats.

6.1.2.2 Algorithm for Computing Severity Values

The following algorithm was used for computing values for the severity of the consequences.

1. Assign a frequency of occurrence (F_i) to each failure mode.
2. Find the failure mode that has the most severe consequences, call its value S_{worst}.
3. For each other failure mode, ask: "How many of these failures would be equally painful to the Worst?" Call this N_i. Alternatively, this can be rephrased as, "Cumulatively, how many of these failures would have an equal impact to the Worst?"
4. Compute the severity for each failure mode as $S_i = S_{\text{worst}}/N_i$.
5. Normalize the severity values so that their range equals the range of the frequency values.
6. Compute the estimated risk using a combining function.
7. Prioritize the risks to show which are the most important.

The results of step 1 are shown in the frequency column of Table 6.1. In our Pinewood example, step 2 identifies "*Placing cars in the wrong lanes thereby declaring the wrong car to be the winner*" to be the most severe failure mode. In step 3, each failure mode is compared to this worst failure mode. For example, our domain expert told us that rerunning 200 heats would be equally painful to declaring the wrong car a winner. Step 4 now produces candidate values for severity.

Severity values should be cardinal numbers, not ordinal numbers. Cardinal numbers indicate size or quantity; ordinal numbers merely indicate rank ordering. (A mnemonic for this is that ordinal is ordering, as in rank ordering.) Cardinal numbers do not just indicate that one failure mode is more severe than another—they also indicate how much more severe. If one failure mode has a weight of 6 and another a weight of 3, then the first is twice as severe as the second. Steps 2–4 in the above algorithm help generate these cardinal numbers.

Step 5 requires normalizing the severity values so that their range equals the range of the frequency values. For Pinewood, the frequency values ranged over five orders of magnitude; therefore the severity values were adjusted so that they also ran over five orders of magnitude, as is shown in the severity column of Table 6.1.

With Table 3.7, we showed that if two items are being multiplied and they have different ranges, the one with the bigger range has more weight. Risk is now being defined as the frequency of occurrence multiplied by the severity of the consequences. Therefore, the range of frequencies must be the same as the range of the consequences.

A technique for step 6 is to use the likelihood of occurrence of the failure event, which is restricted to values of 0–1, and then for each row, assign a scoring function for the severity of consequences. Scoring functions also produce outputs in the range of 0–1. Therefore, the range for both likelihood and severity will be between 0 and 1.

For Table 6.1 we computed the risk as the severity of the consequences times the frequency of occurrence. Other formulae have been used. For example, Ben-Asher [16] writes that people do not equally weigh relative likelihood and severity of the consequences, as shown in Table 6.2. For example, people buy collision insurance for their cars, but they seldom insure their tires against premature wear. Therefore, he computes risk as $R = S^2P$.

Some people have computed risk with the severity plus probability of failure minus the product of severity and probability of failure ($R = S + P - SP$) [17]. But this formula does not perform satisfactorily. For example, if you set the severity to 1 (assuming a range of 0–1), then the probability of failure could be reduced from say 10^{-1} to 10^{-6} without changing the risk. Furthermore, if either the probability or the severity is zero, then the risk should be zero, but this equation does not produce this result. Therefore, we do not use such an equation.

Eric Smith (personal communication) has used exposure instead of probability, that is, Risk = Severity × Exposure. The value for exposure should estimate the subject's likely endangerment to the risk. Exposure does not have probability's time-discounting errors, such as hyperbolic discounting. It would be hard to calculate the probability that particular person would get flu this season, but it would be

Table 6.2 If people do not weigh severity and probability equally

	Severity of consequences (S)	Probability (P)	Product ($S \times P$)	Perceived risk	$R = S^2 \times P$
Low severity, high probability	0.1	0.9	0.09	Low	0.009
Medium severity, medium probability	0.3	0.3	0.09	Medium	0.027
High severity, low probability	0.9	0.1	0.09	High	0.081

easy to estimate that a first-grade teacher would have a greater exposure to viruses than a college professor would.

The Failure Modes and Effects Analysis (FMEA) technique sometimes includes the difficulty of detecting the failure in the product [18], so that ($R = S \times P \times DD$). In some cases this is useful, but we will not use it in this section.

Terrorism risk can be viewed as having three components: the *threat* to a target, the target's *vulnerability* to the threat, and the *consequence* should the target be successfully attacked [19]. The threat is defined as the probability that a specific target is attacked in a specific way during a specific time period. Vulnerability is defined as the probability that damages occur, given a specific attack type, at a specific time, on a given target. The consequence is defined as the expected magnitude of damage, given a specific type of attack, at a specific time, which results in damage to a specific target. Terrorism risk is defined as the expected consequence of an existent threat (for a given target, attack mode, and damage type) which can be expressed as

$$\text{risk} = \text{threat} \times \text{vulnerability} \times \text{consequence}$$

"Rather than seek an optimal method for estimating risk, we seek a method that leads us to make the least egregious errors in decision-making across the range of possible scenarios that might develop in the future" [19].

All of the following equations have been used in the published literature to compute risk.

$$\begin{aligned}
\text{Risk} &= \text{Frequency of Occurrence} \times \text{Severity of Consequences}\\
\text{Risk} &= \text{Likelihood of Occurrence} \times \text{Severity of Consequences}\\
\text{Risk} &= \text{Estimated Probability} \times \text{Severity of Consequences}\\
\text{Risk} &= (\text{Likelihood} + \text{Impact})/2\\
\text{Risk} &= \text{Probability} + \text{Severity} - (\text{Probability} \times \text{Severity})\\
\text{Risk} &= \text{Probability} \times \text{Severity} \times \text{Difficulty of Detection}\\
\text{Risk} &= \text{Probability} \times \text{Severity}^2\\
\text{Risk} &= \text{Exposure} \times \text{Severity}\\
\text{Risk} &= \text{Threat} \times \text{Vulnerability} \times \text{Consequences}
\end{aligned}$$

Each of these equations is good in certain situations. But for the reasons already given in this section, we will only use the top two equations.

Step 7 of the algorithm prioritizes the risks in order to show which are the most important. Often this process is merely that of selecting the risks with the largest values and pointing them out to the decision maker. Later in the Project Risk Management section of this section, we will show a graphical technique for highlighting the most important risks. Section 24 gave a general-purpose technique for prioritizing things other than risks, things like goals, customer needs, capabilities, risks, directives, initiatives, issues, activities, requirements, technical performance measures, features, functions, and value engineering activities [20].

6.1.2.3 Discussion of the Performance Risks

The most important failure mode in this Pinewood Derby risk analysis is *human mistakes in placing cars in the wrong lanes*. In a previous risk analysis of our Pinewood Derby [21, Section 7], we found that in order to satisfy our customer we had to pay the most attention to *mistakes in judging and recording* and *lane biases*. Therefore, we designed our system to pay special attention to these failure modes. First, we designed a computerized judging and recording system that had a very low failure frequency, indeed this failure mode is the least likely one in Table 6.1. Secondly, we designed a round-robin tournament where each car races twice in each lane, therefore lane biases had no effect. Any time the effect of a failure mode was eliminated, we set the severity to zero. We did not remove the failure mode, because we wanted to show that it was considered. Therefore, the failure modes that were very important in our early risk analysis were not important in the risk analysis of this section. In our redesigned derby, humans were the most important element.

Other Columns. A failure modes and effects analysis should include a column indicating who should do what in response to each failure mode. This was not included in this example because all risks were managed by the same person. A "So what?" column would be included for a root-cause analysis. For example, the failure mode *collisions between cars* has the potential effect of "Heat must be re-run." After this, we could ask "So what?" This might produce the response, "This takes additional time." We could again ask, "So what?" producing, "If there were many re-runs the whole Pinewood Derby schedule might slip." etc. Other columns could also be used to document what and when corrective actions were taken.

Risk communication is an important part of a risk analysis. What should be communicated to whom? In general, the risks are communicated to the program manager along with proposals for mitigating these risks. In this case, the Pinewood Derby Marshall served as the program manager, so he was told that *placing cars in the wrong lanes* was the most significant risk. Then he was shown that the race schedules have lane 1 on the left and lane 3 on the right. This matches the perspective of the finish line judges (lane 1 is on their left and lane 3 is on their right), but it is the opposite for the starter at the top of the track. Therefore, we suggested printing regular schedules for finish line judges and printing mirror image schedules in another color for use by the starter.

6.1.2.4 Graphical Presentation

Risk is often presented graphically with frequency of occurrence on the ordinate and severity of consequences on the abscissa. Each of the axes is linear with a range of 0–1, as in Fig. 6.2a. Risks in the upper-right corner are the most serious and should be handled first. The problem with this technique is that all of the risks in Table 6.1 would be squashed onto the x-axis, because they all have frequencies at or below 10^{-1} (except for our anomalous lane bias). If some action resulted in a

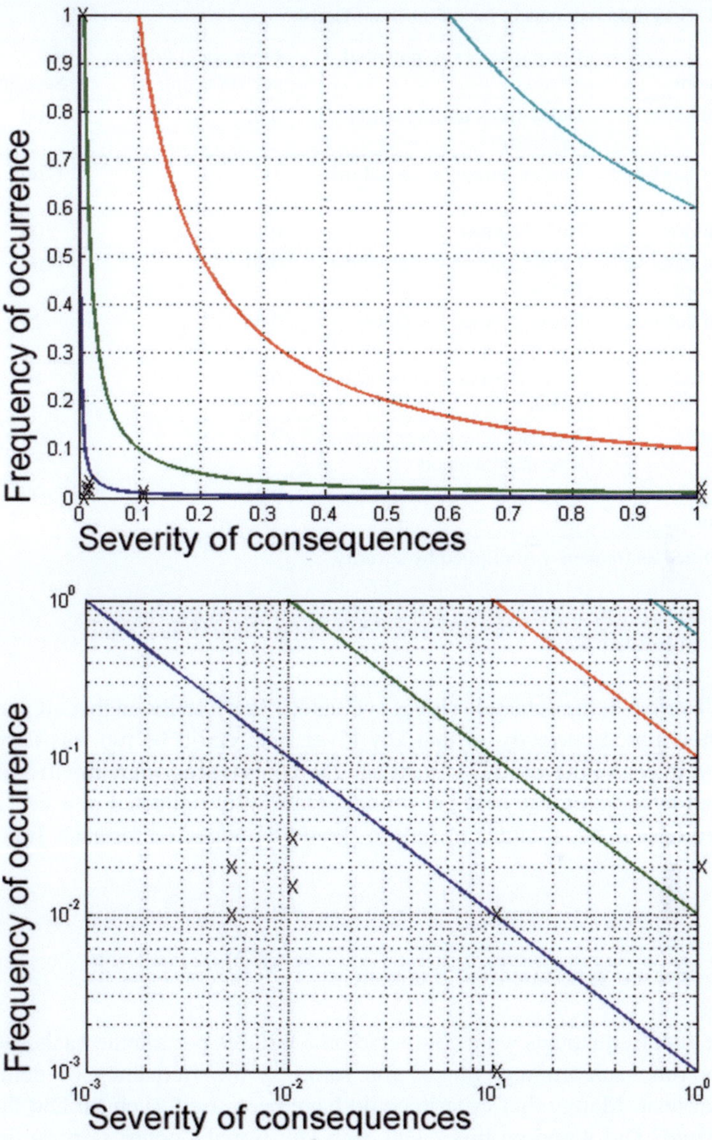

Fig. 6.2 (**a**) Linear plot of $xy = 0.001$, $xy = 0.01$, $xy = 0.1$, and $xy = 0.6$, with normalized data of Table 6.1. (**b**) Log–log plot of $xy = 0.001$, $xy = 0.01$, $xy = 0.1$, and $xy = 0.6$, with normalized data of Table 6.1

reduction of frequency of occurrence from 10^{-2} to 10^{-4} we want the estimated risk to change. And in these graphs, it would not change. Using logarithmic scales as in Fig. 6.2b would help, and indeed this may be the best solution if quantitative data are available.

Table 6.3 Schedule risk matrix for a Pinewood Derby

Failure mode	Consequences or potential effects	Frequency, failures per 1000 min	Severity	Risk
Loss of electric power	Delay races until electricity is restored	0.02	100	2
Computer hardware failure	Cannot compute, cancel the Derby	1	1260	**1260**
Server failure	No schedules	0.1	100	10
Failure of commercial software	Cannot compute, cancel the Derby	0.01	1260	12.6
Failure of custom software	Cannot compute winners	0.05	20	1
Software interface failure	Data is lost and races must be re-run	0.02	10	0.2
Bad weather	Batteries lose power, cannot determine winners	0.008	1000	8
Forgetting equipment	Lose an hour of setup time	0.01	100	1

Risk is defined as frequency multiplied by severity

6.1.2.5 Schedule Risk

Next, in Table 6.3, the schedule risk is evaluated. The likelihood that all parts of the system functioned correctly so that any given heat could be run was determined. Typically, in each Pinewood Derby, about 250 heats (races) were run in a little over 4 h, for an average of one heat per minute. Data were collected in 4 years. So we calculated how many minutes of failure should be expected for each 1000 min of racing.

Rationale for the Schedule Risk Frequencies

Schedule risk frequencies were computed on a failures per minute basis. Because a heat was run each minute, this is the same as the frequency of delaying an individual heat. Many other criteria could have been used, such as: Did the Derby start on time? Did it end on time? Did each Divisional Contest start on time? Did each end on time? The following explains how the schedule risk frequencies were computed.

Loss of electric power. During the 14 years for which we have data, this area of Tucson has lost electric power three times (once when lightning hit a transformer, once when a car hit a pole, and once when the Western Power Grid went down), for a total outage of about 3 h. So for any given 1000 min interval, the power would be expected to be out on average 0.02 min. This means that there was a 2 % chance of delaying any one race during these 4 years.

Personal computer hardware failure. In 4 years, we had three personal computer hardware failures that took 45 h to fix: therefore, there was 1 min of failure per 1000 min of operation.

Server failure. Schedules were printed weeks in advance, so server failure was unlikely to delay a race, 0.1 min of failure per 1000 min.

Failure of commercial software. 0.01 min of failure per 1000 min. This failure frequency was low, because the programs were Unix and DOS-based.

Failure of custom software. 0.05 min of failure per 1000 min.

Software interface failure. In 4 years, there was one interface failure: It took an hour to fix, hence 0.02 min of failure.

Bad weather. In 14 years, Tucson has had one afternoon where the temperature dropped 30° in 2 h. This caused the NiCad batteries to lose power, and the race was shifted to a manual, paper-based system. The switch took 1 h, yielding 0.008 min of failure per 1000 min.

Forgetting equipment. Over these 14 years, on one road trip, equipment was forgotten and a trip back to the lab was needed. This caused a delay of a little over 1 h, yielding 0.01 min of failure per 1000 min.

When assigning the severity values we made evaluations such as "*Forgetting our equipment* 100 times would be equally painful as *Hardware failures* that would cause us to cancel a Derby."

The probability that this system would be operating mistake free for any given heat is (1 – 0.00002) (1 – 0.001) (1 – 0.0001), etc., which equals 0.99878. This is good reliability. It resulted from doing studies like these, finding the weak link and redesigning the system.

As can be seen in Table 6.3, the present weak link is *personal computer hardware failure*. Therefore, when we put on a Pinewood Derby we keep a spare computer in the car. Actually, we are even more paranoid than this. The last time we put on a Pinewood Derby, we designed it with a Round Robin (Best Time) with electronic judging, but we also designed a backup system that required no electricity: it used a Round Robin (Point Assignment) with human judging. All of the equipment and forms for the backup system were in the van.

Failure modes and effects analyses usually do not include acts of nature, like our *bad weather* category. We have included it in our analysis, because we think that if the process is designed with these items in mind, then although we cannot prevent them, the system can better accommodate such failures.

Bahill and Karnavas [15] also did a cost and safety risk analysis for the Pinewood Derby. That analysis is not included in this section. They also considered equity. In this derby, all scouts were exposed to the same risks.

6.1.3 Project Risk Management

Most of this case study has been about analysis of performance and schedule risks. However, project managers are more often worried about project risk, which has been defined as "the potential of an adverse condition occurring on a project which will cause the project to not meet expectations."

For the Pinewood Derby, some project risks are (1) *failure to compute schedules*, (2) *school denies use of facilities*, (3) *Boy Scouts of America (BSA) stops supporting Pinewood Derbies*, (4) *parents or children lose interest*, and (5) *lawsuits*. These are presented in Table 6.4. *Failure to compute schedules* has a large likelihood value, because in our first Pinewood Derby we were not able to compute schedules that met the original requirements. So we had to change the requirements or else run a single elimination tournament, which does not require schedules. *Failure to compute schedules* is an example of a low-level schedule risk that was important enough to be elevated to a project risk. If we have data for failure frequency, we use the term frequency. Otherwise, if we are guessing the future, we use the term likelihood.

The *school denies use of facilities* failure mode includes the school demanding an exorbitant fee, having another big school event scheduled for the same day, and the school not giving access to electric power outlets. We also considered unavailability of equipment. But this is not a big risk, because the pack owns everything we need, except for a PC. We considered loss of knowledge, but our system is well documented. We considered loss of key people, but we have many qualified parents.

For project risk, we do not have accurate numbers, so 0–1 scales *are* appropriate. The likelihood values are *not* probabilities; rather they are relative measures indicating likelihood: The school denying use of their facilities is considered more likely than the *Boy Scouts of America stopping its support of Pinewood Derbies*. But we are not saying that seven out of ten times we expect the school to turn down our request. Id is the identification number that will be used in Figs. 6.3 and 6.4.

Table 6.4 Project risk matrix for a Pinewood Derby

Id	Failure mode	Consequences or potential effects	Likelihood	Severity	Risk	Assigned to
1	Failure to compute schedules	Run a single elimination tournament	0.8	0.4	0.32	Bill
2	School denies use of facilities	Run the Derby in someone's backyard	0.7	0.9	0.63	Terry
3	BSA stops supporting Derbies	Buy commercial car kits	0.1	0.1	0.01	Harry
4	Parents or scouts lose interest	Cancel the Derby	0.1	1.0	0.10	Harry
5	Lawsuits	Buy insurance	0.9	0.9	0.81	Terry

Risk is defined as likelihood multiplied by severity

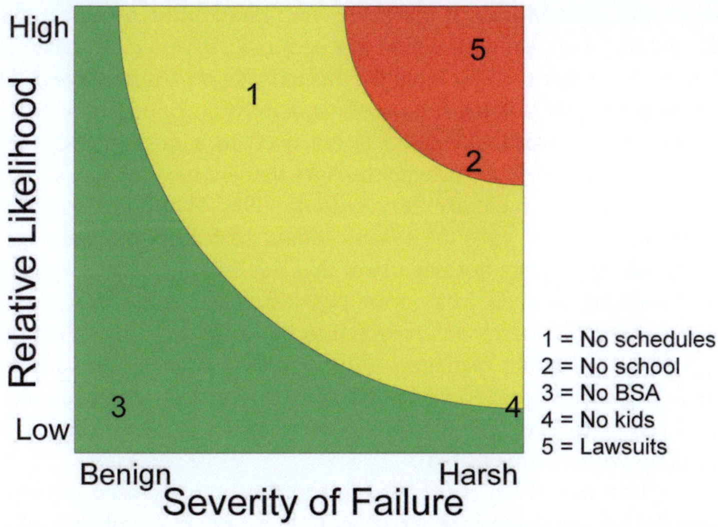

Fig. 6.3 Project risk chart for a Pinewood Derby

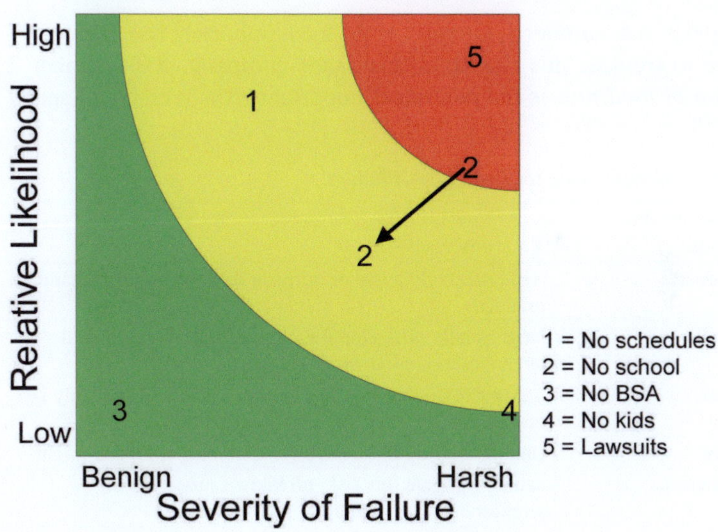

Fig. 6.4 Pinewood Derby project risk plot after risk mitigation

The failure modes are assigned to Bill Karnavas, Terry Bahill, and Harry Williams, the Cub Master. Table 6.4 is analogous to a risk register.

In Fig. 6.3, these data are put into a risk management tool that is commonly used in industry, namely a plot of relative likelihood versus severity of failure. In such plots, the ordinate is often labeled probability or potential probability, but they are *not* probabilities. Uncertainty and unknown unknowns prevent calculation or even estimation of real probabilities. If this scale were probability, then any item with a probability greater than, say, 0.6 would not be a risk; it would be a mandate for a design change. We think the best label for this axis is *relative likelihood.* The word relative emphasizes that it is the relationships between (or ranking of) risks that is being illustrated. The word likelihood is not used in a mathematical sense, but rather in its dictionary sense to indicate the risks that are most likely. These relative likelihood estimates are typically very conservative. They are derived with the following attitude, "If we built the system today, with lofty capabilities, without improving the design, doing nothing about this risk item and if the worst possible circumstances occurred, then what is the possibility that this risk item will harm us?" Initially, some risk items are expected to be in the red zone (the upper-right corner): otherwise, program managers will think that the designers are not trying hard enough to get more performance or less cost. The relative likelihood scale is not linear: a risk with a relative likelihood value of 0.8 is not considered twice as probable as one with a value of 0.4.

The description just given is the way this tool is actually used in industry and government [22]. Statisticians are appalled, because of its use of ordinal numbers instead of cardinal numbers. Often qualitative scales are given to help assign values. The numbers given in Tables 6.5 and 6.6 are typical. Clearly, these are not cardinal numbers.

In Fig. 6.3, risk number 5, *lawsuits*, is the most important risk. Therefore, we buy insurance to transfer this risk to an insurance company. Risk number 2, *school denies use of facilities*, is the next most important risk. We should spend a lot of

Table 6.5 Qualitative scale for frequency of failure

How often will this failure occur?	Your processes ...	Value
Almost never	will effectively avoid or mitigate this risk based on standard practices	0.2
Occasionally	have usually mitigated this type of risk with minimal oversight in similar cases	0.4
Sometimes	may mitigate this risk, but workarounds will probably be required	0.6
Frequently	cannot mitigate this risk, but a different approach might	0.8
Almost always	cannot mitigate this risk, no known processes or workarounds are available	1.0

Table 6.6 Qualitative scale for severity of schedule slippage

If the risk event happens, what would be the severity of the impact on the schedule?	Value
Minimal impact, slight changes could be compensated by available program slack	0.2
Additional activities required, you would be able to meet key dates	0.4
Minor schedule slip, you would miss a minor milestone	0.6
Program critical path affected	0.8
You would fail to achieve a key program milestone	1.0

time making sure it does not happen and preparing backup plans. The third most important risk is number 1, that *failure to compute schedules*, so we should keep an eye on this. The curves should be rectangular hyperbolas, not circles, because they are iso-risk contour lines for severity times likelihood.

The most important project risks are put in the project risk register and are analyzed monthly and at all design reviews. For this Pinewood Derby example, *school denies use of facilities* is an important risk. So we tried to mitigate it. We talked to the school principal and also found an alternate site. (In 1 year, our pack ran the derby at the Northwest Emergency Center.) This ameliorated the risk and at the next review, we presented the risk plot of Fig. 6.4. The forte of this risk management tool is showing changes in risk items from review to review.

In most risk analyses plots that we have seen in industry, most of the risks were in the center of the plot. Most of the risks that would have been in the upper-right corner had already been ameliorated, whereas most the risks in the lower-left corner were considered unworthy of the tracking effort. This distribution could also be caused by a common human decision mistake. People overestimate events with low probabilities, like winning the state lottery, and they underestimate high probability events, such as successfully completing a commercial airplane flight, Section 17.

6.2 Discussion

The highest priority risks are often tracked with technical performance measures (TPMs) introduced in Section 23. TPMs are critical technical parameters that a project monitors to ensure that the technical objectives of a product will be realized. Typically, TPMs have planned values at defined time increments, against which the actual values are plotted. Monitoring TPMs allows trend detection and correction, and helps identify possible performance problems prior to incurring significant cost or schedule overruns. The purpose of TPMs is to (1) provide visibility of actual versus planned performance, (2) provide early detection or prediction of problems requiring management attention, (3) support assessment of the impact of proposed changes, and (4) reduce risk.

Risks will change with time and as they do they will change the frequency and severity of other risks. Once again, we see that the risk process must be iterative.

Preemptive system design can reduce risk. Preemptive design includes these practices: risk management, rapid prototyping, failure modes and effects analysis, sensitivity analysis, identification of cost drivers, reliability analysis, anticipating problems, quality engineering, variation reduction, complexity analysis, six sigma, poka-yoke (translate as fool proofing), design for manufacturability, design for test, design for assembly, and design for cost [23]. The Pinewood Derby design used the first seven of these practices.

The project risk list should not be synonymous with requirements. (1) Requirements have attributes such as priority, stability, and difficulty. However, the difficulty of satisfying a requirement should not be confounded with project risk.

For example, one of the original requirements for the Pinewood Derby was that in a 12-car, 6-heat, round-robin race every car was supposed to race every other car exactly once [15]. This requirement was impossible to satisfy and it had to be relaxed. The performance risk noted that there was a chance that the requirement might not be met. When it was confirmed that the requirement could not be met, it then became a problem, not a risk. Therefore, this problem was handled with requirements management, not with risk management. (2) Sometimes, risks identify unknown effects. If something is spelled out as a requirement, then it is not an unknown. (3) Risks also arise from uncertainty in the environment and external entities. However, a requirement cannot apply to things outside the boundary of its system. (4) Requirements are orders of magnitude more numerous than risks. (5) Risks should be related to the mission, high-level goals or capabilities, and not to numerous low-level requirements.

Figure 6.5 shows a summarizing example of risk management: the hydraulic model for risk of a commercial airplane flight. The total risk is the sum of latent hazards (like poor quality maintenance or poor employee moral) and direct hazards (like lightning, rain, or microbursts). The risk is largest during takeoff and landing. The dispatcher's job is that of balancing risk with resources.

Here is a scenario for risk management in a commercial aircraft situation. You are the dispatcher in Salt Lake City. It is a clear morning. Temperature is in the low 30's °F. A dozen skiers are scheduled to fly to Jackson Hole. But there is a half inch of ice on the Jackson Hole runway. How would you manage this risk? Possible risk management actions are

Transfer: divert to the Idaho Falls airport and bus the skiers over the mountain, thus transferring the risk to Greyhound Bus Lines

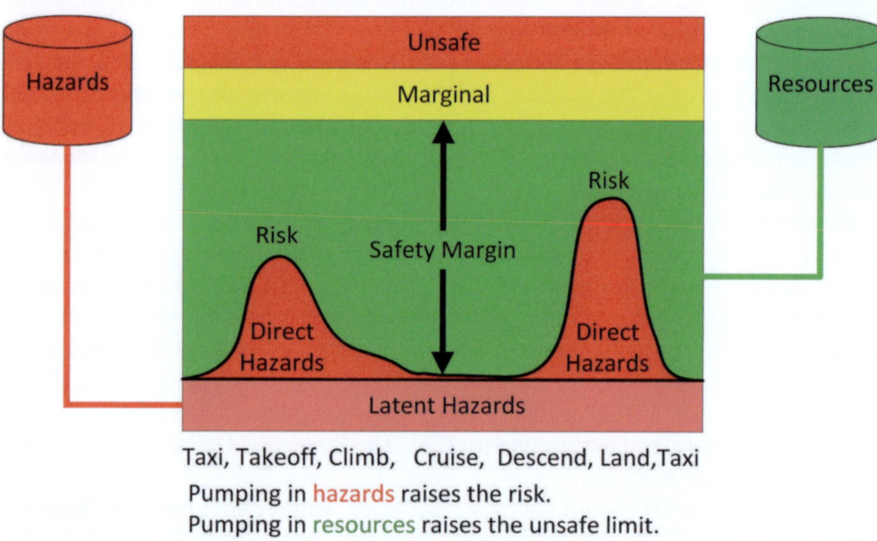

Fig. 6.5 Hydraulic model for risk of a commercial airplane flight

Eliminate: use a helicopter

Accept: send the plane as scheduled

Mitigate

Request removal of ice from runway
Change runways (orthogonal direction)
Change equipment (different type of aircraft)
Change crew (switch the lander from copilot to pilot)

These actions produce the acronym TEAM. How about ignoring the risk? This is not acceptable: there is no I in TEAM.

There is an additional complication: the Icy Runway hazard object interacts with temperature and cross winds. The hazard is amplified by temperatures just above freezing and it is amplified by cross winds. If there are cross winds, the hazard might be mitigated by using a runway colinear with the wind.

The action *eliminate* includes reducing the likelihood of the risk event. However, in this case we cannot do much to reduce the likelihood of occurrence. Mother Nature, and therefore the temperature at Jackson Hole, is not under our control.

This example shows that risk changes with time, that risks are traded off with available resources and common ways of managing risks.

In this section, we treated risk as severity of consequences multiplied by frequency of occurrence. There are many other factors that humans use when assessing risk. In particular, the following are severity amplifiers: lack of control, lack of choice, lack of trust, lack of warning, lack of understanding, manmade, newness, dreadfulness, personalization, ego, recallability, vividness, uncertainty, and immediacy. Examples of these severity amplifiers were given in Section 2.2.7.

There are other project risks that cascade into cost, schedule, or performance risks, such as the political risks resulting from the budgeting process, consequences of constituency pressure, perception problems (especially induced by the Media), and other political facts of life (e.g., the best engineering solution is often not the best political solution). There are illogical and nondeterministic projects risks arising from the political process of budgets and approvals.

The data used in a risk analysis have different degrees of uncertainty: some of the values are known with precision, others are wild guesses. However, in addition to uncertainty, all data have opportunity for errors. For example, [24] calculated that one in a thousand papers published in peer-reviewed journals have been or should have been retracted, because they contained serious errors.

There are three types of errors that can be made in assessing risk: theory, model, and calculation [24]. (1) Using Newtonian mechanics instead of Einstein's relativity theory to describe merging black holes would be an error of using the wrong theory. (2) A modeling error was the direct cause of the failure on the Mars Climate Orbiter [25]. The prime contractor, Lockheed Martin, used English units onboard the spacecraft, whereas JPL used metric units in the ground-based model. (3) Errors in calculating values are common. Several hospital studies have shown that 1–2 % of drug administrations were the wrong dosage. A study of papers from *Nature* and the *British Medical Journal* found that about 11 % of the statistical results were wrong [24]. To validate numerical values in a risk analysis, you should analyze the possibility of errors in choosing the theory, errors in making the model, and errors in estimating and computing numerical values.

In cases where risks can be reduced, the level of effort is driven by a cost/benefit analysis [26]. If we assign expected costs to the risks, then we can see our greatest vulnerabilities. But sheer magnitude is not the whole story. We might also look at the sensitivity of total risk dollars to risk handling dollars. By calculating the cost of implementing a particular strategy versus the reduced expected cost of risk, we can make good decisions. In the Pinewood Derby example of this section, we can see that spending hundreds of dollars on equipment boxes to protect computers that are not very likely to get broken gives a very poor cost reduction to spending ratio. If instead, we spend a couple dollars on duct tape to tape extension cords to the ground to avoid human injury with large potential cost risk, the net savings is great. This also works comparing the cost of insurance against the costs of eliminating a risk. For instance, we may decide that the cost risk reduction of buying insurance does not compare to spending the same money for nets or pads to protect derby cars around elevated areas of track.

History teaches that managers have frequent and common shortcomings when addressing risks. (1) They underestimate the probability of bad things happening. (Prior to the *Challenger* accident, NASA officials put the risk of a shuttle accident at 1 in 100,000: afterwards Richard Feynman computed the odds at 1 in 100.) (2) They underestimate the consequence and chaos caused by the bad things when they do happen. (3) They ignore or fail to recognize that the occurrence of certain risks can trigger the occurrence of other risks. (4) They ignore or fail to recognize that risks may not be independent and may be coupled with other risks. (5) They inadequately investigate unknown unknowns. (6) They suffer from the human imperfections explained by prospect theory in Section 17, such as the fear-related biases like second-guessing, fear of failure, risk aversion, and crisis-mode mentality. Experience teaches that no matter how competent and thorough our technical project team, and no matter how competent and thorough our technical peer review team, Mother Nature still provides a few surprises.

6.2.1 A Risk Template

For a large project, the risk matrix will have to be implemented with a spreadsheet or a database. However, such a wide table does not fit in a textbook. Therefore, in order to produce a risk template, we will start with Table 6.4, project risk matrix for a Pinewood Derby, and turn it on its side. Then we will add more attributes that you might want in your risk matrix (Table 6.7).

After adding more attributes to this matrix, we get the following risk template.

Table 6.7 Transposed project risk matrix for a Pinewood Derby

Id number	1	2	4	5
Failure event	Failure to compute schedules	School refuses facilities	Scouts lose interest	Lawsuits
Consequences or potential effects	Run single elimination tournament	Use someone's backyard	Cancel the Derby	Buy insurance
Likelihood	0.8	0.7	0.1	0.9
Severity	0.4	0.9	1	0.9
Risk	0.32	0.63	0.1	0.81
Assigned to	Bill	Terry	Harry	Terry

Potential failure event that will have a negative impact on the project or the people.

Consequences or potential effects describe what this negative impact might be if the risk is not mitigated or managed: for example, project delay, expenditure to date wasted, reduced performance, or reduced product quality. Once analyzed, risks should be evaluated to determine the likelihood of a risk or threat being realized and the seriousness, or impact, should the risk occur.

Frequency of occurrence (or relative likelihood) of the event occurring can be derived from historical data or it can be estimated based on expert opinion.

Severity of consequences is a measure of negative impact or overall loss of value from a project if the threat emerges. Severity of consequences of a potential failure event can be based on historical data or expert opinion.

Estimated risk is usually defined as the product of relative likelihood and severity of consequences.

Priority: Use numeric values for estimated risk as well as criticality to mission success, customer satisfaction, effect on architecture, business value to prioritize the risks.

Mitigation method: What we are already doing about it.

Countermeasures: Actions to be taken to prevent, reduce, or transfer the risk. This may include production of contingency plans.

Trace to: Most risks can be traced to requirements or brain storming sessions.

Status: Indicates whether this is a current risk or if risk can no longer arise and impact the project. Example classifications are active, inactive, and resolved.

Assigned to: This is the person or team that is responsible for tracking the risk, reporting the risk, and ensuing that proper mitigation of the risk is performed.

Dates: date that the risk was identified, modified, or targeted for resolution.

These then are the attributes for our risk template. They will be the column headings in the risk template spreadsheet: identification number, potential failure event, consequences or potential effects, likelihood of occurrence, severity of consequences, estimated risk, priority, mitigation method, countermeasures, status, assigned to, and dates.

6.2.2 Examples Using the Cookie Acquisition System

Because this is a textbook and not a spreadsheet, the above attributes will be in the rows, not the columns.

Failure event: During test or operation, the audible signal for "cookies are ready" could be too loud.
Potential effects: Someone's hearing could be damaged.
Relative likelihood: Noise induced hearing loss affects 10 % of Americans. Table 1.1 used a range of 0–1 for both likelihood and severity. So we assess this likelihood at 0.01 per year.
Severity of consequences: We compared this risk to all risks that were identified for this system. This was the most serious risk, so we gave it a value of 1.0.
Estimated risk: 0.01.
Priority: high.
Mitigation method: During design, the proposed device is mounted on a test bench and activated in accordance with test instructions. Sound intensity level in decibels (dB) is measured at a distance of 30 cm. The device is put under configuration control to ensure that it is not replaced or altered. This design is conservative in that the device should produce only 80 dB. Exposure to 120 dB or less for only a few seconds is unlikely to cause permanent hearing loss.
Status: active.
Assigned to: Engineer.
Date: Tracking started November 5, 2015.
Failure event: Loss of study time.
Potential effects: This could cause the student to get a low grade on a test.
Relative likelihood: Losing an hour of study time might cause poor performance on an exam with a likelihood of 0.05.
Severity of consequences: 0.5.
Estimated risk: 0.025.
Priority: medium.
Mitigation method: The Engineer is ready to step in and offer advice.
Comment: Although this sounds like a tactical risk, it could become a strategic risk if the student continually comes up with reasons to interrupt his study time.
Status: active.
Assigned to: Hungry Student. The student is an important part of risk management.
Date: Tracking started November 5, 2015.
Failure event: Eating one non-nutritional cookie might instigate eating the whole tray.
Potential effects: This could cause the student to fall off of his diet and start gaining weight.
Relative likelihood: 0.05.
Severity of consequences: 0.1.
Estimated risk: 0.005.
Priority: low.

Mitigation method: The Engineer is ready to step in and offer advice.

Countermeasures: If going to the bakery, the student should take only two dollars. If mom is baking the cookies, she should put out only four cookies and hide the rest. If using Pillsbury dough, use only one-fourth of the package.

Status: active.

Assigned to: Hungry Student.

Date: Tracking started November 5, 2015.

6.2.3 Homework

6.2.3.1 Pinewood Derby Performance Risk

In Table 6.1, why is the severity for Lane Biases equal to zero?

6.2.3.2 Pinewood Derby Risk Charts

Suppose that after the design review of Fig. 6.3 and before the review of Fig. 6.4, Bill shows us that he has computed schedules. Make a copy of Fig. 6.3 and show us how risk 1 would move.

To reduce the risk of lawsuits, we buy insurance. In the case of a pinewood derby, the Boy Scouts of America have an umbrella policy that covers all scouting activities. We have to make sure that we follow the rules for making it an official scooting activity, such as having a roster of attending scouts, having an appropriate number of adults, and training the adults. Assume that we have done these activities correctly. On your copy of Fig. 6.3 show us how risk 5 would move.

6.2.3.3 Other Combining Functions

Create an Excel spreadsheet for the Pinewood performance risk like Table 6.1 and add columns to include the following methods for combining data. Discuss the similarities and differences.

$$\text{Risk} = \text{Frequency of Occurrence} \times \text{Severity of Consequences}$$
$$\text{Risk} = \text{Likelihood of Occurrence} \times \text{Severity of Consequences}$$
$$\text{Risk} = \text{Estimated Probability} \times \text{Severity of Consequences}$$
$$\text{Risk} = (\text{Likelihood} + \text{Impact})/2$$
$$\text{Risk} = \text{Probability} + \text{Severity} - (\text{Probability} \times \text{Severity})$$
$$\text{Risk} = \text{Probability} \times \text{Severity} \times \text{Difficulty of Detection}$$
$$\text{Risk} = \text{Probability} \times \text{Severity}^2$$
$$\text{Risk} = \text{Exposure} \times \text{Severity}$$
$$\text{Risk} = \text{Threat} \times \text{Vulnerability} \times \text{Consequences}$$

6.2.3.4 Linear Versus Log–Log Plots

Examine Fig. 6.2a and state what is the most important risk.
Examine Fig. 6.2b and state what is the most important risk.

6.2.3.5 Schedule Risk Matrix

Suppose that a new computer manufacturer comes on the market and manufactures a computer with superb hardware reliability, which causes us to reassess the frequency of computer hardware failure from 1 to 0.01. Redo Table 6.3. Now what is the most dangerous risk in terms of delaying schedule?

6.2.3.6 Baseball Statistics

What are some possible risks in creating the baseball statistics spreadsheet and in solving the baseball homework problems?

6.2.3.7 Cookie Acquisition System

Create a risk table for the Cookie Acquisition System of Section 3. Most risk tables list many risks. Therefore, there is not enough room to fully explain each risk in the risk tables. So risk analysts usually include only short descriptions (or tags) for the risks in the tables. The full descriptions are given elsewhere. For this homework assignment, we want you to give full explanations for risks. Also, suggest possible mitigation action.

6.2.3.8 Parking Garage System

Create a risk table for the parking garage system of Sections 10–14. Most risk tables list many risks. Therefore, there is not enough room to fully explain each risk in the risk tables. So risk analysts usually include only short descriptions (or tags) for the risks in the tables. The full descriptions are given elsewhere. For this homework assignment, we want you to give full explanations for risks. Also, suggest possible mitigation action.

6.2.3.9 Minimax Design-1

In designing a baseball bat, we know that the distance the ball travels depends on where the ball hits the bat: the ball goes the farthest, if it hits the sweet spot of the bat. Define the *Error* as the absolute value of the distance between the sweet spot and the point of collision. If the collision is not at the sweet spot, then there will be a loss of distance. Define the *Loss* as the number of feet short of 500 where the ball hits the ground. In designing a bat for an amateur use minimax design, that is minimize the *Loss*, if the *Error* is maximum. The ball will not go as far for a perfect hit at the sweet spot, but it will not be a disaster for the more common collision far from the sweet spot. Whereas, in designing a bat for a major leaguer use minimin design, that is minimize the *Loss*, if the *Error* is minimum. It would be a disaster if the collision were far from the sweet spot, but does not matter because he always hits the ball right on the sweet spot. Draw a picture of this design technique and solution.

6.2.3.10 MiniMax Design-2

Your client wants you to help her select a site for a million dollar house. You have narrowed it down to two final alternatives: a million dollar house on a riverbank where a 100-year flood would cause \$900 K damage, and a million dollar house on a mountaintop where a violent thunderstorm would cause \$100 K damage. Assume that such a violent thunderstorm occurs once a decade. Which alternative would you recommend?

6.2.3.11 Risk of Building Below Sea Level

What do you think about the proposal to forbid the construction of houses below sea level in the United States? Would it save money and lives? Who would be in favor and who would be opposed? Which of these has a stronger lobby in Congress? For a particular example, focus your discussion on New Orleans.

6.3 Section 35. Incorporating Solar Power into an Electric Grid

This section explains the risk analysis process. Then it summarizes the main risks associated with incorporating solar photovoltaic (PV) panels into a commercial electric power grid, presents a what-if analysis for extreme scenarios, and explains mitigation strategies to ameliorate these risks. Finally, the section points out some possible unintended consequences of incorporating solar PV systems into a commercial electric power grid. This section is based on Chaves and Bahill [27, 28].

6.3.1 Introduction

Economically viable harvesting of renewable energy is one of the most profound challenges of the twenty-first century. The most promising renewable energy source in the southwest United States is photovoltaic (PV). However, incorporating PV subsystems into an existing electric power grid presents a significant challenge, because of the intermittent and diurnal characteristics of the environment. This, and the uncertainty of dealing with the unknown, means that evolving such a big complex system is risky. Therefore, a risk analysis is a crucial part of system design.

Tucson Electric Power (TEP), the electricity service provider for the Tucson metropolitan area, has been operating a 4.6-megawatt (MW) solar PV panel array at their Springerville Solar Generating Station for the last dozen years. However, this facility provides only 0.2 % of their total generating capacity, and they need to increase their renewable-energy generating capacity significantly in order to comply with the Arizona Corporation Commission's Renewable Energy Standard. This Standard requires that by the year 2025, 15 % of the utility company's retail sales must be supplied from renewable-energy sources.

The risk analysis of this section was conducted under the assumption that most of this Renewable Energy Standard will be satisfied with solar photovoltaic energy. It identifies risks and complications associated with incorporating large-scale photovoltaic-solar systems from the utility company's perspective.

6.3.2 Definition of Risk

The world is full of uncertainty and this makes risk an inherent component in the design of any system. Risk is an expression of the potential harm or loss associated with an activity executed in an uncertain environment.

We quantify risk as the product of the relative likelihood (or frequency of occurrence) of a failure event and the severity of consequences for each occurrence of that failure event, as follows.

$$\text{risk} = \text{relative likelihood} \times \text{severity of consequences}$$

This process uses the *product combining function* of Section 3.3.1.2.

Defining risk as the product of relative likelihood (or frequency of occurrence) and the severity of consequences is the definition usually used in American industry. It is the only approach that we have seen government contractors use. This may be the result of US government documents. For example, the U.S. Food and Drug Administration [www.fda.gov/downloads/Drugs/GuidanceComplianceRegulatoryInformation/Guidances/ucm073511.pdf] wrote, "It is commonly understood that *risk* is defined as the combination of the probability of occurrence of *harm* and the *severity* of that harm." The Department of Defense [5] stated that, "Risks have three components: A future root cause (yet to happen), which, if eliminated or corrected, would prevent a potential consequence from occurring, A probability (or likelihood) assessed at the present time of that future root cause occurring, and The consequence (or effect) of that future occurrence." The Defense Systems Management College [29], which is based on DoDI 5000.2 and DoD 5000.2-R, states, "*Risk* is a measure of the potential inability to achieve overall program objectives within defined cost, schedule, and technical constraints and has two components: (1) the *probability/likelihood* of failing to achieve a particular out-come, and (2) the *consequences/impacts* of failing to achieve that outcome." The Institute of Risk Management [30] states, "Risk can be defined as the combination of the probability of an event and its consequences (ISO/IEC Guide 73)." By design, the CMMI [31] is broad: it states what must be done, never how it should be done. Here it states what is needed to quantify risk: "Parameters for evaluating, categorizing, and prioritizing risks include the following: Risk likelihood (i.e., probability of risk occurrence) Risk consequence (i.e., impact and severity of risk occurrence) Thresholds to trigger management activities." The INCOSE Systems Engineering Handbook [32] states, "Level of risk depends upon both likelihood and consequences . . . risk is expressed as: Risk = Probability of failure (Pf) × Consequence of failure (Cf)." There are differences between these definitions, but for the most part they all state that risk should be quantified as the product of relative likelihood (or frequency of occurrence) of a potential failure event and the severity of the consequences of that event.

Our process for quantifying risk uses the *product* of relative likelihood (or frequency of occurrence) and the severity of consequences [33]. This *product combining function* of likelihood and severity makes intuitive sense. People are familiar with multiplying data; for example, multiplication is used in computing a benefit-to-cost ratio. (A ratio is just multiplication by the reciprocal.) The product combining function is used in many different realms, for example, a person buying a lottery ticket should care about the size of the pot *divided* by the number of people buying tickets; insurance rates on a Corvette are higher than for a typical automobile, because the frequency of accidents is higher and it is an expensive car, so the

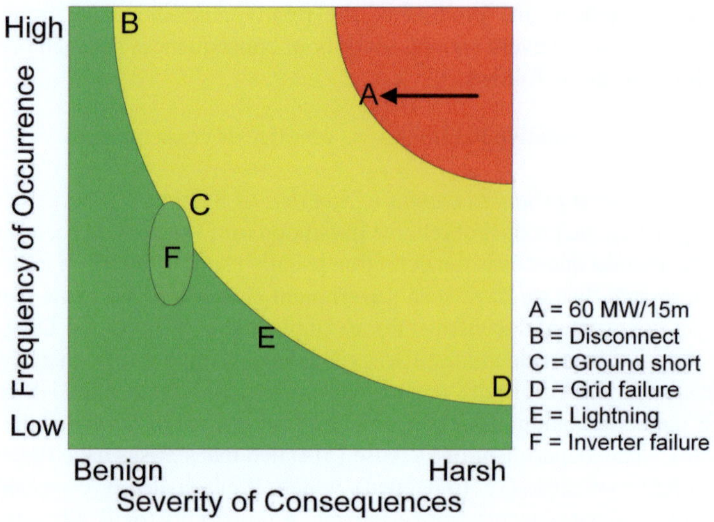

Fig. 6.6 A linear risk chart for some failure events explained in Table 6.8

monetary loss in an accident is higher: it seems intuitive to *multiply* the frequency times the monetary loss.

Of course, a risk analyst would never give a decision maker a single number and say, "This is the most important risk." The risks must be prioritized (Section 24) and discussed with the decision makers. Risk management progress must be understood. Figure 6.6 presents a risk plot that can facilitate these discussions. It is similar to the DoD Risk Reporting Matrix [34, 35].

The arrow in Fig. 6.6 shows that the biggest risk, risk A, has dropped in severity since the last review, due to risk mitigation action. Uncertainty in the likelihood and severity numbers can be shown with ellipses, as is illustrated with risk F. Risks in the red region are high risk and must be managed. Risks in the yellow region are medium risk and should be managed if it fits within the budget. Risks in the green region are low risk and need only to be monitored. The curves are iso-risk contours.

Each row in a risk table (like Table 6.8) describes particular risk. It contains a *potential failure event,* the *consequences* of that failure event, the *frequency of occurrence* (or *relative likelihood*) of the event, the *severity of consequences*, the *estimated risk*, and perhaps a short *identification tag*. The *frequency of occurrence* (or *relative likelihood*) of the event occurring can be derived from historical data or it can be estimated based on expert opinion. Likewise, the severity of consequences of a potential failure event can be based on historical data or expert opinion. The estimated risk is usually defined as the product of relative likelihood and severity of consequences.

However, the terms relative likelihood and frequency of occurrence are not quite synonyms. If there are historical data for an event, then we use the term frequency of occurrence. Otherwise, if we are guessing the future, then we use the term relative likelihood. The word *relative* emphasizes that it is the relationships

Table 6.8 Selected risks for incorporating solar PV subsystems into a commercial electric grid

Potential failure event	Consequences	Frequency of occurrence in the TEP control area (events per year)	Severity of consequences	Estimated risk, defined as frequency times severity	Identification tag
Solar panel output drops by 60 MW in a 15-min interval	Breakers could trip leaving customers without electric power. Voltage on the grid could drop and frequency of coal-fired generators could change	95	200	19,000	A
Feeder circuit disconnects from substation	Feeder circuit voltage could get out of phase with the grid	330	1	330	B
Short to ground on distribution grid	Equipment could be damaged, particularly transformers and capacitor banks	24	10	240	C
Western Electric Power Grid fails	The western United States would be without electric power	0.03	1000	300	D
Lightning strikes the system	Components could be damaged and electric generating capacity would be reduced	0.4	100	40	E
Failure of DC to AC inverters	The customer can no longer sell electricity to TEP	240	0.1	24	F

This is an abridgment of Table 6.11. These data come from TEP managers, TEP documents, academics, and project managers of renewable energy projects

between risks that are being illustrated. The word *likelihood* is not used in a probabilistic sense, but rather in its dictionary sense to indicate the events that are likely to happen. Because humans are poor at assessing probabilities and estimated probabilities interact with stated severities (Section 17), we use frequency of occurrence instead of estimated probability, as discussed in Section 25.

The two types of uncertainty are sometimes called aleatory and epistemic. Aleatory uncertainty describes inherent variability or randomness in the physical system; it would be reported as frequency of occurrence. Epistemic uncertainty is due to lack of knowledge or data about how the system behaves or interacts with its environment. Epistemic uncertainty would be estimated by domain experts and reported as relative likelihood.

The events in Table 6.8 have uncertainty in both the relative likelihood of occurrence and the severity of consequences. However, sometimes we know *when* an event will occur, so the likelihood that the event will occur is 1.0 and we only need to estimate the severity of the consequence. For example, assume that you have bet on tails in a coin-flipping event and you are about to flip the coin. The likelihood that the event will occur is 1.0 and the severity of the consequence is that you will lose half the time. Therefore, the risk of losing your bet in the next moment is 0.5. Most gambling games are of this nature.

On the other hand, sometimes there is no uncertainty in the consequence, only uncertainty that the event will occur in the specified time interval. For example, assume that you are performing an experiment with radium and an alpha particle would ruin your experiment. In radioactive decay of radium-226 into radon-222, we can estimate the likelihood of the event as 3.7×10^{10} decays/s-g. When the event occurs, we know with absolute certainty that the consequence will be an alpha particle; therefore, the severity of this consequence is 1.0. Therefore the risk of getting an alpha particle is 3.7×10^{10} /s-g. If there is 1 g of radium (about 10^{22} atoms) and the experiment lasts 1 s, then the risk is 3.7×10^{10}.

We can compare these three types of events and consequences {(1) uncertainty in both the event and the consequences, (2) uncertainty in only the consequences, and (3) uncertainty in only the occurrence of the event} as long as the time interval is the same. In a fourth case, the likelihood of the event occurring cannot be estimated, for example, when dealing with terrorist attacks on population centers.

In a risk table, a failure event might have multiple consequences, as in Table 6.9. In this case, each consequence will have an expected frequency of occurrence, which will be the product of the frequency of the failure event times the frequency of each consequence.

Table 6.9 A failure event with multiple consequences

Failure event	Consequences	Frequency of occurrence (events per year)
Lightning strikes, the number of lightning strikes per year in Arizona = 10^6	Electrical devices are destroyed, frequency per lightning strike = 10^{-3}	1000
	A house burns down, frequency per strike = 10^{-4}	100
	A person is injured, frequency per strike=3×10^{-6}	3
	A person dies, frequency per strike = 10^{-6}	1

These annual data are for the state of Arizona

A particular consequence might cause a different amount of damage depending on the specific event that caused it and the environment in which it occurred. If data are available for such events, then the severity of the occurrence and the frequency of occurrence can be described with a probability distribution instead of a single value.

It is difficult to get good frequency of occurrence data for physical and cyber infrastructure. For example, terrorism is an asymmetric nonzero sum game with spotty and weak intelligence. The frequency of occurrence of an initiating event (e.g., act of terrorism or a natural disaster) is fraught with epistemic and aleatory uncertainties. On the other hand, the probability distribution of the severity of consequences is easier to assess by experts, using modeling and simulation tools. Indeed, Haimes' approach [36] develops scenarios of future failures, then simulations and expert evidence are used to assess the severity of consequences as functions of different failures.

6.3.3 The Risk Control Process

The following paragraphs explain the activities of the risk control process shown in Fig. 6.7.

Identify risk events: In consultation with the customer, think about everything that could possibly go wrong. You want a lot of unlikely events, everything is fair game. You must identify risks for which curmudgeons would say: "That could never happen."

Analyze risk events: Describe the risks with words and quantitative models. Identify root causes when possible. Look for interactions between risks.

Describe evaluation criteria, namely likelihood of occurrence and severity of consequences. For example, likelihood of occurrence might be described as the number of events expected per year, and severity of consequences might be described as the monetary value of the loss. Other criteria that may be of interest include stage of resolution, estimated effort needed to resolve, and difficulty of detection. These criteria may use scoring functions.

Derive values for the likelihood of occurrence and severity of consequences.

State the combining function to be used. In most cases, this will be multiplication of likelihood by severity. Establish the ranges of these descriptors, which are the weights of importance.

Compute the numerical values for the risks, usually by multiplying likelihood times severity.

Prioritize the risks. Use the risk numeric values as well as criticality to mission success, customer satisfaction, effect on architecture, business value, etc., as listed in Section 24, to prioritize the risks.

Manage risks: Decide how to handle the risks. These are the most common actions. Transfer the risk to a different organization. Eliminating the risk is the utopian

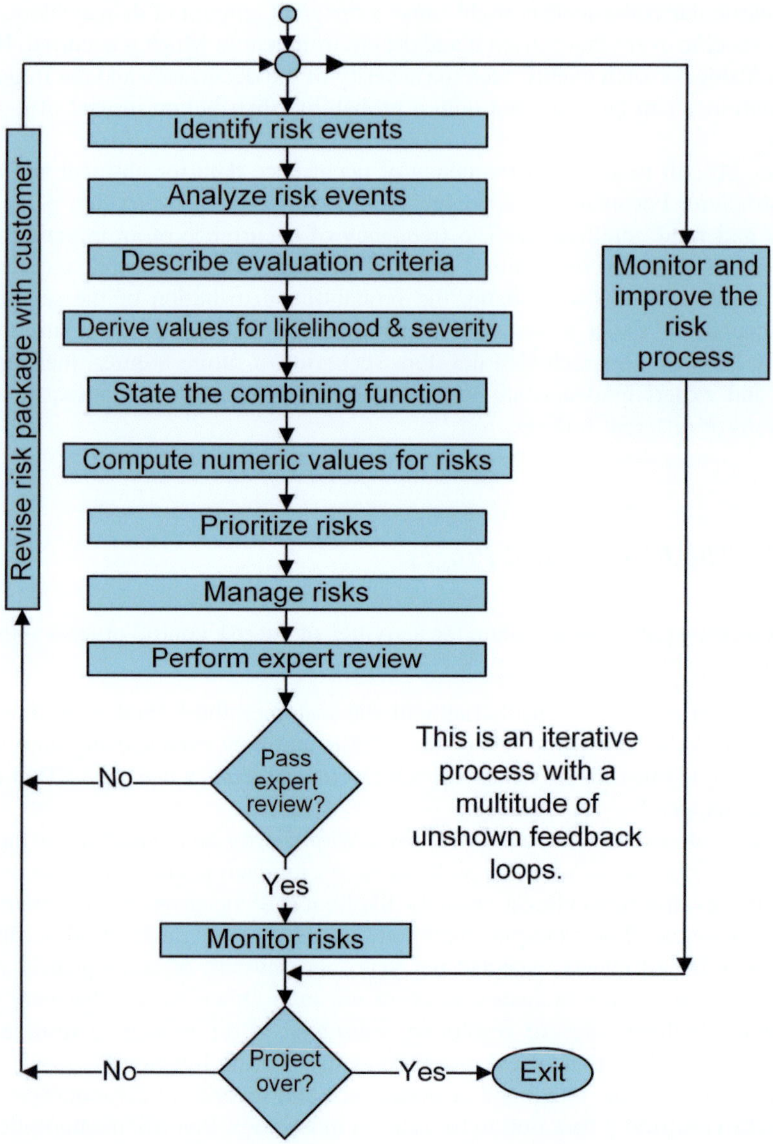

Fig. 6.7 The risk control process

action. Accepting the risk is seldom done. Usually, we mitigate the risk. These actions produce the acronym TEAM. How about ignoring the risk? This is not acceptable: there is no I in TEAM.

Perform expert review or formal inspection of the risk control package.

Monitor risks: The most important risks are put in the project risk register. They get monthly high-level review. All risks are monitored for changes in likelihood, consequences, status, impact, exposure, and other measures that could alter their priorities. Experiments and tests will expose new risks. Technical performance measures (Section 23) may be used to prove that critical risks are under control.

Monitor and improve the company's risk process: The standard company risk control process must be tailored for each project. Tailoring accommodates special schedule and cost constraints of the project and incorporates customer specific needs. The impact of all customization should be studied to determine whether or not they result in improvements to the company's standard risk process. This task of continually monitoring a process, to make beneficial changes to improve the process, is arguably the most important task of any process. However, this task is often forgotten, once the process is designed. This is why the CMMI [31] devotes three generic processes to it. It requires the company to:

Monitor and control the risk process against its plan.

Objectively evaluate adherence of the risk process against its process description, standards and procedures, and address non-compliance.

Review the activities, status, and results of the risk process with higher level management and resolve issues.

The risk control process should be executed by in-house people, not external consultants, because your company's people should understand the product and process better than outside consultants. Your people should feel that they are a part of the risk control process. New data collection processes need not be created. Most of the data you need should come from existing company data collections.

6.3.4 System Characterization

The system hierarchy of Fig. 6.8 is used to help identify system risks.

An electric power grid with solar PV subsystems is defined as follows: "An electric power grid with solar PV subsystems consists of photovoltaic (PV) solar panel arrays, DC to AC inventers, and the hardware that connects them to the electric power grid. It includes both small grid-connected solar systems and utility-scale projects. These systems may be located on residential or commercial property, on rooftops, or in open-land. Net-metering allows customers with grid-connected electric generating systems to buy electricity from the utility company when they need more electricity than they are generating and to sell electricity to the utility company (at a pre-determined price) when they generate more electricity than they need. The utility company uses this solar generated electricity to meet part of their electric demand, and it must be capable of meeting electric demand during the night and during days with reduced solar power output. These systems shall comply with all local and federal laws."

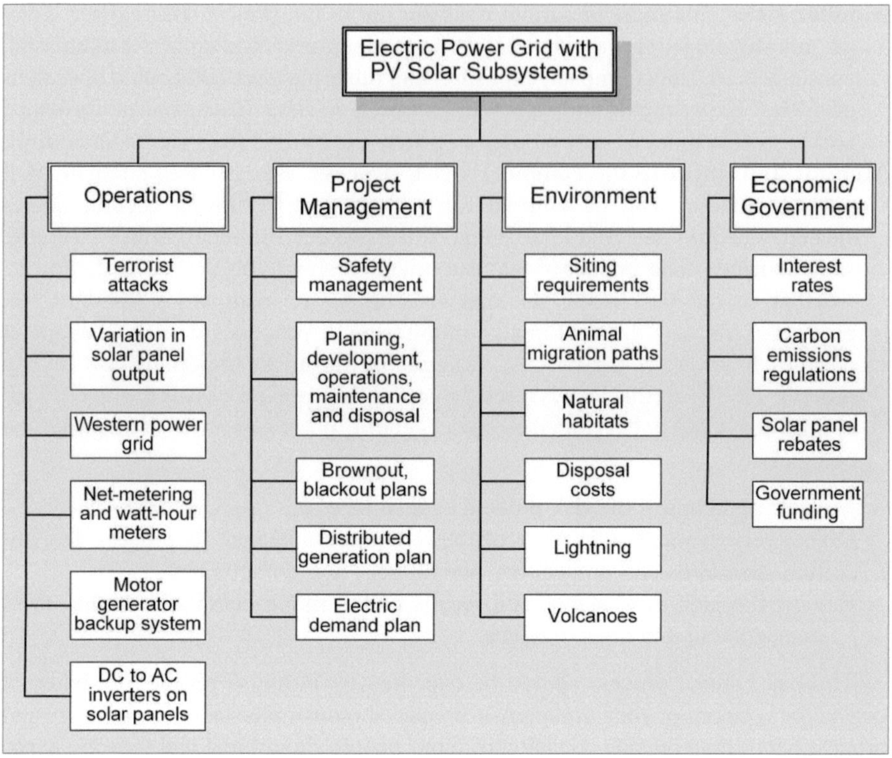

Fig. 6.8 System hierarchy

In this section, we show the risk analysis of the top-level system, the electric power grid with solar PV subsystems. The risk analysis for operating performance is shown in Table 6.11. The risk analysis for the environment is shown in Table 6.14. The risk analysis for project management is shown in Table 6.15. The risk analysis for economic/government is shown in Table 6.16. As a further decomposition example, the risk analysis for a motor-generator backup system is shown in Tables 6.17 and 6.18.

6.3.4.1 Risk Identification and Quantification

For risk identification, first, we obtain significant input from system experts and outsiders to help identify potential failure events and their consequences. We use brainstorming and risk tables. The term risk usually describes risks to the system being designed or its primary users, not to unintended consequences in systems outside the scope of the system being designed. Risk identification is an iterative and hierarchical process. Once a risk table or matrix is obtained, the resulting risks must be discussed with professionals, academics, and other system experts that will help verify, quantify, add, and eliminate risks. The risk tables summarized in this section required multiple iterations. As time goes by and as risk management strategies are implemented, risks and risk severities will have to be revised in order to be a true representation of the existing system.

Next, the effects of these failures should be explained. These failures could affect cost, schedule, performance, operations, the environment, safety, etc. Then the likelihood of each risk occurring should be estimated. If the project has plentiful statistical data, then the frequency of each event might be calculated. But typically such data are not available, so the likelihood of occurrence in some given time interval is estimated.

Our process estimates the frequency of occurrence of each failure event based on observations, statistical analyses of historical events, and expert opinion. Our process quantifies the risk as the product of frequency of occurrence and severity of consequences and emphasizes the importance of normalizing the values of both frequency and severity on the same scale. This normalization guarantees that both frequency and severity are given the same weight when calculating the final risk. This process also recommends the use of log–log plots so that rare events can be tracked without distorting the risk analysis.

Therefore, to avoid skewing the statistics with extreme events, events with low frequency but high severity (such as category 5 hurricanes, volcanic eruptions, acts of war, and terrorist attacks) were removed from the numerical computations and have been marked in the risk tables with a "$0 \times \infty$" symbol. These rare but potentially catastrophic events would have been in the lower-right corners of Figs. 6.2a, 6.3, and 6.6. The upper-left corners of these figures could have contained events with high frequency but low severity (such as birds and airplanes casting shadows on the solar panels, solar corona mass ejections, and solar PV customers connecting to and disconnecting from the electric power grid), but they were removed from the statistics.

6.3.5 *Risk Severities*

The severity of consequences of a risk event is the perceived damage due to its occurrence. Determining severities is an important step, because it allows us to calculate the risks and rank them in order to identify the most critical elements. Severity values can be derived using brainstorming, group decision techniques, modeling, and simulation. However, severity values are subjective and depend on the perception of the analyst. Fortunately, it is possible to reduce analyst-induced bias by sharing the resulting severity values with system experts and other analysts so that they can validate the severity values.

Analyzing risk severities is a common practice. Insurance companies have developed tables to quantify risk so that different risks can be compared. They assess policyholders' risk in order to estimate the total risk of their insured pool and derive the expected payout costs. Understanding risk severities allows them to quantify the risk and act accordingly. By estimating expected payout costs, insurance companies are able to set the price of insurance premiums so that they almost always generate a profit. (It is only the rare but catastrophic events that bankrupt them.) Utility companies also routinely analyze risk severities. If the utility company understands the severity of consequences, they will be able to prioritize risk mitigation strategies.

With Table 3.7, we showed that if two items are being multiplied and they have different ranges, the one with the bigger range has more weight. Risk is being defined as the frequency of occurrence *multiplied by* the severity of the consequences. Therefore, the range of frequencies must be the same as the range of the consequences.

6.3.6 An Algorithm for Computing Severity

The following is a brief restatement of Section 6.1.2.2 for computing values for the severity of consequences.

1. Assign a relative likelihood of occurrence to each potential failure event.
2. Find the failure event that might have the most severe consequences, call its value S_{worst}.
3. For each other failure event, ask, "How many of these failures would be equally painful to the Worst?" Call this N_i. This can be rephrased as, "Cumulatively, how many of these failures would have an equal impact to the Worst?"
4. Compute the severity of consequences for each failure event as $S_i = S_{\text{worst}}/N_i$.
5. Remove the low-frequency and high-severity failure events and the high-frequency low-severity failure events.
6. Normalize the severity values so that their range equals the range of the likelihood values.
7. Compute the estimated risk using a combining function.
8. Prioritize the risks to show which are the most important.

Decision makers should be interested in rare but serious events. Over the last decade, we have witnessed a series of such events. In April 2010, a British Petroleum oil well in the Gulf of Mexico exploded and leaked five million barrels of oil. The terrorist attacks on the World Trade Towers on September 11, 2001, caused severe physical and emotional damage. Hurricane Katrina of August 2005 was the costliest natural disaster in the history of the United States, because so many properties were built below sea level: total property damage was $81 billion. These three events are mentioned because they indicate that the probability density function for the severity of consequences is not Gaussian. The right tail (and perhaps the left tail containing Microsoft, the Internet, Google, and Social Networking) has far more occurrences than a Gaussian distribution would have. Insurance companies and politicians have a hard time dealing with such rare but serious events.

6.3.7 Section Summary

Risk is a multidimensional function, representing the affected multidimensional states of a system and thus it cannot be measured with a single number. Therefore, many additional attributes, such as difficulty of detection and vulnerability, have been included in some risk definitions. If a problem is difficult to detect (in testing, verification, etc.), then we should worry more about it and increase its contribution to risk. If we know that we are vulnerable in a certain area, then we should worry more about that area and increase its contribution to risk. When quantifying risk, the following attributes have also been used: frequency, severity, safety, efficiency, reliability, vulnerability, and resilience.

After completing a risk analysis, you should look at (1) high-risk events, (2) high-severity events (no matter how unlikely), and (3) estimates that have a large uncertainty. These should be discussed with the decision makers. In the next iteration, you should focus resources on these items.

6.3.8 Homework Problems

6.3.8.1 TEP Risk Charts

Suppose that the electric company installed lightning arrestors on all power poles. Make a copy of Fig. 6.6 and show us how risk E would move.

Because the failure of DC to AC inverters is so common, we want to do something to reduce its risk. On your copy of Fig. 6.6, show us how you could move risk F.

6.3.8.2 Power Blackouts

On the Internet, lookup the four major power outages in the United States since 1965. For each list the date, the geographical area affected, the number of people affected, the time until power was partially restored, and the time until power was completely restored. If you feel comfortable, guess at the cause of the failure. See also Fairley, P. 2016. Blackout of the century, *Discover*, **37**(2) pp. 44–51.

6.3.8.3 Generic Tactical Missile

Consider the missile described in Section 7. Estimate the likelihood of failure and the severity of consequences. Compute the risks and comment on your results.

Performance risks

		Numbers are from 1 to 10. 1 is the least likely and the least severe		
Failure mode	Potential effects	Likelihood of failure rank, 10 is most likely	Severity rank, 10 is most severe	Risk (Highest number is greatest risk)
Incorrect target identification	Destroy wrong target	–	–	–
Premature detonation	Personnel injury and missed target	–	–	–
Mislaunch	Missile stays on launcher	–	–	–
Missile seeks wrong target	Destroy wrong target	–	–	–
Missile misses target	Missile must be destroyed	–	–	–
Premature release	Missile falls into sea	–	–	–

Technology risks

Failure mode	Potential effects	Likelihood of failure	Severity	Risk
Cryogenics	Cannot cool detector	–	–	–
Inattention to human factors	Confusing human interface	–	–	–

Schedule risks

Failure mode	Potential effects	Likelihood of failure	Severity	Risk
Miss scheduled demonstration	Cancel contract	–	–	–
Supplier fails to deliver	Schedule slippage	–	–	–

Cost risks

Failure mode	Potential effects	Likelihood of failure	Severity	Risk
Budget overrun	Cancel contract	–	–	–
Supplier goes bankrupt	Find a second source	–	–	–

6.3.8.4 Batteries for a lightweight interceptor missile

Risk management case study

Ten flight batteries are needed for our current missile program. The cost ceiling is \$200,000. The battery average unit production cost ceiling is \$8000. The program Critical Design Review (CDR) is 9 months from now. The Program Office wants the prototype battery test data to support all requirements at the CDR. That means that we must have the prototype battery early enough so that we can run tests on it. No off-the-shelf battery can be identified that meets power and weight requirements. The leading vendor for lightweight batteries has estimated \$150,000 to develop and deliver the CDR prototype battery and ten flight batteries that meet all requirements. The prototype battery and test data would be available in 8 months and the remaining flight batteries would be delivered 2 months prior to their need dates. The vendor expects the cost per production unit to be approximately \$5000 per unit. The battery would be developed using existing proven components and existing technology that has been previously tested. Proven manufacturing processes would be employed. However, this vendor has delivered developmental products as much as 3 months late and has raised average unit production cost and development costs as much as 50 % several months after beginning a project. This vendor has failed to meet cost and/or schedule on three out of four of our company's programs.

Instructions: Identify the risks that are involved. Compute the risk factor as the likelihood times the severity. For severity values, use Table 6.6. However, just show some alternative ways to derive these values, use the following table to get likelihood values.

Qualitative scale for frequency of failure (or likelihood of occurrence)

Rating	Software	Testing	Technology	Hardware	Producibility
9	Theoretical SW concepts beyond known practice. Development of new approach	Theoretical technology, design or concepts beyond known testing processes	Theoretical technology; beyond known capability	Theoretical design, based on advanced research. Requires technology breakthrough	No known capability or technology to produce product
8	New complex SW, new approach, new language. New unproven apps. Extremely large-scale integration	Theoretical technology, design or concepts requiring significant research in test processes	Theoretical technology requiring significant research	Theoretical design. Application of leading edge concepts. Significant research required	Theoretical manufacturing processes, researched but requires significant R&D to develop processes
7	All new SW; beyond experience base. Large integration of new or existing SW	Technology approach has identified conceptual test processes, but not verified or validated	Newer technology; feasible by analogy; Concept untested or not verified	All new complex design with many stringent reqs and/or major integration of	Conceptual manufacturing processes identified. Significant investment to

(continued)

Qualitative scale for frequency of failure (or likelihood of occurrence)					
Rating	Software	Testing	Technology	Hardware	Producibility
				many new HW elements	develop capability
6	Extensive changes in SW. Moderate integration of new or existing SW	Technology approach has limited concept testing. Limited or no in-house experience	Newer technology; feasible by analogy, studies of concept verification	New design and/or major integration of many HW elements	New manufacturing process to industry. Major investment to establish capability and develop experience
5	Major modification of SW or conversion of similar SW. Some integration experience	New technology or similar designs have initial testing. Limited or no in-house experience	New and/or adapted technology with feasibility studies and initial testing	Moderate design change; significant modifications. and/or moderate integration of HW elements	Partially new or modified manufacturing processes to industry and no in-house experience
4	Moderate modification and tailoring of existing SW	Similar designs and technology have been tested by other companies. Limited in-house experience	Proven technology and approach. Concept analysis and testing complete	Redesign, moderate modifications. Integration is primarily of internal functional elements	Proven manufacturing processes but no in-house experience
3	Slightly modified SW, combining existing functions with minor integration	Limited testing done on new or existing components but experienced in test methods	Proven technology and approach; previously validated	Existing proven components; recombined or minor mods in function	Proven manufacturing processes but newly established capability
2	Some modification of existing SW with minimal integration impacts	Components have been tested and validated in similar applications	Proven technology and approach. Used occasionally by a design agent	Existing proven components, repackaged and/or minor usage variation	Proven manufacturing processes used at least twice by design agent
1	Minor revision of existing SW. No impacts to integration tests	System has been tested and validated in similar applications	Proven technology and approach with significant design experience	Functional HW. Mods in form only. Minor usage variation	Proven manufacturing processes used occasionally by design agent

(continued)

Qualitative scale for frequency of failure (or likelihood of occurrence)					
Rating	Software	Testing	Technology	Hardware	Producibility
0	Use of existing SW. No additional test, validation, or integration required	System has been thoroughly tested and validated for current application	Off-the-shelf HW proven to operational environments	Functional HW. Hardware will meet the form, fit, and functional reqs of the application	Off-the-shelf manufacturing processes, which have been used often

6.4 Section 36. Risk Analysis of a Power Grid with PV Solar Panels

This section is based on Chaves and Bahill [27, 28, 37]. There are two categories of risk for incorporating solar photovoltaic subsystems into a commercial electric power grid: risks related to uncontrollable factors such as weather and risks related to software, hardware, and human error. Although many papers on risk do not consider uncontrollable factors or acts of God, because they cannot be predicted, we deem them important given that weather risk is one of the greatest sources of uncertainty for solar power production.

Risks were initially analyzed in different tiers or levels [38]: (1) risks related to the utility company and the power grid (2) operations, project management, environment, and economic/government, and (3) components of these. The risk tiers correspond to the perspectives of the system hierarchy depicted in Fig. 6.8 and Table 6.10.

The risk tiers were analyzed from various stakeholder perspectives. For example, the first tier, "Electric Power Grid with solar PV Subsystems" is clearly a risk to the utility company; however, it can also be a risk to the customer, because

Table 6.10 Top-level risk tiers

Risk tier	Description
Electric power grid with solar PV subsystems	Risks related to not meeting demand, brownouts, blackouts, customer dissatisfaction, etc.
Operations	Risks related to the hardware, software, and bioware of the components of the system during operation: reliability
Project management	Risks that may be encountered throughout the development of the PV project: changes in costs, design issues, permit issues, etc.
Environment	Risks related to the location and surrounding environment of the project: effect on local habitats, weather, environmentalist's opposition, etc.
Economic/government	Risks related to changes in governmental policies or economic conditions

brownouts or blackouts can affect their daily activities and may damage their property. Each tier will have risks that affect various stakeholders (utility company, customers, environment, regulators, etc.); however, the risk tables in this section summarize the risks *from the perspective of the utility company*.

Our project started with a search for risks of using renewable energy resources in an electric power grid [27, 28]. Then to help expand and solidify the risk descriptions, we interviewed TEP managers and directors, academics, and project managers of renewable energy projects. The information provided by them was summarized and analyzed to determine the possible risks. After identifying the risks, risk frequencies were calculated or estimated based on the available information. Finally, the risks were prioritized and discussed with the decision makers.

6.4.1 Description of Identified Risks

We will now describe the most important risks, which are listed in Tables 6.11, 6.12, 6.13, 6.14, 6.15, 6.16, 6.17, 6.18, and 6.19. Our preliminary risk analysis indicated that the greatest risk for an electric power grid with solar PV subsystems was weather causing the solar panels to receive less sunlight than expected. This is a critical factor for a self-sustaining PV system, but it is less important for a large-scale system comprising both renewable (solar) and nonrenewable resources. This risk can be mitigated by using energy storage systems or increasing backup generating capacity. In consequent iterations, this risk was modified in order to encompass output variability: large changes in solar power output (±60 MW) that would correspond to a solar power output variation of ±3 sigma in a 15-min period. This change in power output could introduce transients onto the grid and could produce load shedding.

After our original study was completed, we discovered that areas of Texas and Southern California had the opposite problem: too much solar power. This caused the price of electricity to become negative for hours at a time.

Grid related risks are another risk category. These risks include the grid frequency going out of the ±0.5 Hz limit, feeder circuits disconnecting and shorts to ground. The first two risks are expected to increase as solar PV generation increases, because the solar subsystems may introduce transients or voltages that are out of phase with the grid. The frequency of occurrence of these failures was obtained from TEP.

Hardware risks include failures due to component malfunction or external events such as lightning or dust. The frequency of failures of PV system hardware such as data acquisition systems, junction boxes, PV modules, and general failures due to lightning strikes was based on a report of TEP's experience with the Springerville Generating Station [39]. Severity for hardware failures ranged from a simple system restart to more complex maintenance requirements [39]. In addition to the hardware failures reported by TEP, we included storage system failures, because storage technologies are being considered by TEP and may be implemented in the future. A storage system failure may result in a loss of stored energy and will eliminate the possibility of using this stored energy to meet electric demand. Failure of a backup generator will affect the capability of supplying enough power during peak demand or low power production hours. The severity values and frequencies were estimated based on hardware-specific reliability rates (assuming an expected lifetime of 30 years).

Environmental risks of Table 6.14 include immediate risks to the environment such as habitat destruction, as well as deferred risks (such as the disposal after the system's design life or after irreparable failure). Large-scale solar farms could harm local habitats and modify animal migration routes. Disposing of the solar subsystems at the end of their design life is low risk, because PV panels (as well as the rest of the system hardware) do not contain dangerous or extraneous materials that would complicate system disposal. However, this risk could increase if stronger recycling regulations were enacted. The other environmental risk is unknown hazards and is related to the possibility of discovering that the system contains elements that may become suspects for causing cancer or illnesses, or that the system could produce other unknowns.

Table 6.11 shows that accidents and human mistakes are the risks with the highest severities, given that they can harm people; however, based on TEP's historical record, the occurrence of such events is extremely low and thus their frequencies are almost negligible. Other extreme events such as terrorist attacks on the Western Power Grid and volcanic eruptions were also considered; however, as can be seen in Table 6.11, the estimated risks for these extreme events were filled with our null symbol, $0 \times \infty$, which means that they were excluded from our *numerical* calculations. This is expected to reduce the skewing of numerical calculations that would result by including these rare but serious events, Haimes [36].

Economic risks of Table 6.16 include a change in interest rates. Changes in interest rates were deemed low-severity risks since TEP engages in interest rate swaps, hedging their interest rate exposure and minimizing the impact from future interest rate changes.

Finally, government risks in Table 6.16 include changes in government funding and regulations, such as carbon emission policies and carbon taxes that would have a direct or an indirect impact on the viability and size of PV systems. Government policy changes concerning impacts on wildlife habitat, lands, and water [http://solareis.anl.gov/documents/fpeis/index.cfm] could make TEP's renewable energy portfolio plan obsolete, and could require total replanning of strategies. The early elimination of rebates is another government risk: It would affect customer incentives to convert to solar-powered generation [40].

Tables 6.11, 6.12, 6.13, 6.14, 6.15, and 6.16 contain risk analyses with both PV system-specific risks as well as risks associated with Tucson Electric Power's AC electric power distribution grid. The data for the distribution grid risks were given to us by Tom Hansen, vice president of TEP in 2008, and Bahill derived the rest of the distribution grid data by normalizing the frequency of occurrence and calculating the range: about six orders of magnitude. Since the range for frequency and

Table 6.11 Operating performance risks for incorporating solar PV subsystems into a commercial electric power grid

Potential failure event	Consequences	Frequency of occurrence in the TEP control area (events per year)	Severity of consequences	Risk	Identification tag
Physical or cyber terrorist attack on the Western Power Grid	Load shedding, brownouts, blackouts, transportation gridlocks, hardware damage, chaos and cessation of commerce	0	10^6	$0 \times \infty$	
Nearby volcanic eruption	Clouds of ash and smoke could cover the sky blocking sunlight to solar panels and reducing solar PV output	0	10^5	$0 \times \infty$	
Solar panel output fluctuates by more than 60 MW in a 15 min interval due to clouds, thunderstorms, etc.	Power production plummets tripping breakers and leaving customers without electric power. Voltage on the grid could drop and frequency of coal-fired generators could change: transients are harmful to big electric generators	94.6, This number represents the $\pm 3\sigma$ points for data collected every 15 min for an entire year	200	18,920	A
Feeder circuit disconnects from substation	Feeder circuit voltage gets out of phase with the grid. solar PV	330	1*	330	B
Short to ground on the distribution grid	Equipment is damaged, particularly transformers and capacitor banks	24	10	240	C

(continued)

Table 6.11 (continued)

Potential failure event	Consequences	Frequency of occurrence in the TEP control area (events per year)	Severity of consequences	Risk	Identification tag
Western Power Grid fails (due to other than terrorist activities)	The western United States would be without electric power	0.03	10^4	300	D
Lightning strikes the system	Components may be damaged and electric generating capacity would be reduced	0.39	100	39	E
Grid voltage exceeds ±5 % limits	Customer service deteriorates and solar PV systems trip off-line	24*	1*	24	G
Transient local outages	Outages on transmission or distribution lines trigger shutdown of PV systems	24	1	24	H
Solar panels accumulate layers of dust or other particles	Efficiency of the solar panels will decrease and energy output will be lower than expected	2	10	20	I
Junction box fails	Loss of generated power output	0.27	50	14	J
Data acquisition system fails	Data cannot be read from the solar farm, loss of monitoring	0.14	50	7	K
PV module fails	Loss of production capacity	0.38	10	3.8	L
Grid frequency goes out of ±0.5 Hz limits	Small PV subsystems and big generators trip off-line, perhaps overloading transmission lines. TEP might be fined	0.2	50*	10	M

(continued)

Table 6.11 (continued)

Potential failure event	Consequences	Frequency of occurrence in the TEP control area (events per year)	Severity of consequences	Risk	Identification tag
Software failure	Software failures are ubiquitous and insidious. They can cause a myriad of problems	2	50	100	N
Electric storage system fails	Stored energy is lost. Infrastructure might be damaged. This failure event will become more severe as more batteries are used to smooth the load	0.7	20*	1.4	O

These data are also plotted in Fig. 6.9
Risk is defined as frequency multiplied by severity
Values with asterisks * will increase with an increasing number of solar PV subsystems
In Tables 6.9, 6.10, 6.11, 6.12, 6.13, 6.14, 6.15, and 6.16, estimates are represented with integers or decimals with only one significant figure. Decimal numbers with two or more significant figures were calculated from TEP databases

severity should be about the same [33] numerical values were assigned to the severities as follows.

Severity description	Numerical value
Extreme	1,000,000
Very high	100,000
High	10,000
Medium	1000
Low	100
Very low	10
Minuscule	1

Is it really mandatory to give frequency and severity the same range? Like most systems engineering questions, the best answer is, "It depends." If your customer does not want you to normalize frequency and severity, then don't do it. Maybe he wants to secretly weight one higher than the other.

Table 6.11 summarizes the operating performance risks for solar PV systems and TEP's distribution grid. These risks are related to the functionality of the system.

Table 6.12 The biggest operating performance risks from Table 6.11 (condensed), with the addition of log–log normalization technique data in columns D to F and linear normalization technique data in columns H to J

Operating performance risks with the addition two methods for normalizing risk data										
	A	B	C	D	E	F	G	H	I	J
Row	Potential failure event	Consequences	Frequency of occurrence, (events per year)	Severity of consequences, range of $1–10^6$	Estimated risk	Risk rank order	Identification tag	Severity, range of 1–10	Estimated risk	Risk rank order
4				Log–log technique				Linear technique		
5	Terrorist attack	Blackouts, chaos	0.01	10^6	10^4	2		10	0.10	11
6	Volcanic Eruption	Ash blocks sunlight	0.01	10^5	10^3	3		9	0.09	12
7	Solar power drops 60 MW in 15 min	Power production plummets	94.6	200	18,920	1	A	6	568	2
8	Feeder circuit disconnects from substation	Voltage gets out of phase with grid	330	1	330	4	B	3	990	1
9	Short to ground	Equipment is damaged	24	10	240	6	C	5	120	3
10	Western Power Grid fails	Western United States is without electric power	0.03	10^4	300	5	D	8	0.2	10
11	Lightning strikes system	Equipment is damaged	0.39	100	39	7	E	5	2.0	7
12	Grid voltage exceeds limits	Service suffers. PV systems trip off-line.	24	1	24	8	G	3	72	4
13	Transient local outages	PV systems shutdown	24	1	24	9	H	3	72	5

(continued)

Table 6.12 (continued)

Operating performance risks with the addition two methods for normalizing risk data

	A	B	C	D	E	F	G	H	I	J
Row	Potential failure event	Consequences	Frequency of occurrence, (events per year)	Severity of consequences, range of 1–10^6	Estimated risk	Risk rank order	Identification tag	Severity, range of 1–10	Estimated risk	Risk rank order
14	Dust accumulation	Generated power drops	2	10	20	10	I	5	10	6
15	Junction box fails	Generated power drops	0.27	50	14	11	J	4	1.1	8
16	Data acquisition system fails	Data cannot be read from the solar farm	0.14	50	7	12	K	4	0.6	9

Table 6.13 Correlation coefficients between the frequency, severity, and risk data of Table 6.12

	AA	BB	CC	DD	EE
Row	Technique	Data set	Range, orders of magnitude	Correlation coefficient, r, for estimated risk versus frequency	Correlation coefficient, r, for estimated risk versus severity
3					
4	Linear	10 failure event set	Frequency = 4 Severity = 1	0.964	−0.21
5		12 failure event set	Frequency = 4.5 Severity = 1	0.965	−0.30
6					
7	Log–log	10 failure event set	Frequency = 4 Severity = 4	0.17	−0.08
8		12 failure event set	Frequency = 4.5 Severity = 6	0.10	0.39

Table 6.14 Environmental risk for incorporating solar PV subsystems into a commercial electric power grid

Failure event	Consequences	Frequency of occurrence	Severity of consequences	Estimated risk
More stringent siting requirements by city, county, and state zoning jurisdictions	Plans would become obsolete. Costs would increase. Obstructionist activities and lawsuits would delay the project	2	100	200
Destruction of natural habitats	Strong opposition and lawsuits from environmental groups	1	20	20
Modification of animal migration paths	Migrating species might be affected. Loss of public support, project may require additional environmental studies	0.2	30	15
Unsuspected hazards	Fines, lawsuits, loss of public confidence	0.1	150	15
Higher than expected disposal or recycling cost	Budget overrun, loss of profit	0.1	100	10

Estimated risk is defined as the product of frequency of occurrence and severity of consequences

Table 6.15 Project Management risk for incorporating solar PV subsystems into a commercial electric power grid

Failure event	Consequences	Frequency of Occurrence in the TEP control area (events per year)	Severity of consequences	Estimated risk
Drastic human mistakes	Human fatalities. With 1500 employees, TEP has had no fatalities in the last 25 years of operation	0.02	1,000,000	$0 \times \infty$
Accidents	Injury to humans requiring medical attention	12	10,000	120,000
Project costs becomes higher than projected	The project may have to be delayed or canceled	1	1000	1000
Maintenance costs become higher than expected	Budget overrun, reduction of profits	0.5	1000	500

Estimated risk uses, the product of frequency of occurrence and severity of consequences

Table 6.16 Economic and government risk for incorporating solar PV subsystems into a commercial electric power grid

Failure event	Consequences	Frequency of occurrence	Severity of consequences	Estimated risk
Economic				
Interest rates changes	NPV calculations may become invalid, may affect interest payment on floating rate loans, etc.	12	10	120
Government				
Carbon emission regulations are different than expected	TEP would need to revise their projections and alter their renewable energy acquisition plans	0.5	50	25
Early elimination of rebates	This would increase the net cost of distributed generation (DG) systems to the consumer and may jeopardize TEP's ability to meet the DG requirements	1.2	100	120

Estimated risk is defined as the product of frequency of occurrence and severity of consequences

Failure events in the performance category typically result in system downtime and will affect the quality and reliability of system operations.

Tables 6.11, 6.12, 6.13, 6.14, 6.15, 6.16, 6.17, and 6.18 have five or six columns describing a *potential failure event,* the *consequences or potential effects* of that failure event, the *frequency of occurrence* (or *relative likelihood*) of the event in the relevant environment, the *severity of consequences* for each failure event, the *estimated risk*, and perhaps a short *identification tag*. The *frequency of occurrence* was based on historical data and expert opinion. *Estimated risk* was defined as the product of the *frequency of occurrence* and *severity of consequences.*

Mitigation strategies must be written for all risks. Here are a few examples for Table 6.11.

Risk: Solar panel output fluctuates by more than 60 MW in 15 min. *Mitigation Strategy*: To ameliorate these possibilities, TEP will buy and operate backup generators and negotiate purchase agreements with other suppliers. TEP will plan controlled brownouts with load shedding. Presently this is not much of a problem, because solar power comprises only a few percent of the load. But when solar power approaches one-fourth of the peak power, TEP will need extensive backup systems.

Risk: Feeder circuit disconnects from substation. *Mitigation Strategy*: TEP will use synchronized reclosers.

Risk: Western Power Grid fails (to other than terrorist activities). *Mitigation Strategy*: TEP will have backup generators and plans for controlled brownouts with load shedding.

Risk: Software failure. *Mitigation Strategy*: Every software module in the system shall have redundancy and built-in self-test to help reduce the severity.

The range of magnitudes for frequency of occurrence and severity of consequences must be the same. In Table 6.11, the frequency of occurrence covers four orders of magnitude (from $10^{-1.5}$ to $10^{2.5}$) and the severity of consequences also covers four orders of magnitude (from 10^0 to 10^4). Low-frequency high-severity risks, such as terrorist attacks and volcanic eruptions, and high-frequency low-severity risks, such as the shadows of birds or airplanes blocking the solar panels, are not included in these calculations. This is one example of how including these outliers would screw up the numerical calculations.

The most interesting points of Table 6.11 that are plotted in Fig. 6.9 are (1) risk A, that the "solar panel output fluctuates by more than 60 MW in a 15-min interval," is the riskiest, (2) risk D, that the "Western Power Grid fails," is in the rare but severe corner, and (3) risk B, that a "feeder circuit disconnects from its substation," is in the common but benign corner. Therefore, our overall advice is to (1) apply risk mitigation to risk A, (2) keep an eye on risk D, and (3) ameliorate risk B. Although risk B is not high risk, inexpensive mitigation would improve the overall reliability of the system. Now, please look back at the

Table 6.17 Operating performance risks for a 20 MW backup quick-start natural-gas motor-generator set

Failure event	Consequences	Frequency of occurrence in the TEP control area (events per year)	Severity of consequences	Estimated risk	Identification tag
Backup power generation is unavailable in a timely manner	Natural-gas motor-generator backup sets are complicated machines. They are designed to start up in less than 10 min, but it could take an hour or 2, causing load shedding	2	200	400	P
Unforeseen MG unit outages	Backup generating capacity is inhibited. Given TEP's seasonal capacity requirements, this is only critical during summer months	1	200	200	Q
Grid frequency changes abruptly by more than 0.5 Hz	This would harm the MG set	0.2	300	60	R
MG set introduces transients on grid	Big electric generators can be negatively affected by transients	4	8	32	S
Software failure	Software failures are ubiquitous, but hard to diagnose, particularly when they involve interacting systems. Redundancy and built-in self-test help reduce the severity	0.4	70	28	T

(continued)

Table 6.17 (continued)

Failure event	Consequences	Frequency of occurrence in the TEP control area (events per year)	Severity of consequences	Estimated risk	Identification tag
System connects MG set out of phase with the grid	The MG set could be damaged	2	9	27	U
System connects MG set out of frequency with the grid	The MG set would be damaged	0.5	50	25	V
Human override of the control system causes the MG set to connect at the wrong frequency	The system would disconnect the MG set as quickly as possible. But it may be too late to avoid damage to the MG set	0.2	50	10	W
Lack of fuel	TEP would lose backup capability	0.4	40	16	X
MG backup set hardware failure	TEP would lose backup capability. Then, if the sun were blocked, TEP could not provide full capacity. This would result in initiating a planned phased brownout and load-shedding program	0.2	10	2	Y
TEP's short-range weather prediction system fails	A gas-turbine MG set requires 10 min for startup. Therefore, TEP must have a system that will predict cloud cover 10 min in advance. Otherwise, voltage on	0.8	10	8	Z

(continued)

Table 6.17 (continued)

Failure event	Consequences	Frequency of occurrence in the TEP control area (events per year)	Severity of consequences	Estimated risk	Identification tag
	the grid could decline and the frequency of coal-fired generators could drop				

These values were estimated by Terry Bahill, based on Tom Hansen's data
These data are also plotted in Fig. 6.10. Estimated risk is defined as the product of frequency of occurrence and severity of consequences

Table 6.18 Safety risks for a backup quick-start natural-gas motor-generator set

Failure event	Consequences	Frequency of occurrence in the TEP control area (events per year)	Severity of consequences	Estimated risk
Drastic mistakes	Human fatalities	0.0002	10^6	$0 \times \infty$
Accidents	Injury to humans requiring medical attention	0.12	10^4	1200
Infrastructure damage from violent storms	TEP would be without backup for an extended period of time	0.1	10	1

Estimated risk is defined as the product of frequency of occurrence and severity of consequences

simple risk plot of Fig. 6.6. Figures 6.6 and 6.9 agree except that risk F, "failure of DC to AC inverters," has been transferred from Table 6.11 to a table for risks from the consumers' point of view.

6.4.2 *Normalization of Frequency and Severity*

This section can be skipped without loss of continuity. In this section, we will compare two methods for normalizing risk data. Each cell in Tables 6.12 and 6.13 will be identified with {Table, Row, Column}. For example, the cell in Table 6.13, row 5, column DD will be identified as {6.13, 5, DD}.

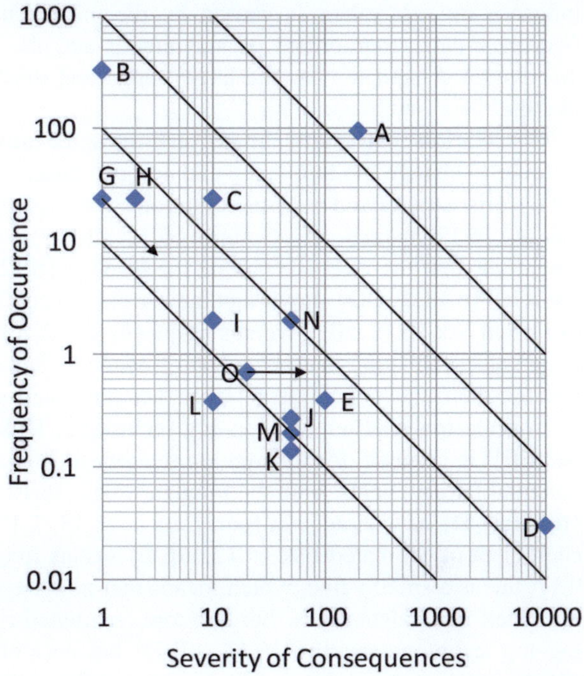

Fig. 6.9 A log–log risk chart for Table 6.11, the TEP electric grid. The *arrows* show that in the next few years risk *G* is expected to move down and to the right and risk *O* is expected to move to the right. The straight lines are iso-risk contours. This process uses log–log plots so that rare events can be tracked without distorting the risk analysis

The linear technique, {6.12, all, C, H, I and J}. Let us now go back to the principle of normalizing frequency and severity so that they have the same range. First, we put the biggest risks from Table 6.11 into cells {6.12, 7–16, A–G}. Then we introduced a new linear technique for computing severity. The severities {6.12, all, H} were assigned using a linear scale of 1–10 [29, 33, 41, Fig. 5-10]. This will be called the *linear technique*. The severities {6.12, all, H} were then multiplied by the frequencies {6.12, all, C} to estimate the risks presented in {6.12, all, I}. These estimated risks were used to compute the risk rank order in {6.12, all, J}.

In Table 6.11 the frequency of occurrence for terrorist attacks and volcanic eruptions were assigned values of zero. In Table 6.12, they were given nonzero values. Rows 7–16 of Table 6.12 are named the 10-failure event data set in Table 6.13. Rows 5–16 of Table 6.12 are named the 12-failure event data set in Table 6.13.

Correlation coefficients, linear technique, 10 failure event set, {6.13, 4, DD and EE}. Next, we used Excel to calculate the correlation coefficients (r) between frequency, severity, and estimated risk. The results of applying Excel to the new linear technique data are shown in Table 6.13. Cell {6.13, 4, DD} shows that for this linear technique the correlation coefficient, r, of estimated risk {6.12, all, I} and frequency {6.12, all, C} is 0.964; this is a large value, which means that frequency is dominating the estimated risk calculations. In contrast, estimated risk {6.12, all, I} versus severity {6.12, all, H} has an r of only −0.21 as shown in {6.13, 4, EE}, which means that severity is having little effect on the estimated risk. This difference in influence on risk values is caused by the mismatch between the ranges of the frequency data (four orders of magnitude) and the severity data (one order of magnitude). *Frequency* has the larger range and *totally dominates* the calculation of risk.

Correlation coefficients, linear technique, 12 failure event set, {6.13, 5, DD}. In order to change the frequency and severity ranges, the frequency of occurrence for terrorist attacks and volcanic eruptions were assigned nonzero values and were put into {6.12, 5 and 6, all}. Column {6.12, all, C} now has a range of 4.5 orders of magnitude, but the linear technique for severity {6.12, all, H}, still has a range of one. Cell {6.13, 5, DD} shows that the correlation coefficient of the linear technique estimated risk {6.12, all, I} versus frequency {6.12, all, C} is now 0.965, which is a large value indicating that once again *frequency dominates* the calculation of risk, because the frequency range is larger than the severity range.

Correlation coefficients, log–log technique, 10 failure events, {6.13, 7, DD and EE}. In contrast, for the log–log severity technique and the original set of ten potential failure events, both the frequency data {6.12, all, C} and the severity data {6.12, all, D} have a range of four. Data in {6.13, 7, DD} show that the correlation coefficient of estimated risk {6.12, all, E} versus frequency {6.12, all, C} is only 0.17; this is a small value, which means that now frequency is not dominating the estimated risk calculations. Furthermore, estimated risk {6.12, all, E} versus the log–log technique severity {6.13, all, D} has an r of only −0.08 {6.13, 7, EE}, which means that severity is not dominating the estimated risk. These low correlation values indicate that the estimated risk is a combination of frequency and severity and *neither one dominates*.

Correlation coefficients, log–log technique, 12 failure events, {6.13, 8, DD and EE}. When frequencies for terrorist attacks and volcanic eruptions were added into {6.12, 5 and 6, all}, the range of the frequency data {6.12, all, C} became 4.5 orders of magnitude and the range of the log–log technique severity data {6.12, all, D} became six. The correlation coefficient of estimated risk {6.12, all, E} and frequency {6.12, all, C} decreased from 0.17 to 0.10 {6.13, 7 and 8, DD}. The correlation coefficient of estimated risk {6.12, all, E} and severity {6.12, all, D} increased from −0.08 to 0.39 {6.13, 7 and 8, EE}. The range for the log–log technique severity data has become larger than the range for the frequency data and *severity is now more important than frequency*.

Is a linear correlation coefficient appropriate? {6.13, 4, DD}. Excel's CORREL and PEARSON functions are the same. They return a value for r that indicates the goodness of a linear fit between two data sets. But what does r mean if the data sets are *not linearly* related? To answer this question, we fit exponential, linear, logarithmic, second-order polynomial and power law trend lines to the data sets. Different functions fit different data sets better or worse. For example, {6.13, 4, DD} shows $r = 0.964$. The functions CORREL, PEARSON, RSQ and the linear trend line from the scatter chart all have the exact same number (to fourteen decimal places). However, the *power function* trend line gave a better match with $r = 0.995$. Small differences like these were common. We used all five of the Excel functions to fit every combination of data sets that we were considering. Functions other than the linear function offered no significant improvements; investigating these other functions merely ensured that we were not trying to fit (for example) a sinusoid with a linear regression line. See Section 3.6.5, Shape of the Data.

Differences between the two techniques. Finally, the most important risks are different for the log–log and the linear techniques. The log–log severity technique {6.12, all, A to G} indicates that the most serious risks are (in order of importance): (1) solar power drops 60 MW in 15 min, (2) terrorist Attack, (3) volcanic eruption, and (4) feeder circuit disconnecting from the substation. Whereas, the linear severity technique {6.12, all, A to C and H to J} indicates that the most serious risks are (in order of importance): (1) feeder circuit disconnecting from the substation, (2) solar power drops 60 MW in 15 min, (3) short to ground, and (4) grid voltage exceeds its $\pm 5\,\%$ limits. To quantify this mismatch we note that the correlation coefficient between the log–log estimated risk {6.12, all, E} and the linear estimated risk {6.12, all, I} is only 0.3; this is a small value. It seems that it is time to discuss these results with the customer.

6.4.3 Other Risks

Table 6.14 summarizes the environmental risks. These risks are related to the environment surrounding the system and affect various stakeholders (utility company, wildlife, humans, and the environment). The first three risks were analyzed from both an environmental and utility company perspective, while the latter two were analyzed strictly from the utility company's perspective due to the possibility of large financial repercussions. In the complete study, Tables 6.14, 6.15, and 6.16 had dozens of rows.

Table 6.15 summarizes project management risks. These risks are associated with the operation and management of grid-connected solar PV farms (either by the utility company or by a third party).

Table 6.16 summarizes economic and government risks. These risks are associated with economic policy changes or changes in government regulations. Although the third risk directly affects customers by increasing the cost of renewable energy systems, the risks in this table were analyzed only from the utility company's perspective.

6.4.3.1 Risk Mitigation with Motor-Generator Sets

To mitigate risk A of Table 6.11, unexpected demand peaks or decreases in power output, motor-generator (MG) sets are used as backup generators. The backup generators that are being considered for PV projects produce 20 MW of power and have startup times of approximately 10 min. To further emphasize the hierarchical nature of risk analyses that was presented in Fig. 6.9, we will now show the main results of a risk analysis performed on the motor-generator backup system.

The risk analysis for a quick-start natural-gas motor-generator set, which is described in Tables 6.17 and 6.18, was conducted in the same manner as the risk analysis for the distribution grid with solar PV systems; however, the severities for this table used the following scale:

Severity description	Numeric values
Very high	1000
High	100
Medium	10
Low	1
Very low	0.1
Minuscule	0.01

As shown in Fig. 6.10 the range of magnitudes for frequency of occurrence and severity of consequences were made about the same: both the frequency of occurrence and the severity of consequences cover almost two orders of magnitude.

Figure 6.10 shows that (1) risk P, "Backup power generation is unavailable in a timely manner" is the riskiest failure event and mitigation efforts should be applied to it, (2) risk R, "grid frequency changes abruptly" is in the rare but severe corner, so we should keep an eye on it, and (3) risk U, "The system connects the MG set out of phase with the grid" is in the common but benign corner: it is not high risk, but inexpensive mitigation would improve the overall reliability of the system, a little money would go a long way.

Because a 20 MW natural-gas MG set would produce 1 % of the total 2000 MW power generated by TEP, we multiplied the frequency of occurrence data of Table 6.11 by 1 % for use in Table 6.17.

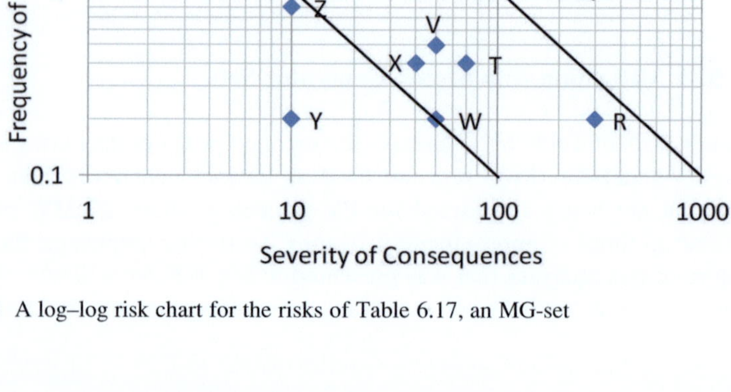

Fig. 6.10 A log–log risk chart for the risks of Table 6.17, an MG-set

Table 6.19 Risks that have already been addressed

Failure event	Consequences	Solution	Problem solver
Home Owners Associations (HOAs) could prohibit or strongly discourage PV systems	HOAs could prevent residents from installing PV systems or could penalize them for doing it	The state of Arizona passed laws making it illegal for HOAs to impose rules against photovoltaic systems	State of Arizona
Accidents or deaths due to installation of PV systems by homeowners	Homeowners could be electrocuted if they installed the systems themselves. High voltages are involved and any mistake could result in death or severe injuries	In order to qualify for TEP's rebate program (which pays for about one third of the installation cost), the system needs to be installed by a *certified* professional; this discourages people from installing the system themselves since they would forego the rebate	TEP with their green incentives program
Electric companies refuse to buy electricity from homeowners	Homeowners would not benefit from net-metering on days when their power output is higher than their power consumption	Federal rules require electric companies to buy electricity from their consumers [IREC, 2009; Arizona Corporation Commission, 2008]	Federal Government, State Governments
The panels contain toxic chemicals or heavy metals	Smashing or crushing a panel would release toxic gases could create a short circuit	No toxic chemicals or heavy metals are contained on the final product	Manufacturers

In addition to the previously described risks, some risks that we identified for the solar PV system had already been mitigated. They are presented in Table 6.19. Discovering risks that have already been mitigated is very important, because it validates the completeness of the search.

6.4.3.2 Identified Risks that Were Not Included

Failure of DC to AC inverters of the solar subsystems is the most common hardware failure. In 2010 TEP had 3500, 3 kW residential inverters connected to their grid. These inventers had typical product warranty periods of about 10 years. We assumed that manufacturers used percentile warranty $p_w = 50\ \%$ and that the underlying failure rate is an exponential distribution, $p_w = e^{-\lambda t}$. For $p_w = 0{:}5$ and a warranty period $t = 10$, $\lambda = 0.0693$. For 3500 inverters connected to the power grid, we can expect around 240 failures per year. When an inverter fails, the homeowner who owns it loses a part of his electric generating capacity and has to buy a new inverter. However, this failure produces only minor inconvenience to TEP. Table 6.19 was constructed from the point of view of the utility company, not the homeowner. Therefore, risk F was removed from Table 6.11.

The original net-metering rules put the electric companies at financial risk by requiring that they buy and sell electricity to homeowners at the same price. But such a practice is economically unfeasible. So electric companies are trying to charge customers a fee for using infrastructure such as the electric grid or else buy and sell electricity at different rates.

A recently discovered risk, that the installers install the solar panels wrong so that they are damaged or get appreciable shade, was also not included in Table 6.19. Acts of war such as cloud seeding and weather manipulation were also considered but not included.

Perhaps the greatest risk, uncontrollability of the electric grid, was not included. Each of the United States electric power grids is a complex system that may be uncontrollable in itself. When many of these are networked together, the problems may become insurmountable. "Mathematical modeling suggests that big blackouts are inevitable" [42].

6.4.4 Sensitivity Analyses

It is important to consider all (or most) possible states of the system and the impact they might have on the output. This requires defining hypothetical situations and examining the consequences and implications on the current system. Both mathematical sensitivity analyses [43–45] and qualitative analysis (a what-if analysis) should be used by describing and exploring possible outcomes and consequences. The next section contains a what-if analysis for an electric power grid with solar PV subsystems.

6.4.4.1 What-If Analysis

Early Elimination of Rebates

Early elimination of rebates would reduce customer incentives to install solar-powered electric generating systems.

Cloudy Days

Weather is the most uncontrollable factor for a PV subsystem. When clouds appear between the solar panels and the sun, there is an immediate and significant drop in power output. What would happen if there were a total blockage of the sun (due to total cloud coverage) when the system load peaked? There are two important

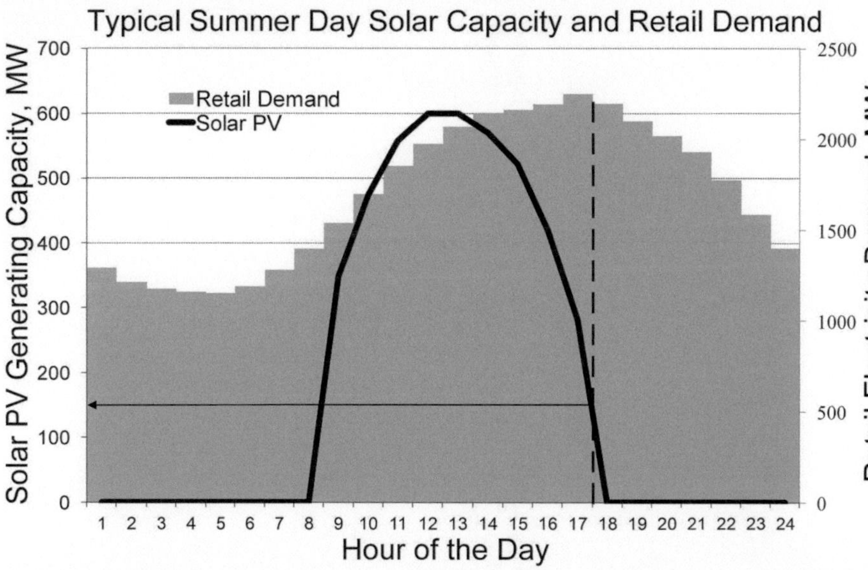

Fig. 6.11 Solar PV generating capacity and retail electric demand projected for a typical summer day in Tucson in 2025. The solar PV generating capacity peaks at 600 MW around noon and drops to 25 % of its peak capacity by 5:30 PM (the *dashed vertical line*). These data are used in Tables 6.20 and 6.21

Table 6.20 TEP power projections for 2025

TEP power projections for 2025	Power (MW)
Solar nameplate power generating capacity	600
Solar power generating capacity at noon	600
Solar power generating capacity at 5 PM	150
Conventional generating capacity	2400
Peak retail load obligations at noon	2000
Peak retail load obligations at 5 PM	2250
Planning reserve	300

factors to consider: first, TEP peak summer loads typically occur in the late afternoon, around 5 PM, and second, during these late afternoon hours, output from PV panels has fallen to one-fourth of their peak output, as shown in Fig. 6.11.

In order to meet the Renewable Energy Standard, TEP predicts that by 2025 its system will have about 600 MW of utility-scale renewable-generating capacity. Assuming that this is all solar PV subsystems, this would produce 600 MW at noon and 150 MW at 5 PM.

The planning reserve will take care of total cloud cover over the solar panels.

Tables 6.20 and 6.21 show that TEP can meet is load obligations under the worst weather conditions (zero solar power output) during peak summer loads. The first row of Table 6.20 is used to compute the needed conventional generating capacity. It states that the peak load in the summer at 5 PM plus the planning reserve must be

Table 6.21 Planning reserve and cloud cover

In 2025, will TEP be able to meet its load obligations under the worst conditions?

Condition	Load obligation (MW)		Planning reserve (MW)	Necessary condition	Conventional generating capacity (MW)		Solar generating capacity (MW)	Equation satisfied?
5 PM, normal	2250	+	300	$\leq$	2400	+	150	Yes
5 PM, total cloud coverage	2250	+	0	$\leq$	2400	+	0	Yes
Noon, normal	2000	+	300	$\leq$	2400	+	600	Yes
Noon, total cloud coverage	2000	+	0	$\leq$	2400	+	0	Yes

less than or equal to the conventional generating capacity plus the solar generating capacity. The other rows show that the total load can be met at the other significant conditions.

Hardware Reliability

Another significant source of risk is failure of the normal day-to-day electric generators. Electric generators (renewable or nonrenewable) could fail during peak load hours. Depending on the total capacity loss and the availability of reserves, it may or may not be possible to meet demand. Given that the incident is not planned, there may be a lag between the time when a generator trips offline and the time when backup or reserve capacity is available to cover this shortfall. Under rare circumstances, a large unit outage during peak load conditions could result in a temporary capacity shortage that requires TEP to shed load and/or call on sharing reserves. In most cases, when generating units trip offline, TEP is able to call on backup capacity from the Southwest Reserve Sharing Group (SRSG). SRSG enables TEP to rely on regional utilities for backup capacity for an hour or less. TEP then uses a combination of its own units (if available) and market resources to replace the needed capacity for the next dispatch hour.

Economic Factors

There are also economic factors that could affect the PV penetration on TEP's system. For example, if the cost of solar subsystems suddenly dropped, then more homeowners might buy solar photovoltaic panel systems. If a large number of households bought solar subsystems and put them on their houses, then, during the day, customers would reduce their dependence on electricity supplied by TEP. If this happened throughout the city of Tucson, it could affect TEP's bottom line. TEP would have underutilized capacity during the day, since they would not be selling as much electricity to residential customers, and thus their revenues would drop. TEP is obligated to provide generating capacity to meet demand during nighttime, during the afternoon (low PV output and high demand hours), and during cloudy days. Additionally, if all of these residential PV systems were grid-tied and customers were taking advantage of net-metering, then during sunny days, TEP will be required to buy all the excess electricity produced by residential customers. As a result, TEP could lose money from decreased revenues and increased net-metering costs. Consequently, TEP could substantially reduce net-metering payments, eliminating one of the incentives for residential customers to acquire PV systems in the first place. This is a negative feedback system with a long time delay (a potentially unstable system).

Analytic Sensitivity Analyses

You should perform a sensitivity analysis anytime you create a model, write a set of requirements, design a system, make a decision, plan a trade-off study, search for cost drivers, or engineer a risk analysis [45]. A sensitivity analysis of risk analyses is simple and it can be done in general terms. In this TEP risk analysis case study, our problem statement was, "*In a risk analysis, what parameters can change the rank order of the most important risks?*"

We will use the definition of risk where the frequency and severity have weights of importance as exponents as discussed in Section 3.3.1.2, like this.

$$\text{risk} = (\text{Frequency of Occurence})^{w_F} \times (\text{Severity of consequences})^{w_S}$$

Estimated risk (R) equals frequency of occurrence (F) raised to the power of weight of importance of frequency (w_F) multiplied by severity of consequences (S) raised to the power of weight of importance of severity (w_S), like this, $R = F^{w_F} \times S^{w_S}$. The first step is to derive the partial derivatives.

$$\frac{\partial R}{\partial F} = S^{w_S} w_F F^{w_F - 1} = S^{w_S} F^{w_F} \frac{w_F}{F} = R \frac{w_F}{F}$$

$$\frac{\partial R}{\partial S} = w_S S^{w_S - 1} F^{w_F} = F^{w_F} S^{w_S} \frac{w_S}{S} = R \frac{w_S}{S}$$

$$\frac{\partial R}{\partial w_F} = S^{w_S} F^{w_F} \ln F = R \ln F$$

$$\frac{\partial R}{\partial w_S} = F^{w_F} S^{w_S} \ln S = R \ \ln S$$

Now the partial derivatives will be multiplied by the normal values of the parameters to get the semirelative-sensitivity functions [45]. The $|_{\text{NOP}}$ symbol means evaluated at the nominal operating point.

$$\widetilde{S}_F^R = \left.\frac{\partial R}{\partial F} F_0\right|_{\text{NOP}} = R_0 w_{F_0}$$

$$\widetilde{S}_S^R = \left.\frac{\partial R}{\partial S} S_0\right|_{\text{NOP}} = R_0 w_{S_0}$$

$$\widetilde{S}_{w_F}^R = \left.\frac{\partial R}{\partial w_F} w_{F_0}\right|_{\text{NOP}} = R_0 w_{F_0} \ln F$$

$$\widetilde{S}_{w_S}^R = \left.\frac{\partial R}{\partial w_S} w_{S_0}\right|_{\text{NOP}} = R_0 w_{S_0} \ln S$$

Note that $\widetilde{S}_{w_S}^{R}$ is the semirelative-sensitivity function of R with respect to w_S, whereas S (*without the tilde*) is the severity of the consequences. Let $F,\ S,\ R,\ w_F$ and $w_S \geq 1$. Which data set (F or S) is most important? If $w_{F_0} > w_{S_0}$ then $\widetilde{S}_F^R > \widetilde{S}_S^R$. This is simply a restatement that the data set with the bigger weight (bigger data range) is more important.

Next, which individual potential failure events give rise to the greatest risks? If the weights of importance are the same, then the potential failure events with the largest sensitivity values create the greatest risks, because the sensitivities are determined by the values of R_0, as shown in the above equations. This means that we should spend extra time and effort estimating the frequency and severity of the highest ranked risks, which seems eminently sensible. Of the potential failure events in Table 6.11 for which data could be collected, the most important are "solar power drops 60 MW in 15 min" and "feeder circuit disconnects from substation." TEP already has good data for the former (several databases with a data set every 15 min for years). Therefore, they should spend more time and resources getting better data for the feeder circuit failure event.

6.4.5 *Risk Management*

Risk analysis is not an isolated step that should be conducted in order to understand the risks associated with the system. Instead, risk analysis should be an integral iterative process that is accompanied by a risk management initiative. After most of the risks inherent in a system have been identified, several risk management or risk mitigation strategies may be implemented to improve the system. The following section describes some risk management strategies that have already been implemented for electric power grids with solar PV subsystems and suggests additional alternatives.

6.4.5.1 Variations in Power Demand and Power Output

Electric power demand varies significantly. TEP's Resource Planning group factors in the variable demand exposure associated with renewable generation resources. Currently, TEP targets a 15 % planning reserve margin to ensure adequate system capacity. This planning reserve margin is used to cover peak load obligations and to mitigate unforeseen system contingencies. TEP is aware of the increased demand risk associated with renewable resources and is considering the potential of customer demand response programs, energy storage technologies, and the need for additional backup quick-start combustion turbines as possible mitigation strategies.

Additionally, electric power output also varies. TEP will create computer models to predict the output of wind and solar systems. Forecasts will be needed annually, monthly, daily, and on a minute-by-minute basis. Annual and monthly forecasts will help with load planning, daily forecasts will help handle the mismatch between peak solar output at noon and peak customer demand at 5 PM, and the minute-by-minute forecasts will allow non-spinning off-line natural-gas motor-generator sets the opportunity to start up in response to rapidly incoming storms.

To accommodate changes in power demand and output power TEP will revise their backup capacity and operating reserve policies. The possibility of designing curtailment options may provide system dispatchers with another tool to maintain system reliability on days with adverse weather conditions. TEP's most recent strategy is to implement a diversified utility-scale renewable portfolio based on a wide range of technologies dispersed over a number of geographical locations. This diversification strategy should reduce output variations and mitigate solar curtailments from cloud cover. In order to maintain future system reliability standards, both TEP and other regional utilities should revise their backup capacity and operating reserve policies as their renewable energy portfolio increases.

TEP will create demand-forecasting models, power output models, and real-time weather-forecasting models in order to identify risky dispatch scenarios that might require higher levels of backup generating capacity. These real-time monitoring systems would take the system to a *prevention* state that would enable system dispatchers to bring on additional generating resources as required.

6.4.5.2 Environmental Risks

The US Department of Energy [46] has written a Solar Energy Development Programmatic Environmental Impact Statement to analyze the environmental impact that solar projects might have and to develop and implement programs that would facilitate responsible solar energy development. All projects will undergo several detailed assessments to comply with governmental and environmental regulations. These studies include architectural studies, environmental studies, and biological studies (all managed by the EPA). They are conducted to ensure that the project's environmental impact is below acceptable levels. Once the Department of Energy develops solar-specific environmental guidelines, the risks associated with developing solar farms in open-land will be reduced, because there will be known requirements that must be met.

What is the carbon footprint for incorporating solar photovoltaic subsystems into a commercial electric power grid? No carbon emissions are produced during solar photovoltaic electric generation; however, the solar panel manufacturing process might not be a zero emissions process. We were concerned that the whole manufacturing, installing, and operating process could have a net positive carbon footprint; however, this is not the case. A study conducted by the Solar Hydrogen Education Program [47] found that the Springerville Solar Generating Station reduced the carbon footprint by 36 t of CO_2 per kW DC installed. Thus, this Station produces 91 % less carbon than a comparable fossil fuel-powered plant. Additionally, the total energy used to manufacture the hardware of the Springerville Solar Generating Station was 12 MWh AC per kW DC (88 % of which corresponds to solar panel manufacturing). Based on expected power production for the Springerville Solar Generating Station, the energy payback time would be 2.8 years, which is less than the 30 year expected life of the solar-powered plant [39]. According to the United Kingdom's Parliamentary Office of Science and Technology [48], the carbon footprint of solar panel manufacturing is expected to be reduced with the development of thin film technologies and the implementation of new, less energy intensive, semiconductor materials. Therefore, incorporating solar photovoltaic systems into a commercial electric power grid will contribute to *global cooling*.

Many technologies are available to simplify project planning and help design environmentally friendly projects. One example is geographic information systems (GIS), which may aid the planning of solar panel location. The GIS analysis may be conducted in various ways in order to reduce environmental impact. For example, GIS may be used to identify solar-feasible greyfields[1] and convenient installation sites such as the roofs of buildings [37] that would not require the modification of open-land in order to locate the PV system, sites that can be easily connected to the grid, and sites that receive enough solar radiation to make the project viable. TEP's current environmental risk mitigation strategy includes constructing utility-scale projects on greyfields such as reclaimed landfills and previous mining sites.

Chaves and Bahill [37] created a tool to help identify the best sites to install solar panels in an urban environment. They used LIDAR data to create a database with the elevation of every house, road, tree, and building. The ground mask eliminated all sites that were at or near ground level. The aspect mask eliminated areas that did not face generally south. The slope mask eliminated sites that were at a slope greater than 35°. The radiation mask eliminated areas that had too little average annual solar radiation; this eliminated areas next to tall building or other objects that create shadows. Finally, the human mask eliminated politically incorrect areas such as outside of the president's window. Next suitable sites had to have an area of at least 1.5 m^2 in order to accommodate a solar panel. Their final map showed that about 2 % of the total area was suitable for solar panel installation.

6.4.5.3 Financial Risks

Solar energy is currently in a developmental state and capital prices are expensive. Depending on the price characteristics of energy and the inherent financial characteristics of the company, developing a solar farm may not be financially viable. However, in some cases, the financial penalty of implementing a solar-powered plant may be reduced by introducing renewable energy tariffs, such as TEP's Renewable Energy Standard Tariff, in order to use these funds to promote the use

[1]Greyfields are underutilized real estate assets or lands. These sites have previously had other uses, such as old mines, mine waste tailings, landfills, mudslide zones, low-level radiation sites, frequently flooded zones, and perhaps as buffer zones for wildlife reserves and wilderness areas.

of renewable energy. Revenues generated by these tariffs are passed on to customers who incorporate renewable-energy solutions in their homes, incentivizing them to switch to greener energy technologies.

Utilities cannot pass on their increased costs directly to their customers since, in most cases, there is a cap on the rate that they can charge for electricity. Therefore, this is not a viable risk management alternative for mitigating the higher power costs from renewable energy sources.

When the Springerville Solar Generating Station was built, there were no federal incentives or rebates to help subsidize the project and TEP does not now have federal incentives that can be used to lower the investment cost of a solar farm. Therefore, TEP uses another alternative for mitigating financial costs: it creates special purpose entities (SPE) to develop and own the solar farms and develop lease agreements with them. An SPE is a legal entity (typically a company or partnership) that is created to serve a particular purpose, in this case, owning the solar farm. This SPE may be owned by one or more entities that are not related to TEP and therefore may qualify for federal or state rebates that are not available to TEP. This may help reduce the investment cost and would reduce the electric generation cost for this facility.

Another financial risk is interest rate risk: the risk of interest rates increasing and affecting the interest payments on PV project loans. However, the interest rate risk may be mitigated by conducting interest rate swaps. When companies enroll in interest rate swaps, they swap floating debt for fixed rate debt (or vice versa) in order to either exploit their competitive advantage in the fixed or floating rate market, or reduce their exposure to interest rate risk.

6.4.6 Unintended Consequences

Implementing new systems, strategies, laws, or controls often has unintended negative or positive consequences. Therefore, it is important that early in the system life cycle the designers try to predict what these consequences may be. The systems engineer is responsible for the big picture of system development. Hence, the system engineer must search for unintended consequences of the system under design [49].

Connecting a solar PV system to an electric power grid could cause problems for the grid. For example, presume that an illumination-controlled building is in the state of Selling AC Electricity, when clouds suddenly cover the sun. The voltage generated by the solar panels will drop as will the illumination in the building. Sensors will sense this drop in illumination and will command the lights to produce more illuminance. The lights will draw more power from the source. This will produce a bigger voltage drop across the source internal impedance, which will further drop the operating voltage. This is a *positive* feedback loop that could cause the grid to become unstable. A second problem with clouds blocking the sun is that the PV system would soon deplete its small local energy store and would switch to the Buying AC Electricity state of Fig. 5.11. This would increase the operating voltage. This is a negative feedback loop, but it contains a significant time delay. Time delays make systems susceptible to instabilities. Therefore, utility companies should create 1-s scale simulations of the interactions between these systems in order to investigate potential instabilities [50].

A third problem with clouds blocking the sun is that, if the voltage of the electric power grid falls out of its $\pm 5\,\%$ limits, then most of the local solar PV subsystems will isolate themselves from the grid. This will further decrease the grid voltage. This is a positive feedback loop. Positive feedback loops can make a system unstable.

Incorporating solar photovoltaic subsystems into a commercial electric power grid has another interesting possible unintended consequence. Solar panels do two things: they absorb solar energy and transform it into electricity, and they also reflect solar energy back into the atmosphere. Both of these actions reduce the solar energy that hits the ground and is absorbed by the Earth. Therefore, solar panels have an unintended consequence of reducing the amount of energy absorbed by the Earth and therefore they contribute to *global cooling*.

Solar panel systems are in positive feedback control loops and negative feedback control loops with large time delays. Both of these could cause instability on the grid. Bahill [49] has listed other unintended consequences of installing solar panels into a commercial electric power grid.

6.4.7 Unlikely Failures

Even the greats make egregious mistakes. The peer-reviewed journal paper that this section was based on [28] was rejected by the prestigious journal *Risk Analysis* in June of 2011. In one section, we had written that an "MG backup set hardware failure would cause TEP to lose backup capability. Then, if the sun were blocked, TEP could not provide full capacity. This change in generated power would result in initiating planned phased brownouts and a load-shedding program." See Section 6.3.8.2. The prime reviewer for *Risk Analysis* wrote, "I am an expert … in electricity grids … the authors display a total lack of knowledge of electricity grids and many of their statements are plain wrong e.g., Section 1.6.1 'The change in power … brownouts'—this is completely untrue." The editors stood behind this reviewer, several times.

Assuming that a risk event is impossible is the *worst* mistake that a risk analyst can make. Four months later, in September of 2011, the Western Power Grid experienced the blackout of the century. More than seven million households, business and hospitals were without electricity for more than 12 h (many for much longer). The economic cost in the dozen affected western states was around \$1 billion. This failure was triggered when a technician "prematurely cranked open a hand-operated switch [42]."

Many risks should just be noted and not quantified until they are more likely or more data about them are available. These include an electromagnetic pulse (EMP) from an atmospheric detonation of a nuclear device, solar flare activities (similar to the 1970 solar flare) and unavailability of critical materials, such as rare earth elements.

It is important to consider unlikely failures. In Section 30 when we introduced the invincibility bias, we recalled that the Spanish Armada was thought to be *invincible* in 1588 and in 1912, the White Star line said that the Titanic was *unsinkable*. Thalidomide was marketed in 1957 in Germany as a sedative that also cured anxiety, insomnia, gastritis, and tension. Its risk was so low that it was sold *over-the-counter*. Prior to the *Challenger* accident, NASA officials estimated the risk of a shuttle accident at 1 in 100,000: afterwards Richard Feynman computed the odds at 1 in 100. In spite of hurricane Katrina in 2005, the US government continues to build dikes and levees to support housing construction below sea level in New Orleans. There were several Wall Street banks that were *too big to fail*. When they did fail, it caused the Great Recession of 2008 and the New York Stock Exchange lost half of its value. The Fukushima Nuclear Power Plant was designed to withstand *any* earthquake that had *previously* hit Japan. Unfortunately, in 2011, it could not withstand the tsunami that accompanied an earthquake. It is very important to seriously consider possible failures that *could never happen*.

6.4.8 Homework Problems

6.4.8.1 Discussion with Customer

Section 6.4.2 ends with statement, "It seems that it is time to discuss these results with the customer." What would you say to your customer in such a discussion?

6.4.8.2 Prioritizing Risks

In Table 6.11, the risks have this numerical order: A, B, D, C, N, and E. Suppose that we told you that after this study was complete, it was decided to mitigate the risks with the following priorities: A then N. Why might this decision have been made?

6.4.8.3 Peak Electric Demand

Why should TEP's peak electric demand occur in the late afternoon, around 5 PM, as shown in Fig. 6.11? This is also the time of peak water usage.

6.4.8.4 Software Mistakes

Table 6.11 for the TEP electric power grid states that software failures (risk N) are ubiquitous and insidious. They can cause a myriad of problems. It lists the frequency of occurrence as two events per year. However, Table 6.17 for the backup motor-generator system (MG-set) states that software failures (risk T) are ubiquitous, but hard to diagnose, particularly when they involve interacting systems. Redundancy and built-in self-test help reduce the severity. It lists the frequency of occurrence as 0.4 events per year. How can you reconcile the different numbers?

6.4.8.5 Planning Reserve

Show how the 300 MW planning reserve in Table 6.20 was calculated.

Of the four conditions given in Table 6.21 which is the worst-case condition? Ignore the planning reserve.

6.4.8.6 Digital Camera

This problem was created by Bob Sklar. A new digital camera is being developed for the consumer market. Like many other amateur "point and shoot" cameras, it will offer compact size, affordable price, high resolution, long-lasting rechargeable battery, flash memory slot, USB computer connection, built-in flash, image stabilization, and face recognition. What distinguishes this new camera from similar priced existing cameras is more powerful processing. This will provide smaller shooting delays (startup to first shot and shot-to-shot) and higher continuous shooting rate (burst speed, frames per second), making it ideal for action photography. However, it will require a new processor, which is both still under development and must be integrated with the camera system.

You will find that the continuous (burst) mode is much faster than the normal mode. In the continuous shooting mode, cameras of this type normally hold many of the camera settings (such as focus) constant between shots, while settings are automatically reset between multiple shots in the normal mode, requiring additional processing time. The faster continuous mode is useful when the distance between camera and subject remains fairly constant.

With an emphasis on the new features, your assignment is to:

- Write some requirements for this new digital camera
- Identify potential performance, cost, and schedule risks
- Prepare a risk register with estimates of likelihood and severity
- Prepare a risk plot (likelihood vs. severity)
- Select one high or moderate risk and prepare a risk waterfall chart showing possible risk mitigation activities including testing

As a possible approach to developing camera requirements, you may adapt requirements from existing digital cameras in the $150–300 MSRP range and then modify shooting speed capabilities to make them significantly more challenging. A factor of two improvement would be significant.

6.4.8.7 SpinCoach

Perform a risk analysis on the SpinCoach.

6.4.8.8 Tucson Electric Power

This table is the risk analysis chart for the data from the vice president of Tucson Electric Power (TEP). Convert it into the numerical type of chart that we have used in class, on homework and in our projects (Table 6.22). What is the biggest risk? How can we mitigate it?

6.5 Summary

Chapter 6 covers risk analysis and management. Starting with the history of risk analysis, we provided examples of how risk analysis is used in real-world contexts. We presented a simple example of the use of risk analysis to identify the most severe risks associated with running a Pinewood Derby. The most serious risks identified were lane biases and placing the cars into the wrong lanes. To ameliorate these risks, the Derby management opted for a round-robin race format. We then presented a much more in-depth example concerned with assessing the risks of incorporating photovoltaic solar panels into an existing electric power grid. This example illuminated common mistakes that people make when performing risk analysis. An example of a common mistake is the use of unequal ranges for the frequency of occurrence and the severity of consequences.

Section 34, devoted to risk analysis, traced the history of risk analysis dating back to 1662. Based on the findings from Section 25 that frequency of occurrence is a superior metric than probability (because humans do a poor job of estimating probabilities), we defined risk as the product of frequency of occurrence multiplied by the severity of the consequences. We provided a few highlights of risk analysis using the Pinewood Derby management example. We also discussed risk management within the context of a commercial aircraft scenario.

The risk assigned to each possible cause is determined by the risk analysis. However, this analysis is not static. Each time the system is changed or a relevant test is performed, the risk analysis needs to be revisited. As risks are eliminated or reduced, other risks tend to increase in relative importance. As a system is deployed, the actual risks become better known. The Pinewood Derby analysis in

Table 6.22 Failures modes and effects analysis for TEP's electric power distribution grid

Failure mode	Effect	Frequency of occurrence in the TEP control area	Severity	Risk
Short to ground on the distribution grid	This could damage TEP's equipment particularly transformers and capacitor banks. The system should be back up within 2 h This is unlikely to damage our customers' equipment	Two times per month	Medium	–
Loss of ability to supply full load	This would trip breakers and leave customers without electric power	One hour per 10 years	Low	–
Accidents	Injury to humans requiring medical attention	Once a year	High	–
Human mistakes	Death of humans	With 100 employees TEP has had no fatalities in 25 years of operation	Extreme	–
Transient loss of energy source such as wind or sunlight due to, for example, clouds	Voltage on grid changes and frequency of coal-fired generators may change	Hundreds of times per day	Miniscule	–
The major weather related risk is intermittency of power produced due to thunderstorms	This would cause the voltage on the grid to change and this might cause the frequency of coal-fired generators to change	A dozen times per day, 5 % of the days of the year	Low	–
Grid frequency goes out of ± 0.5 Hz limits	This will trip generators, perhaps overloading transmission lines and may result in fines. This failure has a low severity with existing equipment. However, it will increase with distributed generation	Twice per week	Low but it will increase	–
Feeder circuit disconnects from substation	Feeder circuit voltage gets out of phase with the grid. This failure has a medium severity for existing equipment, but it will get worse with PV solar panels. It may require synchronized reclosers	Two times per month	Medium, but it will get worse with PV systems	–

Risk is defined as frequency multiplied by severity

Section 34 was written after 4 years of data acquisition and continual analysis and is, therefore, far better than our initial analysis. In the first attempt, the biggest risks were believed to be equipment failures. Inclement weather was not considered because the locale was Tucson, Arizona. However, as the system evolved with new instrumentation, risks changed. The first derbies were not computer-based, so extension cord trip hazards, or computer failures were not relevant. As the system evolved, the risks evolved with it, and so did the risk analysis.

In light of the foregoing, risk managers should manage the identified risks by regularly reviewing and re-evaluating them, because risk identification and data gathering is not an end in itself. Also, a word of caution: More than one project has made the costly mistake of being complacent with their risk ranking, thereby ignoring areas that they initially deemed low risk.

Risk analysis is not an exact science. There are multiple ways to do a risk analysis. Risks change with time, and there is always a level of uncertainty in the data. In this regard, we showed the simplest and most common industry risk analysis technique; namely, plotting the frequency of occurrence versus the severity of consequences. We identified mistakes that are often made when using this technique and ways to ameliorate them. The purpose of risk analysis is to give risk managers insights into the project and to communicate their understanding to others.

The engineering literature defines risk as the product of frequency of occurrence and severity of consequences. This definition of risk originated in the late seventeenth century with Arnauld and Nicole, "In order to decide what we ought to do to obtain some good or avoid some harm, it is necessary to consider not only the good or harm in itself, but also the probability that it will or will not occur." Their book entitled *Logic, or, the Art of Thinking: Containing, Besides Common Rules, Several New Observations Appropriate for Forming Judgment* arose from a series of symposia held at Port-Royal in Paris. This book was used as a university textbook for over 200 years.

The object of risk analysis is not just to assess the likelihood of completing a well-running project on schedule and within budget, it is to also give managers insights into where to invest resources to reduce risk to an acceptable level, how much time they can take out of a schedule without increasing risk, and how much time they can add before the gain is marginal. In this regard, risk analysis is, or should be, a decision tool. Risk analysis is an ongoing process that must be revisited periodically, or whenever significant changes occur to the system or to the problem. It is hard to overemphasize this point because this is where many programs falter. Most programs perform a risk analysis. However, fewer programs build a good risk reduction plan, and fewer still keep it current. Many managers erroneously believe that they are up to date on their programs and that the time spent on risk analysis is not worth the return. This perception is regrettable, because risk analysis is one of the most effective ways to demonstrate to the customer that you have your arms around the customer's problem.

The Pinewood Derby scenario was used to illustrate that risks change over time, and that risks are traded off with available resources. This scenario was used to also illuminate common ways of managing risks. We remind the reader that in this section, we defined risk as severity of consequences multiplied by frequency of occurrence. However, there are several other factors that can amplify human perception of risk: lack of control; choice, trust, or warning; novelty; recall; uncertainty; immediacy, etc. We also presented a risk template that can be employed on projects.

In Section 35, we explained the risk analysis process using the introduction of photovoltaic (PV) solar panels into a commercial electric power grid as an example. Specifically, we employed extreme scenarios to explain mitigation strategies to ameliorate risks. These "what-if" scenarios also identify possible unintended consequences of incorporating solar PV systems into commercial electric power grids.

In Section 36, we described two categories of risks for incorporating solar PV subsystem into a commercial electric power grid; risks associated with uncontrollable factors such as weather; and risks associated with software, hardware, and human error. We introduced the concept of risk tiers and how they can be analyzed from the perspectives of the different stakeholders. We then identified different mitigation strategies for the different risks. Subsequently, we discussed "what-if" analysis for various actions and contingencies for the illustrative example. Next, we discussed risk management from a variety of perspective. We then discussed unintended consequences. We reminded the reader that implementing new systems, strategies, laws, or controls often have unintended consequences, both negative and positive. Therefore, system engineers need to be cognizant of this fact and explore the consequences of implementation decisions early in the system's life cycle. For example, connecting a solar PV system to an electric power grid could cause stability problems for the grid.

We then emphasized two important aspects of risk analyses. The first is that the frequency and severity scales must be normalized in order to give equal weight to both factors. The second is to be wary of problems associated with extreme events, and to avoid basing conclusions about them *exclusively* on the expected value of risk, because this would give equal weight to events with high frequency of occurrence and low severity and events with low frequency of occurrence and high severity.

With regard to the whole solar PV system risk analysis, the risk of clouds blocking the sun and introducing output variability is the biggest risk. Additionally, as solar PV systems become a larger component of an electric company's energy portfolio, it is important to revise the backup capacity policies and consider alternative storage methods in order to reduce the risk of reduced power output during periods with high demand.

After conducting a what-if analysis, even under the worst-case scenario of total sunlight blockage and demand peaking, with appropriate planning, it is possible to develop strategies that will prevent brownouts and power shortages. Based on how much solar PV energy TEP has, it may be important to develop and implement state-of-the-art, 1-minute-scale weather forecasting, which combined with their current demand-forecasting methods, will help them identify risky scenarios and act appropriately. Utility companies should also create 1-second-scale simulations of the interactions between customer PV subsystems and the electric power grid to investigate potential instabilities.

The risk analysis approaches discussed in these sections have a few common weaknesses. (1) Publishing the results of a risk analysis (internally and externally) in a timely manner is a vital part of risk analysis. However, the reality is that publication cycles tend to be slow, but policies and data change rapidly. (2) It is hard to find the time to conduct risk analysis. In the beginning of a risk analysis, the risk analyst (RA) needs time with domain experts to identify and quantify risks. Near the end, the RA needs time with decision makers to explain the results of a risk analysis. It is challenging to schedule such time, given that RAs do not have the clout. (3) It is hard to get relevant data. A RA cannot just, for example, get on the internet and ask for data concerning “Solar panel output fluctuates by more than 60 MW in a 15-min interval due to clouds, thunderstorms, etc.” However, if the RA is willing to change his/her design to match what is available on the internet, then plenty of data are available. (4) Data that the RA does get are usually fraught with uncertainty, mistakes, incorrectly estimated probabilities, and human cognitive biases.

In light of the foregoing, we have two important suggestions. First, when an important risk is discovered, reach outside your organization to find experts that can help. Second, create a culture in your organization that supports open and critical discussion of risks.

Finally, in a risk assessment, an RA could end up spending an inordinate amount of time and money getting better and better data. However, that would not make risk recommendations any more precise. In this regard, Haimes and Chittester [51] drew an apt analogy to the Heisenberg Uncertainty Principle, which states that a person cannot simultaneously measure the position and the velocity of a particle with high precision. Then they expanded this analogy with, “recall Einstein’s statement: ‘So far as the theorems of mathematics are about reality, they are not certain; so far as they are certain, they are not about reality’. Adapting Einstein’s metaphor to risk assessment and management translates into: ‘To the extent risk assessment is precise, it is not real; to the extent risk assessment is real, it is not precise’.”

References

1. Arnauld A, Nicole P (1996) Logic, or, the art of thinking: containing, besides common rules, several new observations appropriate for forming judgment, the first edition was published in 1662, the fifth edition was published in 1683 and was translated from French in 1996 by Jill Vance Buroker, Cambridge University Press, Cambridge
2. Pascal B (1654) Fermat and Pascal on probability. http://www.york.ac.uk/depts/maths/histstat/pascal.pdf. Accessed Jan 2016
3. Witt RC (1973) Pricing and underwriting risk in automobile insurance: a probabilistic view. J Risk Insur 40(4):509

4. Hussey DE (1978) Portfolio analysis: practical experience with the directional policy matrix. Long Range Plann 11(4):2–8
5. Simpleman L, McMahon P, Bahnmaier B, Evans K, Lloyd J (1998) Risk management guide for DOD acquisition. DTIC Document
6. Joksimovic V, Houghton W, Emon D (1977) HTGR risk assessment study. Nuclear Systems Reliability Engineering and Risk Assessment, Society for Industrial and Applied Mathematics, pp 167–190
7. Rasmussen NC (1981) The application of probabilistic risk assessment techniques to energy technologies. Annu Rev Energy 6(1):123–138
8. Whitman RV (1984) Evaluating calculated risk in geotechnical engineering. J Geotech Eng 110(2):143–188
9. Schott H (1997) Risk management: concepts and guidance. DIANE, Collingdale
10. Haimes YY (2009) Risk management. In: Sage AP, Rouse WB (eds) Handbook of systems engineering and management, 2nd edn. Wiley, Hoboken, pp 155–204
11. Kirkwood CW (2009) Decision analysis. In: Sage AP, Rouse WB (eds) Handbook of systems engineering and management, 2nd edn. Wiley, Hoboken, pp 1223–1250
12. Buede DM (2009) The engineering design of systems: models and methods, 2nd edn. Wiley, New York
13. Blanchard BS, Fabrycky WJ (2011) Systems engineering and analysis, 5th edn. Prentice-Hall, Upper Saddle River
14. Bernstein PL (1996) Against the gods: the remarkable story of risk. Wiley, New York
15. Bahill AT, Karnavas WJ (2000) Risk analysis of a pinewood derby: a case study. Syst Eng 3 (3):143–155
16. Ben-Asher JZ (2006) Development program risk assessment based on utility theory. INCOSE Int Symp 16(1):1634–1646
17. Kerzner H (2013) Project management: a systems approach to planning, scheduling, and controlling, 11th edn. Wiley, Hoboken
18. Carbone TA, Tippett DD (2004) Project risk management using the project risk FMEA. Eng Manage J 16(4):28–35
19. Willis HH, Morral AR, Kelly TK, Medby JJ (2006) Estimating terrorism risk. Rand Corporation, Washington, DC
20. Botta R, Bahill AT (2007) A prioritization process. Eng Manage J 19(4):20–27
21. Chapman WL, Bahill AT, Wymore AW (1992) Engineering modeling and design. CRC, Boca Raton
22. British Cabinet Office, National Risk Register (2008) http://www.cabinetoffice.gov.uk/reports/national_risk_register.aspx. Accessed Dec 2015
23. Bahill AT, Botta R (2008) Fundamental principles of good system design. Eng Manage J 20 (4):9–17
24. Ord T, Hillerbrand R, Sandberg A (2010) Probing the improbable: methodological challenges for risks with low probabilities and high stakes. J Risk Res 13(2):191–205
25. NASA (2000) Mars program independent assessment team summary report
26. Madni AM, Sievers MA (2015) Flexible contract-based design framework for exaluating system resilience approaches and mechanisms. IIE annual conference and exposition May 30–June 2, Nashville
27. Bahill AT, Chaves A (2013) 9.4.1 Risk analysis of solar photovoltaic systems. INCOSE Int Symp 23(1):785–802
28. Chaves A, Terry Bahill AT (2014) Comparison of risk analysis approaches and a case study of the risk of incorporating solar photovoltaic systems into a commercial electric power grid. Syst Eng 17(1):89–111
29. DSMC (2001) Risk management guide for DoD acquisition Fort Belvoir, VA [cited 2015 December]. http://www2.mitre.org/work/sepo/toolkits/risk/references/files/DSMC_RiskMgmt_Guide.pdf
30. Institute of Risk Management (2002) A risk management standard, London
31. CMMI for Development (2015) ver 1.3 [cited 2015 December]. http://cmmiinstitute.com/cmmi-models

32. INCOSE (2011) Systems Engineering Handbook
33. Bahill AT, Smith ED (2009) An industry standard risk analysis technique. Eng Manage J 21 (4):16–29
34. Department of Defense Instruction (2008) Number 5000.02: operation of the defence acquisition system. Department of Defense, Washington, DC
35. INCOSE (2015) INCOSE systems engineering handbook: a guide for system life cycle processes and activities. Wiley, New York
36. Henry MH, Haimes YY (2009) A comprehensive network security risk model for process control networks. Risk Anal 29(2):223–248
37. Chaves A, Bahill AT (2010) Locating sites for photovoltaic solar panels 2010 [cited 2015 December]. http://www.esri.com/news/arcuser/1010/solarsiting.html
38. Bahill AT, Szidarovszky F, Botta R, Smith ED (2008) Valid models require defined levels. Int J Gen Syst 37(5):553–571
39. Moore L, Post H, Mysak T (2005) Photovoltaic power plant experience at Tucson electric power. Energy Conversion and Resources. ASME International, Orlando
40. Richerson-Smith D (2010) Energy efficiency and renewable programs, Tucson
41. INCOSE (2006) Systems engineering handbook: a guide for system life cycle processes and activities, INCOSE-TP-2003-002-03, version 3.2.2. In: Haskins C (ed) International council on systems engineering, San Diego
42. Fairley P (2004) The unruly power grid. IEEE Spectr 41(8):22–27
43. Hsu FK, Bahill AT, Stark L (1976) Parametric sensitivity analysis of a homeomorphic model for saccadic and vergence eye movements. Comput Programs Biomed 6(2):108–116
44. Karnavas WJ, Sanchez PJ, Bahill AT (1993) Sensitivity analyses of continuous and discrete systems in the time and frequency domains. IEEE Trans Syst Man Cybern 23(2):488–501
45. Smith ED, Szidarovszky F, Karnavas WJ, Bahill AT (2008) Sensitivity analysis, a powerful system validation technique. Open Cybernet Syst J 2:39–56
46. Bureau of Land Management US (2012) DoE. Final programmatic environmental impact statement (PEIS) for solar energy development in six Southwestern States, vol 1. http://energy.gov/sites/prod/files/EIS-0403-FEIS-Volume1-2012_0.pdf. Accessed Dec 2015
47. Mason J (2004) Life cycle analysis of a field, grid-connected, multi-crystalline PV plant: a case study of Tucson electric power's Springerville PV plant. Final report prepared for Tucson electric power
48. Parliamentary Office of Science and Technology (2006) Carbon footprint of electricity generation. Postnote 1–4 October
49. Bahill AT (2012) Diogenes, a process for identifying unintended consequences. Syst Eng 15 (3):287–306
50. Bahill AT (2010) Design and testing of an illuminance management system. ITEA J 31 (1):63–89
51. Haimes YY, Chittester CG (2005) A roadmap for quantifying the efficacy of risk management of information security and interdependent SCADA systems. J Homeland Secur Emerg Manag 2(2):1–21

Chapter 7
Comparing Four Processes

7.1 Section 37. Requirements, DAR, Trade-offs, and Risk

The first three chapters of this book described the system design process, human decision-making, and technical processes that are commonly used in a wide variety of fields. The next three chapters described and discussed three specific applications of these processes (i.e., requirements discovery, trade-off studies, and risk analysis) that employ different vocabularies but share a common underlying structure.

Requirements discovery, decision analysis and resolution (DAR), trade-off studies, and risk analysis are four distinct activities in the systems life cycle. Even though they employ the same underlying structure in their processes, they appear to be different because they employ different vocabularies, inputs, and outputs. We abstracted and grouped the common activities in these four processes. In the interest of clarity, these processes, shown in the following figures, suppress explicit representation of temporal sequences. Many of these activities are performed in parallel, and give rise to a multitude of feedback loops, which, once again, are not shown in the interest of clarity.

7.1.1 The Requirements Discovery Process

Figure 7.1 presents the requirements discovery process. Uncertainty exists in several aspects of this process. To begin with, not all stakeholders can be identified at the start of the project. The identification of customer needs requires iterations, because vagueness exists in the initial identification of needs. The same goes for the problem statement. We can expect the initial statement of the problem to be reused and refined as customer needs are more clearly articulated. Therefore, requirements discovery is an iterative process in which each iteration improves the identification of requirements. Upon customer consensus, the requirements are finalized,

A.T. Bahill, A.M. Madni, *Tradeoff Decisions in System Design*,
DOI 10.1007/978-3-319-43712-5_7

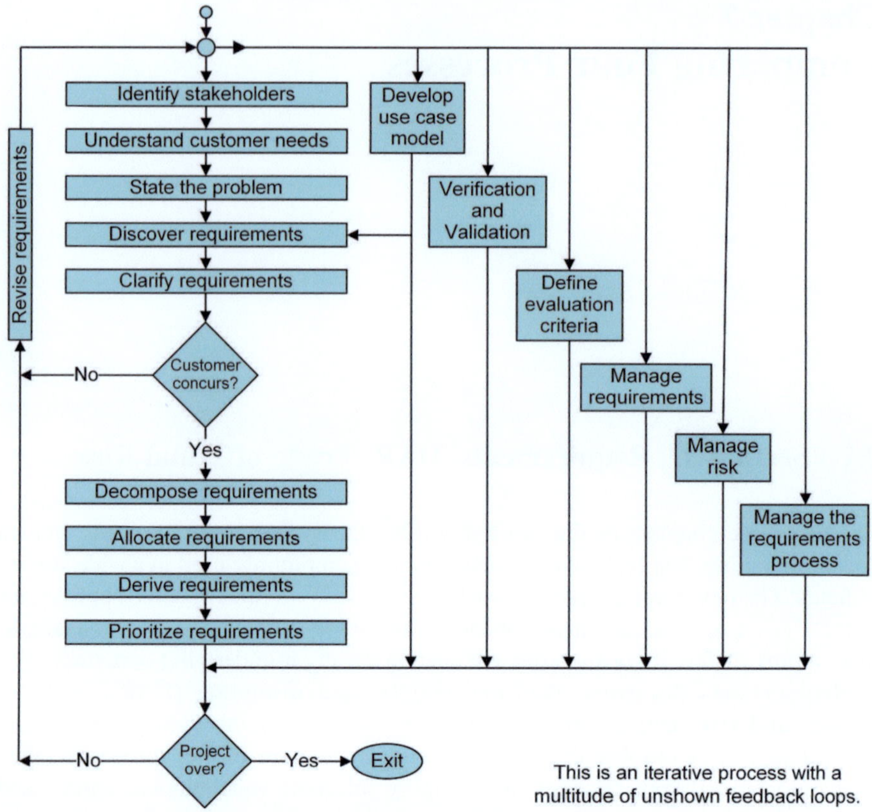

Fig. 7.1 Requirements discovery process of Fig. 4.7

decomposed, allocated, further refined (derived), and prioritized. Iteration is the primary means to reduce uncertainty in the requirements discovery and derivation subprocess. Query reformulation is the primary means for reducing uncertainty in the problem statement subprocess. Program schedule and technology maturation is addressed through incorporating schedule buffers and safety margins.

7.1.2 The DAR Process

Decision analysis and resolution (DAR) is the big-picture within which trade-off studies fit (Fig 7.2). In this big-picture, the top-level activity is DAR and trade-off studies are a technique of DAR. Managers select the decision problems that require DAR and incorporate them into their management plans. DAR is a common process. Common processes are company tools that the user retrieves, tailors, and uses. DAR is invoked throughout the whole program life cycle, whenever a critical

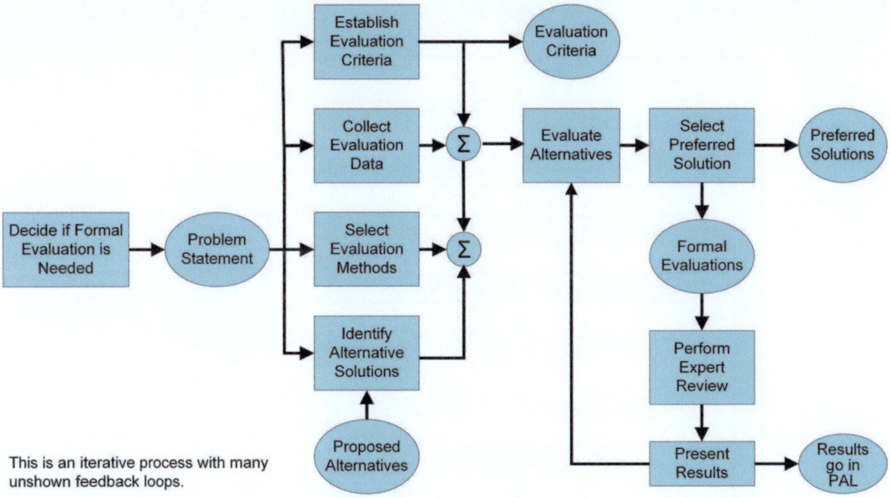

Fig. 7.2 DAR process of Fig. 5.2

decision is to be made. In all decisions, something is gained and something is lost. As an engineer or manager making decisions, you need to know what they are and make trade-offs deliberately.

7.1.3 The Trade-off Study Process

Figure 7.3 presents the trade-off study process. It is invariably the case that the initial problem statement is imprecise and the trade-off space initially defined is incomplete. These sources of uncertainty need to be addressed before proceeding with trade-off studies. Probing the statement of the problem, reformulating queries, and identifying new variables that need to be included in the trade-off space are the means for reducing ambiguity in the problem statement and trade-off space completeness.

7.1.4 The Risk Analysis Process

Figure 7.4 presents the risk analysis process flow. Once again, to prepare for risk requires the identification of risk events. Risks are explored by the risk analyst working with domain experts using storytelling, mental exploration, and envisioning methods to identify risk events. Once a comprehensive set of risk events are identified, the risk analysis process proceeds in accordance with the steps laid out in this figure.

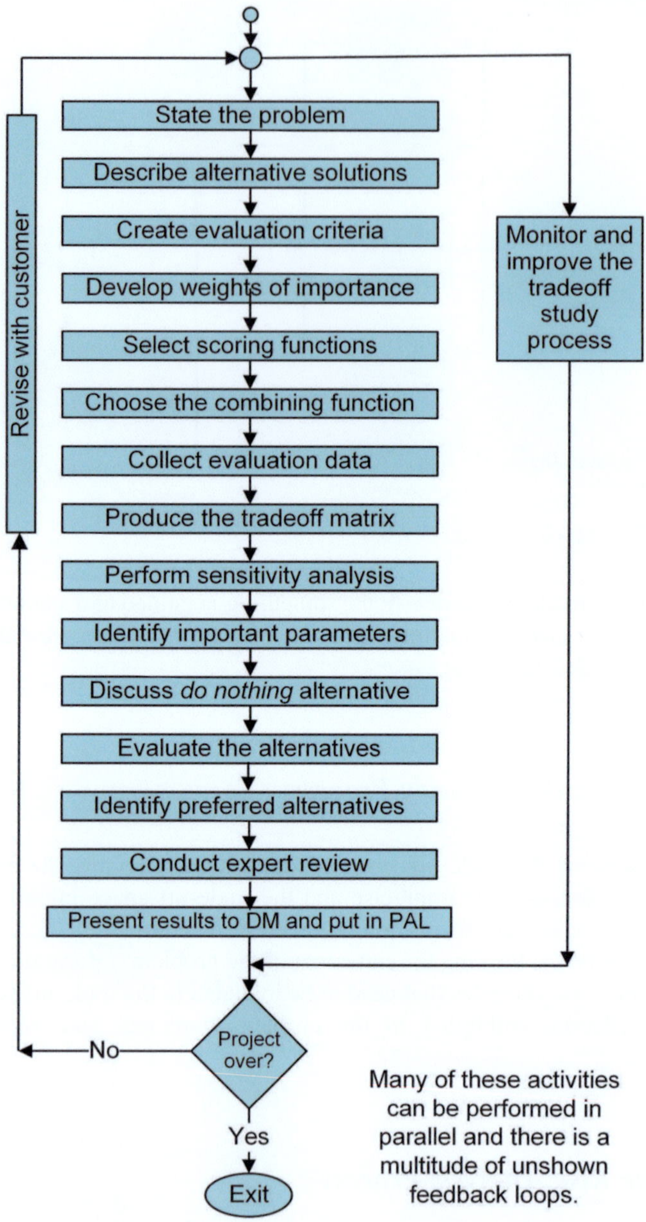

Fig. 7.3 Trade-off study process of Fig. 5.32

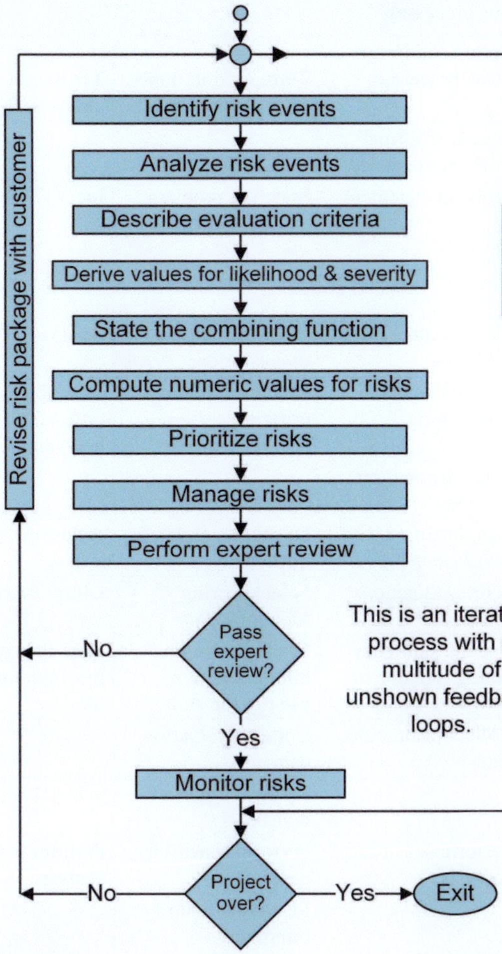

Fig. 7.4 Risk analyses process of Fig. 6.7

7.1.5 Comparison of Requirements, DAR, Trade-off, and Risk Processes

Table 7.1 compares the requirements discovery process, the decision analysis and resolution process, the trade-off study process, and the risk analysis process.

We did not expect a one-to-one mapping between these disparate fields. However, it is quite striking how closely they correspond to each other from the perspective of process structure. The comparison of these three well-established processes has an added benefit—it facilitates process improvement. For example, comparing the rows in this table might suggest that the requirements process should include an activity for formal inspection or expert review, that all of the processes

Table 7.1 Comparison of four processes

Term used in requirements discovery, Fig. 7.1	Term used in the DAR process, Fig. 7.2	Term used in trade-off studies, Fig 7.3	Term used in risk analysis, Fig. 7.4
	Decide if formal evaluation needed		
Identify stakeholders, understand customer needs, state the problem, develop use case models	Problem statement	State the problem	Identify risk events
Discover requirement	Identify alternative solutions	Describe alternative solutions	Analyze risk events
Define evaluation criteria (attributes)	Establish evaluation criteria	Create evaluation criteria	Describe evaluation criteria, mainly likelihood and severity
	Select evaluation methods		
Requirement priorities	Weights of importance	Develop weights of importance	Adjust the range of the criteria
Utility functions	Scoring functions	Select scoring functions	Utility functions
Use the Boolean AND function	Use the sum combining function	Choose the combining function, usually the sum	State the combining function, usually the product
	Collect evaluation data	Collect evaluation data	
		Produce trade-off matrix	
Identify cost drivers	Perform sensitivity analysis	Perform sensitivity analysis	Perform sensitivity analysis
		Identify the most important parameters	
	Examine the status quo, *do nothing* alternative	Discuss the *do nothing* alternative	Identify acceptable risks
	Evaluate alternatives	Evaluate the alternatives	Compute numeric values for likelihood and severity
Clarify, decompose, allocate, and derive requirements			
Revise requirements with customer		Revise with customer	Revise risk package with customer
Manage risk			Manage risks
Prioritize the requirements set (find the most important requirements)	Select preferred solution or decision	Identify preferred alternatives	Prioritize risks (Find the greatest risks)

(continued)

Table 7.1 (continued)

Term used in requirements discovery, Fig. 7.1	Term used in the DAR process, Fig. 7.2	Term used in trade-off studies, Fig 7.3	Term used in risk analysis, Fig. 7.4
Test, validation, and verification	Evaluation, validation and verification		
Review with customer and perform formal inspections	Perform expert review	Conduct expert review	Perform expert review
			Monitor risks
Manage requirements (put results in a requirements database)	Present results to DM and put results in PAL	Present results to DM and put in PAL	Put results in risk register
			Track outliers (both high-frequency but low-severity and low-frequency but high-severity)
Manage the requirements process		Monitor and improve the trade-off study process	Monitor and improve the risk process
Inputs			
Requirements come mainly from the customer and the systems engineer	Evaluation criteria and proposed alternatives come from the design engineer	Evaluation criteria and proposed alternatives come from the design engineer	Risks are identified by the risk analyst
Outputs			
Requirements specification	Preferred alternatives	Preferred alternatives	Risk register

This table uses the vocabulary used in the fields of requirements discovery, decision analysis and resolution (DAR), trade-off studies, and risk analysis

Table 7.2 This row from Table 7.1 might be the most important row

Term used in requirements discovery	Term used in the DAR process	Term used in trade-off studies	Term used in risk analysis
Prioritize the requirements set (find the most important requirements)	Select preferred solution or decision	Identify preferred alternatives	Prioritize risks (find the greatest risks)

should start with understanding the customer's needs, and that the DAR process should include a feedback path that goes all the way back to the problem statement. In other words, activities in each row can be evaluated in terms of their relevance (or not) for the other processes. This activity is intended to introduce consistency and uniformity into the four processes.

Pablo Picasso might have said that this row (Table 7.2) is the essence of this textbook.

We will now compare the key products of requirements discovery, DAR, trade-off studies, and risk analyses. These products *appear* to be different because they use different vocabularies and structure.

In the following paragraphs, we discuss the foregoing within the context of the Cookie acquisition system that allows a student on an afternoon study break to acquire chocolate chip cookies using one of several approaches. This illustrative example is used to convey the key terms associated with our three processes.

7.1.6 A Requirement

See Table 7.3.

7.1.7 An Evaluation Criterion

This is an evaluation criterion from the Cookie acquisition system of Section 30.5.

Name of criterion: Audible signal indicating cookies are ready.
Description: An audible signal shall inform the student when the cookies are ready. This signal should have a nominal intensity level of 80 ± 2 dB.
Weight of importance: 9.
Basic measure: Intensity level of an audible signal.
Measurement method: During design and construction, the proposed device is mounted on a test bench and is activated in accordance with test instructions. Sound intensity level in decibels (dB) is measured at a distance of 30 cm. At final test and during operation, an actual oven activates the audible signal and the sound intensity level in decibels (dB) is measured at a distance of 30 cm.
Units: decibels (dB).
Scoring function input: The measured sound intensity should lie between 70 and 90 dB.
Scoring function: SSF5 (76, 78, 80, 82, 84, 0.5, −2, RLS [70–90]). Here we used the mandatory requirement thresholds of 78 and 82 dB as the baseline values, because we expect the values to improve through the design process (Fig. 7.5).
Scoring function output: 0–1.
Trace to Section 3, ReqF5.
Owner: Engineering.
Date of last change: 11/18/2015.

Table 7.3 A requirement from the Cookie acquisition system of section 26.7

Attribute	Explanation
Identification tag (Id)	ReqNF4
Name	Audible signal for cookies are ready
Text	An audible signal shall inform the student when the cookies are ready. This audible signal shall have an intensity level of 80 ± 2 dB
Priority	9
Verification method	During design and construction, this requirement will be verified by test. The proposed device will be mounted on a test bench and will be activated in accordance with test instructions. Sound power level in decibels (dB) will be measured at a distance of 30 cm. At final test and during operation, this requirement will be verified by demonstration. An actual oven will activate the audible signal and the sound power level in decibels (dB) will be measured at a distance of 30 cm
Difficulty	It will be easy to satisfy this requirement
Link to TPM?	There is no technical performance measure (TPM) for this requirement
DeriveReqt:	This requirement refines `ReqF5: Cookie shall emit an audible signal when the timer has elapsed`
Owner	Engineer
Date of last change	November 26, 2015

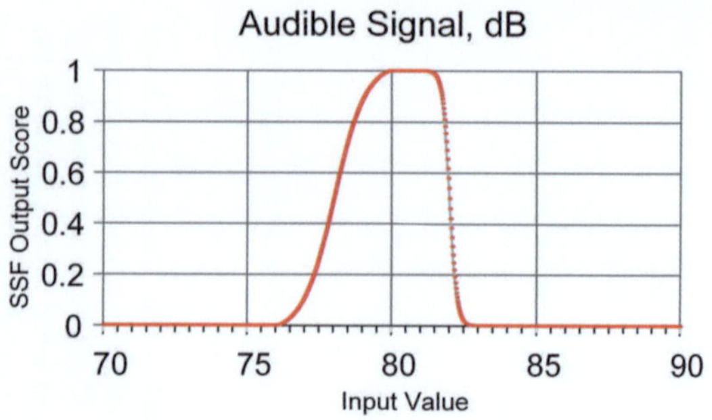

Fig. 7.5 Audible signal scoring function of Fig. 5.3

7.1.8 A Risk

This is a risk from the Cookie acquisition system of Section 34.6.

Failure event: During test or operation, the audible signal for "cookies are ready" could be too loud.

Potential effects: Someone's hearing could be damaged.

Relative likelihood: Noise-induced hearing loss affects 10 % of Americans. Table 1.1 used a range of 0–1 for both likelihood and severity. So we assess this likelihood at 0.01 per year.

Severity of consequences: We compared this risk to all risks that were identified for this system. This was the most serious risk, so we gave it a value of 1.0.

Estimated risk: 0.01.

Priority: high.

Mitigation method: During design, the proposed device is mounted on a test bench and activated in accordance with test instructions. Sound intensity level in decibels (dB) is measured at a distance of 30 cm. The device is put under configuration control to ensure that it is not replaced or altered. This design is conservative in that the device should produce only 80 dB. Exposure to 120 dB or less for only a few seconds is unlikely to cause permanent hearing loss.

Status: active.

Assigned to: Engineer.

Date: Tracking started November 5, 2015.

7.1.9 Comparison of the Requirement, Evaluation Criterion, and Risk

Table 7.4 compares these three descriptions of the auditory output.

Once again, it is not surprising that these products are similar, because they come from similar processes. However, we might compare these three products and decide if we should change our templates for them.

Table 7.4 Comparison of a requirement, an evaluation criterion, and a risk for the Cookie acquisition system

Requirement	Evaluation criterion for a trade-off study/DAR	Risk
Name: Audible signal indicating cookies are ready	Name of criterion: Audible signal indicating cookies are ready	Failure event: During test or operation, the audible signal for "cookies are ready" could be too loud
Text: This audible signal shall have an intensity level of 80 ± 2 dB	Description: This audible signal should have an intensity level of 80 ± 2 dB	Potential effects: Someone's hearing could be damaged
Priority: 9	Weight of importance: 9	Relative likelihood: 0.01
		Severity of consequences: 1.0
		Estimated risk: 0.01
		Priority: high
Verification method	Basic measure:	Mitigation method:
	Measurement method:	
	Units: dB	
Difficulty: easy to satisfy		
	Scoring function: SSF5 (76, 78, 80, 82, 84, 0.5,–2, RLS [70–90])	
		Status: active
Link to TPM? no	Trace to Section 3, ReqF5	
DeriveReqt: This requirement refines ReqF5:		
Owner: Engineer	Owner: Engineering	Assigned to: Engineer
Date of last change: 11/5/15	Date of last change: 11/5/15	Date: 11/5/15

For example, requirements have an attribute listing the difficulty of satisfying the requirement. Should the trade-off study process and the risk process also have such an attribute? We do not think that the trade-off study should. However, one school of risk analysis has three columns in their risk tables: relative likelihood, severity of consequences, and *difficulty of detection*. We did not use difficulty of detection, because we found that it added complexity without adding comparable value. Therefore, we did not change the templates by adding an attribute for difficulty.

The trade-off study process requires scoring functions. Should the requirements process and the risk process also require scoring functions? Some requirements have scoring functions, but we do not want to *require* scoring functions for *all* requirements. We have not found risk analyses that used scoring functions. Therefore, we did not change the templates by adding a requirement for use of scoring functions.

The risk analysis process has an attribute named *status*. Requirements are not likely to come and go like risks. But it would be easy to add a column in the requirements database to cover this possibility. In a trade-off study, it is not likely that evaluation criteria would come and go. But again, it is possible that a whole trade-off study could have status, possible values would be: under construction, alternatives are being evaluated, and the decision has been made.

Both the requirements process and the trade-off study process have an attribute to trace where the item came from. The risk process should have a similar attribute. It would be easy to satisfy, if the risk traced to a requirement or a particular brain storming session.

7.2 Section 38. The SIMILAR Process

Using Figs. 7.1, 7.2, 7.3, and 7.4 and Tables 7.1 and 7.4 we created a general process that can be used for system design, requirements discovery, decision analysis and resolution, trade-off studies, and risk analysis. It is based on the SIMILAR process of Chap. 1, Fig. 1.1 that comprises seven key activities: **S**tate the problem, **I**nvestigate alternatives, **M**odel the system, **I**ntegrate, **L**aunch the system, **A**ssess performance, and **R**e-evaluate. These seven activities are conveniently summarized using the acronym, SIMILAR, see Fig. 7.6. We use this process

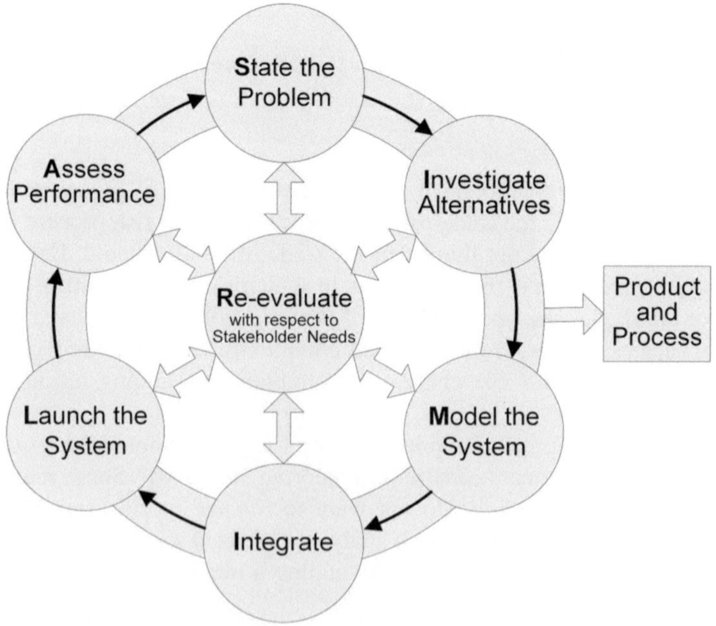

Fig. 7.6 The SIMILAR process of Fig. 1.1

to provide the overall context for problem solving and system design. At the outset, we note that the activities in the SIMILAR process are performed iteratively and in parallel. Each activity in this process is described next.

7.2.1 *State the Problem*

The problem statement contains many tasks that are performed iteratively, many of which can be performed in parallel. Figures 7.1, 7.2, 7.3, and 7.4 and Tables 7.1 and 7.4 contain the following tasks that fit into the state the problem activity. Stating the purpose of the system is the first and foremost task. Next in importance is understanding customer needs. To help you do this you should identify stakeholders such as end users, operators, maintainers, suppliers, acquirers, owners, customers, bill payers, regulatory agencies, affected individuals or organizations, sponsors, manufacturers etc. Because the DAR process and the trade-off study process are so similar in Table 7.1, we will now combine them. So from this point on we will only use three processes: requirements, trade-off study, and risk. Where do the inputs for these three processes come from? Requirements come mainly from the customer and the systems engineer. Evaluation criteria and proposed alternatives for trade-off studies come from the design engineer. Risk events are identified and described by the risk analyst. Another task in this activity is describing how the system works using stories and use case models. The use case models provide preliminary requirements and test cases. The problem should be stated in terms of *what* needs to be done, not *how* to do it. The problem statement may be in prose form or in the form of a model. Finally, you should initiate risk analysis. Yes, this should be done for all of the processes including requirements discovery and trade-off studies.

"The beginning is the most important part of the work," Plato, *The Republic*, fourth century B. C.

7.2.2 *Investigate Alternatives*

You should investigate alternative requirements, designs, and risk events using criteria such as performance, cost, schedule, and risk. For quantitative analyses, identify attributes of requirements, evaluation criteria for trade-off studies and the likelihood of occurrence and severity of consequences for risk events. Assign them weights of importance to show priorities. Scoring (utility) functions are mandatory for trade-off studies, but are optional for requirements and risks. Select evaluation methods, that is, methods for combining the data. State the combining function that will be used. Usually this will be the Boolean AND function for requirements, the sum of weighted products for trade-off studies and the product function for risks. You collect data and use it to assign values to attributes for requirements, weights and scores for trade-off studies, and likelihood and consequences for risk analyses.

Finally, study the data that you just collected: you must study the shape and texture of the data. The shape of the data will determine the desired combining function. The texture will identify outliers.

7.2.3 Model the System

Abstract models are typically created for most requirements, alternative designs, and risk events. These models are appropriately elaborated (that is, expanded) throughout the system life cycle. A variety of models can be used. Requirements can be modeled with use case models, textual shall statements, tables, spreadsheets, and specialized databases. The requirements must be clarified, decomposed, allocated, and derived. Trade-off studies are usually modeled with trade-off matrices implemented with spreadsheets. The alternative designs within them are modeled with UML diagrams, SysML diagrams, analytic equations, computer simulations, and mental models. Risks are modeled with tables containing values for the likelihood of occurrence and severity of consequences and figures displaying these data. Everything must be prioritized. You should prioritize the requirements set to find the most important requirements. For trade-off studies, you identify preferred alternatives with a trade-off matrix. You adjust the ranges for likelihood and severity for risk events in order to find the greatest risks. The results of a sensitivity analysis can be used to validate a model, flag unrealistic model behavior, point out important assumptions, help formulate model structure, simplify a model, suggest new experiments, guide future data collection efforts, suggest accuracy for calculating parameters, adjust numerical values of parameters, choose an operating point, allocate resources, detect critical criteria, suggest tolerance for manufacturing parts, and identify cost drivers.

7.2.4 Integrate

Integration means bringing elements together so that they work as a whole to accomplish their intended purpose and deliver value. Specifically, systems, enterprises, and people need to be integrated to achieve desired outcomes. To this end, interfaces need to be designed between subsystems. Subsystems are typically defined along natural boundaries in a manner that minimizes the amount of information exchanged between the subsystems. Feedback loops between individual subsystems are easier to manage than feedback loops involving densely interconnected subsystems. Evaluation criteria should trace to requirements. Risks should trace to requirements or particular brain storming sessions. Requirements should trace to higher level requirements and should link to risks. Requirements and risks might link to technical performance measures (TPMs).

7.2.5 *Launch the System*

System launch means either deploying and running the actual system in the operational environment, or exercising the model in a simulated environment to produce necessary outputs for evaluation. The purpose of system launch is to provide an environment that allows the system or its model to do what it is being designed to do. The outputs of these processes are a requirements specification, preferred alternatives, and the risk register. One should continually monitor the requirements (in the requirements database), alternative designs (in the PAL), and risks (in the risk register) looking for possible changes and bring these to the attention of the decision makers. One should continually monitor the market place looking for new requirements, products, designs, and risks, and bring these to the attention of the decision makers.

7.2.6 *Assess Performance*

Test, validation, and verification are important tasks for all processes. There should be regularly scheduled and performance initiated expert reviews. The results of these reviews are presented to the decision maker (DM) and are put in the process assets library (PAL). Evaluation criteria, measures, metrics, and TPMs are all used to quantify system performance. Evaluation criteria are used in requirements discovery, trade-off studies, and risk analyses. Measures and metrics are used to help manage a company's processes. TPMs are used to mitigate risk during design and manufacturing.

7.2.7 *Re-evaluate*

The distinction between an engineer and a mathematician is the use of feedback in the design. For a century, engineers have used feedback to control systems and improve performance. It is one of the most fundamental engineering concepts. Re-evaluation is a continual process with multiple parallel feedback loops. Re-evaluation means observing outputs and using this information to modify the inputs, the system, the product and the process.

The SIMILAR process (Fig. 7.6) clearly shows the distributed nature of the re-evaluate function in the feedback loops. However, it is important to realize that not all the loops will always come into play. The loops that are used depend on the problem to be solved and the problem context. Re-evaluation includes formal inspections, expert reviews, and reviews with the customer. A very important and often neglected activity in any process is monitoring and improving the process itself. This activity is shown explicitly in Figs. 7.1, 7.3, and 7.4 with the following tasks. Manage the requirements process. Monitor and improve the trade-off study process. Monitor and improve the risk process.

7.3 Section 39. The Overarching Process

Our general process for everything is shown in Fig. 7.6 with the contents of the circles being described in Table 7.5. This process can be used for discovering requirements, doing trade-off studies, and performing risk analyses. Common activities are listed down the center of the table. Similar activities with different vocabularies are listed in the three columns. Only in the modeling area are the three activities different enough to warrant separate rows.

Many of these activities can be done at the same time. For example, creating evaluation criteria, developing weights of importance, selecting scoring functions, choosing combining functions, and collecting data can all be done in a parallel, but in iterative fashion. Accordingly, the process of Table 7.5 was not represented with an activity diagram or a functional flow block diagram.

Some actions that were listed in Figs. 7.1, 7.2, 7.3, and 7.4 are not in the general process of Table 7.5. They have been subsumed in the activities. For example, the action of studying the shape and texture of the data is included in the choose combining function activity. The action of studying the do nothing alternative is in the create trade-off matrix and evaluate alternatives activity. The identify important parameters action is in the perform a sensitivity analysis activity. Finally, the action to track outliers is in the monitor and manage risks activity.

In Fig. 7.7, we have taken the information in Table 7.5 and recast it as a figure. This figure presents the overarching process as a superset of the three processes. Viewed another way, each of the three processes is a tailoring of the overarching process. The value of the overarching process is that it serves as a reference model for uniform handling of requirements, trade-offs, and risks.

7.4 Section 40. Denouement

At this point, you have read this book. You should now have a good understanding of processes. In particular, you should understand the system design process. With some additional experience with UML, SysML, and system architecture, you can be on your way to being a system designer. At this point, you should also have an understanding of how to write and manage requirements. Upon becoming familiar with commercial requirements database tools, you can be on your way to becoming a requirements engineer. By now, you should understand risk analysis and the common mistakes that people make when performing risk analyses. With additional breadth of experience, you can be on your way to becoming a risk analyst. An

Table 7.5 General process for everything

A general process for requirements, trade-off studies, and risk analyses		
Requirements discovery	DAR/trade-off study	Risk analysis
State the problem		
Declare the purpose of the system Identify stakeholders Understand customer needs State the problem Create use case models Initiate risk analysis		
Investigate alternatives		
Discover requirements	Describe alternative designs	Identify and analyze risk events
Create evaluation criteria Develop weights of importance Select scoring functions Choose combining functions Collect data		
Assign values to attributes	Assign values to weights and scores	Assign values to likelihood and severity
Model the system		
Clarify, decompose, allocate, and derive requirements		
	Create a trade-off matrix and evaluate alternatives	
		Compute numerical values for risks, plot risk events and track outliers
Prioritize requirements	Prioritize alternatives	Prioritize risks
Perform sensitivity analyses		
Integrate		
Identify links to risks	Identify links to requirements	Identify links to requirements
Check for links to TPMs		Check for links to TPMs
Launch the system		
Run the system and observe outputs		
Monitor and manage requirements	Monitor and manage trade-off studies	Monitor and manage risks
Monitor the marketplace and the environment		
Assess performance		
Conduct formal inspections and expert reviews Test, validate, and verify Present results to DM and put them in the PAL		
Re-evaluate		
Review with customer and revise		
Monitor and improve the process		

DM decision maker, *PAL* process assets library, *TPM* technical performance measure

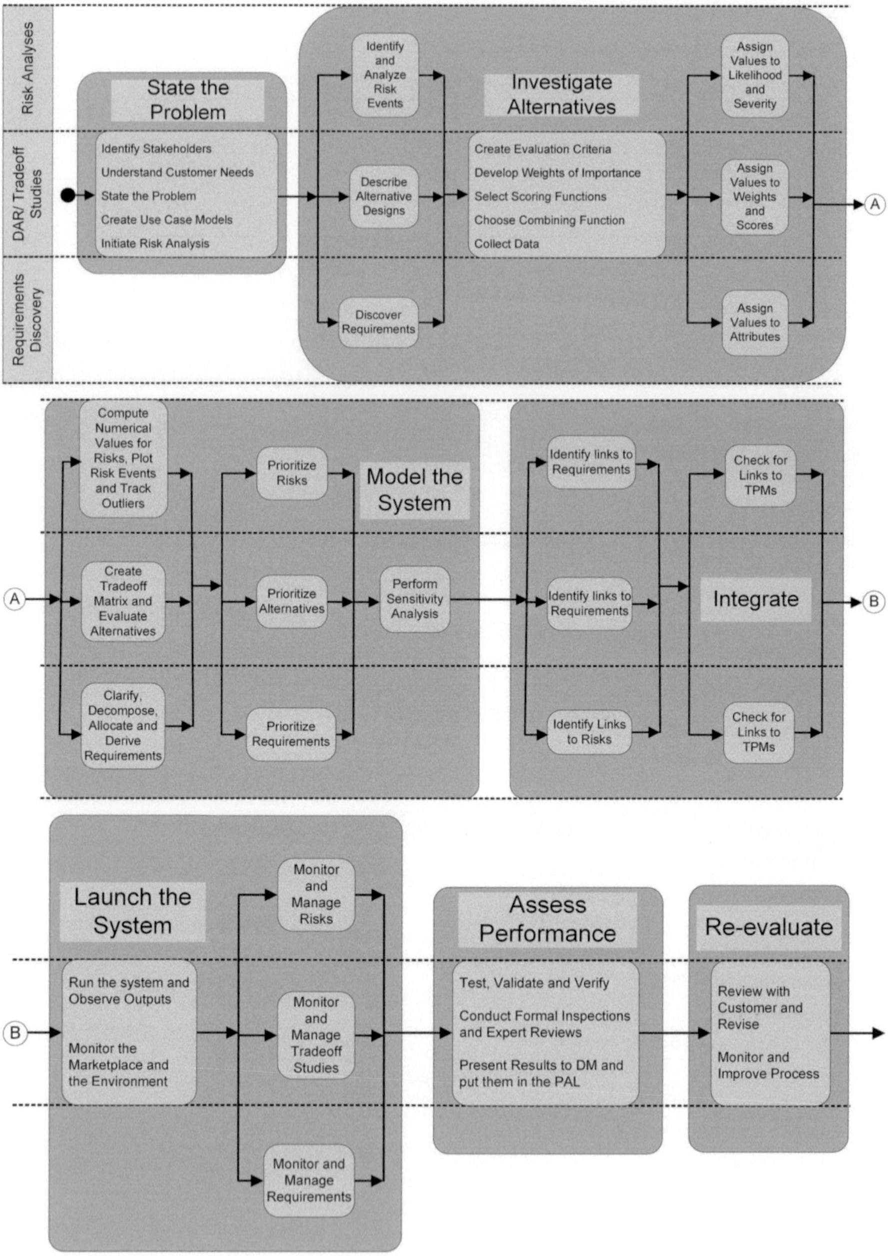

Fig. 7.7 The overarching process

example of this breadth would be learning to apply risk analysis to both threats and opportunities. And, finally, you are reasonably conversant with conducting a trade-off study. You are knowledgeable about the mistakes that people make when conducting trade-off studies. As importantly, you now know how to do a sensitivity analysis as part of a trade-off study. You now know more than most people about the promise and pitfalls of trade-off studies, and how to avoid the pitfalls. With additional experience, you can be on your way to becoming expert in performing trade-off studies.

7.5 Section 41. Homework

7.5.1 Human Behavior

Mark with an asterisk or highlighting the activities in Table 7.5 or Fig. 7.7 where the concepts of prospect theory and behavioral economics should particularly be applied, and explain why you think so.

7.5.2 What Have You Learned?

Describe what you have learned in this course.

7.5.3 Suggestions for Improvement

Suggest improvements for this textbook. Mail then to terry@sie.arizona.edu.

Index

A.T. Bahill, A.M. Madni, *Tradeoff Decisions in System Design*,
DOI 10.1007/978-3-319-43712-5

MIX
Papier aus verantwortungsvollen Quellen
Paper from responsible sources
FSC® C105338

If you have any concerns about our products,
you can contact us on
ProductSafety@springernature.com

In case Publisher is established outside the EU,
the EU authorized representative is:
Springer Nature Customer Service Center GmbH
Europaplatz 3, 69115 Heidelberg, Germany

Printed by Libri Plureos GmbH
in Hamburg, Germany